Advanced Rock Dynamics and Applications

Advanced Rock Dynamics and Applications

Editor:

Navodita Bhatnagar

Delve Publishing

www.delvepublishing.com

Advanced Rock Dynamics and Applications

Editor: Navodita Bhatnagar

708 3rd Avenue, 6th Floor
New York, NY 10017
United States of America
www.delvepublishing.com

ISBN: 978-1-68095-840-9

Library of Congress Control Number: 2016951593

About the Editor

Navodita Bhatnagar

Navodita Bhatnagar finished her masters in technology and currently pursuing her PhD in environmental sciences from Institute of Technology, Carlow, Ireland. She's a keen enthusiast of environment protection and management. An alumunus of Indo-German centre for sustainibility which is essentially a collaborative interface to gather scientists in the field of environment sciences to come together and work for the common cause which is environment protection.

Contents

List of Contributors

Adeline Delonca
UMR 7359 georesources, Lorraine University, CNRS, CREGU School of Mines Campus Artem, CS2434, 54042 Nancy Cedex, France

Thierry Verdel
UMR 7359 georesources, Lorraine University, CNRS, CREGU School of Mines Campus Artem, CS2434, 54042 Nancy Cedex, France

Yann Gunzburger
UMR 7359 georesources, Lorraine University, CNRS, CREGU School of Mines Campus Artem, CS2434, 54042 Nancy Cedex, France

Minggao Tang
State Key Laboratory of Geohazard Prevention and Geoenvironment Protection (Chengdu University of Technology), Chengdu, 610059, China
College of Environment and Civil Engineering, Chengdu University of Technology, Chengdu, 610059, China

Qiang Xu
State Key Laboratory of Geohazard Prevention and Geoenvironment Protection (Chengdu University of Technology), Chengdu, 610059, China

Yusheng Li
College of Environment and Civil Engineering, Chengdu University of Technology, Chengdu, 610059, China

Runqiu Huang
State Key Laboratory of Geohazard Prevention and Geoenvironment Protection (Chengdu University of Technology), Chengdu, 610059, China

Niek Rengers
State Key Laboratory of Geohazard Prevention and Geoenvironment Protection (Chengdu University of Technology), Chengdu, 610059, China

Xing Zhu
State Key Laboratory of Geohazard Prevention and Geoenvironment Protection (Chengdu University of Technology), Chengdu, 610059, China

Eugie Kabwe
Department of Mining and Mineral Resources Engineering, University of Science and Technology Beijing, Beijing, China

Yiming Wang
Department of Mining and Mineral Resources Engineering, University of Science and Technology Beijing, Beijing, China

Weiren Lin
Kochi Institute for Core Sample Research, Japan Agency for Marine-Earth Science and Technology, Nankoku, Japan.

Osamu Tadai
Department of Marine & Earth Sciences, Marine Works Japan Ltd., Nankoku, Japan.

Manabu Takahashi
Research Institute of Earthquake and Volcano Geology, National Institute of Advanced Industrial Science and Technology, Tsukuba, Japan.

Daisaku Sato
Graduate School of International and Area Studies, Hankuk University of Foreign Studies, Seoul, Korea.

Takehiro Hirose
Kochi Institute for Core Sample Research, Japan Agency for Marine-Earth Science and Technology, Nankoku, Japan.

Wataru Tanikawa
Kochi Institute for Core Sample Research, Japan Agency for Marine-Earth Science and Technology, Nankoku, Japan.

Yohei Hamada
Kochi Institute for Core Sample Research, Japan Agency for Marine-Earth Science and Technology, Nankoku, Japan.

Kentaro Hatakeda
Department of Marine & Earth Sciences, Marine Works Japan Ltd., Nankoku, Japan.

Giulia Sofia
Department of Land, Environment, Agriculture and Forestry, University of Padova, 35020, Legnaro (PD), Italy

John K. Hillier
Department of Geography, Loughborough University, Loughborough, LE11 3TU, UK

Susan J. Conway
Department of Physical Sciences, The Open University, Milton Keynes, MK7 6AA, UK
Laboratory of Planetology and Geodynamics - UMR CNRS 6112, 2 rue de la Houssinière - BP 92208, 44322 Nantes Cedex 3, France

Andrew Valentine
Department of Earth Sciences, Universiteit Utrecht, Postbus 80.021, 3508TA Utrecht, the Netherlands

Lara Kalnins
Department of Earth Sciences, Science Labs, Durham University, Durham, DH1 3LE, UK

P. Bottelin
ISTerre, University of Grenoble 1, CNRS, BP 53, F-38041 Grenoble Cedex 9, France

D. Jongmans
ISTerre, University of Grenoble 1, CNRS, BP 53, F-38041 Grenoble Cedex 9, France

D. Daudon
3SR Laboratory, University of Grenoble 1, G-INP, CNRS, BP 53, F-38041 Grenoble Cedex 9, France

A. Mathy
Alpine Geotechnical Society (SAGE), 2 rue de la Condamine, Gières, France e

A. Helmstetter
ISTerre, University of Grenoble 1, CNRS, BP 53, F-38041 Grenoble Cedex 9, France

V. Bonilla-Sierra
3SR Laboratory, University of Grenoble 1, G-INP, CNRS, BP 53, F-38041 Grenoble Cedex 9, France

H. Cadet
Alpine Geotechnical Society (SAGE), 2 rue de la Condamine, Gières, France

D. Amitrano
ISTerre, University of Grenoble 1, CNRS, BP 53, F-38041 Grenoble Cedex 9, France

V. Richefeu
3SR Laboratory, University of Grenoble 1, G-INP, CNRS, BP 53, F-38041 Grenoble Cedex 9, France

L. Lorier
Alpine Geotechnical Society (SAGE), 2 rue de la Condamine, Gières, France

L. Baillet
ISTerre, University of Grenoble 1, CNRS, BP 53, F-38041 Grenoble Cedex 9, France

P. Villard
3SR Laboratory, University of Grenoble 1, G-INP, CNRS, BP 53, F-38041 Grenoble Cedex 9, France

F. Donzé
3SR Laboratory, University of Grenoble 1, G-INP, CNRS, BP 53, F-38041 Grenoble Cedex 9, France

G. Ciabocco
Department of technology and engineering and environment sciences and the Forest of Tuscia University, Viterbo, Italy

L. Boccia
Department of agricultural engineering and territory – University of Naples Federico II, Italy

M. N. Ripa
Department of technology and engineering and environment sciences and the Forest of Tuscia University, Viterbo, Italy

V. Wirz
Department of Geography, University of Zurich, Zurich, Switzerland

S. Gruber
Department of Geography and Environmental Studies, Carleton University, Ottawa, Canada

R. S. Purves
Department of Geography, University of Zurich, Zurich, Switzerland

J. Beutel
Computer Engineering and Networks Laboratory, ETH, Zurich, Switzerland

I. Gärtner-Roer
Department of Geography, University of Zurich, Zurich, Switzerland

S. Gubler
Federal Office of Meteorology and Climatology MeteoSwiss, Zurich, Switzerland

A. Vieli
Department of Geography, University of Zurich, Zurich, Switzerland

Lailin Li
Oil Exploration and Development Research Institute of Daqing, Daqing, China.

Haibo Zhao
Oil Exploration and Development Research Institute of Daqing, Daqing, China.

Xuehui Han
China University of Petroleum (East China), Qingdao, China.

D. T. Milodowski
School of GeoSciences, University of Edinburgh, Edinburgh, UK

S. M. Mudd
School of GeoSciences, University of Edinburgh, Edinburgh, UK

E. T. A. Mitchard
School of GeoSciences, University of Edinburgh, Edinburgh, UK

F. Beaud
Department of Earth Sciences, Simon Fraser University, 8888 University Drive, V5A 1S6, Burnaby, BC, Canada

G. E. Flowers
Department of Earth Sciences, Simon Fraser University, 8888 University Drive, V5A 1S6, Burnaby, BC, Canada

J. G. Venditti
Department of Geography, Simon Fraser University, 8888 University Drive, V5A 1S6, Burnaby, BC, Canada

R. H. Guthrie
MDH Engineered Solutions, SNC-Lavalin Group, Calgary, AB, Canada
P. Friele
Cordilleran Geoscience, Squamish, BC, Canada

K. Allstadt
Earth and Space Sciences, University of Washington, Seattle, WA, USA

N. Roberts
Centre for Natural Hazard Research, Simon Fraser University, Burnaby, BC, Canada

S. G. Evans
Landslide Research Programme, University of Waterloo, Waterloo, ON, Canada

K. B. Delaney
Landslide Research Programme, University of Waterloo, Waterloo, ON, Canada

D. Roche
Kerr Wood Leidal Associates Limited, Burnaby, BC, Canada

J. J. Clague
Centre for Natural Hazard Research, Simon Fraser University, Burnaby, BC, Canada

M. Jakob
BGC Engineering Inc., Vancouver, BC, Canada

L. Superchi
Dept. of Geosciences, University of Padova, Italy

M. Floris
Dept. of Geosciences, University of Padova, Italy

M. Ghirotti
Dept. of Earth and Geo-Environmental Sciences, University of Bologna, Italy

R. Genevois
Dept. of Geosciences, University of Padova, Italy

M. Jaboyedoff
Institute of Geomatics and Risk Analysis (IGAR), University of Lausanne, Switzerland

D. Stead
Dept. of Earth Sciences, Simon Fraser University, Burnaby, Canada

L. Zhang
State Key Laboratory of Hydroscience and Engineering, Tsinghua University, Beijing, China

G. Parker
Department of Civil & Environmental Engineering and Department of Geology, Hydrosystems Laboratory, University of Illinois, Urbana, IL, USA

C. P. Stark
Lamont-Doherty Earth Observatory, Columbia University, Palisades, NY, USA

T. Inoue
Civil Engineering Research Institute for Cold Regions, Hiragishi Sapporo, Japan

E. Viparelli
Dept. of Civil & Environmental Engineering, University of South Carolina, Columbia, SC, USA

X. Fu
State Key Laboratory of Hydroscience and Engineering, Tsinghua University, Beijing, China

N. Izumi
Faculty of Engineering, Hokkaido University, Sapporo, Japan

Preface

Rock dynamics is a branch of rock mechanics and engineering which essentially deals with the response of rock masses to various dynamic loading conditions including vibrations from earthquakes and explosions. The studies also encompass formation and propagation of dynamic loads, rock failures under dynamic and static loads, responses and transmission. In the last few decades, experimental and computational techniques have seen a tremendous development thus allowing capturing the progress of fracture in microsecond steps, understanding the initiation, propagation, and branching of fracture, thus leading better research and development of new theories. Rock dynamics is useful in civil and geological engineering studies as well but the presence of fourth dimension, that is time, makes it a bit difficult to understand and apply.

Rock dynamics and applications is an important part of rock mechanics and rock engineering, when rock materials and rock masses are subjected to dynamic loads including earthquake and explosions. The studies of rock dynamics involve the formation and transmission of dynamic loads, responses and failures of rock materials and rock masses under the dynamic loads, as well as dynamic rock failure processes of under static loads. Due to the additional "dimension" of time, dynamics, at least in the discipline of rock mechanics, remains a challenging topic to understand and to apply. Much new researches are needed and are indeed on-going. This book is organized to summarize the state-of-the-art in rock dynamics and applications, highlighting the current scientific research activities and engineering application challenges, covering all the aspects ranging from analytical and numerical works to laboratory testing, and to engineering. The book includes research dealing with rock dynamics theory, rockfall hazards, earth surface dynamics and geomorphometry. Rockfalls and landslides have been given special focus by including case histories which also discuss about risk assessment. I hope it will be insightful for the readers and help them understand the subject in depth.

Editor
Navodita Bhatnagar

INTRODUCTION

Rock dynamics deals with time responses of rock materials and masses under dynamic loads Dynamic loads include explosion, impact, and seismic, that are generally in the form of stress waves. Since rock masses generally contain fractures in different scales, the interaction between stress waves and rock materials and fractures affects not only stress wave attenuation, but also rock mass instability. The wave transmission and transformation across rock fractures and the fracture and failure of rock materials are traditional issues in the rock dynamic study. Stress wave propagation across non-filled fractures usually focuses on the coupling of wave attenuation and fracture geometrical properties (spacing, frequency, orientation, aperture, and roughness). To deal with filled fractures, stress wave attenuation due to dynamic compaction of filling gouges (e.g. sand or clay) shall be considered, as well as the attenuation at fracture interfaces. Dynamically induced rock instability is commonly associated with frictional slip on rock fractures

Dynamical triggering of fracture slip may take place close to or far from the main shock. The fracture damage depends on the incident wave energy and the stored strain energy at the fracture plane. If energy release from the damage is sufficiently large, it can cause seismic events, and further induce aftershocks in close proximity. Laboratory experiments offer a direct observation on energy release pattern. Stress wave with high amplitude can influence rock material strength increases and rock material fails with more fractures However, it is not clear yet the cause of high density of fracturing. The Split Hopkinson Pressure Bar (SIIPB) technique is a popular tool to explore the mechanical causes of rate effects on rock strength and failure pattern. The SIIPB application is developing to investigate rate effects on fracture branching, multiple fracture initiation and crack propagation velocity. The numerical modelling of stress wave propagation across rock fractures and rock fracture and failure needs to be micromechanics and discrete based.

The Particle Manifold Method (PMM) is a recently developed by introducing particle concept into the Numerical Manifold Method. In the PMM, the geometry of the physical field is represented by particles. Through different particle models, micromechanics can be investigated by the PMM. This keynote addresses recent advancements on rock dynamic research, emphasizing on the laboratory experiments and the numerical modelling. It covers stress wave propagation across a single and a set of filled parallel rock fractures, dynamically induced seismicity, mechanism of rock fracturing, discontinuous micromechanics modelling of wave propagation and dynamic fracturing.

Rock dynamics is important because many rock mechanics and rock engineering problems involve dynamic loading ranging from earthquakes to vibrations and explosions. The Rock dynamics deals with the distribution and propagation of loads, dynamic responses, and processes of rocks and rate-dependent properties, coupled with the physical environment. Rock dynamics has a wide range of applications in civil, mining, geological and environmental engineering.

Rock dynamics, as a branch of rock mechanics, deals with the responses of rock (materials and masses) under dynamic stress fields, where an increased rate of loading (or impulsive loading) induces a change in the mechanical behavior of the rock materials and rock masses. Differing from static mechanics, dynamic stresses are in the forms of stress waves propagating in the loaded medium with time, and therefore the response of rock is influenced by, and interacts with, the stresses in motion. Rock dynamics deals not only with the end effects of the forces, but also the processes of the forces acting on the rocks. In these processes, both forces and objects are in motion. Rock dynamics specifically examines the processes of dynamic motions of both the forces and the rocks, at different scales varying from micro particles to rock blocks. Rock dynamics as a science subject covers a wide scope related to forces, and responses of rock, in the time domain. It deals with the distribution of stress fields, responses and properties of rocks, and dynamic behavior coupled with the physical environment. Sources of dynamic loads include explosion, impact, and seismic events. These loads are typically given in the form of time histories of particle acceleration, velocity, or displacement.

The distribution of a dynamic stress field is in the form of a stress wave moving in the loaded medium, including the propagation behavior of the stress wave. Stress wave propagation in rock masses is governed by wave transmission and transformation across the discontinuities (rock joints) in the rock masses. The response of rock materials and rock masses under dynamic stress field includes displacements of rock at particle scale, material fracturing and failure, and large movements at discontinuities. Rock fracturing, for example, is a dynamic micro-scale process leading to macro-scale deformation and failure. Rock dynamic behavior is often coupled with, and frequently induced by, the physical environment, e.g. water and temperature. Changes of physical environment may alter the stress fields as well as the properties of the rock materials and rock masses, hence leading to dynamic responses of the rocks.

When a hard working pillar, or rock bridge in the mine fails, or a highly stressed fault slips, load is redistributed from that yielded 'structural element' of the rock mass. Other parts of the mine must do that 'work'. Whether other components elsewhere in the mine subsequently fail is a function of their capacity and the load path and ultimate loading which evolves as the system converges to a new equilibrium. Subsequent failures are rarely due to impulsive loads or vibrations from the seismic wave emanating from the original failure. Though seismic waves do damage excavations in mines, the magnitude and occurrence of the hazard is less than the exposure of personnel to increasing rock mass damage and deformation due simply to load redistribution caused by mining and under-supporting of excavations. Most often, the 'dynamic' damage seen in working areas of a mine which is hazardous to personnel, is the rapid dilation of rock in the walls of the excavation or dilation adjacent to and associated with dislocation of discontinuities. The dynamic damage occurred as an essential consequence of additional stress, which came on to that part of the mine following an episode of excavation or failure elsewhere.

The most common example is in tunnels adjacent to work areas, which experience a very rapid increase in the depth of damage in the walls when stress becomes more adverse following some other change in the mine. If that additional damage were to occur slowly, the installed ground support would only need to accommodate the strains and retain the loose material, subject to the loads that evolve in the loose material due to gravity. In this case, the engineer would have considerable opportunity to intervene before the situation become dangerous. When the new damage occurs quickly enough and in a large enough increment so that there are additional accelerations imparted to the newly loosened material, the loads that ground support must sustain may transiently be higher than that due to gravity alone, and there is no opportunity to intervene part way through the damage process. An example of a grossly distorted tunnel, which failed in a dynamic way. In this case, the stress in the tunnel wall increased during a several hour period, after a nearby pillar failed. Several hours after the nearby pillar failure which did generate a large seismic wave, the failure in the photograph occurred violently, ejecting a significant volume of rock into the drive. The ejection was not caused by a stress wave emanating from elsewhere it occurred because of the rate of dilation of the rock in the tunnel wall, as it failed.

CHAPTER 1

ROCKFALL HAZARD ASSESSMENT USING EMPIRICAL METHODS

Adeline Delonca, Thierry Verdel, and Yann Gunzburger

INTRODUCTION

Rockfall instabilities are a major hazard for people, human activities and infrastructure (Bell and Glade, 2004; Moreiras, 2006). It is thus essential to assess the rockfall hazard in areas over which they are likely to occur and to propose a map-ping of the hazard to manage the risk in urban areas. Natural hazard zoning has been introduced in many countries all over the world. For instance, the PPR (Plans de Prevention des Risques Naturels Previsibles) in France (Besson et al., 1999) and the cartes de dangers in Switzerland (Leroi et al., 2005) evaluate the hazard level in affected zones, according to a predefined set of hazard classes. Based on this hazard maps, different areas are identified: areas where construction is restricted, areas where a monitoring system or a protection system is required for reducing the risk and areas where no restrictions apply for constructions (Fell et al., 2008). Be-cause of the implications in the territory management, the rockfall assessment must be as accurate as possible (neither underestimated nor overestimated) and reliable.

Rockfall hazard can be defined as the probability that a specific location at the toe of a studied slope will be reached by a rockfall of a given magnitude (Jaboyedoff et al., 2001). Whatever the magnitude of the expected rockfall, the probability can be divided into two terms: the failure probability and the propagation probability (Jaboyedoff et al., 2005).

Various methods simulate the trajectory of rock masses fragments after rupture and evaluate the propagation probability (Dorren, 2003). However, to date, there is no fully reliable method to estimate the failure probability (Hantz, 2007). Even though some existing methods are quantitative and based on historical inventory (Dussauge-Peisser et al., 2002; Hungr et al., 1999; Luckman, 1976), in most cases such an inventory is not available. Then, qualitative assess-ment methods are used to assess the rockfall probability failure. These methods are mainly based on expert judgment (Abella and Van Westen, 2008; Budetta, 2004; Effendiantz et al., 2004; Hantz et al., 2003; Jaboyedoff et al., 2001). Therefore, experts in charge of the hazard assessment have a key role. The method considered could likewise have an influence in the hazard assessment. Different research works (Abbruzzese and Labiouse, 2010; Bormioli et al., 2011) have compared different mapping methods and evaluated the in-fluence of the method chosen on the obtained results. These studies highlighted the statistical significance of the chosen method: the use of one method over another gives different levels of failure probability for the same sector.

In this paper, an experiment is undertaken in order to evaluate the influence of the method and the level of expertise on the rockfall failure probability assessment. To simplify the reading, and the understanding of the paper, the term "rock-fall hazard" will be used when referring to the rockfall failure probability. It corresponds to the combination of the occurrence probability, the temporal probability and the magnitude (volume). The rockfall hazard is assessed on a test site by three populations with different levels of expertise. Two different methods for rockfall hazard assessment are used:

- a qualitative one, the Laboratoire des Ponts et Chaussées (LPC) method, which is the method mainly used in France, and
- a quantitative one, based on the slope mass rating (SMR).

The experiment was conducted in two phases: the first one was realized in May 2012 and the preliminary results were presented in EUROCK Congress (Delonca et al., 2013). The

Citation: Delonca, A., Verdel, T., and Gunzburger, Y.: Rockfall hazard assessment using empirical methods, Nat. Hazards Earth Syst. Sci., 16, 1657-1672, doi:10.5194/nhess-16-1657-2016, 2016.

second one was realized in May 2013, in order to confirm the first results, by increasing the size of the three populations and adding statistical procedure to the study. For each phase, the same protocol has been followed: three populations with different levels of expertise have assessed the level of rockfall hazard by considering two different methods of analysis, on three sectors of the same test site. The experimental protocol considered in the study is presented in more details in the first part of the paper. Then, the results are presented and discussed for the two phases together.

The objective of the study is to consider different methods of rockfall hazard assessment and to evaluate the differences in terms of levels of rockfall hazard. The levels of rockfall hazard considered in the study correspond to classical rock-fall hazard levels: very low, low, moderate and high (Copons et al, 2008; Bauer, 2011; OFAT, 1997). These levels are commonly used to build hazard maps for risk management in urban areas. Thus, it is possible to compare these levels, which are obtained using different methods. Moreover, the influence of the level of expertise on the result, and so on the obtained hazard levels, is also investigated. Indeed, it may interest engineers and researchers in charge of hazard mapping or those concerned with the development of rockfall assessment methods. Note that the objective of the paper is not to evaluate the "true" level of rockfall hazard but to compare the evaluation process, considering different levels of expertise and different methods.

EXPERIMENTAL PROTOCOL

Selection of the Test Site

The experiment was realized on a cliff situated in Liverdun, a town close to Nancy in France. The test site has been chosen because of its history. It is an old quarry that became a climbing site, which has been closed after repetitive rock-falls. Moreover, it has been previously classified as presenting a high-level rockfall hazard (Moiriat et al., 2008). The methodology used to evaluate the rockfall hazard on the site is the following: (1) bibliographic review of all the documents available and (2) field recognitions. The first step leads to evaluate the risk area at a departmental scale. The second step leads to prioritize the level of rockfall hazard on the area of study. Therefore, the evaluation is made at a regional scale. The site is a 50 m long cliff of Jurassic limestone that is made up of massive blocks at the base and small blocks in the upper part (Delonca et al., 2013).

Three sectors were identified (Delonca et al., 2013):

- Sector 1 presents a massive wall with well-defined stratification beds. A pluridecametric fracture isolates a rock panel several hundreds of cubic meters in size. At the top, small unstable blocks are present (Fig. 1a).
- In sector 2, two major fractures form a rock wedge of a few cubic meters in size. Small unstable blocks are present in the upper part (Fig. 1b).
- In sector 3, limestone beds are overhanging, and some blocks at the top are unstable (Fig. 2).

The prevailing failure mechanism on the three sectors is de-scribed in the following (after Hantz et al., 2003): for sec-tor 1, the main failure mechanism is column toppling; for sector 2, it is wedge slide; for sector 3, it is overhang failure. The latter type of failure is also present on the two others sectors. The overhang failure is the result of a traction failure for some of the small blocks at the top of the three sectors. Moreover, there also exists a failure mechanism that involves shearing of joints on the three studied sectors.

Population

Three populations were involved in the experiment.

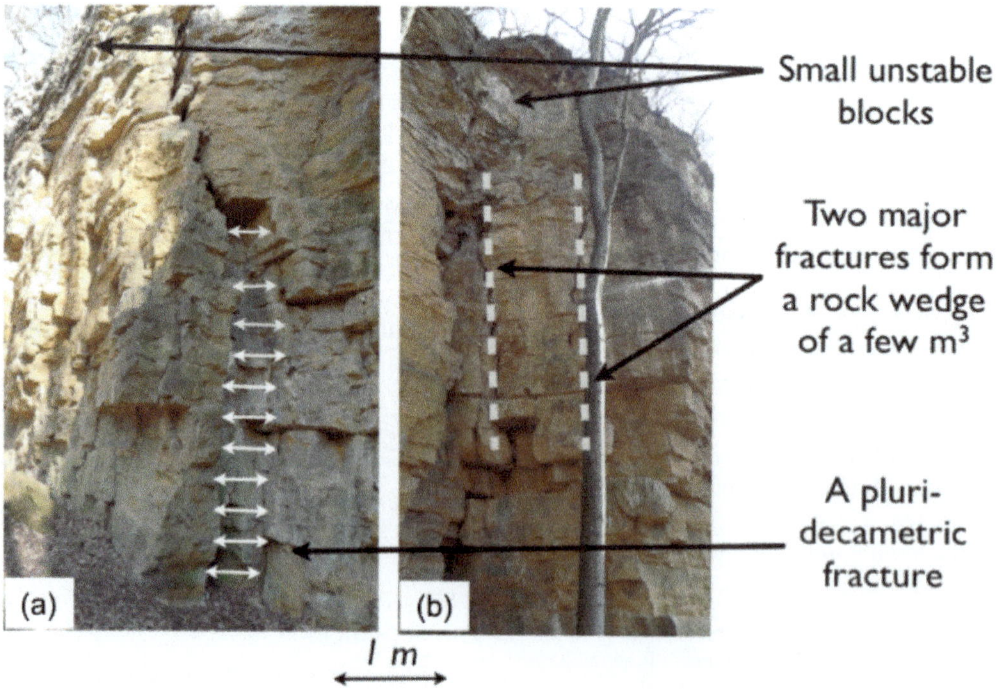

Figure 1. Sector 1 (a) and sector 2 (b) and their main characteristics (Delonca et al., 2013).

Figure 2. Sector 3 and its main characteristics (Delonca et al., 2013).

- A total of 38 first-year MSc students in the geosciences, confronted with a hazard assessment study for the first time. These students may later be asked to prepare hazard or risk maps, for example during an internship.
- A total of 10 researchers in the geosciences, working in the field of hazard and risk assessment but not accustomed to regular rockfall assessment studies.
- A total of 8 confirmed experts benefiting from a long experience in rockfall hazard assessment and risk studies.

The comparison of the rockfall hazard evaluated by each population is supposed to highlight the influence of the level of expertise on the rockfall hazard assessment. Note that the expert assessment is not necessarily the "best" one or the "true" one.

Methods used for the Rockfall Hazard Assessment

At first, an a priori assessment was requested from each population, without making use of any method. Then, two rockfall hazard assessment methods are used: the LPC method, which is a qualitative method frequently used in France, and a SMR-based method that relies on more quantitative information.

A Priori Assessment

The first assessment of the rockfall hazard was requested from each individual of the three populations. Four failure probability levels were proposed: (i) zero, (ii) low, (iii) medium and (iv) high. A presentation of rockfall hazard theory and a presentation of the methods and their parameters were delivered before this work as well as a document containing additional information on the history of the site. The objective of this first assessment was to compare it with the assessment later carried out using the two other methods and to estimate the differences between an a priori assessment and one based on the use of a guided method.

LPC Method

The LPC method is detailed in Laboratoire des Ponts et Chaussées (Effendiantz et al., 2004). It is a qualitative method that is frequently used in France for official hazard mapping. It consists of two main steps: (1) the collection of preliminary data and (2) the use of these data to localize and characterize the potential instability.

The collected preliminary data incorporate all relevant information regarding the study site, including the following.

- Documentary information: the objective is to exploit the archives to avoid duplication of studies already realized.
- Historical information: to study the past of the site to determine whether events have already occurred on the site.
- Geological information: to analyze the lithological and stratigraphic data and the regional geological history to establish the geological context of the study site.
- Structural information: to identify the structural characteristics of the study site on different scales and to propose global and local structural models.
- Morphological information: to identify the main historical steps leading to the actual morphology.
- Hydrogeological, hydrological and climatic information: to characterize the fluid intakes, their nature and their importance. It is also to identify the flows inside the massif and unfavorable climate patterns (e.g., freezing, thawing and important thermal contrasts).

Parameters	Favorable to the stability					Unfavorable to the instability
CODE						
		1	2	3	4	
GEOMECHANICAL PARAMETERS						
Frequency of the discontinuities	Absent					Frequent
Surfaces of the discontinuities	Rough					Smooth
Discontinuity apertures	Closed					Open
Filling of the discontinuities	Sealed					Not sealed
Orientation of the discontinuities	Favorable to the stabilization of the sector *Scheme of the sector :*					"Favorable" to the destabilization of the sector
HYDROLOGICAL AND CLIMATIC PARAMETERS						
Interstitial water	Never exposed					Exposed several days per month
Rainfall / snowmelt	Never exposed					Exposed several days per month
High temperature variation / freeze-thaw cycles	Never exposed					Exposed several days per month
Vegetation	Stabilizing					Destabilizing

Figure 3. Quantification of LPC parameters.

Information regarding vegetation: to identify and characterize the main vegetation on the site and its influence on stability or instability processes.

- Potential mechanism of rupture: to identify the potential mechanisms of rupture associated with the studied zone.
- Sectoring: to identify homogeneous areas from the hazard characterization point of view. This step is performed in the study of large zones.

Following the collection of these preliminary data, the potential instabilities are localized and characterized. To do so, three new stages are carried out.

- Localize and identify potential instabilities. In this stage, the expert in charge of the study observes the following parameters:
- Nature of the studied area: description of the instable rock mass, lithology, hydraulic and hydrologic data.
- Geometric parameters: geometry and dimensions of the potentially unstable compartments.
- Geomechanical parameters related to discontinuities: e.g., spacing, roughness, apertures, filling and orientation of critical discontinuities.
- Triggering factors: e.g., interstitial water, rainfall, high temperature variation, freeze-thaw cycle and vegetation

This stage helps to determine the unstable volume (magnitude) and the potential mechanism of failure. In the experiment proposed in this paper, these parameters were coded to process them statistically (Fig. 3). This codification, which has been validated by expert users of the LPC method for the purposes of this experiment, does not belong to the original method.

- Define the rockfall hazard. The analysis of the previous parameters helps to define potentially unstable volumes and the potential mechanism of rupture associated with these volumes. For each of these, a coupled "temporal probability / occurrence probability" is assessed. These two terms are qualitatively defined as follows.
- The occurrence probability is subjectively assessed from the parameters presented in Fig. 3. It answers the question, "Can the rockfall occur?" The qualitative scale of the level of occurrence probability is presented in Table 1.
- The temporal probability corresponds to the annual frequency or return period of the rockfall on the study site. It is defined as the time expected to failure. In other words, the temporal probability is de-fined by the probability that the failure occurs be-fore an expected delay. The periods are defined on a scale from "imminent" to "long term". The definitions of the terms are presented in Table 2.

Then, for each volume under consideration, these two probabilities are combined in a matrix (Table 3). The plotted couples are then used to provide the best response to the risk: quick action for the shortest terms and the strongest issues and the planning of preventive actions for the longest terms.

- Qualification of the probability of propagation. With the LPC method, it is proposed to carry out a qualitative trajectory analysis before considering a more advanced study using numerical simulations. We do not detail this last step because it has not been considered in the experiment.

Table 1. Qualitative scale of the level of occurrence probability, after the guide of the Laboratoire des Ponts et Chaussées (2004)

Very high	The occurrence of the phenomenon is normal. Its non-occurrence will be exceptional.
High	The occurrence of the phenomenon is more probable than its non-occurrence.
Moderate	The occurrence of the phenomenon is equivalent to its non-occurrence.
Low	The non-occurrence of the phenomenon is more probable than its occurrence.
Very low	The non-occurrence of the phenomenon is normal. Its occurrence will be exceptional.

Table 2. Qualitative scale of the level of temporal probability, after the guide of the Laboratoire des Ponts et Chaussées (2004)

Imminent	The time is measured in hours, days, weeks, or months
Very short-term	Approximately 2 years
Short-term	Approximately 10 years
Medium-term	Approximately 30–50 years
Long-term	Approximately 100–150 years

The use of the temporal probability / occurrence probability matrix is not easy for beginners. Therefore, we have pro-posed a new matrix, developed with experts, to assess the rockfall hazard level. The temporal probability and the occurrence probability are combined to assess the level of predisposition to instability (Table 4). Finally, the predisposition to instability is coupled with the volume (magnitude of the phenomenon) to determine the rockfall hazard (Table 5). Three levels of rockfall hazard are defined: low, medium and high. Several volumes can be identified on a single sector, and then several values for the level of rockfall hazard can be assessed. The global rockfall hazard level of the site corresponds to the worst hazard level on the study site.

SMR-Based Method

The SMR index, proposed by Romana (1985), is a geomechanical classification commonly used for the characterization of rock slopes (Corominas Dulcet and Mavrouli, 2009; Irigaray et al., 2003) and derived from the rock mass rating (RMR) as follows:

$$\mathrm{SMR} = \mathrm{RMR_b} + (F_1 \cdot F_2 \cdot F_3) + F_4, \qquad (1)$$

where $\mathrm{RMR_b}$ is the basic RMR index resulting from Bieniawski's rock mass classification without any correction (Bieniawski, 1972). It is obtained by adding rating values for the following five parameters.

- The strength of the intact rock.
- The rock quality designation (RQD) (Deere and Miller, 1966), which gives a quantitative estimate of the rock mass fracturing based on the study of cores obtained by drilling. The RQD is defined as the percentage of intact pieces of length greater than 10 cm over the total length of the hole. It can also be estimated from surface measurements.
- The spacing of discontinuities.
- The condition of discontinuities: the roughness, weathering and opening of the discontinuities are assessed.
- The water inflow through discontinuities and/or the pore pressure ratio.

F_1, F_2, F_3 and F_4 are defined as follows:

- F_1 characterizes the angle (A) between the slope face strike and joint azimuth. It ranges from 0.15 to 1.00, ac-cording to the relationship F_1 D (1 sin(A))^2. A value of 1 indicates that the joint azimuth and face strike are parallel.
- F_2 refers to the joint dip angle. For a plane sliding mechanism, its value ranges from 0.15 to 1, according to the relationship F_2 D tan(B_j)^2, where B_j denotes the joint dip angle. For a toppling mode of failure, F_2 is equal to 1.00.
- F_3 reflects the relationship between the slope and joint dips. This parameter uses the Bieniawski adjustment factors that range from 60 to 0 points and reflects the probability that outcropping blocks will be subjected to planar and wedge failure mechanisms.
- F_4 is an adjustment factor to take into account the method of excavation (natural slope, presplitting, smooth blasting, blasting or mechanical and deficient blasting). Its value ranges from 8 to 15, and it is chosen empirically.
- The SMR calculation leads to five stability classes (Romana, 1985). Then, based on the work of El-Shayeb et al. (1997) or El-Shayeb (1999), a level of the site activity is evaluated.

Table 3. Matrix used to assess the coupled "temporal probability/occurrence probability", after the guide of the Laboratoire des Ponts et Chaussées (2004)

Temporal probability	Imminent	Very short-term	Short-term	Medium-term	Long-term
Occurrence					
Very high					
High					
Moderate					
Low					
Very low					

Table 4. Qualitative scale of the predisposition to instability

Very High	The fall of the rock mass will happen imminently.
High	The fall of the rock mass is more probable than its stability. The temporal probability estimate is approximately 2 years.
Moderate	The probability of the fall of the rock mass is equivalent to its stability. The temporal probability estimate is approximately 10 years.
Low	The stability of the rock mass is more probable than its fall. The temporal probability estimate is approximately 30-50 years.
Very Low	The fall of the rock mass will be exceptional, or the temporal probability estimate is approximately 100-150 years.

There are four different levels of activity corresponding to the following.

- Sleeping: weathering traces are faded, and there is no alteration of the rock mass.
- Inactive: few weathering traces and superficial alteration.
- Low active: recent weathering traces and deeper rock mass alteration.
- Active: numerous weathering traces and deep rock mass alteration.

The combination of the SMR class and the site activity pro-vides a level of predisposition to the instability (Table 6), which is coupled to the volume of unstable masses to obtain the rockfall hazard level (Table 7).

Note that the level of activity, assessed considering the weathering traces is considered as the history of the site (El-Shayeb et al, 1997). Thus, it includes the temporal activity of rockfalls. Indeed, it is considered that some weathering traces can be used to evaluate the imminent and very short-term qualitative scale of the level of temporal probability (Table 2 – LPC method). Thus, even though this approach is less global than the one proposed by the LPC, it allows the rock-fall hazard level to be evaluated. The different terms used in both the LPC and SMR-based methods are presented in Fig. 4.

Complementary Data

The on-site observations took place at the foot of the cliff. Each individual had a compass, a sclerometer and a geologist's hammer with him and had to report his observations and measurements on a separate form for each sector. On this form, each method was presented again, and all the used parameters were listed (cf. Fig. 3 for the LPC method).

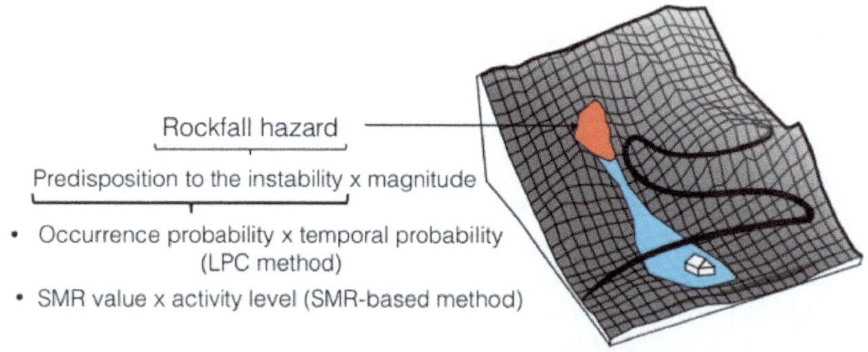

Figure 4. Terms considered in the LPC and SMR-based method.

RESULTS

To statistically treat the results, the level of the rockfall hazard was coded as follows.

- Low-rockfall hazard level: code value equal to 1.
- Moderate-rockfall hazard level: code value equal to 2.
- High-rockfall hazard level: code value equal to 3.

This coding allows the level of rockfall hazard to be quantified and reflects its intuitive increase. Other encodings have been tested, as discussed in Sect. 4.3.

Methods for the Statistical Treatment

Each individual had to make three types of rockfall hazard assessment: an a priori assessment, an assessment with the LPC method and an assessment with the SMR-based method.

Table 5. Assessment of the rockfall hazard with the LPC method

Predisposition to instability	Very low	Low	Moderate	High	Very high
Volume					
< 0.001 m^3	L	L	L	M	M
0.001 to 0.01 m^3	L	L	M	M	H
0.01 to 1 m^3	L	M	M	H	H
1 to 100 m^3	M	M	H	H	H
> 100 m^3	M	H	H	H	H

L: low; M: moderate; H: high

Table 6. Assessment of the predisposition to instability with the SMR method

SMR Activity	VF	F	M	U	VU
Sleeping	VL	L	L	M	M
Inactive	L	L	M	M	H
Low active	L	M	M	H	H
Active	M	M	H	H	H

VF: very favorable; F: favorable; M: moderate; U: unfavorable; VU: very unfavorable; VL: very low; L: low; M: moderate; H: high.

After the experiments were carried out, a descriptive statistical analysis was performed to compare the levels of the rock-fall hazard assessment. Then, the results were analyzed using the chi-square independence test and the analysis of variance (ANOVA) method (Scheffé, 1959) to assess the significance of the influence of the level of expertise and of the influence of the method used in determining the rockfall hazard level. Finally, the means of the LPC and SMR-based method parameters were compared to identify those that most influence the rockfall hazard assessment.

Preliminary Descriptive Analysis

An initial statistical treatment has been performed for the three sectors. Figure 5 and Table 8 show the results and exhibit the following.

- Regarding the a priori assessment, there is a wide dispersion (standard deviation higher than 0.5) of the results in terms of rockfall hazard levels for all populations in all sectors, except for the experts in sector 3. Except for sector 3, where the populations give similar results, the students more often give a higher value for the rockfall hazard level, followed by the experts and finally the researchers. This can be explained by the level of in-experience of the students, who were initially impressed by the presence of cracks and overhanging blocks on the cliff.
- Regarding the LPC method, the results are similar for all population groups, with less dispersion than with the a priori assessment. Students achieved the highest mean score whatever the sector. Experts and researchers are closer to each other, particularly in sectors 2 and 3. For comparison purposes, a random sampling (using uniform distribution) of the levels of predisposition to in-stability and magnitude provided a 6 / 25 D 24 % low-rockfall hazard level, 9 / 25 D 36 % moderate-rockfall hazard level and 10 / 25 D 40 % high-rockfall hazard level (i.e., a mean value of 2.16); the observed values are clearly different from what a random assessment would give.
- With the SMR-based method, the results exhibit more dispersion for the three sectors, with the standard deviations being higher than 0.5 except for the students in sector 2. We also note that the mean rockfall hazard level is always lower than that from the LPC method. For comparison, a random sampling on the SMR matrix (predisposition to instability – magnitude) gives the following results (number of favorable cases on number of possible cases): a 6 / 16 D 37.5 % low-rockfall hazard level, 7 / 16 D 43.75 % moderate-rockfall hazard level and 3 / 16 D 18.75 % high-rockfall hazard level, corresponding to a mean value of 1.8. Most of the values obtained in the experiment are higher than what would be randomly obtained.

The influence of the level of expertise and of the chosen method on the level of rockfall hazard could not be clearly determined from this preliminary descriptive analysis. This is the reason why statistical tests have also been performed.

Influence of the Level of Expertise and the Method on the Level of Rockfall Hazard

Chi-Square Independence Test (χ^2)

The chi-square independence test allows the dependence be-tween two qualitative variables to be investigated.

Let $x_1, \ldots, x_i, \ldots, x_p$ and $y_1, \ldots, y_j, \ldots, y_q$ be the terms (categories) of two qualitative variables X and Y. A sample of n individuals from whom the values of the two variables were simultaneously taken yielded the following results: n_{ij} is the number of individuals who presented both the x_i value of X and the y_j value of Y. $n_{i:}$ and $n_{:j}$ are, respectively, the total of line x_i and the total of column y_j. It is then possible to build the contingency table of the observed values (Table 9).

Table 7. Assessment of the rockfall hazard with the SMR method

Predisposition to instability Volume	Very low	Low	Medium	High
<0.01 m^3	VL	L	L	M
0.01 to 1 m^3	L	L	M	M
1 to 10 m^3	L	M	M	H
>10 m^3	M	M	H	H

VL: very low; L: low; M: moderate; H: high.

Table 8. Mean and standard deviation of the rockfall hazard levels for the three sectors

		A priori assessment		LPC method		SMR-based method	
		Mean	Standard deviation	Mean	Standard deviation	Mean	Standard deviation
Sector 1	Students	**2.15**	0.5	**2.66**	0.3	**2.43**	0.52
	Researchers	**1.6**	0.58	**2.4**	0.5	**2.2**	0.62
	Experts	**1.75**	0.61	**2.27**	0.6	**1.88**	0.6
Sector 2	Students	**2.64**	0.53	**2.89**	0.35	**2.66**	0.43
	Researchers	**2.1**	0.68	**2.8**	0.38	**2.3**	0.52
	Experts	**2.5**	0.5	**2.75**	0.4	**2.5**	0.5
Sector 3	Students	**2.48**	0.51	**2.82**	0.38	**2.37**	0.51
	Researchers	**2.50**	0.5	**2.6**	0.46	**1.8**	0.56
	Experts	**2.75**	0.42	**2.62**	0.44	**1.99**	0.53

Under the hypothesis that variables X and Y are independent, we can also build a contingency table of theoretical values equal to $\frac{n_{i.} \cdot n_{.j}}{n}$ at the intersection of row i and column j . It is then possible to calculate the following quantity.

$$D = \sum_{i=1}^{p} \sum_{j=1}^{q} \frac{\left(n_{ij} - \frac{n_{i.}n_{.j}}{n}\right)^2}{\frac{n_{i.}n_{.j}}{n}}, \quad (2)$$

which obeys a χ^2 distribution with (p −1)(q −1) degrees of freedom.

This test is performed side by side, between (i) the method and the level of rockfall hazard (here considered as a categorical variable) and (ii) the level of expertise and the level of rockfall hazard. As an example, Tables 10 and 11 present the contingency tables of the observed (o_{ij}) and theoretical values (t_{ij}), respectively, for the chi-square test to be performed between the level of expertise (variable X) and the level of rockfall hazard (variable Y) for sector 1.

Table 9. Contingency table of the observed values of the X and Y variables

	y_1	...	y_j	...	y_q	Total
x_1	n_{11}	...	n_{1j}	...	n_{1q}	n_1
⋮	⋮	⋮	⋮	⋮	⋮	⋮
x_i	n_{i1}	...	n_{ij}	...	n_{iq}	$n_{1.}$
⋮	⋮	⋮	⋮	⋮	⋮	⋮
x_p	n_{pi}	...	n_{pj}	...	n_{pq}	$n_{p.}$
Total	$n_{.1}$	...	$n_{.j}$	...	$n_{.p}$	n

The χ^2 distance is then calculated following Eq. (2):

$$D = \sum_{i=1}^{p} \sum_{j=1}^{q} \left((o_{ij} - t_{ij})^2 / t_{ij}\right) \quad (3)$$

Table 12 shows the distances computed for the two independency test carried out on each sector together with their p values. A p value lower than 0.05 indicates that the independency hypothesis can be rejected at a 5 % risk level. This table highlights the following:

- a non-dependency between the level of expertise and the level of rockfall hazard for sectors 2 and 3;
- a dependency between the level of expertise and the level of rockfall hazard for sector 1;

- a dependency between the method and the level of rockfall hazard for the three sectors.

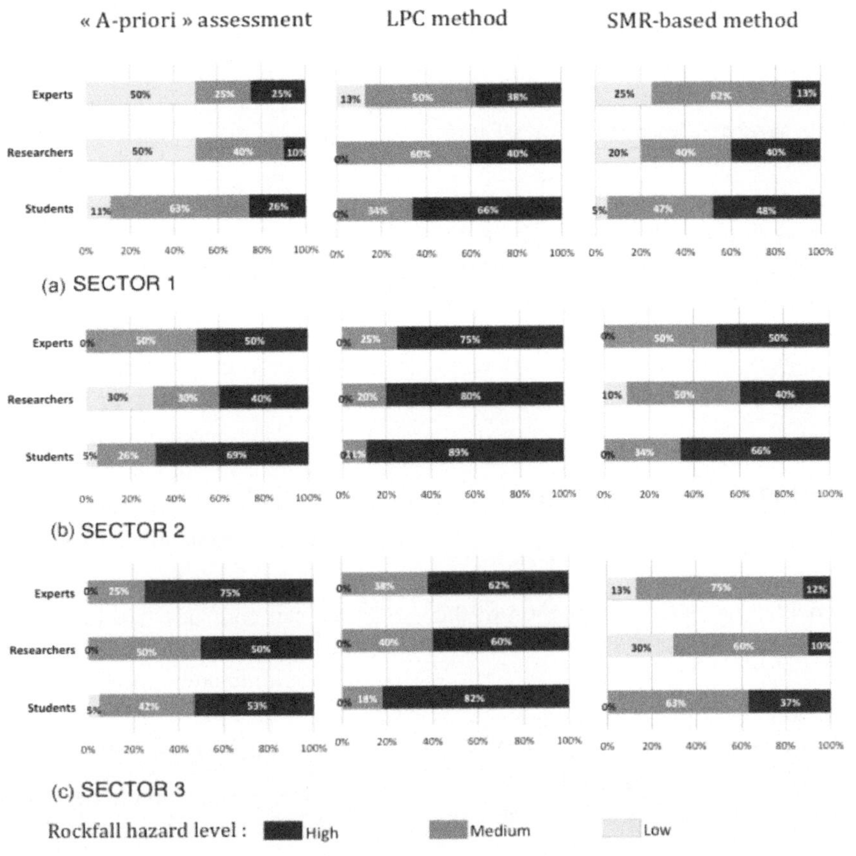

Figure 5. Rockfall hazard level for (a) sector 1, (b) sector 2 and (c) sector 3.

Table 10. Contingency table of the observed values for the chi-square test performed between the level of expertise and the level of rockfall hazard (sector 1)

Observed values (o)	Low-rockfall hazard level	Medium-rockfall hazard level	High-rockfall hazard level	Total
Students	6	55	53	114
Researchers	7	14	9	30
Experts	7	11	6	24
Total	20	80	68	168

Analysis of Variance

ANOVA is a statistical test indicating the influence of qualitative variable(s) on a quantitative variable to be assessed. It is based on the comparison of the mean values of the quantitative variable for each category of the qualitative variable.

Before performing the ANOVA, the normality of the level of rockfall hazard (the quantitative variable) and its homoscedasticity have to be verified. The levels of rockfall hazard are discrete data, which make the condition of normality difficult to assume. However, authors such as Legendre and Borcard (2008) and Underwood (1996) state that, in a balanced experimental design, ANOVA has problems with the heterogeneity of the variance when the variance is in high contrast to others, but the ANOVA is little affected by nonnormality. This condition is considered as not discriminating to achieve the ANOVA.

To test the homogeneity of the variances, there are several tests, including Levene's test, that are insensitive to the normality of the data (Legendre and Borcard, 2008). If Levene's test is statistically significant, the assumption of the homogeneity of variances should be rejected. We used the software program R to perform this test. For the three sectors, Levene's test has shown that we can consider the variances to be homogeneous.

In the ANOVA, the influence of a factor is considered significant when the p value associated with this factor is lower than a given risk such as 5 or 1 % (error chosen here). The ANOVA results are presented in Table 13. To help the reader in the analysis, the significant factors are grayed. The analysis of variance shows the following:

- a very significant influence of the factor "method" on the level of rockfall hazard assessments for the three sectors;
- a very significant influence of the factor "expertise" on the level of rockfall hazard assessments for sector 1;

- the non-significant influence of the factor "expertise" on the level of rockfall hazard assessments for sectors 2 and 3 at a 1 % risk (but a significant influence at a 5 % risk);
- the absence of interaction between these two factors on the level of rockfall hazard assessments (for all three sectors) at a 1 % risk or even at 10 % risk.

These results confirm the results given by the test of independence carried out previously. They first exhibit the main importance of the method chosen, as found previously for the three sectors. Then, the great influence of the level of expertise for sector 1 is highlighted. Finally, both tests highlight the negligible influence of the level of expertise for sectors 2 and 3, regardless of the level of confidence we are working with.

Influence of the Parameters of the LPC and SMR-Based Methods

To better understand the influence of the LPC and SMR-based methods on the level of rockfall hazard, a comparison of the mean values of the parameters used in these methods has been conducted (Table 14).

For each parameter presented, a two-by-two comparison of the mean values has been carried out. More precisely, we have compared the mean values of the levels given for each parameter by (i) the students and the researchers, (ii) the students and the experts and (iii) the researchers and the experts. The objective was to identify the mean values that are statistically similar and those that are different. This analysis could explain the influence of the chosen method previously high-lighted, depending on the sector and the level of expertise.

The first step consists of checking the hypothesis of equal variance for each couple of the population. If the variances can be considered equals, it is then possible to compare the mean values. Let n_1 and s_1^2 be the size and the variance of the first population sample (students, researchers or experts), and let be n_2 and s_2^2 be the size and the variance of the second population sample. Under the hypothesis of equal population variances, the two estimates of the variance should be equal. This hypothesis has been tested (F test) and verified, and the comparison of the means can then be performed.

Let m_1 and m_2 be, respectively, the mean values of the first and second population samples under test. Both are unbiased estimates of the population means (μ_1 and μ_2). Under the assumption that $\mu_1 = \mu_2$, we compute the t ratio (Eq. 3), which is supposed to belong to a Student's t distribution with $(n_1 + n_2 - 2)$ degrees of freedom. This value has then to be compared to the interval $[-t_{\alpha/2} t_{\alpha/2}]$, where is the risk and (1) the confidence level:

$$t = (m_1 - m_2)/\sigma^{*2}\sqrt{(1n_1) + (1/n_2)},$$

(4)

where

- m_1 is the average of the first population tested (students, researchers or experts), and m_2 is the average of the second population tested;
- n_1 is the sample size of the first population, and n_2 is the sample size of the second population;
- σ^{*2} is the unbiased estimate of the common variance of the two populations based on s_1, s_2, n_1 and n_2.

If t belongs to the interval $[-t_{\alpha/2} t_{\alpha/2}]$, being the chosen risk, we can conclude that the two population means are not significantly different. We considered a 5 % risk, so the interval is thus approximately equal to [2,2]. Table 15 presents the t ratio for the different parameters tested.

Table 16 summarizes the results and shows the following:

- the predisposition to the instability (LPC method only) is the same, whatever the level of expertise and the sector;
- the magnitude is similarly assessed for sectors 2 and 3, regardless of the level of expertise, for both methods (LPC and SMR-based methods);
- the magnitude is assessed differently by the students on one side and the experts on the other side for sector 1;
- The SMR classes in the SMR-based method is the same for sectors 2 and 3, but not for sector 1, where there is a significant difference between the students and the experts;
- the level of activity is assessed differently for the three sectors. Then, seeing that the combination of the SMR classes and the site activity provides the level of predisposition to the instability, the predisposition to the instability is assessed differently for the three sectors.

Table 11. Contingency table of the theoretical values for the chi-square test performed between the level of expertise and the level of rockfall hazard (sector 1)

Theoretical values (t)	Low-rockfall hazard level	Medium-rockfall hazard level	High-rockfall hazard level	Total
Students	13.57	54.29	46.14	114
Researchers	3.57	14.29	12.14	30
Experts	2.86	11.43	9.71	24
Total	20	80	68	168

Table 12. Chi-square distance between the observed and theoretical values for all the cases tested, under the hypothesis of independency

D / p value	Sector 1	Sector 2	Sector 3
Level of expertise/level of rockfall hazard	16.98/0.04	9.25/0.99	6.35/0.53
Chosen method/level of rockfall hazard	19.79/0.03	17.09/0.01	25.53/0.003

Note that another analysis (principal component analysis) has been realized to compare directly the parameters used to evaluate the occurrence probability of the LPC method (Fig. 3). However, the analysis did not produce any concluding results.

DISCUSSION

Choice of Hazard Assessment Methods

During the experiment, three methods were used to assess the rockfall hazard. First, each participant was asked to assess the level of rockfall hazard from his own experience with-out using any tool, form or measurement. Second, a qualitative method (LPC method) and, third, a quantitative method (SMR-based method) were used. In both of these latter cases, a form was provided to guide the process of assessment.

We chose to use the LPC method due to its wide use in France for most official hazard mapping. This methodology aims to formalize the practice gradually developed over 2 decades through many field studies produced by both operational services of the French government and local authorities. Many experts contributed to the preparation of this guide, and some of these experts also participated in our experiment. The methodology proposed by the LPC method is somehow classical and follows the following steps as previously described:

- a literature review, which avoids repeating studies or investigations already done;
- an historical review, which makes an initial zoning of hazard levels, taking into account past events;
- a geomorphological analysis, during which the level of the rockfall hazard is actually assessed.

These steps can be found in many other methods that exist in different countries. In Spain, for instance, Copons and Vilaplana (2008) proposed a similar qualitative method for hazard mapping. Several authors also tried to introduce more quantification to hazard assessment for parameters that are usually qualitatively addressed, such as in the Matterock method developed in Switzerland (Baillifard et al., 2003), the RES method (Hudson, 1993; Mazzoccola and Hudson, 1996; Rozos et al., 2008) and the RHAP method (Ferrero et al., 2011; Lombardia, 2000).

Then, even though the LPC method is not an internationally recognized method, we may consider that it treats the hazard analysis in a common manner, similar to other methods. In contrast, the SMR method, also used in the present work, is commonly used in rock slope stability analysis along roadways (Budetta, 2004; Corominas Dulcet and Mavrouli, 2009; Irigaray et al., 2003; Tomas et al., 2012). It is derived from the globally used RMR method for the stability study of excavated slopes or underground excavations.

Even though it is a qualitative method, the LPC method helps the user to identify the block masses that are potentially instable and the volume concerned and leads to the assessment of the level of rockfall hazard in a similar manner by the three tested populations. However, the introduction of a paper form to assist in the process of the assessment of the level of rockfall hazard could have introduced a bias favoring the LPC method in terms of less dispersion in the assessment results when compared with the results of the SMR-based method. As a result, such a form could be promoted to engineers making use of the LPC method to enforce a common interpretation of some aspects of the method.

The SMR-based method shows results similar to those obtained by the LPC method for students. However, for re-searchers and experts, the dispersion in the results is more important. The assessment of the site activity is not precise enough to allow the same level of activity to be assessed by all the individuals. Surprisingly, the use of a partly quantitative method does not prevent differences in the final rock-fall hazard level, mainly because of the assessment of the site activity, which is a subjective component of the method. The terms used to describe the activity may be clarified to reduce this effect. Moreover, the prevailing failure mechanism on the three sectors plays an important role in the use of the SMR-based method. Indeed, the SMR method is routinely used to assess rockfall susceptibility of failure mechanisms involving shear on joints (including flexural toppling) though it seems inapplicable to failures caused by traction. In the test site, the main failure mechanisms are column toppling, wedge slide and overhang failure. However, there also exists a failure mechanism that involves shearing of joints on the three studied sectors. Because it is not the main failure mechanism on the test site, the SMR-based method has been considered anyway into the analysis.

Table 13. Results of ANOVA: influence of the level of expertise and the choice of the method on the level of rockfall hazard for the three sectors. The bold values highlight the significantly influencing factors, for which the ***p*** values are lower than 1 %

		Degrees of freedom	Sum of squares	Mean square	Fisher–Snedecor test value	*p* value
Sector 1	Method	2	8.73	4.36	11.79	$\mathbf{1.68 \times 10^{-5}}$
	Expertise	2	6.21	3.11	8.4	$\mathbf{3.40 \times 10^{-4}}$
	Method–expertise	4	1.35	0.34	0.91	0.46
	Residuals	159	58.83	0.37		
	Total	167	75.12	8.18		
Sector 2	Method	2	7.17	3.59	13.03	$\mathbf{5.56 \times 10^{-6}}$
	Expertise	2	2.31	1.15	4.2	0.017
	Method–expertise	4	1.79	0.45	1.63	0.17
	Residuals	165	45.40	0.28		
	Total	173	56.77	5.47		
Sector 3	Method	2	8.43	4.22	16	$\mathbf{4.64 \times 10^{-7}}$
	Expertise	2	1.71	0.86	3.24	0.042
	Method–expertise	4	1.82	0.46	1.73	0.15
	Residuals	162	42.75	0.26		
	Total	170	54.71	5.8		

Table 14. Code of the parameters used in the LPC and SMR-based methods

Parameters studied	Code
Predisposition to instability (LPC)	1 (very low level) to 5 (very high)
Magnitude (LPC)	1 ($<0.001\,m^3$) to 5 ($>100\,m^3$)
SMR classes (SMR-based)	calculated in the SMR-based method
Activity (SMR-based)	1 (sleeping) to 4 (active)
Predisposition to instability (SMR-based)	1 (very low level) to 4 (high level)
Magnitude (SMR-based)	1 ($<0.01\,m^3$) to 4 ($>10\,m^3$)

Choice of the Test Site

For the experiment, one test site was selected, among which three sectors could be defined. These three sectors provide repeatability in the analysis and produce more relevant results. However, because there are three sectors on the same site, all of them are composed of the same lithology. Hence, the results obtained in the present study only apply to this case study, and the experiment should be repeated on other sites with different lithology and geological structures to prove that the findings are generic. The test site was chosen be-cause of its history, as it was a climbing site which has been closed after repetitive rockfalls. However, no quantitative in-formation concerning the occurrence of the phenomenon was available. The site is not monitored, and the sole data considered in the analysis were the visual evidence of events. The scenario considered here is that an expert is in charge of the evaluation of a new site, and he does not have any temporal data available to him. Therefore, to propose a more complete analysis, it would be interesting to consider another test site where the estimation of temporal frequency is possible. A national French project is underway that may provide a frame-work to repeat this experiment on different study sites, with more students, researchers and experts involved, as well as other methods to be tested. However, no sites have yet been chosen to carry out this experiment again.

Coding of the Level of Rockfall Hazard

As explained previously, to carry out the statistical analysis, the three rockfall hazard levels were transformed into quantitative variables: the low-rockfall hazard level was coded to 1, the moderate-rockfall hazard level to 2 and the high-rockfall hazard level to 3, so as to reflect the increase of the rockfall hazard in a familiar and convenient way. However, to confirm that this coding has no influence on the results, we performed additional ANOVAs with different encodings. The following cases were used, respectively, for low-, moderate- and high rockfall hazard levels:

- 0, 5 and 10;
- 1, 2 and 5;
- 1, 4 and 5;
- 3, 2 and 1 (converse coding).

Table 15. Values of the t ratio, to compare the average 2 by 2 of the parameters of the LPC and SMR-based methods. The bold values correspond to the population means that cannot be significantly differentiated (or can be considered as equal)

Sector 1			
Ratio *t*	Students–researchers	Students–experts	Researchers–experts
Predisposition to instability (LPC)	**0.83**	**−0.49**	**−1.02**
Magnitude (LPC)	**0.55**	3.4	**1.96**
SMR classes (SMR-based)	**1.6**	2.87	**1.22**
Activity (SMR-based)	3.81	**1.57**	**−1.55**
Predisposition to instability (SMR-based)	**1.39**	**−1.08**	**−1.61**
Magnitude (SMR-based)	**0.11**	2.51	**1.39**
Sector 2			
Ratio *t*	Students–researchers	Students–experts	Researchers–experts
Predisposition to instability (LPC)	**−0.6**	**−1.52**	**−0.97**
Magnitude (LPC)	**−0.98**	**0.49**	**1.02**
SMR classes (SMR-based)	**0.82**	**0.46**	**−0.19**
Activity (SMR-based)	5.9	**0.12**	−3.53
Predisposition to instability (SMR-based)	2.57	**−0.95**	−2.47
Magnitude (SMR-based)	**−1.74**	**0.39**	**1.76**
Sector 3			
Ratio *t*	Students–researchers	Students–experts	Researchers–experts
Predisposition to instability (LPC)	**0.25**	**0.71**	**0.77**
Magnitude (LPC)	**1.64**	**1.46**	**0.23**
SMR classes (SMR-based)	**−0.47**	**0.06**	**0.39**
Activity (SMR-based)	4.05	**1.97**	**−1.53**
Predisposition to instability (SMR-based)	3.41	**0.7**	**−1.85**
Magnitude (SMR-based)	**0.56**	**1.41**	**0.60**

Table 16. Sample mean and standard deviation of the values of the LPC and SMR-based parameters. The bold values correspond to the population means that cannot be significantly differentiated

			Sector 1		Sector 2		Sector 3	
			Mean	Standard deviation	Mean	Standard deviation	Mean	Standard deviation
LPC method	Predisposition to instability (1 to 5)	Students	**3.24**	**1.09**	**3.36**	**0.63**	**3.96**	**0.78**
		Researchers	**2.88**	**0.92**	**3.5**	**0.52**	**3.88**	**0.33**
		Experts	**3.5**	**1.22**	**3.83**	**0.75**	**3.71**	**0.48**
	Magnitude (1 to 5)	Students	3.72	0.68	**3.48**	**0.59**	**3.56**	**0.71**
		Researchers	3.55	0.88	**3.7**	**0.48**	**3.11**	**0.6**
		Experts	2.5	1.04	**3.33**	**0.82**	**3**	**1.1**
SMR-based method	SMR (value of the SMR)	Students	59.69	16.45	**44.28**	**20.54**	**47.09**	**16.35**
		Researchers	49.07	16.86	**38.08**	**18.59**	**50.47**	**21.47**
		Experts	36.88	18.75	**40.02**	**18.79**	**46.64**	**12.22**
	Activity (1 to 4)	Students	3	0.71	3.2	0.57	3.2	0.57
		Researchers	1.88	0.78	1.8	0.63	2.22	0.67
		Experts	2.5	0.55	3.17	0.75	2.71	0.49
	Predisposition to instability (1 to 4)	Students	3.08	0.49	3.44	0.50	3.44	0.51
		Researchers	2.77	0.67	2.9	0.56	2.66	0.70
		Experts	3.33	0.52	3.66	0.51	3.29	0.49
	Magnitude (1 to 4)	Students	3.04	0.73	**2.64**	**0.81**	**2.72**	**0.25**
		Researchers	3.02	1.22	**3.2**	**0.78**	**2.55**	**0.88**
		Experts	2.17	0.75	**2.5**	**0.54**	**2.29**	**0.76**

It was found that the coding choice does not affect the conclusions regarding the significant influence of the method on all sectors and the non-significant influence of the level of expertise on two (nos. 2 and 3) of the three sectors.

Note that the quantification of the variable is used only for the statistical analysis.

CONCLUSIONS

The novel experiment developed in this paper had the goal of statistically evaluating the influence of the level of expertise and the choice of the method used for the rockfall hazard assessment. Three levels of expertise (students, researchers and experts) and two methods (LPC and SMR-based, plus an a priori assessment) were used on three sectors at one test site.

The main result obtained is that the influence of the level of expertise is not significant on the rockfall hazard assessment, which means that geoscience students and geoscience researchers who are not experienced at rockfall evaluations provide the same assessment as engineers experienced in rockfall hazard assessment.

More precisely, the qualitative analysis of the results, as well as the chi-square independence test and the ANOVA test highlighted the following:

a statistically significant influence of the factor "method" on the level of the rockfall hazard assessment for all three sectors;

- a statistically significant influence of the factor "expertise" on the level of the rockfall hazard assessments for sector 1;
- a statistically non-significant influence of the factor "expertise" on the level of the rockfall hazard assessments for sectors 2 and 3.

The influence of the level of expertise on the rockfall hazard assessment for sector 1 can be explained by the characteristics of this sector. A pluri-decametric fracture isolates a rock panel several hundreds of cubic meters in size. After discussions with the three populations, it appears that this particular configuration is a challenge to the students, who tend to over-estimate the rockfall hazard. For researchers and experts, the stability of the sector is not an issue in the present state, and the hazard level should be low. Here, we highlight the difficulty of assessing the temporal probability on a specific site.

Moreover, the statistical analysis showed a high dispersion in the rockfall hazard assessment for the a priori assessment, especially for the students and researchers, explained by their inexperience in this type of exercise. This result, coupled with the previous one, highlights the "plus-value" of the experts: they do not overestimate the level of rockfall hazard in complex cases, and their assessment is more homogeneous and based on similar cases they have already studied due to their experience. Indeed, for sector 1, which exhibits the above mentioned pluri-decametric fracture, the results show a higher ratio of low-rockfall hazard levels for the experts. This suggests that the experts link the rockfall hazard assessment to the temporal probability assessment. For them, there is no risk of failure within a short time.

Other assessments of the rockfall hazard level will be realized on other sites to enrich this experiment and confirm or challenge the conclusions. The rockfall hazard level will be assessed by other people (students, researchers and experts), and other methods will be used.

If confirmed, the conclusions drawn in the presented paper would suggest that the use of a qualitative approach could be as relevant as a quantitative approach for rockfall hazard assessment even though it leaves room for some subjectivity. Nevertheless, the use of a standard form as proposed in this study (as exists for many other methods, either basically quantitative or qualitative) could be promoted to engineers making use of the LPC method to enforce a common interpretation of some aspects of the method. Such a form may contribute to a reduction of the subjectivity expected in a qualitative method.

ACKNOWLEDGEMENTS

The authors are indebted to all the people (students, researchers and experts) who participated in the experi-ment and the region of Lorraine for its financial support.

REFERENCES

1. Abbruzzese, J. M. and Labiouse, V.: Comparison of two rock fall hazard mapping methodologies based on the French and Swiss guidelines, in Rock Slope Stability RSS 2010 Symposium, Paris, France, 2010.
2. Abella, E. A. C. and Van Westen, C. J.: Qualitative landslide suscep-tibility assessment by multicriteria analysis: a case study from San Antonio del Sur, Guantánamo, Cuba, Geomorphology, 94, 453–466, 2008.
3. Baillifard, F., Jaboyedoff, M., and Sartori, M.: Rockfall hazard map-ping along a mountainous road in Switzerland using a GIS-based parameter rating approach, Nat. Hazards Earth Syst. Sci., 3, 435– 442, doi:10.5194/nhess-3-435-2003, 2003.
4. Bauer, M., and Neumann, P.: A Guide to Processing Rock-fall Haz-ard from Field Data, 3rd International Symposium on Geotechni-cal Safety and Risk (ISGSR 2011), edited by: Vogt, Schuppener, Straub and Bräu, Bundesanstalt für Wasserbau. 2011.
5. Bell, R. and Glade, T.: Multi-hazard analysis in natural risk assess-ments, WIT Transactions on Ecology and the Environment, vol. 77, 2004.
6. Besson, L., Durville, J. L., Garry, G., Graszk, E., Hubert, T. and Toulemont, M.: Plans de prévention des risques naturels (PPR) - Risques de mouvements de terrain., 1999.
7. Bieniawski, Z. T.: Rock Mass Rating (RMR), Rock Mass Classifi-cation, 34–46, 1976.
8. Bormioli, D., Damato, D., Lanteri, L., Morelli, M., Pispico, R., and Troisi, C.: Comparison of methods for the speditive rockfall haz-ard assessment: Activity 1 of the Massa (medium and small size rockfall hazard assessment) project, in Geoitalia 2011 VIII Fo-rum Italiano di Scienze della Terra Torino, 19–23, 2011.
9. Budetta, P.: Assessment of rockfall risk along roads, Nat. Hazards Earth Syst. Sci., 4, 71–81, doi:10.5194/nhess-4-71-2004, 2004.
10. Copons, R. and Vilaplana, J. M.: Rockfall susceptibility zoning at a large scale: From geomorphological inventory to preliminary land use planning, Eng. Geol., 102, 142–151, 2008.
11. Corominas, J. and Mavrouli, O. C.: Evaluation of the susceptibility to failure of rocky slopes based on the SMR index, 2009.
12. Deere, D. U. and Miller, R. P.: Engineering classification

and index properties for intact rock, Technical report no. AFNL-TR-65-116, Air Force Weapons Laboratory, Albuquerque, NM, 1966.

13. Delonca, A., Verdel, T., and Gunzburger, Y.: Impact of expertise on rock fall assessment – An original experimentation, in Rock Mechnaics for Resources, Energy and Environment, 117–121, Kwasniewski and Lydzba, EUROCK, Wrolclaw, 2013.

14. Dorren, L. K.: A review of rockfall mechanics and modelling ap-proaches, Prog. Phys. Geogr., 27, 69–87, 2003.

15. Dussauge-Peisser, C., Helmstetter, A., Grasso, J.-R., Hantz, D., Desvarreux, P., Jeannin, M., and Giraud, A.: Probabilistic ap-proach to rock fall hazard assessment: potential of histori-cal data analysis, Nat. Hazards Earth Syst. Sci., 2, 15–26, doi:10.5194/nhess-2-15-2002, 2002.

16. Effendiantz, L., Guillemin, P., Rochet, L., Pauly, C., and Payany, M.: Les études spécifiques d'aléa lié aux éboulements rocheux, Laboratoire central desponts et chaussées, Paris, 2004.

17. El-Shayeb, Y, Verdel, T., and Didier, C.: Fuzzy Reasoning for the analysis of risk in geotechnical engineering. Application to a French case, Proceedings of the 1997 International Symposium on Intelligent Design in Engineering Applications (IDEA'97), 11 September, edited by: Elite Foundation, Aachen, Germany, 101–105, 1997.

18. El-Shayeb, Y.: Apport à la logique floue à l'évaluation de l'aléa "Mouvement de Terrain des sites géotechniques": propositions pour une méthodologie générale, Institut National Polytechnique de Lorraine, Ecole des mines de Nancy, 1999.

19. Fell, R., Corominas, J., Bonnard, C., Cascini, L., Leroi, E., Savage, W. Z. and others: Guidelines for landslide susceptibility, hazard and risk zoning for land use planning, Eng. Geol., 102, 85–98, 2008.

20. Ferrero, A. M., Migliazza, M., Roncella, R. and Segalini, A.: Rock cliffs hazard analysis based on remote geostructural surveys: The Campione del Garda case study (Lake Garda, Northern Italy), Geomorphology, 125, 457–471, 2011.

21. Hantz, D.: Contribution à l'évolution de l'aléa éboulement rocheux - Approche multidisciplinaire et multiéchelles, Mémoire de diplôme d'Habilitation à Diriger des Recherches, 2007.

22. Hantz, D., Vengeon, J. M., and Dussauge-Peisser, C.: An his-torical, geomechanical and probabilistic approach to rock-fall hazard assessment, Nat. Hazards Earth Syst. Sci., 3, 693–701, doi:10.5194/nhess-3-693-2003, 2003.

23. Hudson, J. A.: Rock Engineering Systems: Theory and Practice, El-lis Horwood Ltd., 1993.

24. Hungr, O., Evans, S. G., and Hazzard, J.: Magnitude and frequency of rock falls and rock slides along the main transportation corri-dors of southwestern British Columbia, Canad. Geotech. J., 36, 224–238, 1999.

25. Irigaray, C., Fernández, T. and Chacón, J.: Preliminary rock-slope-susceptibility assessment using GIS and the SMR classification, Natural Hazards, 30, 309–324, 2003.

26. Jaboyedoff, M., Baillifard, F., Hantz, D., Heidenreich, B. and Maz-zoccola, D.: Prévention des mouvements de versants et des in-stabilités de falaises?: confrontation des méthodes d'étude des éboulements rocheux dans l'arc alpin, Programme INTERREG II, Suisse, 2001.

27. Jaboyedoff, M., Dudt, J. P., and Labiouse, V.: An attempt to refine rockfall hazard zoning based on the kinetic energy, frequency and fragmentation degree, Nat. Hazards Earth Syst. Sci., 5, 621–632, doi:10.5194/nhess-5-621-2005, 2005.

28. Leroi, E., Bonnard, C., Fell, R., and McInnes, R.: Risk assessment and management, edited by: Hungr, O., Fell, P., Couture, R., Eberhardt, E., Landslide Risk Management, Taylor & Francis Group, London, 159–198, 2005.

29. Lateltin, O.: Prise en compte des dangers dus aux mouvements de terrain dans le cadre des activités de l'aménagement du territoire, Recommandations, OFEFP, vol. 42, 1997.

30. Luckman, B. H.: Rockfalls and rockfall inventory data: Some obser-vations from surprise valley, Jasper National Park, Canada, Earth Surf. Process., 1, 287–298, doi:10.1002/esp.3290010309, 1976.

31. Mazzoccola, D. F. and Hudson, J. A.: A Comprehensive Method of Rock Mass Characterization for Indicating Natural Slope Insta-bility, Q. J. Eng. Geol. Hydrogeol., 29, 37–56, 1996.

32. Moiriat, D., Colin, S. and Dufrenoy, R.: Aléa chute de blocs sur le territoire de Meurthe et Moselle (54) – Etat des connaissances et cartographie au 1/50000, BRGM/RP-56628-FR., 2008.

33. Moreiras, S. M.: Frequency of debris flows and rockfall along the Mendoza river valley (Central Andes), Argentina: Associated risk and future scenario, Quaternary International, 158, 110–121, doi:10.1016/j.quaint.2006.05.028, 2006.

34. Office fédéral de l'aménagement du territoire OFAT: Prise en compte des dangers dus aux mouvements de terrain dans le cadre des activités de l'aménagement du territoire, 1997.

35. Regione Lombardia: Procedure per la valutazione e la zonazione della pericolositá e del rischio da frana in Regione Lombardia. Bollettino Ufficiale della Regione Lombardia 51, Milano (in Ital-ian), 2000.

36. Romana, M.: New adjustment ratings for application of Bieniawski classification to slopes, 49–53, International Society of Rock Me-chanics, Zacatecas, 1985.

37. Rozos, D., Pyrgiotis, L., Skias, S., and Tsagaratos, P.: An imple-mentation of rock engineering system for ranking the instability potential of natural slopes in Greek territory. An application in Karditsa County, Landslides, 5, 261–270, 2008.

38. Scheffé, H.: The Analysis of Variance, John Wiley & Sons, 1959. Tomas, R., Cuenca, A., Cano, M., and García-Barba, J.: A graphical approach for slope mass rating (SMR), Eng. Geol., 124, 67–76, 2012.

39. Underwood, A. J.: Experiments in ecology: their logical design and interpretation using analysis of variance, Cambridge University Press, 1996.

CHAPTER 2

PARTIAL REACTIVATION OF A HUGE DEEP-SEATED ANCIENT ROCK SLIDE: RECOGNITION, FORMATION MECHANISM, AND STABILITY

Minggao Tang, Qiang Xu, Yusheng Li, Runqiu Huang, Niek Rengers, and Xing Zhu

INTRODUCTION

Before the old town of Fengjie was submerged in the water of the Three Gorges Reservoir, the Sanmashan slope area was recommended as the resettlement site for building a new Fengjie in 1997 (Fig. 1). During the construction process of the new Fengjie town, a large-scale discontinuous weak zone was found in the Sanmashan slope area in 1998 (Fig. 2). To determine the safety of the slope by geotechnical analysis, it was necessary to determine the origin and characteristics of the weak zone.

The identification and formation mechanism of deepseated ancient rock slides is an important research subject (Crosta, 1996; Crosta and Zanchi, 2000). It is well known that both a rock slide and tectonic faulting can form an important weak zone (Illies, 1981; Chang et al., 1981; Crosta, 1996; Crosta and Zanchi, 2000; Cruden and Martin, 2007; John and Douglas, 2012). However it is often not easy to decide which process has taken place. Michael et al. (2012) found that a landslide scarp can easily be misinterpreted as a fault. However, tectonic activity can play a key role in the development of a large-scale rock slide (Brideau et al., 2009). In this paper we describe the study of a very large, interesting deep-seated rock slide, called the Sanmasan landslide, in the new Fengjie town of the Three Gorges Reservoir area, Chongqing, China (Figs. 1 and 2).

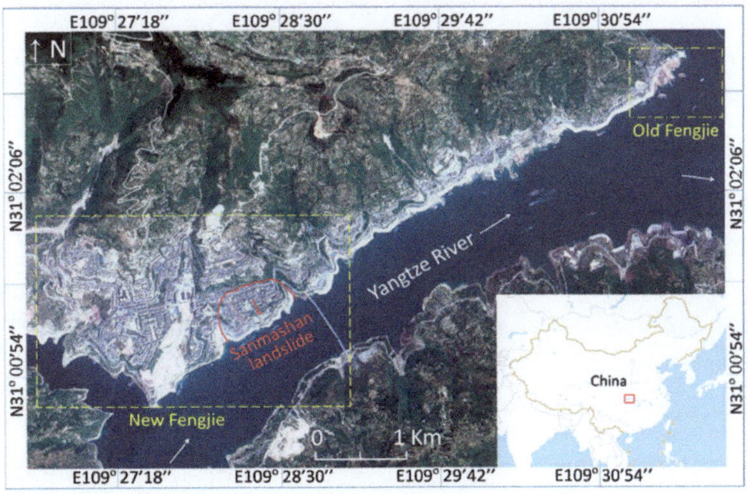

Figure 1. Location map of the Sanmashan landslide (31◦0101800 N, 109◦2801600 E) (Quickbird image, 2 May 2010). It is located in the new Fengjie town in the Three Gorges Reservoir area, Chongqing, China. The old Fengjie town was submerged after the Three Gorges Reservoir filled with water in 2003.

Citation: Tang, M., Xu, Q., Li, Y., Huang, R., Rengers, N., and Zhu, X.: Partial reactivation of a huge deep-seated ancient rock slide: recognition, formation mechanism, and stability, Nat. Hazards Earth Syst. Sci., 16, 1719-1735, doi:10.5194/nhess-16-1719-2016, 2016.

Before the start of our research there were two viewpoints or hypotheses about the origin of the weak zone: a huge ancient slide or a fault graben. Some authors (BIS-CWRC, 1997, 1999) suggested that it was caused by geological tectonic activity, as the slope area is surrounded by four normal faults (see Supplement 1, from the previous study), due to the formation of a local fault graben, and this viewpoint was quoted in later research papers (Li et al., 2002; Wang et al., 2006; Qi et al., 2012). Other authors (Nanjiang Geological Team, 1999) speculated that the weak zone was the result of a very large ancient rock slide. These two different viewpoints confused the land use planners and those responsible for geological hazard prevention in the new Fengjie town. A large number of resettlemed residents and buildings on the Sanmashan area were located in the new Fengjie town (Fig. 3), and filling–drawdown cycles of the reservoir water table and rainfall may affect the stability of the landslide with disastrous consequences (Alonso and Pinyol, 2010; Barla et al., 2010; Paronuzzi et al., 2013; Tang et al., 2015a). Therefore, the main question to be answered by our research is as follows: was this large-scale discontinuous weak zone caused by a huge ancient rock slide or by a local fault graben structure? If it was concluded that it is a huge landslide, how did that ancient slide happen? Is it stable and safe under the filling–drawdown cycles of the Three Gorges Reservoir?

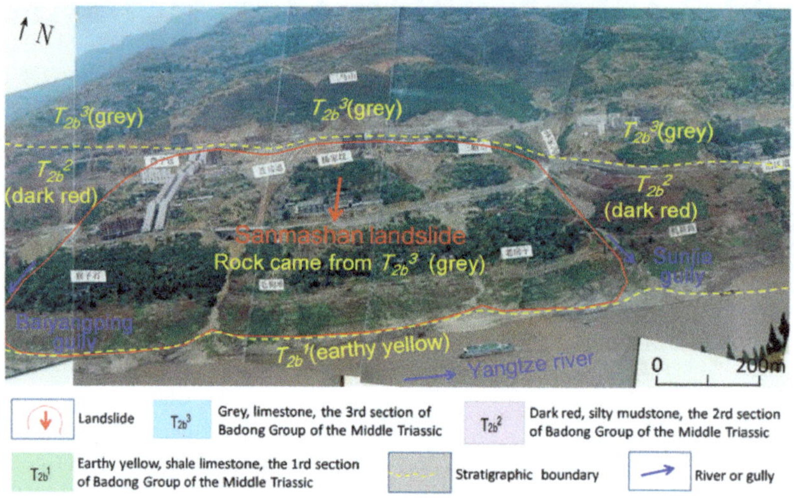

Figure 2. Panoramic photograph of the Sanmashan landslide before the Three Gorges Reservoir filled with water and before the relocation of the new Fengjie town (photograph was taken by Nanjiang Geological Team in 1998). The attitude of the bedrock bedding is 335–350° < 19–30°, but the rock in the Sanmashan landslide is dislocated. The elevation of Yangtze river water level is about 80 m a.s.l. View direction is NW.

Figure 3. The residential buildings on the Sanmashan landslide in the new Fengjie town after the Three Gorges Reservoir filled with water (photograph taken 6 July 2013). The elevation of the Yangtze river water level is about 147 m a.s.l.

To solve these questions, the authors have carried out geological investigations and mapping, performed a deep exploration borehole, and carried out rock and soil mechanics tests, a large centrifuge model experiment, and a stability analysis.

GEOLOGICAL BACKGROUND

Regional Geological Tectonics

There are three major geomorphological levels in China (Li et al., 2001). The new Fengjie town in the Three Gorges Reservoir area is close to the eastern margin of the second level. This region is part of the Yangtze platform, a tectonic unit in which the basement rocks are mainly composed of early Proterozoic metamorphic volcaniclastic rocks and intrusive magmatic rocks. The overlying sedimentary rocks, deposited during the Triassic, were folded and uplifted during the Yanshan phase at the end of the Jurassic. The upper Yangtze platform fold belt, the marginal depression of the Sichuan Basin, and the Dabashan platform fold belt converge in this region (No. 107 Geological Team of Sichuan Province, 1980) (Figs. 4 and 5).

Strata, Rocks, and Soils

The rocks in the new Fengjie town area are of Triassic age (No. 107 Geological Team of Sichuan Province, 1980; Li et al., 2006).

- The Early Triassic Jialingjiang group (T_{1j}) was deposited in a shallow lagoon facies, and contains sedimentary limestone, dolomitic limestone, and limestone.
- The Middle Triassic Badong group (T_{2b}) was formed in inland lakes and lagoon facies, with both clastics and carbonate deposition. The rocks consist of mudstone, silty mudstone, marly limestone, and limestone. This group (T_{2b}) is divided into four units, T_{2b}^1, T_{2b}^2, T_{2b}^3 and T_{2b}^4 (Fig. 6).
- The overburden soils contain landslide deposit (Q^{del}), residual and deluvium deposit (Q^{edl}), alluvial and diluvial deposit (Q^{apl}), and artificial fill (Q^{ml}).

ANALYSIS OF THE LOCAL GEOLOGICAL STRUCTURE

Geological Field Study and Mapping

After reviewing the available geological maps, aerial photographs, previous papers, and reports on the area (No. 107 Geological Team of Sichuan Province, 1980; Chen and Zhang, 1998; Nanjiang Geological Team, 1999), we conducted field work and geological mapping and collected the following information: the stratigraphic distribution in the Sanmashan area (shown in Fig. 7); the outcrop features of bedding and rock types in the Sanmashan area (shown in Fig. 6); and the cross section of the key geological structure and lithologies, the Zhuyi anticline in the new Fengjie town area (shown in Fig. 8). The palaeostress reconstruction of the tectonic stress field in the Sanmashan area, the attitude of the two limbs of the Zhuyi anticline (Fig. 9) and its conjugate shear joints (Fig. 10) are analyzed in Fig. 11.

Analysis and Discussion of Local Geological Structure

It is well known that the formation, orientation, and evolution of faults and joints in rock are controlled by the palaeotectonic stress field. Conversely, the tectonic stress field can be deducted from the mechanical properties of faults and joints and their structural geological orientation. The Zhuyi anticline at the Sanmashan area is a double inverted anticline (an overturned and closed fold with a dipping axial plane and overturned strata) with strong crushing of strata in the core of the Zhuyi anticline (Figs. 7–10). This shows that compressive stresses have resulted in closed folding and compressional thrust structures. With the help of the stereographic projection analysis of the conjugate shear jointing and the overturned bedding, the regional maximum compressive stress direction in the new Fengjie town can be determined as 344– 352.5° (north–south) (Figs. 11 and 4). The character of faults (shear or tension) is determined by the tectonic stress field (Illies, 1981; Chang et al., 1981). Normal faults need attentional stress field, so a local fault graben (BIS-CWRC, 1997, 1999) (see Supplement 1) could not have formed under the regional compressive tectonic stress field conditions in this area.

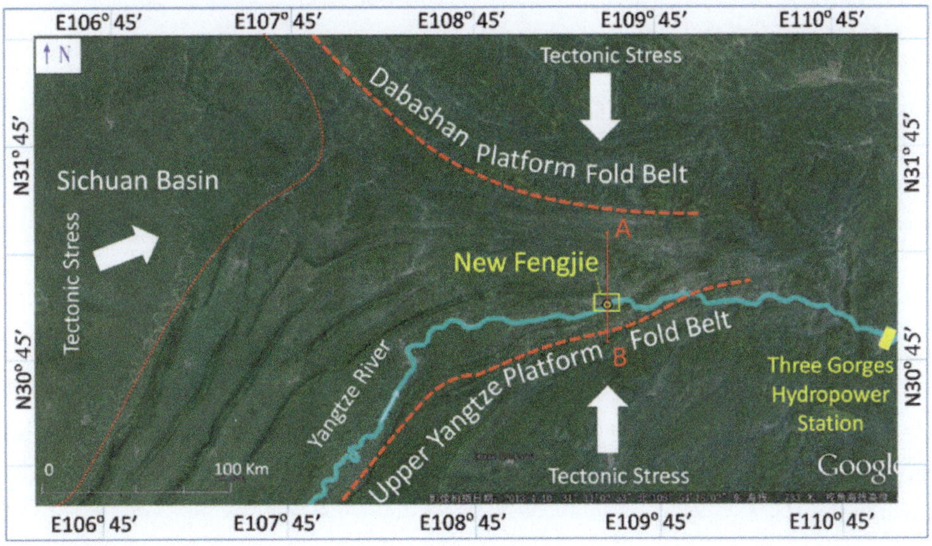

Figure 4. Regional tectonic domains and geostress fields (image from Google earth).

We especially investigated the four normal faults in the field, F_8, F_{20}, F_{19}, and F_{26}, which form the local fault graben (BIS-CWRC, 1997, 1999) (see Supplement 1). They share some common characteristics which do not fit with a local fault graben for the following reasons. (1) The extension lengths of the faults are not long enough (F_8: 1110 m, F_{19}: 150 m, F_{26}: 100 m), the fault fracture zone is not wide (F_8: 3.5 m, F_{19}: 0.1–1.1 m, F_{26}: 1–2 m), but the fault displacement is very large (F_8: 150–180 m, F_{26}: 25–35 m). According to geological mechanics (Illies, 1981; Chang et al., 1981), four normal faults of this type, F_8, F_{20}, F_{19}, and F_{26} (BIS-CWRC, 1997, 1999), cannot form in a compressive tectonic stress field. (2) The footwall rock mass of the faults is preserved intact, but the hanging wall is almost entirely disintegrated, which does also not support the formation of normal faults around a local fault graben. (3) The thickness of T_{2b}^2 in the local fault graben is only 200 m (BIS-CWRC, 1997, 1999). According to earlier research (No. 107 Geological Team of Sichuan Province, 1980; Li et al., 2006), the thickness of T_{2b}^2 is regionally about 500 m, which means that about 300 m thickness of T_{2b}^2 has been lost in the fault graben (BIS-CWRC, 1997, 1999). (4) Faults F_{20} and F_{26} have not been found on the spot. Faults F_8 (which is actually a rear sliding rupture of the ancient slide, shown in Fig. 14c) and F19 (a sliding zone of the ancient slide, shown in Fig. 14f) show the characteristics of pressure-shear sliding rather than tension as would be expected at a normal fault. Thus an analysis of the results of our field data collection does not agree with the hypothesis of a local fault graben as the cause of the large discontinuous weak layer in the Sanmashan area.

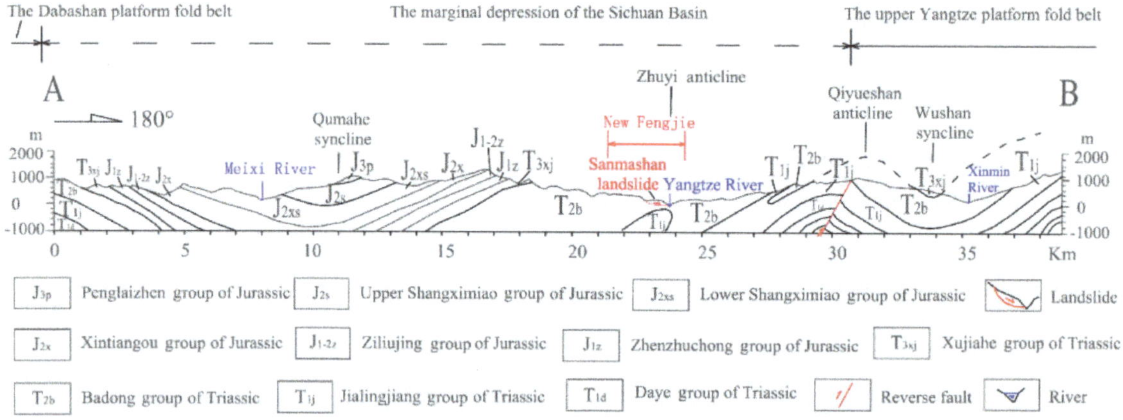

Figure 5. Schematic N–S geological cross section of background with regional tectonic domains. The location of the A–B profile is shown in Fig. 4.

Figure 6. Rock outcrops in the new Fengjie town area. D01: the first section of the Middle Triassic Badong group (T_{2b}^1 beds). D02: the second section of the Middle Triassic Badong group (T_{2b}^2 beds). D03: the third section of the Middle Triassic Badong group (T_{2b}^3 beds). D04: the fourth section of the Middle Triassic Badong group (T_{2b}^4 beds). The outcrop locations of D01, D02, D03 and D04 are shown in Fig. 7.

These data provide rather a large amount of evidence for the assumption that this area is a huge deep-seated ancient rock slide.

IDENTIFICATION OF THE SANMASHAN LANDSLIDE

Methodology

Detailed geological mapping, the investigation of an exploratory tunnel and a foundation pit, and the logging of 21 deep bore holes were carried out at the site. The results are shown in a detailed geological map of the new Fengjie town area at a scale of 1 : 10 000 (Fig. 7), and an engineering geological map of the Sanmashan landslide at a scale of 1 : 1000 (Fig. 12). The outline of the landslide, the geomorphological features and the sliding traces and structural geological features of the Sanmashan landslide are shown in Figs. 12–14. Eighteen samples were collected from the sliding plane, the slide mass and bedrock for the geotechnical laboratory tests.

Geomorphological Features and Sliding Traces

A steep (70) cliff, over 130 m in height, forms the back scarp of the landslide (Figs. 13 and 14b), a gully called Baiyang-ping forms the western lateral boundary, and a gully called Sunjia forms the eastern lateral boundary, while the Yangtze River (the Three Gorges Reservoir) forms the front of the landslide. The terrain slope angle of the landslide is 15– 20 . The length of the landslide is over 780 m, the width is about 1020 m, and the maximum thickness is over 150 m. The average thickness is about 125 m, and the volume is about 100 million m^3 (Figs. 12 and 13). The geomorphological characteristics of the Sanmashan landslide are as follows.

During the construction of the new Fengjie town in 2000, the sliding plane in the upper part of the landslide and the scarp were exposed in a foundation excavation (Fig. 14c). The sliding surface is very obvious and very smooth, showing the characteristics of squeezing and sliding.

The sliding plane at the eastern lateral boundary has a thickness of 1.1 m (Fig. 14e). It is exposed in the exploratory tunnel near the eastern lateral boundary, it is very smooth, and also shows the characteristics of squeezing and sliding (Fig. 14f). These sliding traces are very clear, but have been misidentified as faults $\boldsymbol{F}_{19}$ in earlier research (BIS-CWRC, 1997, 1999).

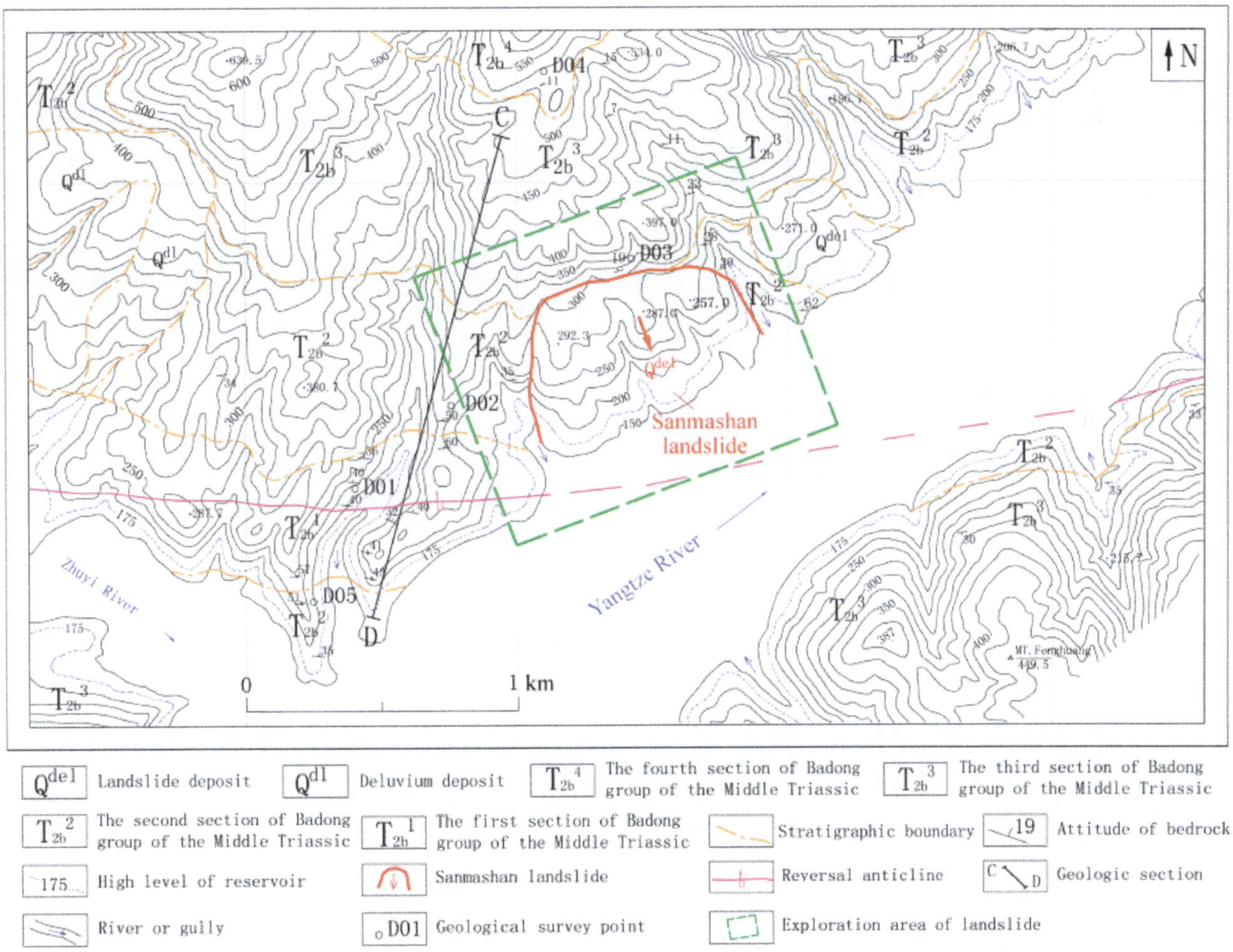

Figure 7. Geological map at 1: 10 000 scale of the new Fengjie town area. The location is the area of the yellow box in Figs. 1 and 4. The topographic map was prepared in 1990 before the relocation of the new Fengjie town. The area of the green box in this map is an area of major site investigations (see Fig. 12).

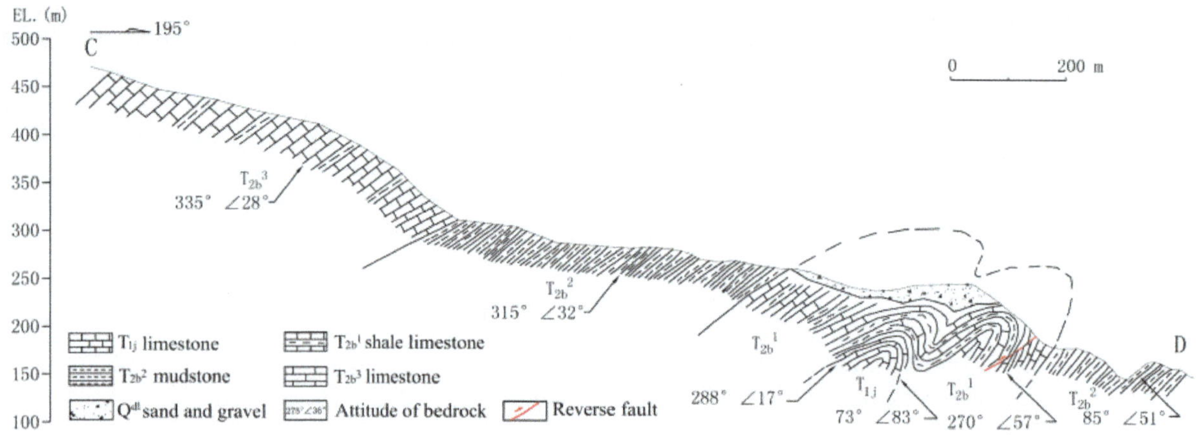

Figure 8. Geological cross section of the Zhuyi anticline. The location of the C–D cross section is shown in Fig. 7. It is a double overturned anticline, see also Figs. 9 and 10.

The western lateral boundary, called Longzibaogou, leads into the Yangjiaping gully (Figs. 12 and 14). The sliding surface there was found and observed in a horizontal borehole (Fig. 14g); the slickensides are clear and smooth.

The elevation above sea level of the toe of the landslides about 90 m. There are three secondary (partial) landslides at the foot of the Sanmashan landslide called the Houzishi, Zhiwuyou, and Laofangzi landslides (Figs. 12–14).

Geological Structure of the Landslide

The Sanmashan landslide can be subdivided into four domains (I–IV) based on differences in morphology, geological structure, and inferred failure mechanics (Figs. 12 and 14). Seven engineering geological units were identified in these four domains (Fig. 13).

- Units 1 and 2 are the landslide deposits (Q^{del}) (Fig. 13) with a thickness of up to 150 m. (1) The first unit is dense to very dense, and contains grey or dark grey crushed stone in a matrix of silty clay and sand. The crushed stone is derived from limestone (T_{2b}^{3}). A large foundation pit was constructed in this unit in the center of the landslide (Fig. 14i). The slide zone of the Zhiwuyou secondary landslide is exposed in the eastern wall of this pit (Fig. 14j) and the rock masses in the western wall have anomalous attitudes with respect to each other (Fig. 15). (2) The second unit is even more dense, reddish or dark red/brown and contains crushed stone in a matrix of clay and silty clay. The crushed stone is derived from mudstone and silty mud-stone (T_{2b}^{2}).
- Units 3, 4, and 5 are the bedrock (T_{2b}) (Fig. 13). (3) The third member of the Badong group (T_{2b}^{3}/ contains grey or dark grey limestone, with a local thickness of up to 205 m. (4) The second member of the Badong group (T_{2b}^{2}), which consists of kermesinus or dark red/brown mudstone and silty mudstone, has a local thickness of up to 550 m. (5) The first section of the Badong group of the Middle Triassic (T_{2b}^{1}) consists of yellowish or light grey shale limestone, with a local thickness of about 70 m.
- Unit 6 is alluvium and proluvium (Q^{apl}) (Fig. 13). (6) It is loose to slightly dense and contains subrounded to rounded gravel, cobbles, and mixed boulders of limestone and silty mudstone in a matrix of sand, deposited at the front of the landslide. The observed unit thickness is up to 15 m.
- Unit 7 is artificial fill soil (Q^{ml}) (Fig. 13). (7) It is loose to dense, consists of concrete, clay, and sand within mixed with crushed stone, and is randomly filled with depressions or spread out on top of the landslide surface. The observed unit thickness is up to 10 m.

Table 1. Physical, mechanical, and hydraulic properties for the five main geologic units

Geologic unit	Modulus/E (MPa)	Poisson's ratio/ ν	Unit weight/ r (kN m^{-3})	Cohesion/ c (kPa)	Internal friction angle/ ϕ (°)	Saturated permeability/ K_s (m s^{-1})
Q^{del} (T_{2b}^{3})	85	0.35	22.5	28	33	1×10^{-5}
Q^{del} (T_{2b}^{2})	50	0.40	21	30	18.2	5×10^{-8}
Limestone (T_{2b}^{3})	4800	0.25	26	200	40	10^{-9}
Mudstone (T_{2b}^{2})	1500	0.35	23	100	25	10^{-9}
Shale limestone (T_{2b}^{1})	3000	0.30	24.8	120	35	10^{-9}

Figure 9. The bedding plane on the south limb of the Zhuyi anticline. The ripple marks show that the sequence is overturned.

Figure 10. Conjugate shear joints of in rock outcrop on the south limb of the Zhuyi anticline.

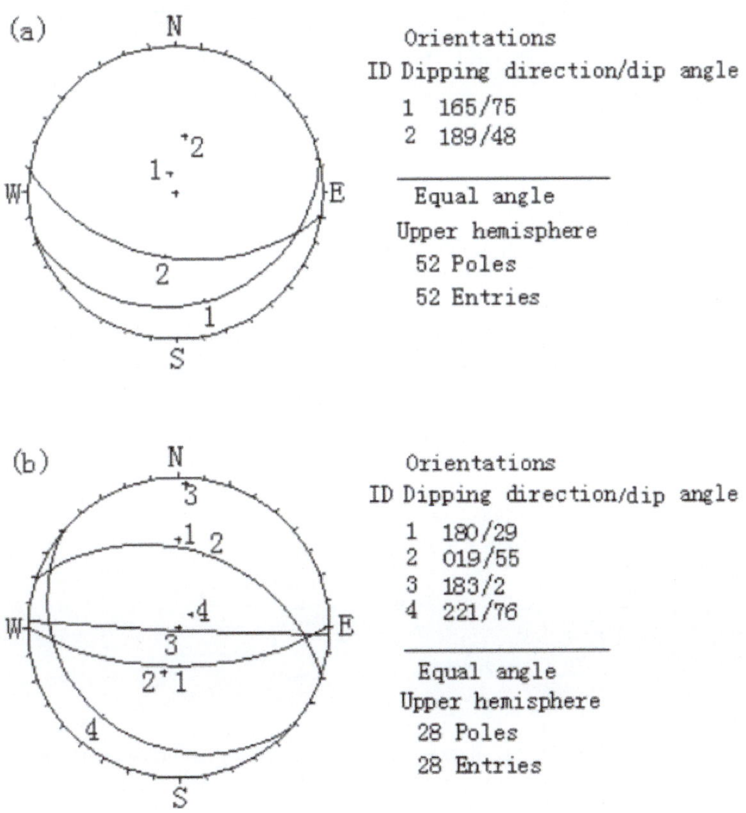

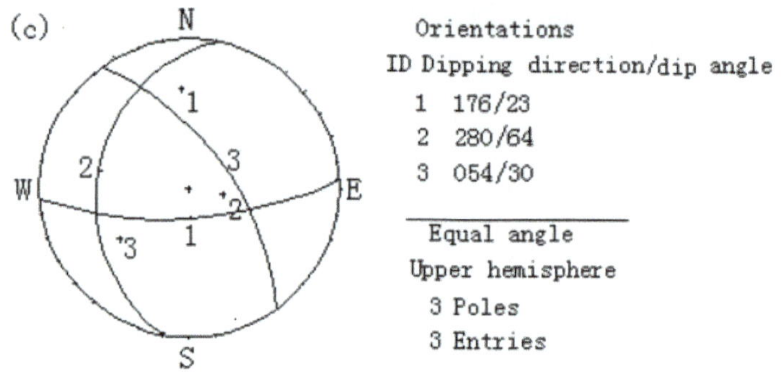

Figure 11. Stereographic projection of the orientation of the two limbs of the Zhuyi anticline and the conjugate shear joints. (a) The attitude of the north limb of the Zhuyi anticline, measuring point (D01) and (D02) (location see Figs. 6 and 7). (b) The attitude of the south limb of the Zhuyi anticline, measuring point at location of Fig. 9. (c) Conjugate shear joints, measuring point at location of Fig. 10.

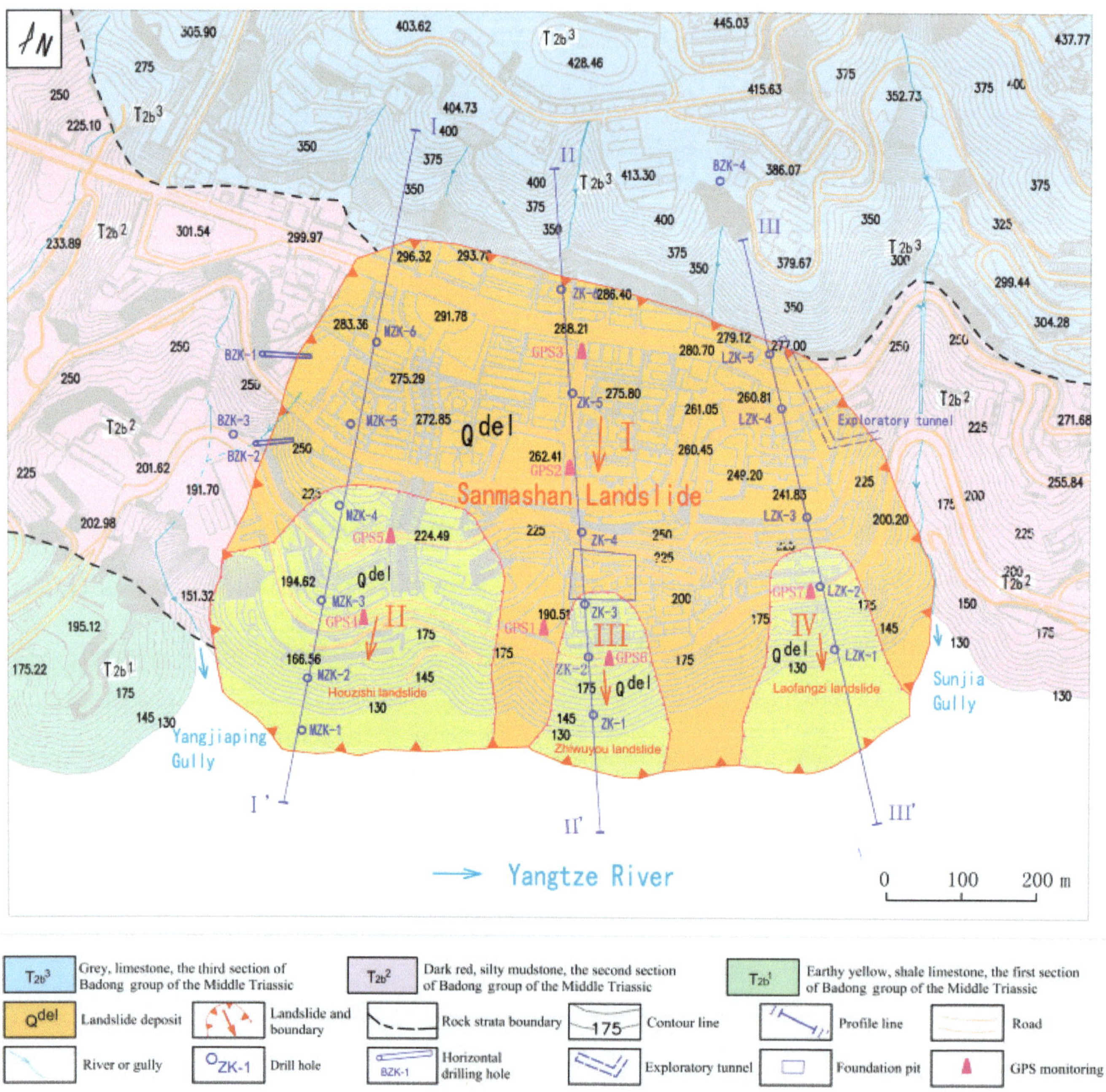

Figure 12. Geological map at a scale of 1 : 1000 of the Sanmashan landslide (the topographic base map is surveyed in 2004). The location of map is the area of the green box in Fig. 7. The area of the Sanmashan landslide is subdivided into four domains (I–IV). The three secondary partial landslides are outlined at the foot of the Sanmashan landslide.

Properties of the Materials in the Landslide

Eighteen samples were collected from the sliding zone, the slide mass, and the bedrock below the Sanmashan landslide. The physical and mechanical parameters of the main six geologic units were determined by laboratory testing. Cohesion and internal friction angle were obtained from triaxial saturated consolidated undrained shear tests. The saturated permeability coefficient of the slide mass was acquired by double-ring infiltration testing in the field. The saturated permeability coefficients of the bedrock are estimated values. The test results are shown in Table 1 and were used for the following analysis.

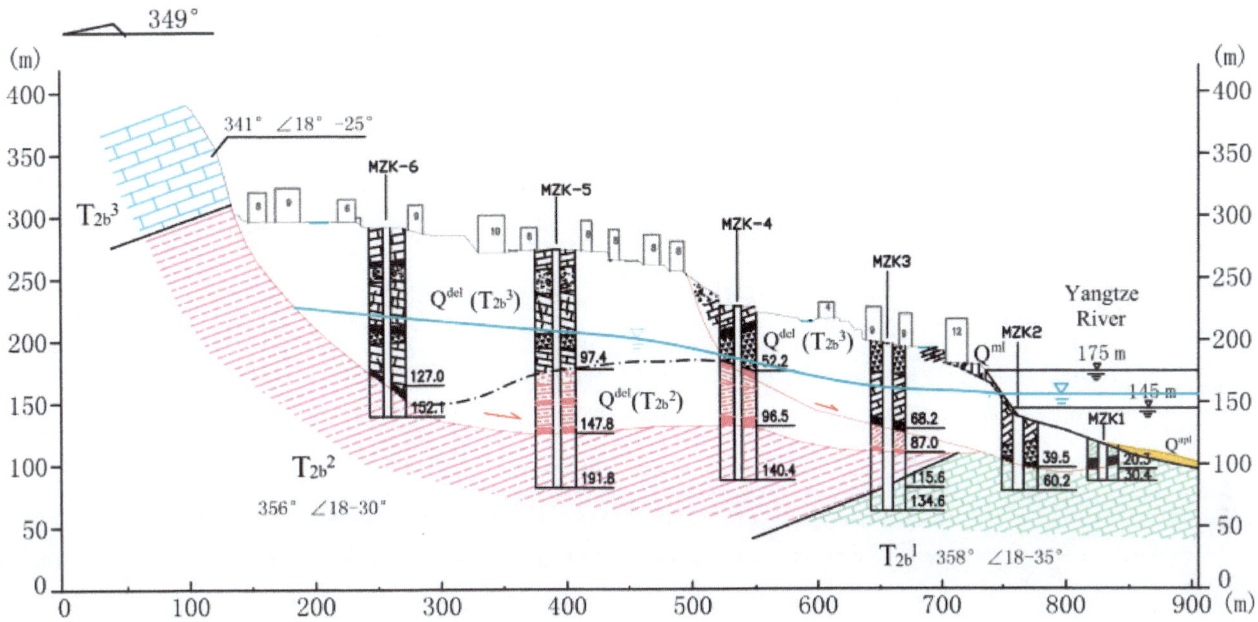

(a) Cross section I–I′

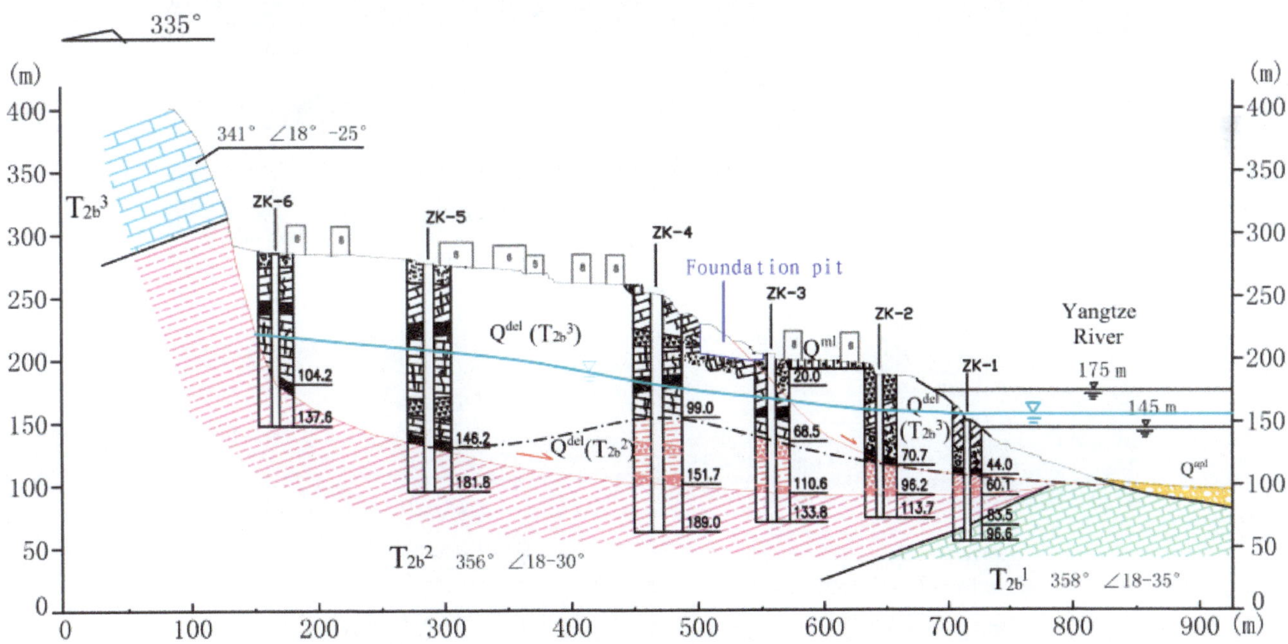

(b) Cross section II–II′

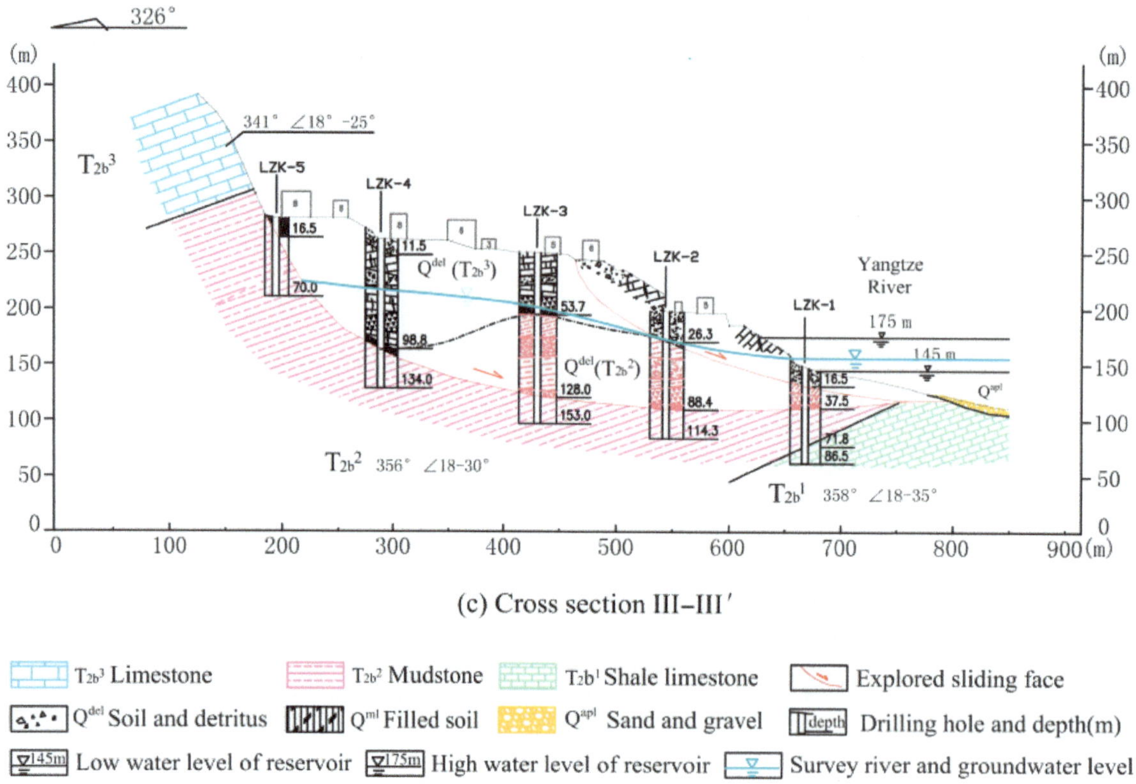

Figure 13. Geological cross sections through the Sanmashan landslide. The location of the cross sections is shown in Fig. 12. (a) Cross section I–I'. (b) Cross section II–II'. (c) Cross section III–III'.

Figure 14. Various views of the Sanmashan landslide. (a) Quickbird image map (September 2004). Area is subdivided into four domains (I–IV). (b) The crown and scarp of the Sanmashan landslide. (c) Slickenside discovered in a foundation excavation in 2000 (taken by Zhengquan Cui). (d) Landslide seen from the east. (e) Sliding zone near the eastern boundary. (f) The sliding zone in the exploratory tunnel. (g) The sliding scarp in a horizontal drilling hole near the western boundary. (h) The western slide boundary and the toe of the Houzishi secondary landslide in 2000. (i) Large foundation pit (in 2013), in which the sliding zone of the Zhiwuyou secondary landslide is exposed (see also (j) and Fig. 15).

Deformation Characteristics

According to residents, a small shallow soil slide and the formation of cracks had already occurred at the front of the secondary (partial) landslide during rainfall before 2003.

The filling–drawdown cycles of the Three Gorges Reservoir from June 2003 to December 2015 are shown in Fig. 16.

(i) On 1 June 2003, the first reservoir filling phase started. On 10 June 2003, the reservoir level reached an elevation of 135 m. (ii) On 20 September 2006, the Three Gorges Reservoir second phase of the reservoir filling started. On 27 October 2006, the reservoir level reached an elevation of 156 m. (iii) On 28 September 2008, the third phase of the reservoir filling started. On 14 November 2008, the reservoir level reached an elevation of 171.81 m. On 26 October 2010, the reservoir level reached its maximum design elevation level of 175 m.

During the first phase of the reservoir filling, several cracks were found in the surface of the Houzishi landslide in December 2004. The first crack occurred at an elevation of 150 m, had a length of 17 m and a width varying from 5 to 28 mm. GPS monitoring of the deformations was carried out by the local government at locations shown in Fig. 12. GPS1, GPS2, and GPS3 were located near the central axis of the Sanmashan landslide, GPS4 and GPS5 on the Houzishi landslide, GPS6 on the Zhiuwuyou landslide, and GPS7 on the Laofangzi landslide. The monitoring data were collected twice a month starting on 5 July 2006, but this frequency was increased in the rainy season. The results of the GPS monitoring are shown in Fig. 17.

The displacement curves show that the total displacements of GPS1, GPS2, and GPS3 remain near zero, which show that the middle and back parts of the Sanmashan landslide are stable. However, the total displacements of GPS4 and GPS5 were 19.3 and 16.4 mm from 5 July 2006 to 31 December 2011 on the Houzishi landslide. The total displacements of GPS6 and GPS7 were only 6.7 and 9.8 mm for the Zhiwuyou and Laofangzi landslides respectively. Analysis of the relation with reservoir level and rainfall, the displacements began to increase after the filling phases of 20 September 2006 and 28 September 2008. Since May 2009, however, none of the GPS monitored displacements have increased further. We assume that the Sanmashan landslide as a whole is inactive nowadays, but a partial reactivation of the landslide can be induced by reservoir water fluctuations or rainfall.

(a)

(b)

Figure 15. (a) The structural geological features of the landslide body in four walls of the foundation excavation. (b) The rock mass inside the Sanmashan landslide body in the western wall of the foundation excavation shows chaotic attitudes as generally generated by landslides.

FORMATION MECHANISM AND PROCESS OF THE ANCIENT SLIDE

Figure 13 shows that the original surface of the ancient slide was a gently inclined slope consisting of limestone of T_{2b}^{3} in its upper part and mudstone and siltstone of T_{2b}^{2} in its lower part. It is a hard-on-soft rock slope that provides a basis for the sliding; a free face is provided by the downcutting of the Yangtze River. These factors finally led to the formation of the ancient Sanmashan slide. Analysis of our field observations shows that there is a high steep cliff in the crown of the landslide and that the surface of rupture and sliding is concavely curved. So the ancient slide of Sanmashan is most likely a rotational rock slide (Varnes, 1978; WP/WLI, 1993; Cruden and Varnes, 1996; Huang, 2012; Hungr et al., 2014). In order to confirm this assumption, we have carried out a large centrifuge model experiment, which was briefly dis-cussed and published earlier by Tang et al. (2015b). According to international standards (WP/WLI, 1993), we further analyze and discuss the formation process and failure mechanism of the Sanmashan ancient slide, which is essential in order to recognize a potential hazard.

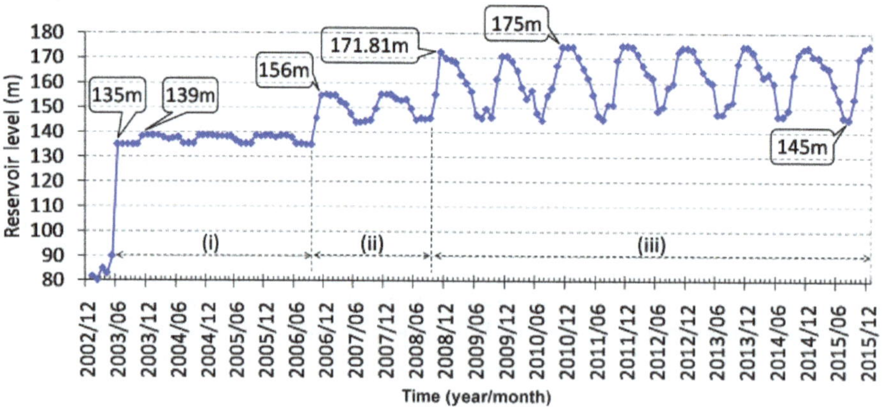

Figure 16. Filling-drawdown cycles of the reservoir from 2003 to 2015.

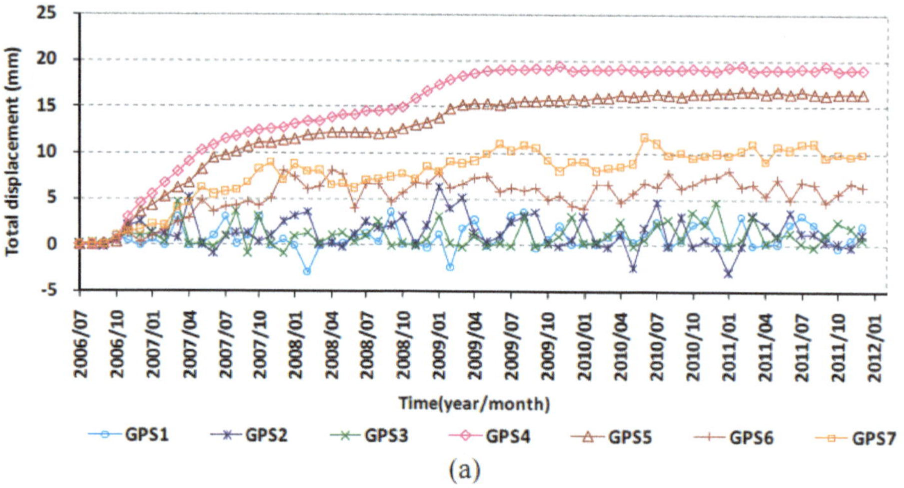

(a)

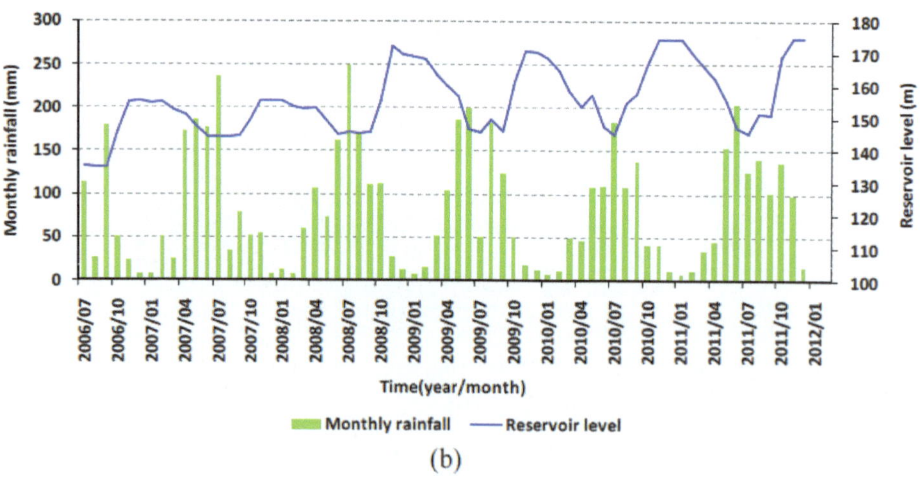

(b)

Figure 17. (a) The result of GPS monitoring and (b) reservoir level and rainfall from 5 July 2006 to 31 December 2011. The location of GPS monitoring points are shown in Fig. 12.

Geotechnical Centrifuge Test

The centrifuge technology is widely applied in geotechnical and geological engineering, especially in the simulation of slope deformation and failure (Taylor, 1995; Nakajima and Stadler, 2006)

Table 2. The similarity coefficients for the main physical quantities in the centrifuge model

Physical quantity/ symbol	Unit	Similarity coefficient
Acceleration/a	g	$N_a = 30$
Dimension/L	m	$N_l = 1/3000$
Density/ρ	$g\,cm^{-3}$	$N_\rho = 1$
Modulus/E	MPa	$N_E = N_\rho\, N_a\, N_l = 1/100$
Internal friction angle/φ	°	$N_\varphi = 1$
Cohesion/c	kPa	$N_c = N_\rho\, N_a\, N_l = 1/100$
Poisson's ratio/ν	–	$N_\nu = 1$

Table 3. Physical and mechanical parameters of the materials in the centrifuge model test

Model materials	Modulus (MPa)	Poisson's ratio	Density ($g\,cm^{-3}$)	Cohesion (kPa)	Internal friction angle (°)
Model unit 1[a]	30	0.30	2.48	1.2	35
Model unit 2[b]	15	0.35	2.30	1.0	25
Model unit 3[c]	48	0.25	2.60	2.0	40

[a] Model unit 1 simulates Shale limestone T_{2b}^1; [b] model unit 2 simulates mudstone T_{2b}^2 and [c] model unit 3 simulates limestone T_{2b}^3.

A centrifuge is a piece of equipment that rotates an object around a fixed axis (it spins in a circle), applying a potentially strong centrifugal (outward) force perpendicular to the axis of spin. Geotechnical centrifuge modelling is used for physically testing soils and rock models. The centrifugal acceleration is applied to models to scale the gravitational acceleration and enable prototype scale stress in scale models. Problems can be studied such as occur in the design of foundations for buildings and bridges, the design of earth dams, tunnels, and analysis of slope stability, including effects such as blast loading and earthquake shaking. Large centrifuges are used to simulate high gravity or acceleration environments. For our model tests we used the centrifuge at the State Key Laboratory of Geohazard Prevention and Geoenvironment Protection in Chengdu, China (Fig. 18). With a maximum acceleration of 500 gt it is the largest geotechnical centrifuge in Asia.

Tested Model

The geological model of the original slope (Fig. 19) was generalized into a geotechnical test model (Fig. 20). Based on the previous research (Yang, 1988; Li et al., 2001), three periods of incision of the Yangtze River are assumed. According to the similarity theory (Taylor, 1995) and the dimension relationship between geological model and experimental model we determined the similarity coefficient (Table 2). From the physical and mechanical parameters of the geological units of the Sanmashan landslide (Table 1), we determined the physical and mechanical parameters of the model materials with the similarity coefficient (Table 3).

Figure 18. The geotechnical centrifuge used for the model test.

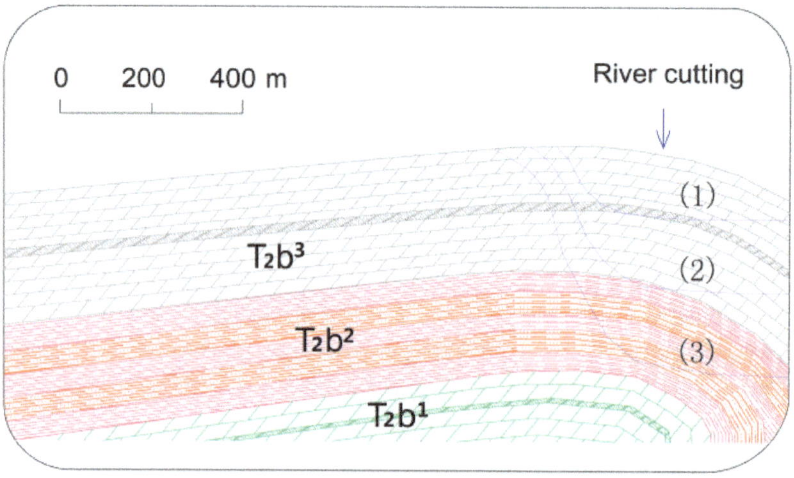

Figure 19. The assumed geological structure of the original river bank slope.

Figure 20. The centrifuge test model.

Analysis and Discussion of Results of Centrifuge Test

The model test results were shown in an earlier study (Figs. 97.8–97.10, Tang et al., 2015b) (see Supplement 2). The deformation and failure process during the centrifuge test has the following characteristics. First, the soft rock in the lower part of the model slide body is deformed in shear.

Figure 21. The model after slide. The rupture surface of the slide is roughly curved, successive rotational slides occurred around one axis.

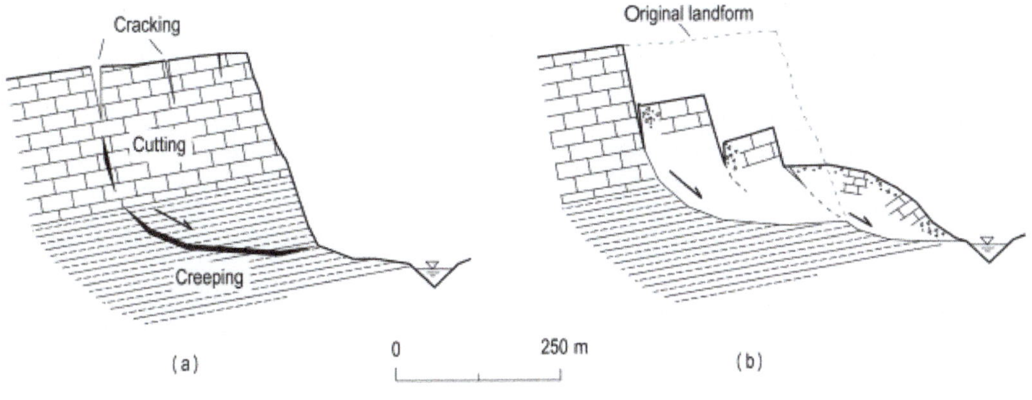

Figure 22. Schematic cross section of the Sanmashan ancient slide. (a) Slope deformation, (b) slope after slide.

Then, tension cracking occurs in the hard rock in the upper part. Finally, the rupture plane is formed when the tension cracks extend downwards towards the zone of shear deformation and the slide body moves. The deformation and failure process can be characterized as creep–crack–cut. The sliding surface is curved, but non-circular. A steep main scarp cuts through the hard rock in the upper part of the slope and forms the proximal part of the rupture surface, daylighting at the crown. The slide movement is roughly rotational around an axis that is parallel to the ground surface (Fig. 21).

According to international standards (WP/WLI, 1993), the type of the Sanmashan ancient slide should be classified as a successive rotational rock slide (Fig. 22), as we already concluded after our field survey (Fig. 13). This type of slide generally occurs as a result of river incision, in subhorizontal layered rock masses on a gently inclined slope with hard rock overlying soft rock. Other examples are the Chana landslide, Longyangxia along the Yellow River (Fig. 23) and the Yanchihe landslide in Hubei Province, China (Zhang and Huang, 1990).

Figure 23. The Chana landslide, Longyangxia, China.

INFLUENCE OF SEEPAGE ON THE STABILITY OF THE SANMASHAN LANDSLIDE

On the basis of the results of the research described above, we analyzed the influence of seepage on the stability of the Sanmashan landslide.

Methodology

Modelling Software and Stability Analysis

We used the software Geo-Slope 2007 (Geo-Slope International Ltd., 2007). The variation of the groundwater level induced by the reservoir filling or rainfall and the related change in pore water pressure are simulated using the code SEEP/W. We carried out the limit equilibrium slope stability analysis with the software SLOPE/W. During the analysis, the output of the seepage simulation was used as input for the stability analysis, for which we adopted the Morgenstern and Price (1965) method to calculate the factor of safety (FOS).

Model Assumptions

Three cross sections (Fig. 13) were chosen to conduct the two-dimensional combined seepage–slope stability analysis. For each cross section, five main hydrogeological– geomechanical units are identified.

(1) The upper unit of the slide mass is dense to very dense and contains grey or dark grey crushed stone in a matrix of silty clay and sand. The crushed stone is derived from lime-stone (T_{2b}^{3}). (2) The intermediate unit has a higher density, is reddish or dark red/brown in color, and contains crushed stone in a matrix of clay and silty clay. The crushed stone is derived from mudstone and silty mudstone (T_{2b}^{2}). Units (3)–(5) are the underlying stable and undeformed bedrock of the geological units T_{2b}^{3}, T_{2b}^{2} and T_{2b}^{1}. Each unit has been subdivided into a finite element mesh that combines triangular and quadrilateral elements of different sizes (Fig. 24).

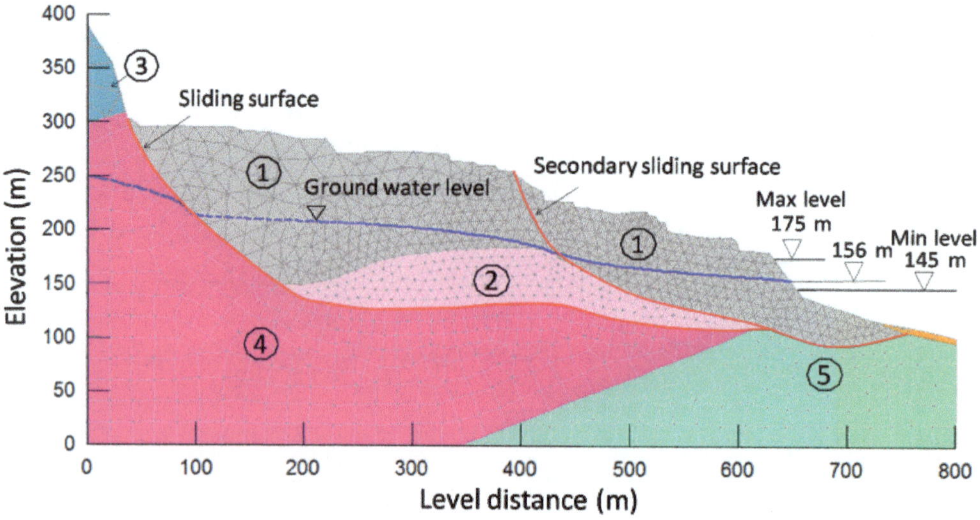

Figure 24. Model of the two-dimensional numerical stability analysis of the Sanmashan landslide (cross section I–I' in Fig. 12).

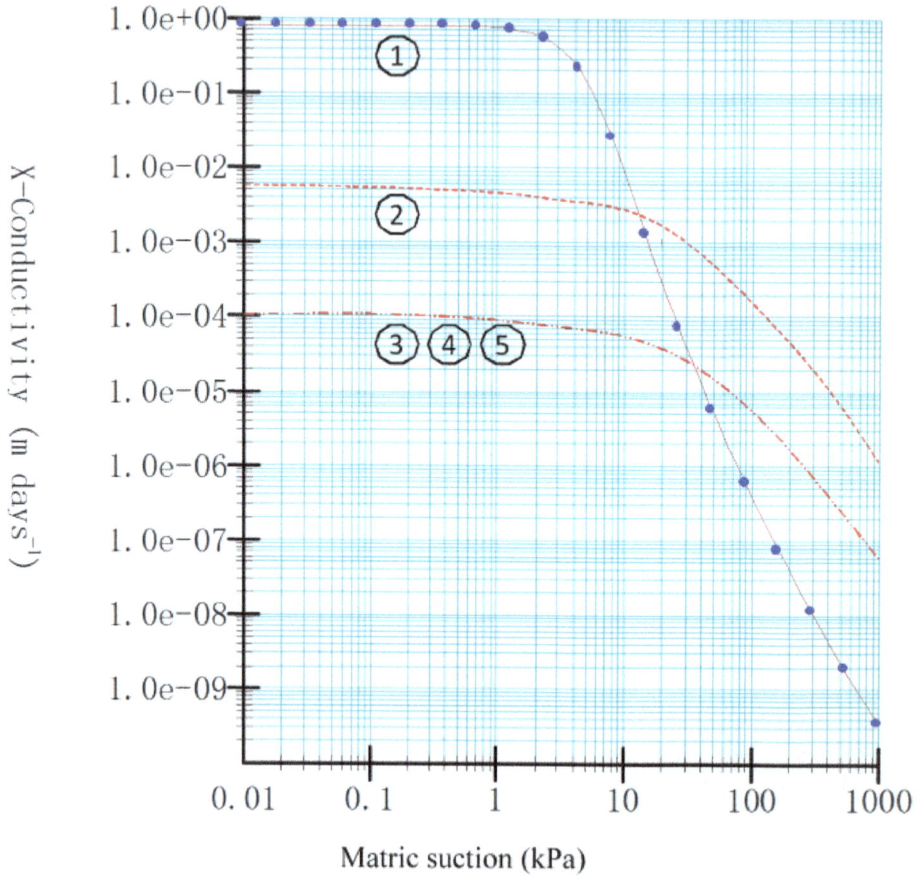

Figure 25. Hydraulic conductivity functions adopted for the main hydrogeological units: (1) Q^{del} (T_{2b}^{3}), (2) Q^{del} (T_{2b}^{2}), (3)–(5) bedrock types T_{2b}^{3}, T_{2b}^{2}, and T_{2b}^{1}.

The physical and mechanical parameters of the analytical model were determined according to the material testing results. The permeability of the different units is an important parameter affecting the stability analysis. We have carried out an inversion analysis on the data used in a previous study (Wang et al., 2013). The parameters of the model materials of stability analysis are listed in Table 1. The hydraulic conductivity functions that we adopted for the main hydrogeological units are shown in Fig. 25.

Historical Reservoir Level Variations and Rainfall Data

The repeated filling–drawdown cycles and rainfall are the main factors that affect the stability of the Sanmshan land-slide (Zangerl et al., 2010; Paronuzzi et al., 2013; Galeandro et al., 2013). Based on a statistical analysis of the historical data, the reservoir level and rainfall values are determined, as follows.

Table 4. The water level fluctuation rates chosen for the filling– drawdown cycles of the reservoir level (Stages A–H see Fig. 27)

Stage	Time	Reservoir level (m)	Average fluctuation rate (m day^{-1})
AB	1 Jan–15 May	175–160	−0.11
BC	15 May–10 Jun	160–145	−0.58
CD	10 Jun–4 Jul	145–145	0.00
DE	4 Jul–29 Jul	145–155	+0.40
EF	29 Jul–15 Aug	155-145	−0.59
FG	15 Aug–18 Oct	145–175	+0.47
GH	18 Oct–31 Dec	175–175	0.00

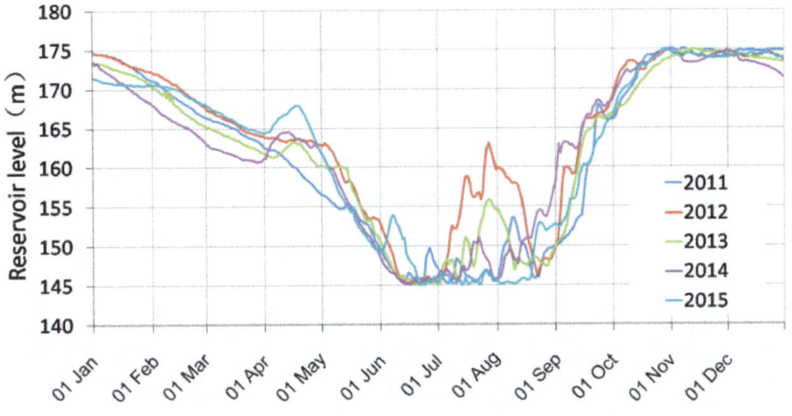

Figure 26. The reservoir level data from 2011 to 2015.

Figure 17 shows that the reservoir level has varied annually between the elevations of 175 and 145 m since January 2011. Through the statistical analysis of the historical data from 2011 to 2015 of the reservoir level (Fig. 26), the annual fluctuation rate of the cycles is determined (Fig. 27 and Table 4). It was used for the boundary conditions in the numerical modelling.

The historical rainfall records of 1964–2013 for the new Fengjie town have been collected and analyzed. The variation of average monthly rainfall during the years 1964–2013 shows that heavy rainfall is mainly concentrated in June and July (Fig. 28). Based on the probability statistics of the maximum daily rainfall (Table 5), the maximum rainfall intensity for different return periods has been calculated (Fig. 29 and Table 6). A daily rainfall of 140 mm in return periods of 50 years (2 % probability) has been adopted as the boundary condition in numerical modelling.

Table 5. Annual maximum rainfalls in the period 1964–2013

Date	Annual maximum rainfall (mm day^{-1})	Date	Annual maximum rainfall (mm day^{-1})	Date maximum	Annual rainfall (mm day^{-1})
24 May 1964	121.40	25 Aug 1981	56.30	29 Jun 1998	141.30
17 Aug 1965	88.60	8 Sep 1982	110.30	7 Jul 1999	161.40
7 Oct 1966	37.70	6 Oct 1983	96.10	16 May 2000	79.70
21 Jul 1967	70.10	13 Jun 1984	113.20	12 Jul 2001	100.80
14 Sep 1968	71.90	21 Jun 1985	95.90	21 Sep 2002	73.20
5 May 1969	108.80	15 Jun 1986	96.00	19 Jul 2003	72.20
28 May 1970	84.10	14 Sep 1987	52.20	20 Sep 2004	86.80
3 May 1971	97.80	8 May 1988	81.90	28 Aug 2005	82.40
21 Jun 1972	57.50	11 Jul 1989	127.90	5 Jul 2006	75.10
10 Sep 1973	66.90	1 Jul 1990	65.50	30 Jul 2007	66.80
5 Aug 1974	45.70	30 Jun 1991	70.40	20 Jun 2008	81.70
26 Jun 1975	103.20	11 May 1992	77.00	29 Aug 2009	93.80
21 Jul 1976	56.70	18 Jul 1993	88.50	7 Jun 2010	49.20
4 Oct 1977	50.10	25 Jun 1994	91.30	23 Jun 2011	95.10
24 Jun 1978	71.90	25 Jul 1995	69.50	5 Jul 2012	60.30
12 Sep 1979	106.70	28 Aug 1996	63.10	25 May 2013	56.90
31 May 1980	71.30	5 Jul 1997	49.70		

Table 6. The maximum rainfall intensity for different return periods

Frequency/probability (%)	20 %	10 %	5 %	2 %	1 %
Return periods (years)	5	10	20	50	100
Rainfall intensity (mm day^{-1})	101	115	126	140	151

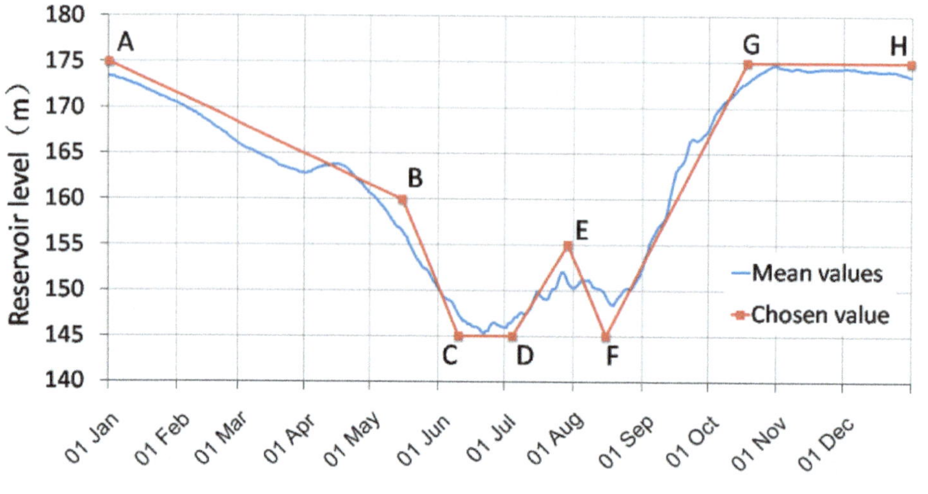

Figure 27. The mean values of the reservoir level during the filling– drawdown cycles in 2011–2015. The chosen values for the analysis (see also Table 5).

Stability Analysis

Factor of Safety (FOS) Variations Induced by Reservoir Filling–Drawdown Cycles

With a drawdown of the reservoir level from 175 to 145 m (Fig. 27), the delay in the lowering of the groundwater level after the lowering of the reservoir water level will cause a decrease in the landslide stability. After stabilization of the reservoir water level at 145 m, further lowering of the groundwater level in the sliding body will gradually lead to an increase in the stability of the landslide (Figs. 30 and 31).

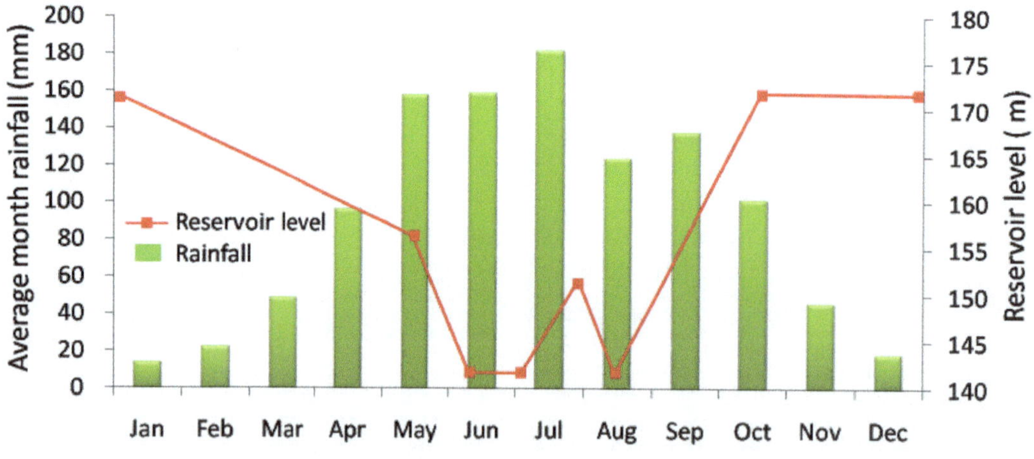

Figure 28. The average monthly rainfall from 1964 to 2013 and the chosen value of the filling–drawdown cycles of the reservoir level for the numerical stability analysis.

During a rise in the reservoir water level from 145 to 175 m (Fig. 27), the stability of the landslide will increase as the rise in the groundwater level in the slide mass lags behind the rise of the reservoir water level. After reaching the reservoir water level of 175 m, the stability of the landslide will again gradually decrease with a further rise of the groundwater level (Figs. 30 and 32).

Due to the filling–drawdown cycles of the Three Gorges Reservoir (Fig. 27), but without the rainfall effects, the minimum stability coefficient (FOS) of the Sanmashan landslide is 1.617, and that of the partial landslide at the front is 1.114.

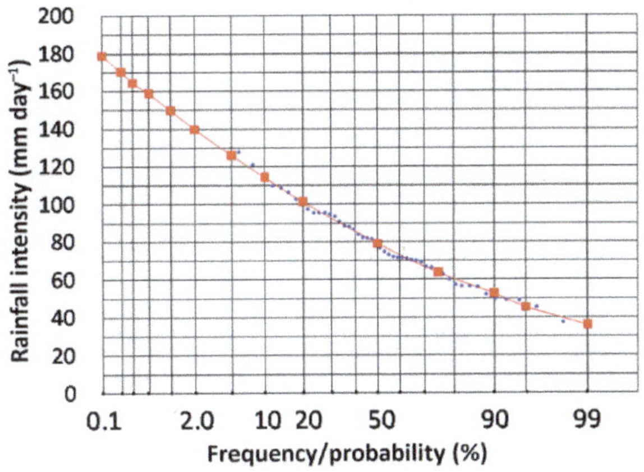

Figure 29. Maximum rainfall intensity in mm day [1] for different return periods.

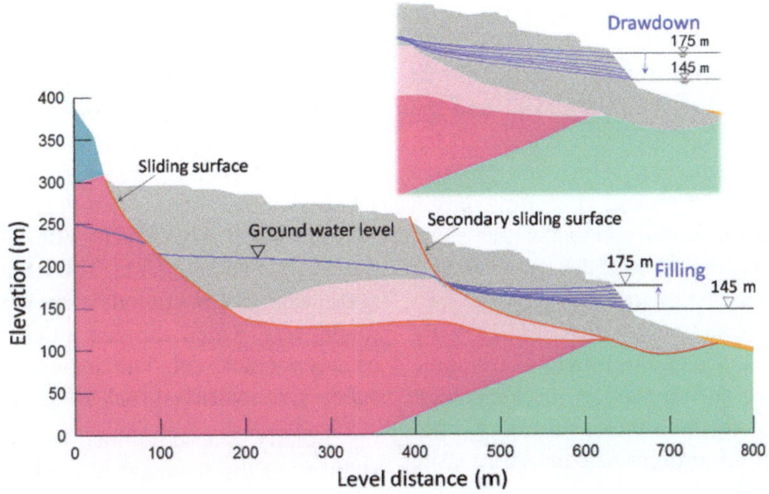

Figure 30. Ground water level changes during the filling– drawdown cycles of the Three Gorges Reservoir.

Rainfall Influence on the Sanmashan Landslide Stability

A daily rainfall of 140 mm in return periods of 50 years (2 % probability), during a period of reservoir water level drawdown from 175 to 145 m, has little effect on the stability of the whole landslide (decrease in FOS of 2.4 %), but has a large influence on the secondary partial landslides at the front (decrease in FOS of 4.2 %) (Fig. 32).

During rainfall and filling–drawdown cycles in the Three Gorges Reservoir, the minimum stability coefficient (FOS) of the Sanmashan landslide is 1.578 and of the partial landslide at the front it is 1.067.

CONCLUSIONS

Extensive field studies and mapping, deep borehole drilling, investigation of the already existing tunnels and a large foundation excavation pit as well as centrifuge model testing and numerical stability analysis were carried out to investigate the stability of a deep-seated ancient rock slide in the Three Gorges Reservoir area.

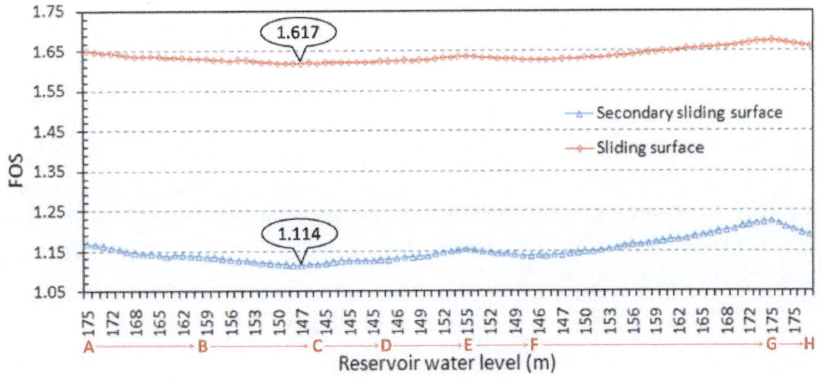

Figure 31. Factor of safety (FOS) variation during the filling– drawdown cycles of the Three Gorges Reservoir (Points A–H see Fig. 27).

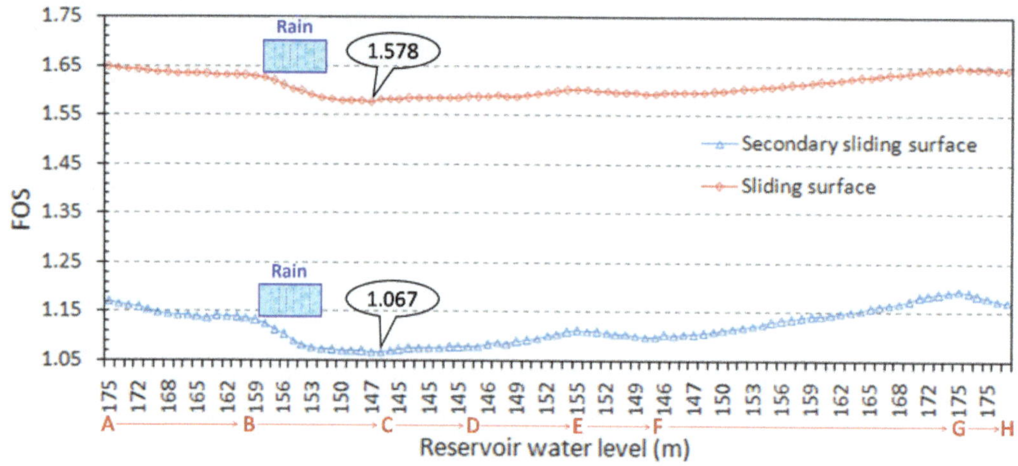

Figure 32. Factor of safety (FOS) variation due to rainfall and filling–drawdown cycles of the Three Gorges Reservoir (Points A– H see Fig. 27).

Three tectonic units, the upper Yangtze platform fold belt, the marginal depression of the Sichuan Basin, and the Dabashan platform fold belt, converge on this region. In the new Fengjie town, the site of the landslide, a N–S compressive geostress field is the tectonic background. For this reason we discarded the theory that the deformations in the slope were related to a fault graben, as that would have needed a tensional geostress field. Our investigations have also shown that the four normal faults, which in earlier studies were used to support the hypothesis of a fault graben, share some common characteristics, which do not support the local fault graben theory. Thus we come to the conclusion that a local fault graben does not explain the presence of the large-scale weak layer in the Sanmashan area.

Meanwhile, there are numerous unique geological features and interesting slickenside relics of the ancient slide discov-ered in the field. From the structural geological evidence, geomorphological evidence and slickenside traces, we conclude that this area is a huge deep-seated ancient rock slide, called the Sanmashan landslide.

A steep cliff over 130 m in height forms the back scarp of the Sanmashan landslide. Two gullies, Baiyangping as a western and Sunjia as an eastern boundary, as well as the Yangtze River in the front, form the boundaries of the landslide. The terrain slope angle of the landslide is 15– 20. The length of the landslide is over 780 m, the width is about 1020 m, the maximum thickness is over 150 m, the average thickness is about 125 m, and the volume is about 100 million m^3. The Sanmashan landslide can be subdivided into four domains based on differences in morphology, geological structure, and inferred failure mechanics. There are three secondary landslides at the foot of the Sanmashan land-slide called the Houzishi, Zhiwuyou, and Laofangzi land-slides. There was a sign of a partial reactivation of one of the secondary landslides in October 2006 and October 2008 at the moment of reservoir filling, but after May 2009 none of the GPS monitored displacements of the landslide showed any further increase. This suggests that the Sanmashan land-slide is presently inactive without the likelihood of reactivation by fluctuations of the reservoir water or rainfall.

The original surface of the ancient slide was a gently inclined slope consisting of limestone of T_{2b}^{3} in its upper part and mudstone and siltstone of T_{2b}^{2} in its lower part. It is a hard-on-soft rock slope, which provides a basis for the sliding, while the free face is provided by the downcutting of the Yangtze River. These factors finally led to the formation of the Sanmashan ancient slide. The formation process of the ancient slide was demonstrated in a large geotechnical centrifuge model test, which showed a deformation and failure process of creep–crack–cut. The ancient slide can be characterized as successive rotational rock slide.

Finally, the combined analysis of seepage and slope stability of the Sanmashan landslide has been carried out, the results of which show that during rainfall and reservoir filling–drawdown cycles the Sanmashan landslide as a whole is stable and dormant, while the secondary (partial) landslides in the toe area are basically stable but can be reactivated. This research provides an important guidance for strict land use planning in the sliding zone.

ACKNOWLEDGEMENTS

This research was financially supported by the National Basic Research Program of China (No. 2013CB733202), the Key Project of TGR (No. SXKY3-1-1-200901), and the State Key Laboratory of Geohazard Prevention and Geoenvironment Protection (No. SKLGP2015Z008). The authors sincerely thank Xue-bin Huang, Wenming Cheng, and Kaixiang Xu for their assistance. Wei Zhang, Ruihua Xiao, Lei Zhang, and others contributed to the investigation and drilling. Guang Zheng and Kai Wang contributed the centrifuge model testing. Fan Yang and Yangjian Cao contributed the stability analysis. The manuscript was directed by Zhuoyuan Zhang and Guangzong Peng. The authors wish to thank the anonymous referees for their helpful suggestions and constructive comments, which have contributed greatly in improving the quality of the manuscript.

REFERENCES

1. Alonso, E. E. and Pinyol, N. M.: Criteria for rapid sliding I. A re-view of Vaiont case, Eng. Geol., 114, 198–210,

2010.

2. Barla, G., Antolini, F., Barla, M., Mensi, E., and Piovano, G.: Monitoring of the Beauregard landslide (Aosta valley, Italy) using advance and conventional techniques, Eng. Geol., 116, 218–235, 2010.
3. BIS-CWRC – Bureau of Investigation and Survey, Changjiang Water Resource Commission: The geology argument report of the relocation site of the Fengjie County in Three Gorges Reservoir (Detailed investigation stage), Wuhan, 1997.
4. BIS-CWRC – Bureau of Investigation and Survey, Changjiang Water Resource Commission: Engineering Geology Investigation Report of protecting project of bank slopes of the new Fengjie County (feasibility study stage), Wuhan, 1999.
5. Brideau, M., Yan, M., and Stead, D.: The role of tectonic damage and brittle rock fracture in the development of large rock slope failures, Geomorphology, 103, 30–49, 2009.
6. Chang, W. Y., Li, Y. H., Ma, F. C., and Zhong, J. Y.: On the mechanical mechanism of the formation of graben, Chin. Scient. Geol. Sin., 1, 1–11, 1981.
7. Chen, M. S. and Zhang, S. H.: Characteristics and forming mechanism of opposite protruding arcuate structural zones in the Three Gorges Area of the Yangtze river, south China, Geol. Miner. Re-sour. S. China, 1, 48–55, 1998.
8. Crosta, G. B.: Landslide, spreading, deep-seated gravitational deformation: Analysis, examples, problems and proposals, Geograf. Fis. Dinam. Quatern., 19, 297–313, 1996.
9. Crosta, G. B. and Zanchi, A.: Deep-seated slope deformations: Huge, extraordinary, enigmatic phenomena, in: Landslides in Re-search, Theory and Practice, edited by: Bromhead, E., Dixon, N., and Ibsen, M. L., Thomas Telford, London, 351–358, 2000.
10. Cruden, D. M. and Martin, C. D.: Before the Frank Slide, Can. Geotech. J., 44, 765–780, 2007.
11. Cruden, D. M. and Varnes, D. J.: Landslide types and processes, in: Landslides investigation and mitigation, Transportation re-search board, chap. 3, Special Report 247, edited by: Turner, A. K. and Schuster, R. L., US National Research Council, Washing-ton, D.C., 36–75, 1996.
12. Galeandro, A., Doglioni, A., Guerricchio, A., and Simeone, V.: Hy-draulic stream network conditioning by a tectonically induced, giant, deep-seated landslide along the front of the Apennine chain (south Italy), Nat. Hazards Earth Syst. Sci., 13, 1269–1283, doi:10.5194/nhess-13-1269-2013, 2013.
13. Geo-Slope International Ltd.: Seepage and stability modeling with SEEP/W and SLOPE/W, Users Manuals, Calgary, Alberta, Canada, 2007.
14. Huang, R. Q.: Mechanisms of large-scale landslides in China, Bull. Eng. Geol. Environ., 71, 161–170, 2012.
15. Hungr, O., Leroueil, S., Picarelli, L.: The Varnes classification of landslide types, an update, Landslides, 11, 167–194, 2014.
16. Illies, J. H.: Mechanism of garben formation, Tectonophysics, 73, 249–266, 1981.
17. John, J. C. and Douglas, S.: Landslides: Types, Mechanisms and Modeling, Cambridge University Press, Cambridge, 2012.
18. Li, H. L., Yi, S. H., and Deng, Q. L.: Development characteris-tics and their spatial variations of Badong formation in the Three Gorges Reservoir Region, J. Eng. Geol., 14, 578–581, 2006.
19. Li, H. Z., Zhou, Y., and Pan, Y. Z.: The stability analysis and protection measures of the Houzishi landslides in the Three Gorges Reservoir Region, Hubei Geol. Miner. Resour., 16, 97– 104, 2002.
20. Li, J. J., Xie, S. Y., and Kuang M. S.: Geomorphic evolution of the Yangtze Gorges and the time of their formation, Geomorphology, 41, 125–135, 2001.
21. Michael, W. H., Phillip, J. S., and Gregory, T. F.: When Land-slides are Misinterpreted as Faults: Case Studies from the West-ern United States, Environ. Eng. Geosci., 11, 313–325, 2012.
22. Morgenstern, N. R. and Price, V. E.: The analysis of the stability of general slip surfaces, Geotechnique, 15, 79–93, 1965.
23. Nakajima, H. and Stadler, A. T.: Centrifuge modeling of one-step outflow tests for unsaturated parameter estimations, Hydrol. Earth Syst. Sci., 10, 715–729, doi:10.5194/hess-10-715-2006, 2006.
24. Nanjiang Geological Team: Investigation report of Sanmashan land-slide in New Fengjie Town, Chongqing, 1999.
25. No. 107 Geological Team of Sichuan Province: 1 : 200000 the re-gional geological report of Fengjie, China University of Geo-sciences, Wuhan, 37–39, 1980.
26. Paronuzzi, P., Rigo, E., and Bolla, A.: Influence of filling-drawdown cycles of the Vajont reservoir on Mt. Toc slope stability, Geomor-phology, 191, 75–93, 2013.
27. Qi, S. W., Wu, F. Q., Yue, Z. Q., and Liu, C. L.: Characteristics and mechanism of deep weathering of argillaceous limestones at Fengjie County, Three Gorges Region, Central China, Bull. Eng. Geol. Environ., 71, 285–295, 2012.
28. Tang, M. G., Xu, Q., and Huang, R. Q.: Site monitoring of suc-tion and temporary pore water pressure in an ancient landslide in the Three Gorges reservoir area, China, Environ. Earth Sci., 73, 5601–5609, 2015a.
29. Tang, M. G., Xu, Q., Huang, X. B., Xu, K. X., Cheng, W. M., and Wang, K.: Recognition and Genetic Mechanism of Sanmashan Deep-Seated Landslide, Three Gorges Reservoir Area, China, in: Engineering Geology for Society and Territory – Volume 2: Landslide Processes, edited by: Lollino, G., Giordan, D., Crosta, G. B., Corominas, J., Azzam, R., Wasowski, J., and Sciarra, N., Springer International Publishing, Switzerland, 587–591, 2015b.
30. Taylor, R. N.: Centrifuges in Modeling: Principles and Scale Effects, in: Geotechnical Centrifuges Technology, edited by: Tay-lor, R. N., Blackie Academic & Professional, London, 19–33, 1995.

31. Varnes, D. J.: Slope movement types and processes, in: Landslides – Analysis and control, Special Report 176, edited by: Schuster, R. L. and Krizek, R. J., National Research Council, Transportation Research Board, Washington, D.C., 11–33, 1978.
32. Wang, J. G., Zhou, Y., and Huang, Z. P.: Groundwater analyses of Houzishi landslide in the Three Gorges Reservoir, J. Rock Mech. Eng., 25, 2757–2762, 2006.
33. Wang, K., Tang, M. G., and Xu, Q.: Study on the Inversion Method of Selecting Seepage Parameter for Sanma Mountain Landslide, Yellow River, 35, 124–126, 2013.
34. WP/WLI – International Geotechnical Society's UNESCO Working Party on World Landslide Inventory: A multi-lingual landslide glossary, Bitech Publishers, Vancouver, 51–58, 1993.
35. Yang, D. Y.: Genetic mechanism of the alluvial terraces along the Three Gorges course of the Changjiang River, Acta Geogr. Sin., 42, 120–126, 1988.
36. Zangerl, C., Eberhartd, E., and Perzlmaier, S.: Kinematic behavior and velocity characteristics of a complex deep-seated crystalline rockslide system in relation to its interaction with a dam reservoir, Eng. Geol., 112, 53–67, 2010.
37. Zhang, Z. Y. and Huang, R. Q.: Epigenetic recreation of rock-mass structure and time-dependent deformation, in: Proc. of the 6th Congress of IAEG, A. A. Balkema, Rotterdam, the Nether-lands, 2065–2072, 1990.

CHAPTER 3

ROCKBURST THEORY AND TYPES OF ROCK SUPPORT IN ROCKBURST PRONE MINES

Eugie Kabwe, Yiming Wang

Keywords: Cone Bolts, Jacketted Rock Bolt, Rockburst, Rock Support, Yield-Lok

INTRODUCTION

The overall stress field attains high values when mining reaches greater depths. This stress is further redistributed due to the activity resulting in excessive stress concentrations in certain sections of the rock mass. Accumulation of stress in the rock leads to destructive stress that causes fracture which can be a seismic source. The passing dynamic stress wave as a result of fracturing in the rock mass is defined as a seismic event. A seismic event can also be a sudden inelastic deformation within a given size of rock [1] [2] . Effectively, a seismic event is the vibration of rock breaking. Seismic events are a normal response of a rockmass to the stress changes caused by the creation of mining excavations. The term seismicity is given various definitions with different authors. One definition can be stated seismicity as the rock mass response to deformation and failure [2].

ROCKBURST

The rockburst phenomena, came into sight in the early 1900s in the mines of South Africa [3] . Today several deep underground mines in all around the world face the problem of rockbursts. A rockburst is a mining-induced seismic event that affects and cause destruction to excavations in the rock [2] . It is vital to differentiate between seismic event and rockburst. A seismic event does not necessarily cause damage in openings, while a rockburst will, with varying severity of damage. A very good definition of rockburst given by the Canadian Rockburst Research Program is "damage to an excavation that occurs in a sudden or violent manner and is associated with a seismic event" [3] [4] . The role of seismic events in a rockburst phenomenon can be examined as to the actual course of burst; if it's the seismic event or the stored stress energy in the rock around an excavation released during breaking. A seismic event can be located at a certain distance away from the rockburst location, or can also overlap with the rockburst phenomenon as in the case of strain bursts. Hence the cause of rockbursts can generally be classified as self-initiated and remotely triggered [5] .

Self-initiated rockbursts arise from the stresses close to the boundary of an excavation which exceed the strength of a rock mass and failure can be in unstable manner. The stress redistribution that is the outcome of a nearby mining increases the concentration of the stresses at the spot. Degradation of strength with time and loss of confinement can also deteriorate the capacity of the rock mass. One or all of these conditions lead the strength of the rock mass to be exceeded by the stress and consequently lead to failure. Loss of structural stability, a factor independent of the strength of the rock mass, can also cause a self-initiated rockburst as can be demonstrated in the sudden buckling of column or slab of rock [6] .

Remotely triggered rockbursts are caused by large magnitude seismic events. Remotely triggered rockbursts commonly occur in some hard rock mines, usually after the mine has been significantly mined out and where faults intersect stopes or large mined-out areas and sill pillars. Large vibrations and the accompanying dynamic stresses as a result of incoming seismic waves can lead to both fracturing of the rock mass and structural instability. Furthermore the

Citation: Kabwe, E. and Wang, Y. (2015), Rockburst Theory and Types of Rock Support in Rockburst Prone Mines. Open Journal of Safety Science and Technology, 5, 104-121. doi: 10.4236/ojsst.2015.54013.

mining stage can influence the phenomena of rock bursting. In the initial stage of the mine's life highly localized stress concentrations near drifts that are relatively isolated from each other are the usual causes of the events. In the later life stage of the mine multiple opening and numerous stopes cause additional rockburst mechanisms [6] .

Rock bursts are generally divided into three classifications:

- Strain bursts: These are caused by high-stress concentrations at the edge of mine openings that exceed the strength of the rock. Events can range from small slivers of rock being ejected from the walls to the collapse of a complete wall as it tries to achieve a more stable shape.
- Pillar bursts: Severe rock bursts, involving thousands of tons, have been caused by the complete collapse of support pillars. These tend to occur in extensively mined-out areas, and the resulting damage can be severe.
- Fault slip bursts: Recognized in the 1980s, this type of rock burst occurs when slippage suddenly occurs along a geological weakness plane. This is the same mechanism as for an earthquake.

Damage may occur as rock bulking by fracturing, ejection of rock due to seismic energy, or rockfalls by seismic quaking (Figure 1) [7] . Each mechanism may result in different levels of damage to an excavation and its support system. The damage severity depends on many factors, including:

- Failure potential near the opening;
- Support effectiveness;
- Local rock stiffness;
- Magnitude of seismically induced stress, rock accelerations, or velocities;
- Opening geometry, size, and orientation;
- Geological structure.

Three levels of rock burst damage severity are defined in Figure 1. Damage level determination is usually based on observations of previous damage, where such observations are available, plus analytical methods [8] . Severity levels are summarized in Table 1. Only a limited number of ground support elements are suitable for dynamic loading conditions, and under the most severe conditions even these are restricted by maximum practical support limitations [8] .

ROCK REINFORCEMENT SYSTEMS

The main purpose of rock reinforcement in underground mining is to ensure stability of the rock mass around excavation openings while maintaining access to the mine, allowing the rock mass to support itself with its natural strength, while keeping the intended functionalities.

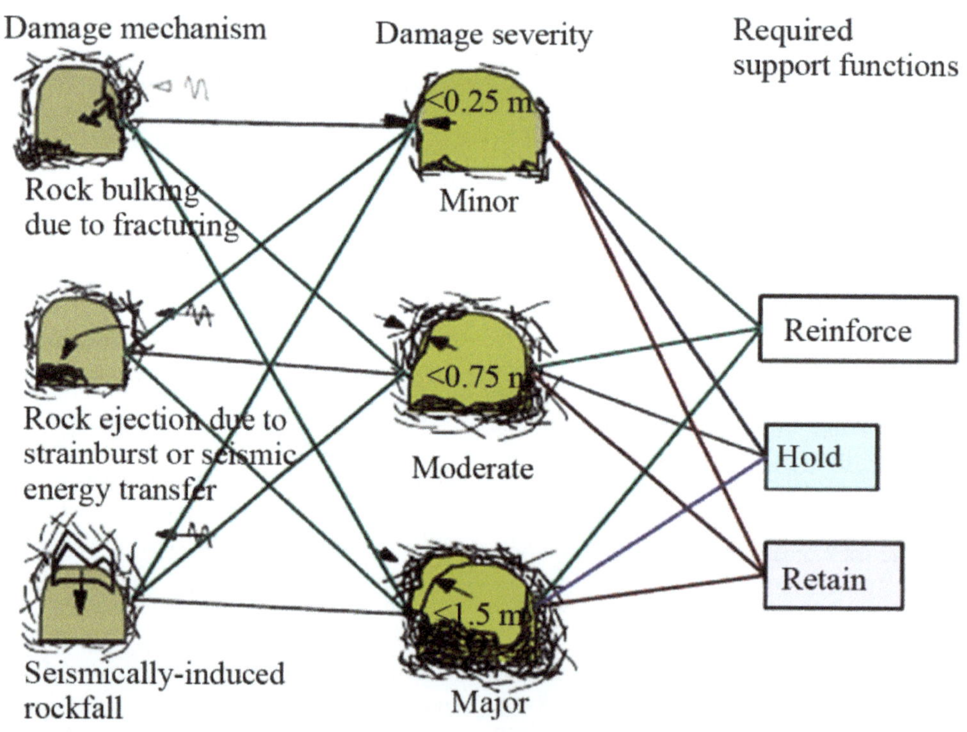

Figure 1. Type of damage mechanism, damage intensity and applicable support functions (After Kaiser et al. (1996)).

Table 1. Rock-burst damage mechanisms and the nature of anticipated damage (After Kaiser et al. (1996))

Damage Mechanism	Cause of Rock—Burst Damage	Damage Severity	Thick-ness m	Weight kN/m^{2*}	Closure†mm	Ve,‡m/s	Energy kJ/m^2
Bulking with-out ejection	Rock with high stress and small amount of stored stress energy	Negligible Reason-able Foremost	<0.25 <0.75 <1.5	<7 <20 <50	15 30 60	<1.5 <1.5 <1.5	Not critical Not critical Not critical
Bulking caus-ing ejection	Rock with high stress with substantialextrastress energy	Negligible Reason-able Foremost	<0.25 <0.75 <1.5	<7 <20 <50	50 150 300	1.5 - 3 1.5 - 3 1.5 - 3	Not critical 2 - 10 5 - 25
Ejection by remote seismic event	Seismic energy trans-fer to jointed or broken rock	Negligible Reason-able Foremost	<0.25 <0.75 <1.5	<7 <20 <50	<150 <300 >300	>3 >3 >3	3 - 10 10 - 20 20 - 50
Rockfall	Inadequate strength, forces increased by seismic acceleration	Negligible Reason-able Foremost	<0.25 <0.75 <1.5	<7 g/(a + g) <20 (a + g) <50 (a + g)	- - -	- - -	- - -

Support systems may be classified into two broad categories: internal and external, which can be either active or passive. A support becomes active when stresses are induced in it at the time of installation. Therefore, such supports reinforce the rock mass structure by exerting an "induced" stress on the ground immediately after its installation [8] . The common examples are pre tensioned rock bolts, hydraulic props and power-driven supports for long wall faces. Active supports are applied in situations where an excavation is believed to cause excessive deformation in the ground [9] . An example of this situation is separation of rock wedges from the rock mass. Passive supports do not reinforce the rock mass immediately after installation, but their effect is seen as subsequent mining activities take place. Examples of passive supports are steel arches, timber sets, composite packs and un-tensioned bolts. External supports are generally of the passive type, they are placed around the boundary of the excavation to help restrain the movement of the rock walls and avoid the failure of rock mass. Steel arches, wooden cribs and fiber reinforced shotcrete are some types of external supports. Backfill is another common type of external, passive support used in hard rock mining. Internal supports continue to see technological developments in the mining industry. The basic mechanism of internal support is to bound rocks together to maintain the overall stability of the rock mass around an excavation. Internal supports which are pre-tensioned at the time of installation are considered to be of the active type. Swellex, Split Sets, grouted bars and mechanical anchors are some of the common examples of internal supports. Rock bolts are most preferred during support installations in underground excavations than other types of supports because they provide unbarred excavation openings and have minimum repairs and set up, they require less storage space, they are very easy to install and require less installation time.

Instability of the rock mass is usually triggered by the mining activities, alteration of the in-situ stresses, causing either extreme high stresses that are way above the rock strength, this result in shear failure or relatively extensive spalling or low stresses causing progressive failure along planes of low strength. In cases of increasing depth or at later age of the mine, seismicity and rockbursts arise due to slippage on natural or mining-induced planes of weakness and fracture of the intact rock itself (strain burst), usually close to excavation boundaries [5] . For a successful support and reinforcement systems, the maximum effect is achieved by the interaction between the bolt and shotcrete, thus retaining and holding. (Figure 2) shows the concept between rock mass, and the reinforcement components: bolts, mesh and shotcrete.

The five most important characteristics of support in burst-prone grounds are: 1) Peak load capacity of the support; 2) Displacement at peak capacity; 3) Ultimate load bearing capacity or yield support; 4) Ultimate displacement capacity or stretch limit; 5) Energy dissipation capacity of support. The choice of a support system will depend on the geo-mechanical conditions, stress field, geology, the expected loading and nature of the damage.

Dynamic Supports for Burst Prone Ground

Reinforcing mechanisms generally restrict and control the bulking of the rock mass. Typically, reinforcing elements like grouted rebars act as stiff support elements, however split set bolts, yielding Swellex or Cone bolts [10] , act as ductile or yielding elements under high stress conditions. The holding function of supports is aimed to hold the retaining elements of the system and loose rock to stable ground.

Dynamic support design effectively implies rock-burst-resistant support design and is a function of the size of the seismic event and the location of the event hypocenter relative to the mine infrastructure in question. Local site factors such as the orientation of the infrastructure can also play a significant role in support and ground performance. Support ductility is required to achieve displacement capacity and energy absorption capacity [8] .

The main issue with underground excavation is the stabilization of the rock around excavations at depth. The main difference between deep rock and surface rock is the increment of the in situ stresses. Significantly this increase in the rock stresses, result in rockburst occurrence in hard rocks, large squeezing deformations in soft and weak rocks. From observation in most underground mines, rockburst phenomena begins at depths from 600 m to 800 m and more pronounced below 1000 m. Most metal mines in Sweden, Canada, Australia and South Africa, at the moment mining operations are at depths between 1000 m and 3000 m. At such depths, conventional support devices [11] [12] fail to cope with the severe rock conditions.

Conventional rock bolts are categorized in three types looking at their anchoring mechanisms: 1) Two-point fixed mechanical bolts; 2) Fully encapsulated rebar bolts; 3) Frictional bolts. Fully encapsulated rebar bolts are fused to the grout/rock lengthwise with a link between the bolt ribs and the grout. Rebar has a high load-bearing capacity, though, rebar cannot accommodate large rock dilations, would only endure a deformation of 2 - 3 cm when exposed to a fracture this indicates that it's a tough but rigid rock bolt (Figure 3). Interaction of frictional rock bolts with the rock is through friction amid the hole wall and the cylinder-shaped surface of the bolt.

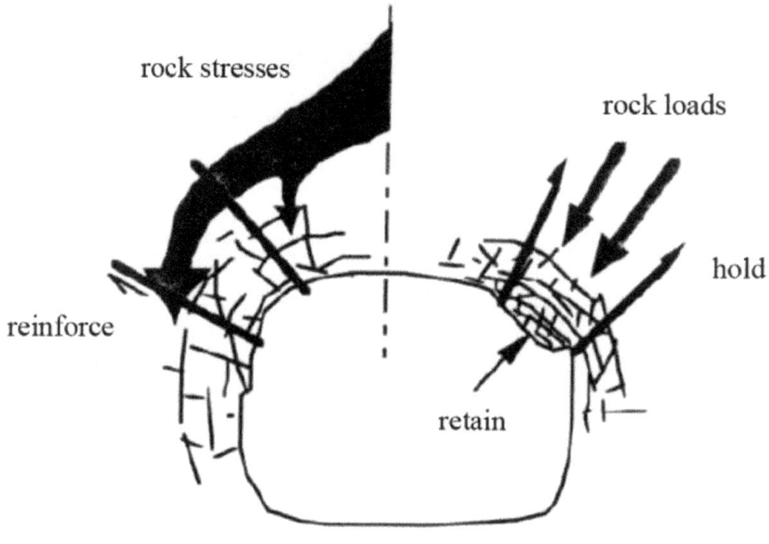

Figure 2. Interaction of reinforcement components (McCreath and Kaiser, 1992).

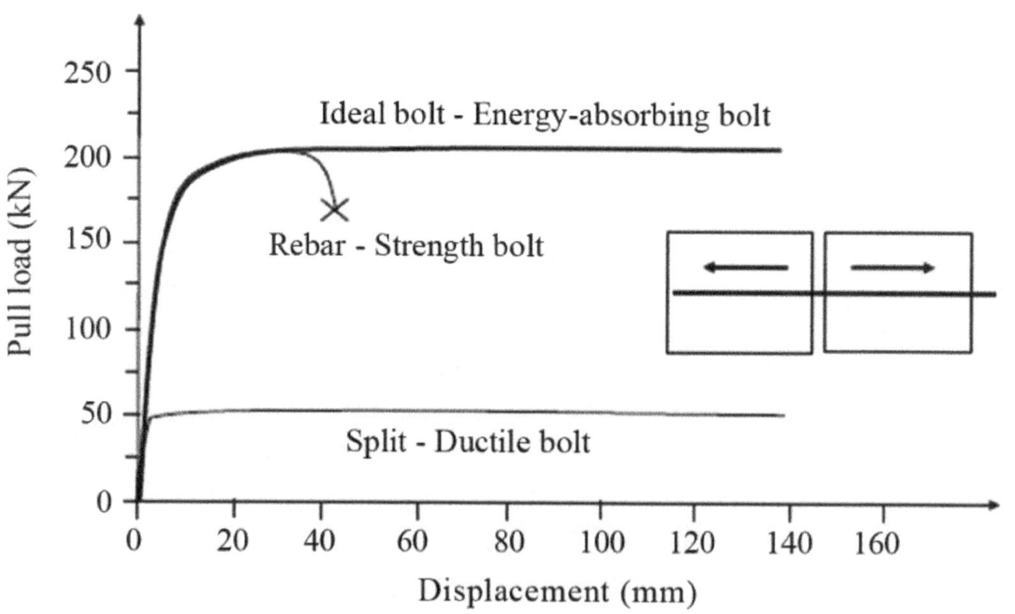

Figure 3. Ideal bolt properties in relative to other bolts (After Li CC (2013)).

The bolts can endure large rock deformations with a low load-bearing capacity. A standard Split Set bolt cannot carry a load greater than 50 kN (Figure 3). A perfect support scheme needs to be tough but rigid like are bar and endure deformation like a Split Set bolt (Figure 3), with the ability to handle load similar to the strength of the rock bolt material [12] .

It is not suitable to use rebar bolts as rock reinforcement devices at depths with high in situ rock stresses. When used with weak and soft rocks, it is often observed that either the face plate of the rebar is greatly loaded, or the thread of the bolt is pulled to failure [12] [13] . In the case of hard rocks, the rock spalls and peels behind the bolt plate, leaving a short part of an extruding bolt out of the rock surface. In Kiirunavaara mine in Sweden, old rebar bolts are occasionally visible on the work faces of successive cut slices. It is seen that some of those bolts fail all because of minor shear and initial movement at rock fractures [14] .

The untimely failure of the rebar bolts (Figure 3), indicates that a rebar is too stiff to withstand rock dilations in high stress rock masses [15] . To withstand large rock dilations, Split Sets are used as reinforcement devices in many deep mines. They are capable of enduring large rock deformations, but their load-bearing capacity is very low. Both rebar and Split Sets are low energy-absorbing devices. Ortlepp pointed out that support devices applicable in deep mines should be able to handle high loads and also withstand large deformations, in that they should be able to absorb a large amount of energy prior to failure [12] [16] . The first energy-absorbing device designed to counteract rock burst problems is the cone bolt [12] [17] .

Ideal Bolt

Rock bolts can be strength bolt, ductile bolt, or an energy-absorbing bolt. Strength bolts are those that provide a support load equal or close to the intrinsic strength of the bolt material. Ductile bolts are able to accommodate large rock deformations. Energy-absorbing bolts are characterized by their high load capacity and also their large deformation capacity [12] . The installed rock support system must be able to absorb dynamic energy while at the same time be able accommodate large rock deformation due to rock failure.

Strategically, installed support system is focused on the control of rock performance after failure, not on the prevention of rock failure occurrence [18] . In fact, it is actually un-economical to prevent rock damage from occurring by the increase in rock support capacity. In rock damage control, the support behavior must be essentially altered to allow for yielding. This section reviews the current support applicable in burst prone ground that have energy absorption or yielding capacity. Enforcement of stability for large rock blocks, and the holding together of weak lamented rock are achieved by rock bolts of the correct dimension. Widespread research and advance work on yielding rock support has been conducted, energy-absorbing bolts have been successively developed and applied. Currently, dozens of energy-absorbing bolt types and their yielding mechanism can be illustrated as structural components sliding and steel deformation (Figure 4) [18] .

Swellex Rock Bolt

Swellex is a friction bolt; the dowel varies up to a length of 12 m (Figure 5). The Swellex Mn2 4 is a 3 m long and a 25 mm diameter tube before folded during manufacturing. Pm24C Swellex bolts can be used to provide deep tendon support under dynamic loading (Table 2). When inserted into a drill hole (32 - 39 mm) no extensive driving force is required. It has expansion capability of 42 mm when activated by injection of high water pressure, causing a tight contact with the drillhole walls after the inflation. The concept of the bolt entails that the rock is secured by immediate and full support action. The moment the bolt is expanded in the hole, it interacts with the rock to maintain its integrity. The excellence of the bolt installation is confirmed when the pump stops, and is independent of rock mass conditions or operator experience [19] . The Swellex rockbolts are intended to enhance the efficiency of each bolt, so that the bolting operation ties with the required safety levels as intended [20] .

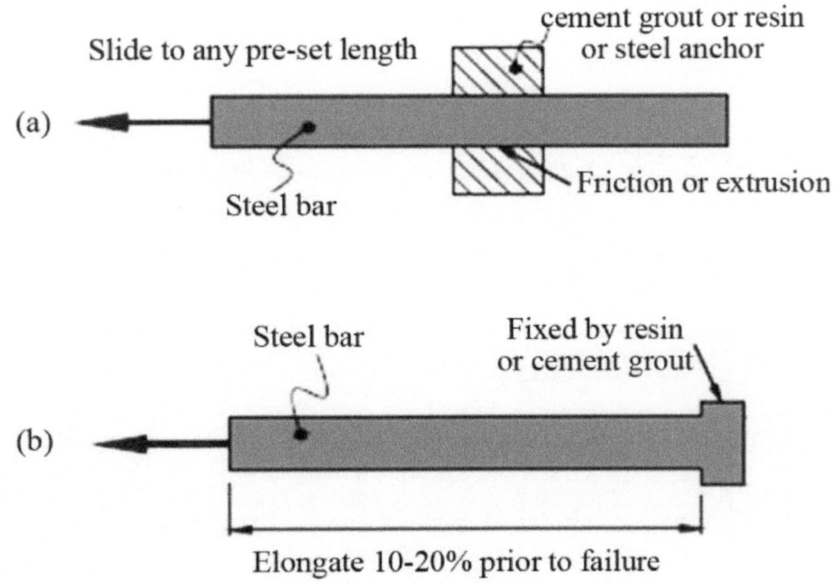

Figure 4. Illustration of Yielding mechanism of a rock bolt (After Wang et al. (2013)).

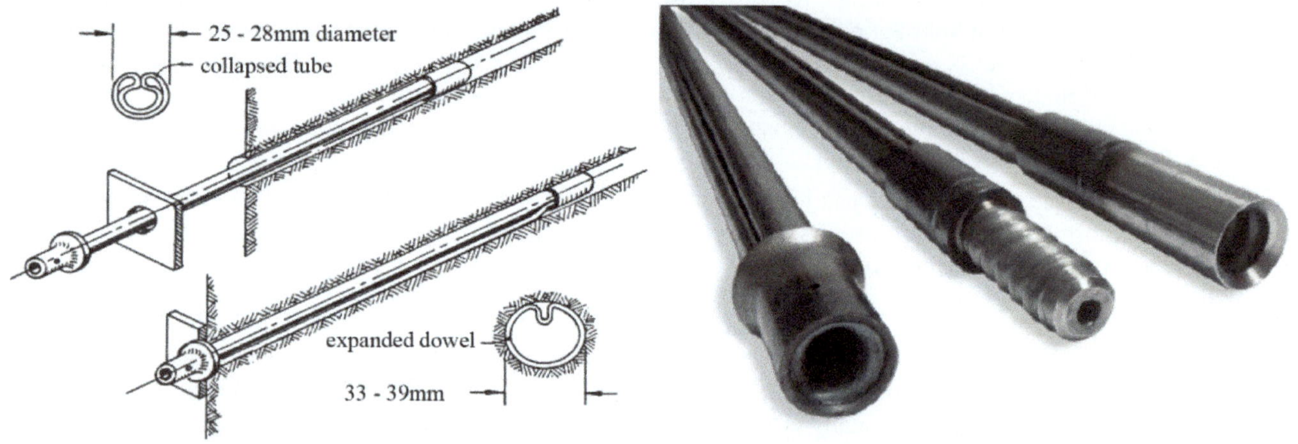

Figure 5. Friction anchored rockbolt—Swellex.

Roofex

Roofex is made up of a steel bar encased in a smooth plastic cover fixed inside the borehole with cement or resin grout (Figure 6). The bolt has an energy absorber, it works like a sliding element over the steel bar. Enabling the bolt to extend outwards during sudden movements from rock bursts while at the same time providing constant load capacity (Figure 7). This capability makes the Roofex rock bolt especially appropriate for developing new underground excavations in rock burst prone areas [20] . The bolt can be made in different lengths usually used in mining, and capacity of displacement can be designated at the time of bolt manufacturing.

Cone Bolts

The cone bolt is made up of a smooth steel bar coated with a thin layer of waxy lubricating material, so that it easily detaches from the grout under a tensile load, the bar has a flattened conical bell-shaped forged section at one end (Figure 8). The bolt is entirely encapsulated with cement grout or resin in a borehole. Installed in a specifically formulated Coneloc or Fasloc resin, clearly mixed by the resin mixing blade. A plastic sleeve is installed over the shaft of the bolt to debond the resin from the bar, thus allowing the cone to yield or plough through the resin. Cone bolts show high-ductility and energy absorption characteristics. Cone bolts should provide high-shear resistance [8] .

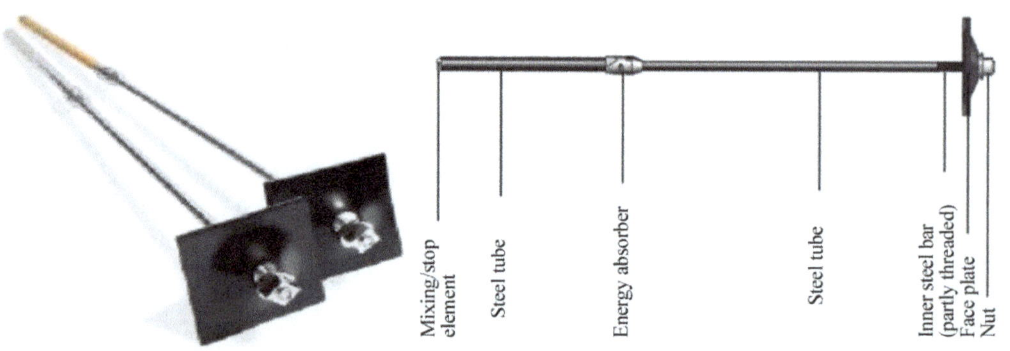

Figure 6. Roofex rock bolt.

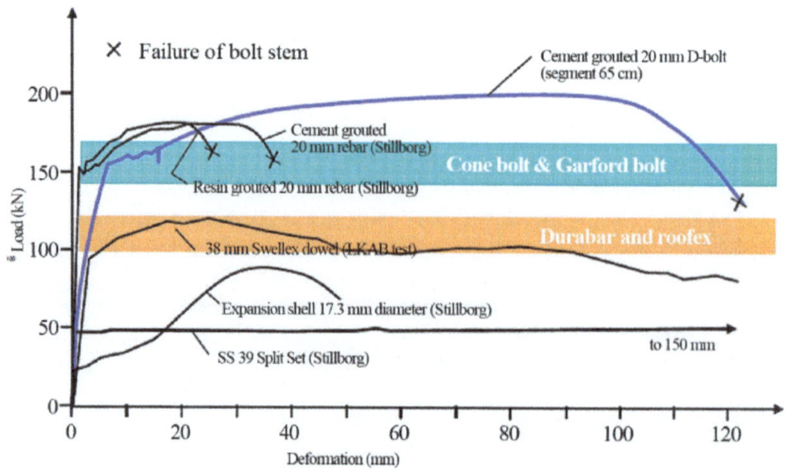

Figure 7. Displacement characteristics of rock bolts (after Li, 2013).

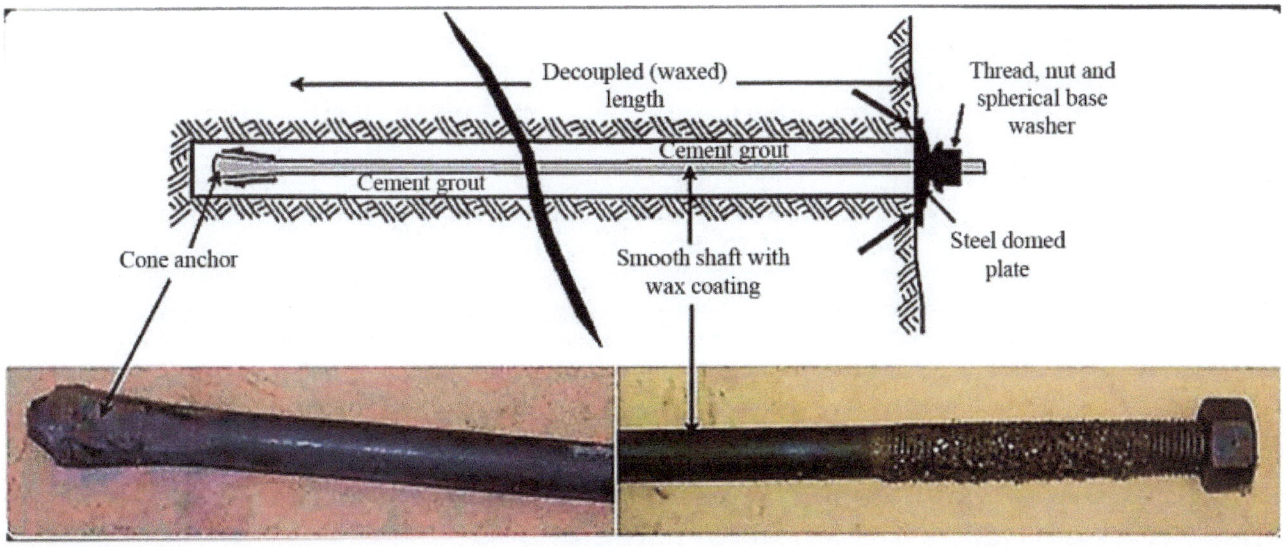

Figure 8. Cone bolt.

A pull load is induced in the bolt shank when the rock mass dilates in between the cone and the bolt plate. The bolt has been designed in that the conical end section works through the grout material when exposed to a pull load exceeding a predefined value. This enables the bolt to absorb energy from the rock, initially it was designed for use with cement grout later adapted to resin grout [12] .

Modified Cone Bolt

The Modified Cone Bolt cone bolt is basically a smooth bar with a threaded section at one end, forged cone and mixing blade at the other end which works as a resin mixer (Figure 9). These bolts are usually applied in many burst-prone mines especially in Canada as the main dynamic rock support systems. Ground observations carried out in mines revealed susceptibilities in the exterior anchoring point of the bolt especially at the bolt plate. The anchoring may be lost if the rock fractures behind the plate and falls down [21] , this causes the bolt to completely lose its reinforcement ability. Observed that the resin takes time and at times fails to harden, indicating that the cone bolt mixing quality of resin is not always definite and to the standard [22] .

Table 2. Swellex properties

Type	Swellex load capacity		
	Minimum breaking load, expanded profile (kN)	Minimum yielding load, expanded profile (kN)	Minimum elongation (%)
Pm16	160	140	10
Pm24	240	200	10
Pm24C	240	200	10
Pm24H	200	190	10
Mn24	220	180	20
Sp24	240	200	6

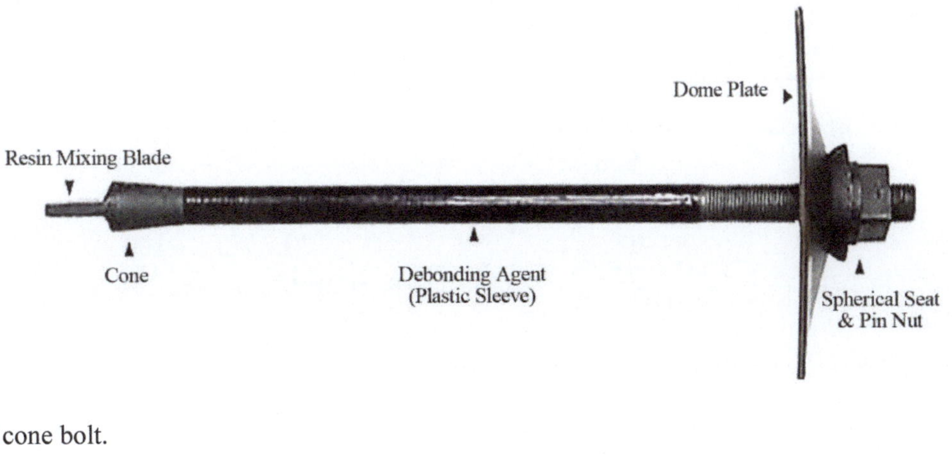

Figure 9. Modified cone bolt.

D-Bolts

The D-bolt was developed to stabilize both burst-prone and squeezing rock conditions that are a major factor in underground excavation. The bolt is made up of a smooth steel bar with anchored sections along its length. These can firmly be fixed within a borehole with the aid of cement grout or resin, the smooth segments between the anchors may easily bend in reaction to rock stretching (Figure 10). Reinforcement performance of the other segments can never be affected if one of the other segment fails. The bolt is designed to completely utilize the strength and the deformation capability of the bolt material lengthwise. The bolt has large load-bearing and deformation capacities. Through static pull trials and dynamic drop trials it was seen that the bolt extent lengthens from about 14% - 20% at a load equal to the strength its material, thus absorbing a larger quantity of energy (Figure 11). The impact normal load of a 20 mm D bolt was about 200 - 230 kN, with some percentage of the load transmitted to the bolt plate [12] [23] .

Garford Bolt

The Gar ford dynamic bolt is comprised of a compact steel bar, an anchor and a coarse-threaded steel cover at one end (Figure 12). Its characterized by its anchor this permits the bolt to expanse by a great extent when the rock dilates. Varden et al. presented some tests and selection of the gar ford dynamic bolt (Figure 13) for use under dynamic conditions and explained that the bolt comprises of a 20 mm mild steel solid bar and a resin mixing device of 350 mm long, 43 mm diameter rough threaded steel cover folded at the end. The dynamic segment is an original sliding anchor mechanism which is constrained on to the bolt beneath the mixing device.

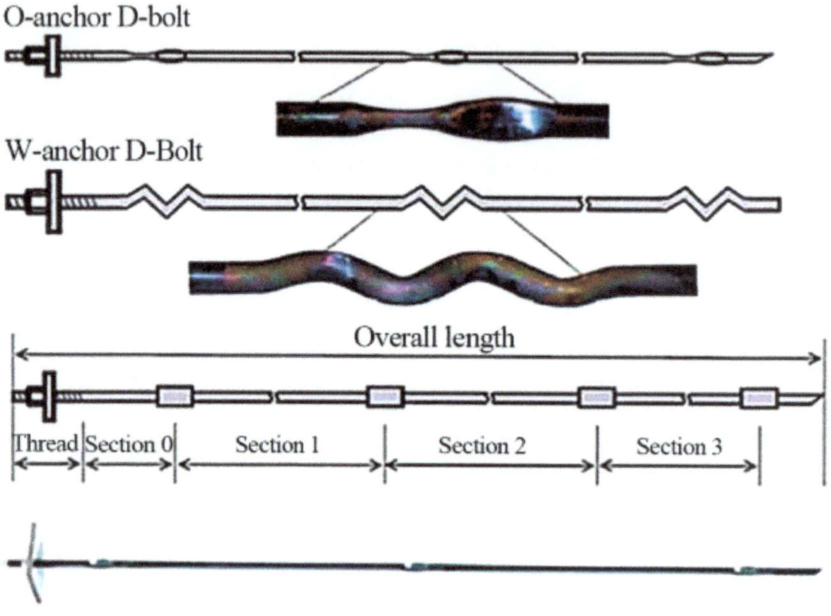

Figure 10. D-bolt.

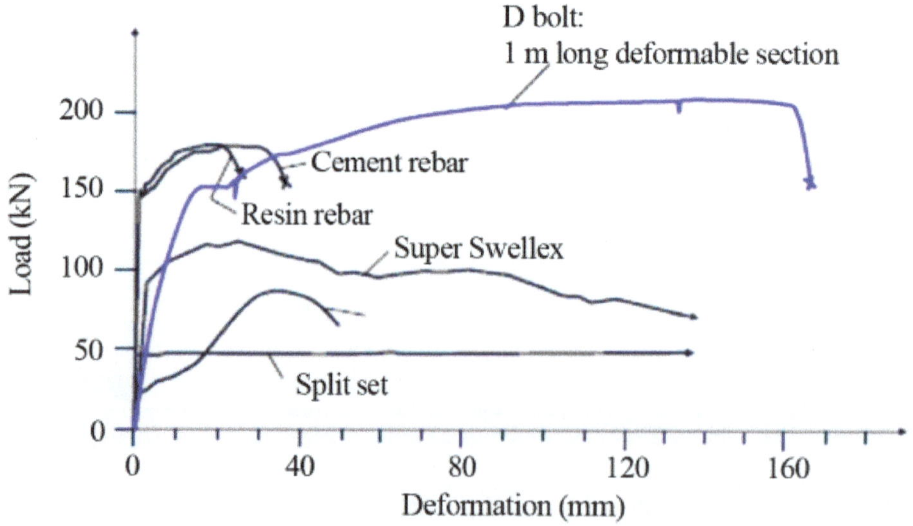

Figure 11. Load-displacement of D bolt under pull loading test.

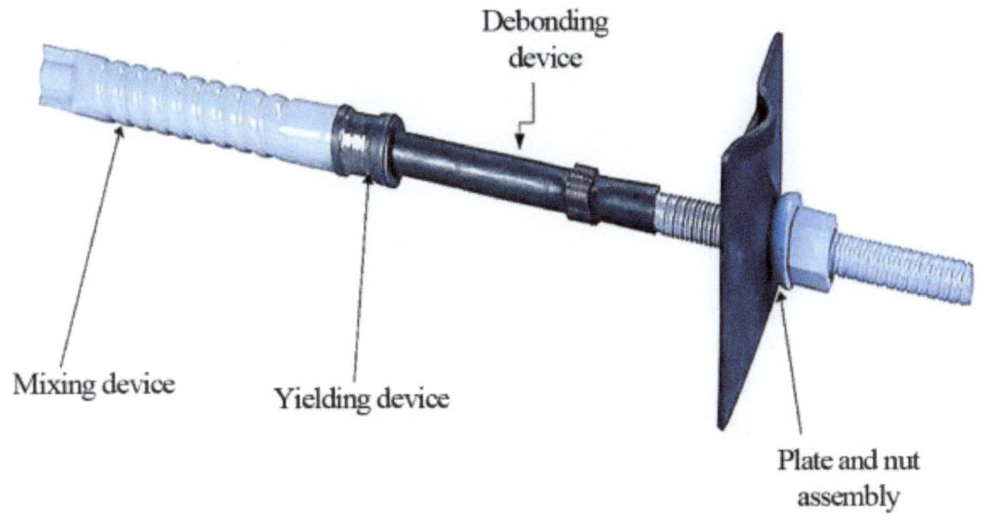

Figure 12. Garford bolt.

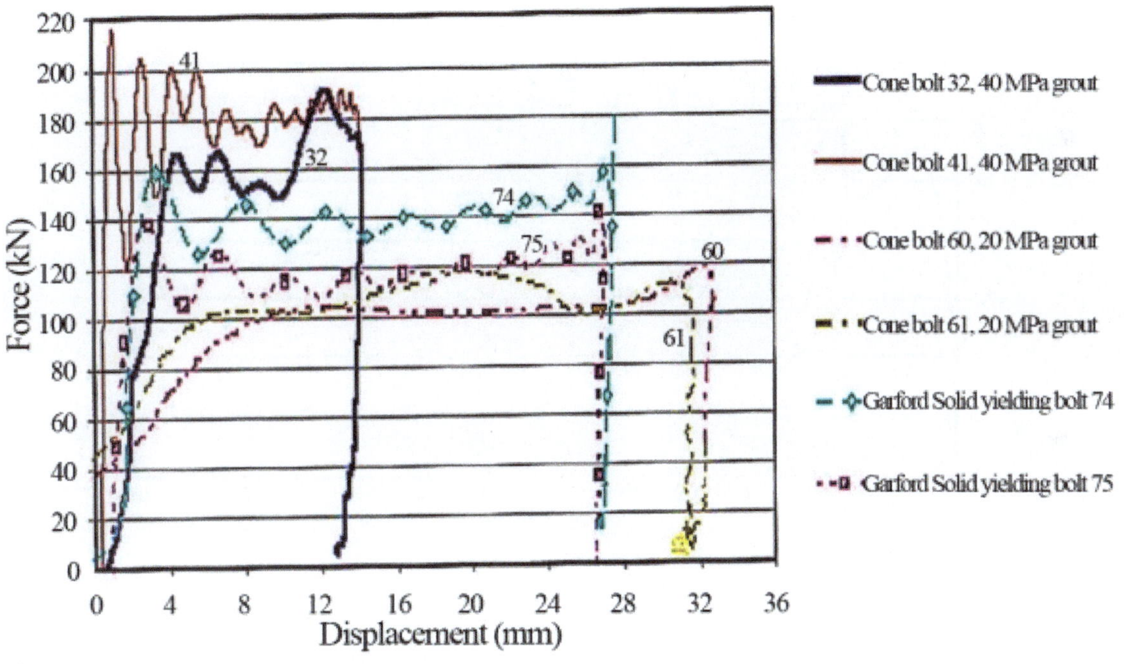

Figure 13. Dynamic force displacement curves (After Varden et al. 2008).

The rest of the bolt is enclosed in a polyethylene sheath, this provides a debonding action, which debonds the bolt below the dynamic section. Under a dynamic condition the bolt is pushed through the narrow opening and it stretches [24] .

Dura Bar

This is an energy-absorbing yielding rockbolt. It is anchored at two-points. The inner anchor of the bolt is a crinkled segment of the smooth bolt stem that slips at a predefined pull load. The external anchor is the bolt plate, the rod is bent to form a wave and act as a ductile anchor. The collar end is completed by an eye for cable lacing or a threaded segment for nut and washer (Figure 14). The entire length of the bar is coated with wax excluding the collar end for improved de-bonding. The hole is injected using cement grout the anchor of Dura bar is a creased segment of the bar. During face plate loading, the anchor slips alongside the creased outline in the cement grout at a definite pull load [25] . (Figure 15) shows displacement of the bar under a dynamic test.

Dynatork Bolt

Its Dywidag systems international dynamic bolt, it's made with a coiled mixing blade to ensure ultimate mixing of DSI resin in the borehole (Figure 16).

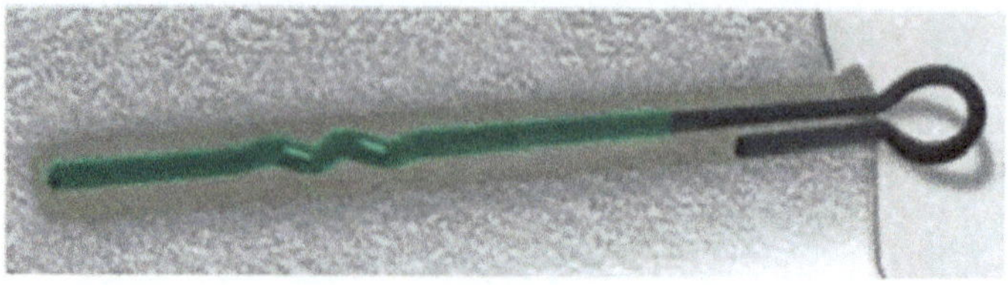

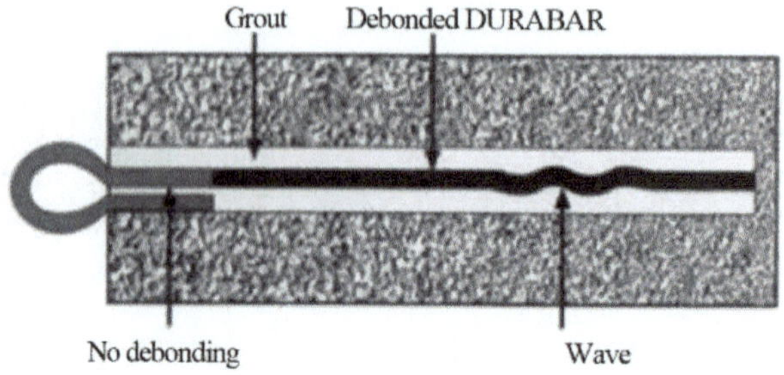

Figure 14. Dura bar.

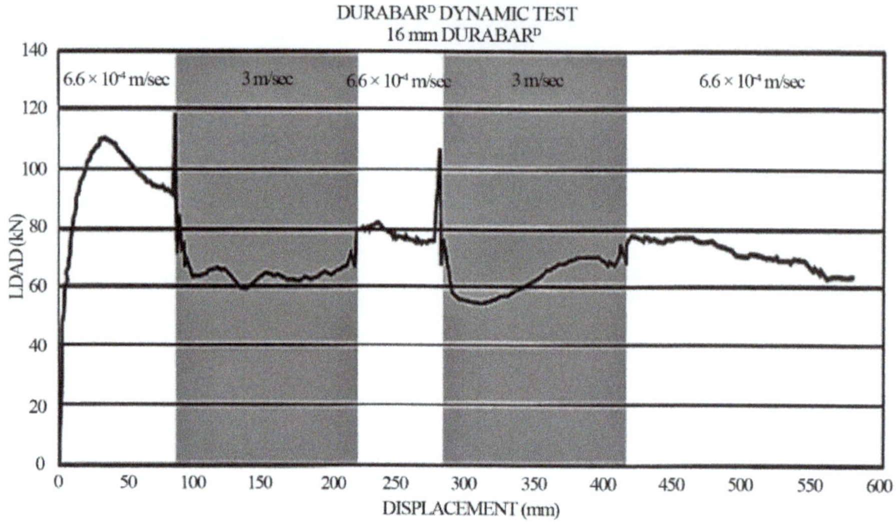

Figure 15. Dynamic test results of DURABAR yieldable bolt (CANMET-MMSL, 2012).

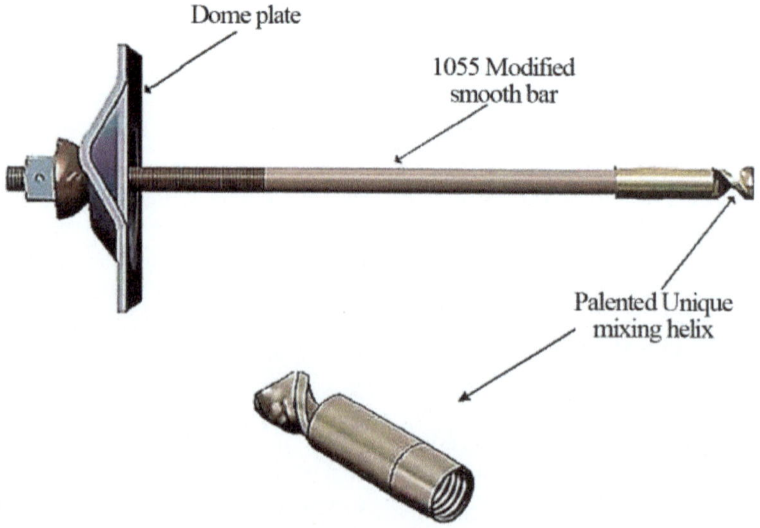

Figure 16. DSI dynamic bolt.

This blade design, according to DSI has a double function, mixing the resin correctly and providing perfect docking for support in static conditions. The bolt is engineered to yield and aid transmission of dynamic load into the resin during rock dilation in rock burst conditions, absorbing the energy through controlled deformation (Figure 17). The energy is expected to be absorbed during the ploughing effect of the cone through the resin [25] .

Split Set Bolts

Comprise of a slotted, hollow highly-strengthened steel tube and faceplate. The bolts are fitted by pushing them into an undersized hole. The circular coil force created by compression of a C-shaped tube delivers frictional anchorage along the complete length of the tube (Figure 18). The toe end of the tube is slanted inward to facilitate quick and easy installation. It can be particularly useful in mild rock-burst conditions where the bolt will slip and not break, when used with wire mesh it will hold shattered rock caused by a mild burst.

These bolts can also be used to pin up mesh in highly fractured and yielded ground prior to the application of shotcrete. Provided the load demand does not exceed capacity, this system can work well for short-term support. However, susceptibility to corrosion rules out this system for most long-term support requirements. The bolts are installed as dowels. Split Set bolt pull-out capacity is rated in metric tons per meter (t/m) and is derived from in-situ pull tests. Nominal pull-out strengths vary from 1 to 2 t/m [8] .

The Yield-Lok

This bolt can be fully or partially grouted with resin or cement mortar. The yielding performance is based on the interactions between the polymer coating, the head and the grouting media (Figure 19). The angled head of the polymer coating helps to scrap resin cartridge packing when the bolt is inserted into the resin and this improves anchorage. Mixing of the resin is enabled by deformations on polymer coating similar to rebar, thus the bolt can be fully inserted without rotation, and then spun afterwards to mix the resin. The Bolt can be constructed in dynamic condition a case of YL-Dynamic Bolt or for specified loads in static condition a case of YL-Static Bolt (Table 3) [25] [26] .

When the rock bolt is tensioned, it immediately provides primary support upon being installed. During static loading conditions, the bolt performance is actually similar to a rebar bolt (Figure 20). The head transmits the effects on the polymer coating during dynamic loading, this brings about restricted compression, thermal softening and flow of the polymer around the upset head, creating a pulling effect. The energy from dynamic loading is absorbed by pulling the upset through the polymer, some of the energy is used up during friction between the smooth bar and the polymer coating [25] .

The feature of yielding elements can be well controlled, ensure the steady performance of the bolt in the full length of polymer coating. The purpose of grouting material is to provide confinement of the polymer coating, instead of working as a yielding element as in the case of cone bolts. It means that if the bolt is completely grouted and confined, the performance of the bolt is not dependent on the grouting material. Since the mechanism of displacement is confined within the polymer, de-bonding agents like grease cannot help achieve the definite effect and steady performance of the bolt [26] .

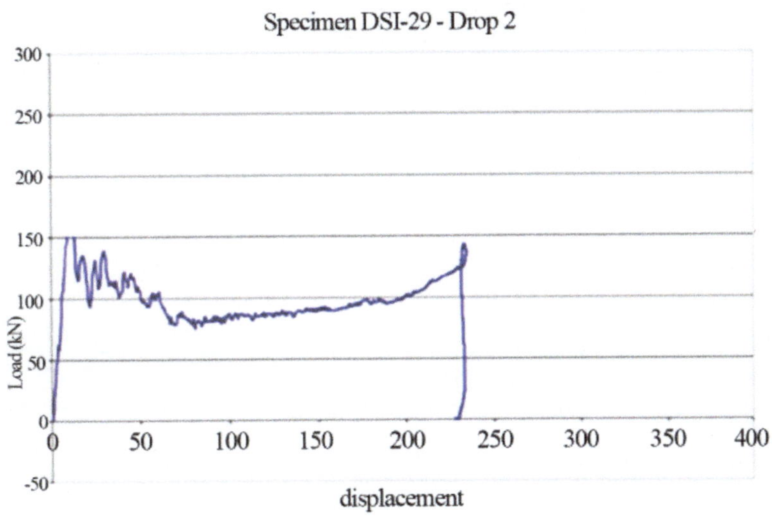

Figure 17. Dynamic laboratory test results of DSI-Dynatork bolt (CANMET-MMSL, 2012).

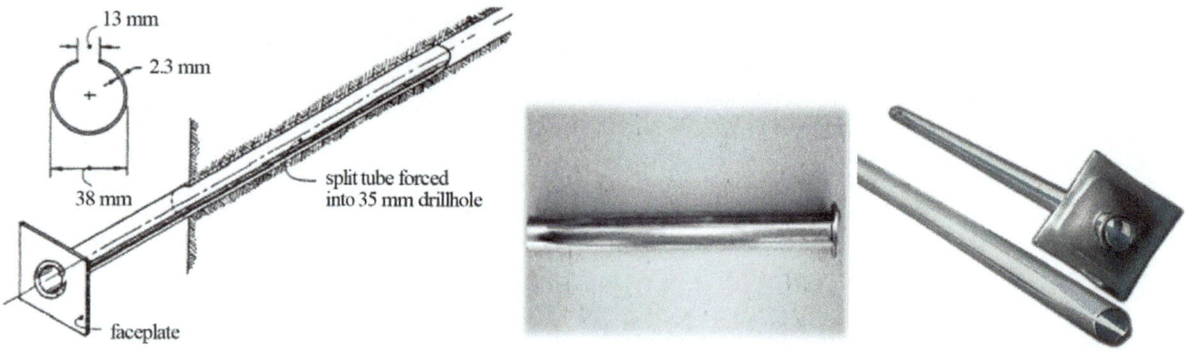

Figure 18. Friction anchor or Split Set.

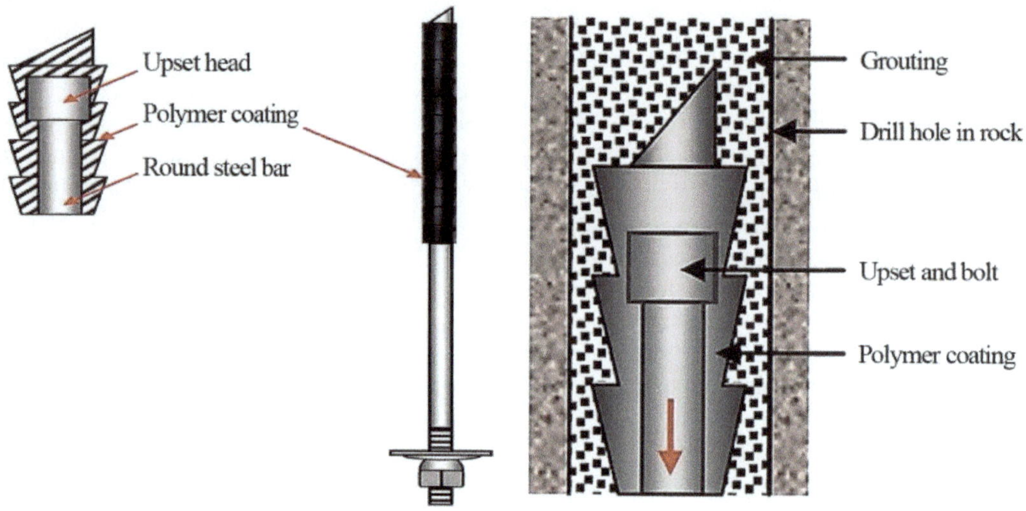

Figure 19. Yield-Lok.

Jacketted Rock Bolt

The function of jacketted rock bolts proposed by Chileshe (2015) was to balance the stress and strain, imposed by surrounding rock on grout and/or grouted rock bolts. If the high strain energy in the excavation periphery were to be accommodated over the whole length of a grouted rock bolt to the toe by a load balancing rock bolt, localized strain and stress excess would be mitigated. The low shear stress induced in the minimal strain zone in the deepest toe part of the grouted drill-hole would be increased, while that at the collar of the hole would be reduced. The result would be equitably distributed average shear stress on the rock bolt. The likelihood of grout or grout-rock or grout-bolt or rock bolt failure would be significantly reduced thus increasing the effectiveness of the grouted rock bolt. The overall effect of the load-balancing rock bolt would be to increase the effective anchorage length, especially with longer rock bolts [27] .

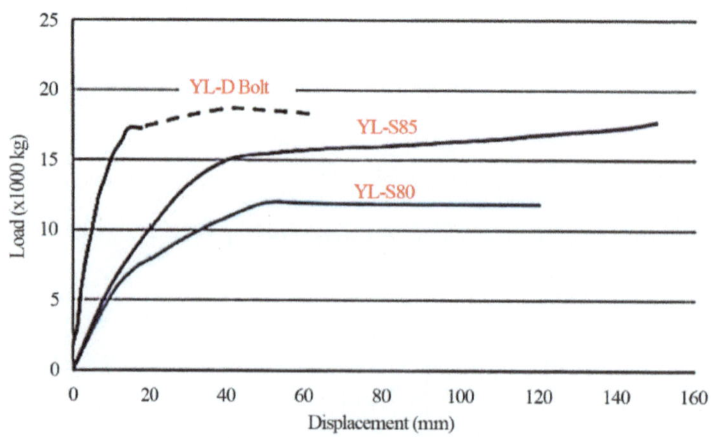

Figure 20. Static test results of Yield-Lok Dynamic and static bolts (CANMET-MMSL, 2012).

Table 3. Yield-Lok Bolt properties

Type	YIELD-LOK Bolt		
	Minimum breaking load, expanded profile (kN)	Minimum yielding load, expanded profile (kN)	Minimum elongation (%)
	178	133	8
	Dynamic yield load (kN)	Dynamic yield displacement	Displacement capacity, max (mm)
YL—dynamic bolt	71 - 89	200 mm per effect with energy input 16.4 kJ	762
YL—static bolt	Static yield load (kN) 116 - 133	-	(500 mm) or more up to length of polymer coating

The bolt is assembled in concentrically tubular steel jackets (Figure 21). Each jacket is 0.6 m to 1.1 m in length, and fitted within each other. The jackets overlap by, say, 0.2 - 0.4 m, thus forming a total length and diameter compatible with a typical rock bolt hole. The tightly or spirally or groove-fitted or lined or indented overlaps would provide friction between jackets and thus transmit energy from one jacket to the other in a frictional balancing system. The capacity to slip is essential. The entire jacketted assembly is grouted into the drill hole or simply in force-fitted direct contact with the rock [27] .

Alternatively, to accommodate extreme cases of gross deformation, such as rock-bursts, the innermost tube can be replaced with a similarly sized solid rod (Figure 22) of a length that allows extension, even protrusion out of the hole [27] .

New Yielding Bolt Proposed by (Wang et al., 2012)

This bolt has similar appearance with Roofex, but a different structure of energy-absorber, its more simpler and stable, this increases its resistance. The bolt uses the elastic properties of steel through its operations. The bolt is intended to disperse and control large amount of energy liberated from the rock mass during the process of deformation. The new bolt is made up of a drawing die and drawing rod (Figure 23(a)). The rod is a metal bar with different diameters coated with a thin layer of wax, for easy detachment from the grout under a pull load, the die which also acts as an anchoring point is fixed inside the borehole with cement grout or resin. Operations of the bolt involve the metal being pushed through a die by tension applied at the end of the die. The elastic movement is sorely due to compression, from the metal with the die [15] [23] .

The deformation of the rock induces a tension in the bolt, this causes sliding between the drawing die and rod. When the drawing die orifice diameter gets smaller than the thick end of drawing rod, pressure is induced. When the tension exceeds the defined drawing force, the rod is slowly pulled out (Figure 23(b)). The new bolt is designed to act like a stiff rock reinforcement element until the defined drawing load is exceeded. In the process of drawing, the bolt is capable of absorbing a large amount of energy (Figure 24). In addition to the continuous resistance and the large deformation, the effect of bolt operation is an increase in the rod's tensile strength. Upon installation, suitable pre-stress is enforced to support the rock mass this enhances the supporting effect considerably [15] [23] .

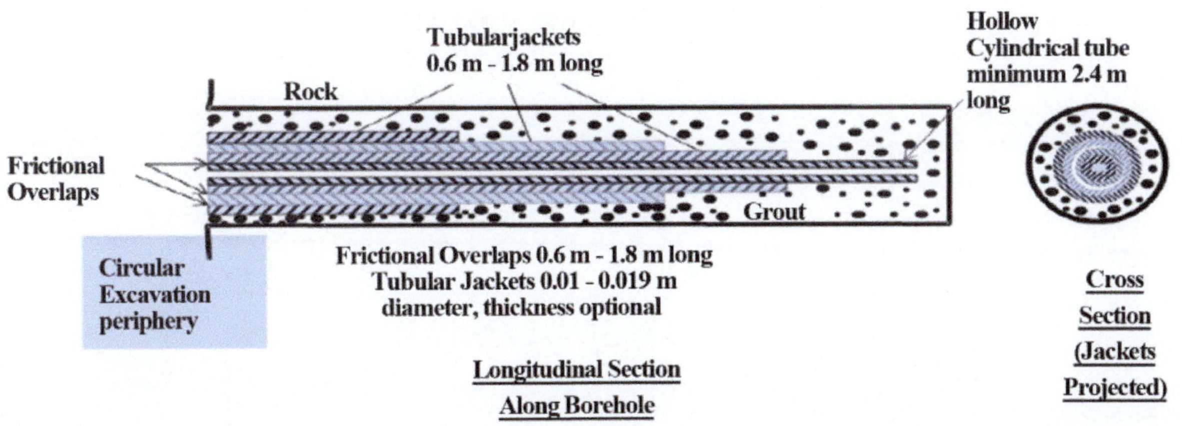

Figure 21. Jacketted Rock Bolt.

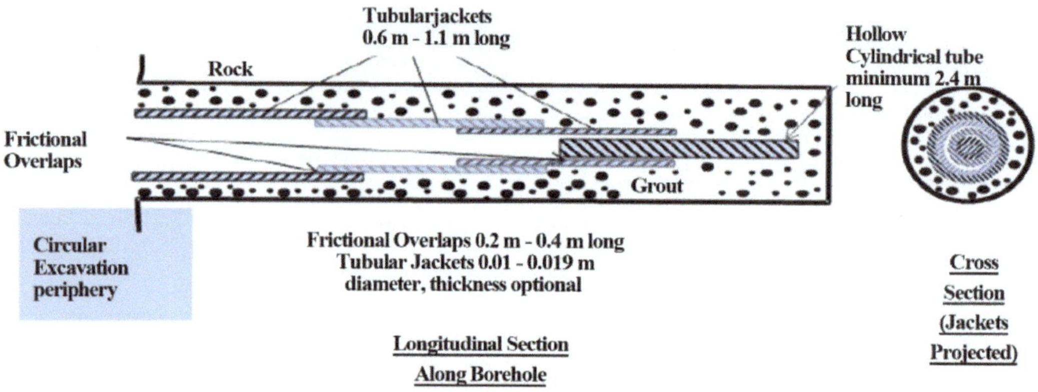

Figure 22. Jacketted Rock Bolt.

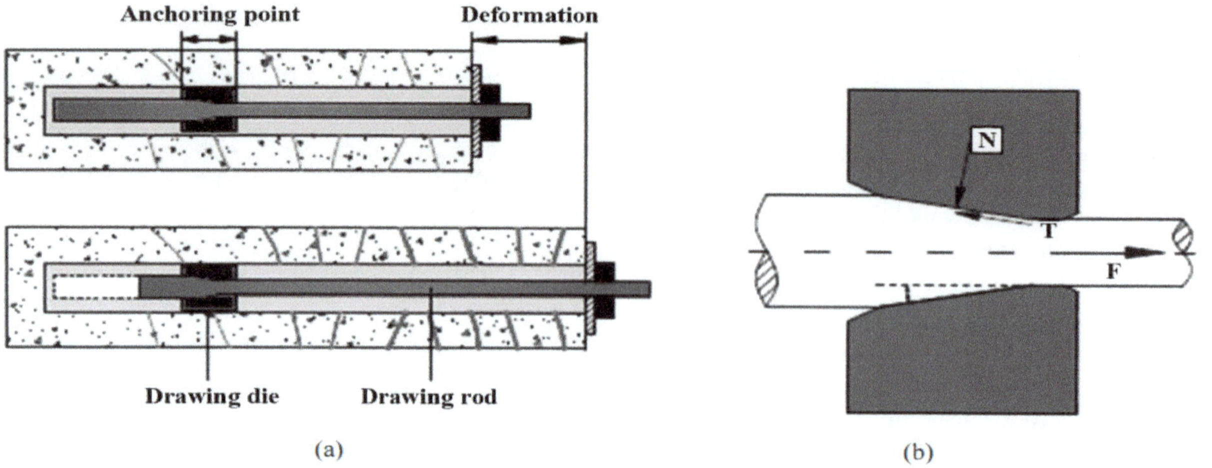

Figure 23. (a) & (b) Mechanism of the new bolt (After Wang et al. (2012)).

INTEGRATED SYSTEM SUPPORT

To form an integrated system support to be employed in high stress ground conditions rockbolts can be combined with surface support elements. Installing rock bolts as the only support system reduces the anticipate defficiency of rockburst support. This was observed underground in real rockburst conditions and was also established by a simulated rockburst trial conducted in a South African mine [28] . In the simulated rockburst trial, the support comprised of rebars, mesh, and lacing. Before blasting, mesh and lacing was removed and only there bars were left. Rebars were affected by the dynamic loading but not a single bolt failed due to dynamic loading of maximum peak particle velocity of 3.3 m/s and an ejection velocity of 2.5 m/s which was verified. Milev and Spottiswood (2005) [28] clarified that the cause was due to rapid weakening of peak particle velocity from the surface to the more competent rock mass [18].

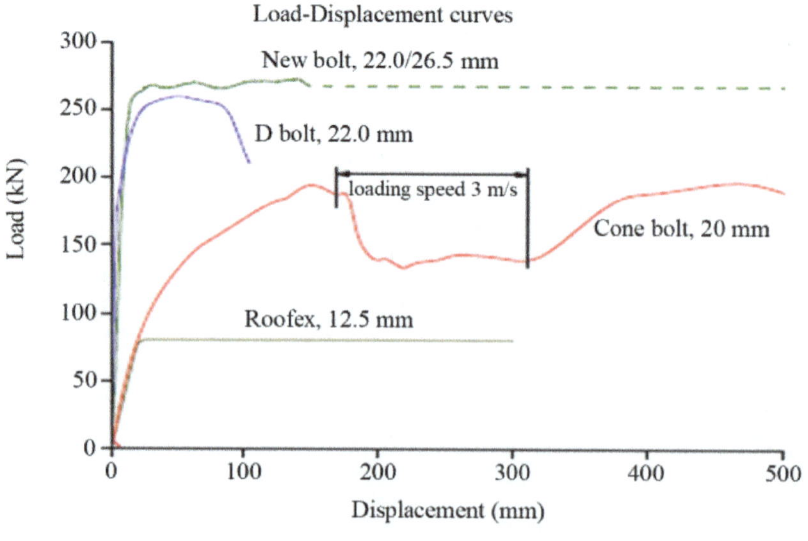

Figure 24. Load versus displacement behaviors of various boltunder pull loading test.

Support displacement capacity is critical in areas of generated dynamic loading [29] . Cablebolts can be effective where fractured ground has relaxed and become vulnerable to vibration. Support loads generated over larger displacements absorb more of the kinetic energy of such a groundmass. Stiff supports will rupture before the mass can be decelerated and stabilized. Cable bolts are rock support elements which can reinforce and hold the rock mass. Cablebolts can survive higher values of ultimate displacement before failure; even though the energy that they could absorb is moderate [8] . It may be advisable to combine long partially bonded cables, with modified geometry or plated end lengths, with closely spaced primary support such as grouted rebar [30] . Under dynamic loading the rebar will maintain a reinforced film at the excavation surface. The large displacement of this film can be accommodated by the cables, maintaining ultimate holding capacity after the disturbance has passed. Without the rebar, the surface skin may loosen and disintegrate, rendering the cable bolts ineffective [31] .

Tannant and Kaiser (1995) [32] describe an innovative support element for support of blocky ground under dynamic loading. A cable is anchored in a borehole using a Swellex friction bolt. A length of cable extends beyond the down hole extent of the Swellex and has an optional button or swage at the end. Even if the Swellex breaks in several places under loading, the cable remains frictionally anchored within the Swellex segments. The load-displacement characteristics are ideal for dynamic conditions; a consistent yield load and a large displacement capacity before rupture [33] .

Wire mesh is used to contain ejected rock and hold back protruding pieces on the excavation surface; it's usually used in combination with shotcrete. Steel fibre reinforced shotcrete and mesh reinforced shotcrete is extensively used and preferred in underground mining. In poor quality and loose rock masses, where shotcrete to the rock surface is poor due to adhesion, the mesh delivers a significant reinforcement. Thus when stabilizing and building walls for underground fill, weld mesh is applied in the stabilization of the surface [34] .

Steel arches were generally used to support haulages especially in coal mines where they are often essential to withstand large deformations. Deformations can be accommodated by employing yielding arches comprising of elements intended to slip at predetermined loads [35] . Hydraulic props are set with a pre-load to provide active support, The load bearing capacities of individual props may vary from as little as 5 t for a very light prop to more than 100 t for a 0.3-m-diameter prop. Rapid-yielding hydraulic props are widely used to provide concentrated, active support of the hanging wall close to the face. Their rapid-yielding capability allows the energy released by rock bursts to be absorbed rapidly and safely, thereby minimizing the damage caused [35] .

DISCUSSION

Charette and Hill (2011) [36] analysed basic static and dynamic loading data of six grouted rock bolt types capable of long term support. They concluded that the rebar and D-Bolt had a higher static load performance, while the MCB cone, Yield-Lok, D-Bolt and Roofex provided large deformation capacities at various loading levels. The MCB33 cone is best coupled with rebar reinforcement, while the D-Bolt and Yield Lok are designed to be used as single pass reinforcement bolts. Laboratory testing appeared to show that rebar and MCB cone bolt worked best together, at bolt displacements of less than 25 mm, while a dynamic event exceeding 7 kJ occurring would fail fully-grouted rebars (displacement larger than 25 mm). Thereafter, only the MCB cone bolt is able to provide static support [27] .

Yielding bolts application such as cone bolts or friction sets is a priority in rockburst prone mines, Rebars are considered to be too stiff and that they can't be used for rockburst support. Rebar bolts can provide suitable support to uneven blocks to prevent rock falls, considered suitable in shallow excavations with low in situ rock stresses, and the main risk is from rock falls aided by gravity. The key task of the bolt in a shallow tunnel is to equilibrate the weight of loosened blocks that are reinforced by the bolt. Thus, the bolt strength is more significant than its deformation capacity in low in situ stress conditions [12] [13] .

Combination of mesh or shotcrete with rebar bolts, can control the fracturing and rock bulking perfectly in hard rock mines, if the stress is somewhat low to moderate. When a rockburst strikes, rebars usually break at the threaded section and lose their function of holding but still maintain their reinforcing function to some point. However, if yielding bolts with straps are added to the rock support system, then a two tiered mechanism is formed. Rebars streng then the rock mass to ensure that it does not go through pre-mature fracturing. During rock mass failure, yielding bolts ensure the broken rocks are correctly retained. When developing a standard bolt applicable for rock burst prone ground, significant factor is its high initial load capability and stiffness for static loading as well as its high yielding and energy absorbing capacity for dynamic loading [18] .

Application of shotcrete in burst-prone grounds is considered unlikely because it has been observed that shotcrete became part of the ejected rock when a rockburst occurs. Shotcrete improves the installed rockbolt-mesh support system by enhancing its link between the bolt and mesh. Prevent main blocks in blocks from moving and thus enhances the overall integrity of the rock mass and it needs to be used effectively in the support system. In preventing shotcrete being ejected off with rocks, mesh-over-shotcrete is applied. The problem of ejected rock can be solved by mounting yielding bolts and mesh/mesh straps over shotcrete. Thus, it is very vital to appreciate the three vital support function of for each support element [18] . As stated above its best to have a combined reinforcement to withstand ground deformations from seismic activity and rockburst, but cone bolts proves to be the best yielding bolt in line with its deformation capacity, but however can also work best in an integrated system of supports.

CONCLUSION

Yielding rock bolts are most preferred during support installations in rockburst prone ground because they provide obstructions in excavation openings, they are easy to repair, install and set up, they occupy small space during storage and they also require less installation time. However, with these advantages of rock bolts over other supports, rock bolts are not very effective installed by themselves, and they are best in an integrated system. Support components have many functions in terms of reinforcement, holding and retaining in that they can be strong on one side and weak on the other. It's very important that support components are combined in order to increase their capabilities for support in rockburst-prone mines. A very good support system is one which has many and different elements to form a combined system.

ACKNOWLEDGEMENTS

The authors would like to thank the department of mining and mineral resources engineering of the University of Science and Technology Beijing.

REFERENCES

1. 1. Mendecki, M.A. (1999) A Guide to Routine Seismic Monitoring in Mines. Creda Communications, Cape Town.
2. Larsson, K. (2004) Mining Induced Seismicity in Sweden. Master's Thesis, Luleå University of Technology, Luleå.
3. Blake, W. and Hedley, D.G. (2003) Rockbursts, Case Studies from North American Hard-Rock Mines. Society for Mining, Metallurgy and Exploration Inc., Littleton.
4. Per John, L. (1983) Hard Rock Pillar Strength Estimation an Applied Empirical Approach. Masters's Thesis, The University of British Columbia, Vancouver.
5. Woldemedhin, Y.B. and Mwagalanyi, H. (2011) Investigation of Rock-Fall and Support Damage Induced by Seismic Motion at Kiirunavaara Mine. Master's Thesis, Luleå University, Luleå.
6. Malmgren, L. (2005) Interaction between Shotcrete and Rock: Experimental and Numerical Study. PhD Thesis, Luleå University of Technology, Luleå.
7. Cai, M. and Kaiser, P.K. (2012) Design of Rock Support System under Rockburst Condition. Journal of Rock Mechanics and Geotechnical Engineering, 4, 215-227.
8. Bawden, F.W. (2011) Ground Control Using Cable and Rock Bolting. In: Darling, P., Ed., SME Mining Engineering Handbook, Society for Mining, Metallurgy and Exploration Inc., Littleton, 616-617.
9. Stillborg, B. (1994) Professional Users Handbook for Rock Bolting. Trans Tech Publications, Clausthal-Zellerfeld.
10. Rajua, D., Mitria, H. and Thibodeaub, D. (2011) Mine Safety Enhancement by Designing Dynamic Rock Supports. 1st International Symposium on Mine Safety Science and Engineering, Sudbury, November 2011, 1591-1602.
11. Hoek, E., Kaiser, P.K. and Bawden, W.F. (1995) Support of Underground Excavation in Hard Rock. Balkema, Rotterdam.
12. Li, C.C. (2010) A New Energy-Absorbing Bolt for Rock Support in High Stress Rock Masses. International Journal of Rock Mechanics & Mining Sciences, 47, 396-404. http://dx.doi.org/10.1016/j.ijrmms.2010.01.005
13. Li, CC. (2007) A Practical Problem with Threaded Rebar Bolts in Reinforcing Largely Deformed Rock Masses. Rock Mechanics and Rock Engineering, 40, 519-524. http://dx.doi.org/10.1007/s00603-006-0094-7
14. Li, C.C. (2010) Field Observations of Rock Bolts in High Stress Rock Masses. Rock Mechanics and Rock Engineering, 43, 491-496. http://dx.doi.org/10.1007/s00603-009-0067-8
15. Wang, G., Wu, X.Z. and Jiang, Y.J. (2012) A New Yielding Bolt for Rock Support in High Stress Rock Masses. Applied Mechanics and Materials, 204-208, 366-369. http://dx.doi.org/10.4028/www.scientific.net/AMM.204-208.366
16. Ortlepp, W.D. (1992) The Design of Support for the Containment of Rockburst Damage in Tunnels—An Engineering Approach. In: Kaiser, P.K. and McCreath, D.R., Eds., Rock Support in Mining and Underground Construction, Balkema, Rotterdam, 593-609.
17. Jager, A.J. (1992) Two New Support Units for the Control of Rockburst Damage. In: Kaiser, P.K. and McCreath, D.R., Eds., Rock Support in Mining and Underground Construction, Balkema, Rotterdam, 621-631.
18. Cai, M. (2013) Principles of Rock Support in Burst-Prone Ground. Tunneling and Underground Space Technology, 36, 46-56. http://dx.doi.org/10.1016/j.tust.2013.02.003
19. Smith, M. (2007) Rock & Soil Reinforcement. Ulf Linder, &OUML;rebro.
20. Smith, M. (2007) Underground Mining Methods. Ulf Linder, &OUML;rebro.
21. Li, C.C. and Marklund, P.I. (2004) Field Tests of the Cone Bolt in the Boliden Mines. In: Hamre, L., Rohde, J.K.G., Berg, K.R. and Nilsen, B., Eds., Fjellsprengninsteknikk/Bergmekanikk/Geoteknikk, Norsk Jord og Fjellteknisk Forbund, Oslo, 12.
22. Simser, B., Andrieux, P., Mercier-Langevin, F., Parrott, T. and Turcotte, P. (2007) Field Behaviour and Failure Modes of Modified Conebolts at the Craig, LaRonde and Brunswick Mines in Canada. In: Hadjigeorgiou, J., Stacey, D. and Potvin, Y., Eds., Challenges in Deep and High Stress Mining, Australian Centre for Geomechanics, Perth, 347-354.
23. Wang, G., Wu, X.Z., Jiang, Y.J., Huang, N. and Wang, S.G. (2013) Quasi-Static Laboratory Testing of a New Rock Bolt for Energy-Absorbing Applications. Tunnelling and Underground Space Technology, 38, 122-128. http://dx.doi.org/10.1016/j.tust.2013.05.010
24. Varden, R., Lachenicht, R., Player, J., Thompson, A. and Villaescusa, E. (2008) Development and Implementation of the Garford Dynamic Bolt at the Kanowna Belle Mine. 10th Underground Operators' Conference, Launceston, 14-16 April 2008, 95-102.
25. Guntumadugu, D.R. (2013) Methodology for the Design of Dynamic Rock Supports in Burst Prone Ground. PhD Thesis, Department of Mining and Materials Engineering, McGill University, Montreal.
26. Wu, R., Oldsen, J. and Campoli, A. (2011) Application of Yield-Lok Bolt for Bursting and Convergence Grounds in Mines. 30th International Conference on Ground Control in Mining, Morgantown, 26-28 July 2011, 1-6.
27. Chileshe, P.R.K. (2015) Review and Design Concept for Grouted Multi-Jacketted Friction Balancing Rock Bolt. International Journal of Scientific & Technology Research, 4, 92-100.

28. Milev, A.M. and Spottiswood, S.M. (2005) Strong Ground Motion and Site Response in Deep South African Mines. The Journal of the South African Institute of Mining and Metallurgy, 105, 515-524.

29. Kaiser, P.K., McCreath, D. and Tannant, D. (1996) Rockburst Research Handbook. CD-ROM Version, Ontario.

30. Kaiser, P.K. and Maloney, S.M. (1997) Scaling Laws for the Design of Rock Support. Pure and Applied Geophysics, 150, 415-434. http://dx.doi.org/10.1007/s000240050085

31. Hutchinson, D.J. and Diederichs, M.S. (1996) Cablebolting in Underground Mines. Bi Tech Publishers, Richmond.

32. Kaiser, P.K., McCreath, D.R. and Tannant, D.D. (1995) Rockburst Support Handbook. Geomechanics Research Centre, Sudbury.

33. Thompson, A.G., Villaescusa, E. and Windsor, C.R. (2012) Ground Support Terminology and Classification: An Update. Geotechnical and Geological Engineering, 30, 553-580. http://dx.doi.org/10.1007/s10706-012-9495-4

34. Wood, D.F. (1992) Specification and Application of Fiber Reinforced Shotcrete. In: Kaiser, P.K. and McCreath, D.R., Eds., Rock Support in Mining and Underground Construction, Balkema, Rotterdam, 149-156.

35. Brady, B.H.G. and Brown, E.T. (2004) Rock Mechanics for Underground Mining. 3rd Edition, Springer Science and Business Media, Dordrecht.

36. Charette, F. and Hill, M. (2011) Single Pass Bolting Approach for the Stabilisation of Rock Masses in High Stress Conditions. Metsco Proceeding CIM/ICM Conferences & Exhibitions, Montreal, 2011.

CHAPTER 4

AN EXPERIMENTAL STUDY ON MEASUREMENT METHODS OF BULK DENSITY AND POROSITY OF ROCK SAMPLES

Weiren Lin, Osamu Tadai, Manabu Takahashi, Daisaku Sato, Takehiro Hirose, Wataru Tanikawa, Yohei Hamada, Kentaro Hatakeda

Keywords: Rock, Density, Porosity, Caliper Method, Buoyancy Method, Helium-Displacement Pycnometer, Mucury Intrusion Posimetry

INTRODUCTION

Density and porosity are fundamental and important physical properties of rocks in various geological problems, and affect the other physical properties such as elastic wave velocity, permeability, resistivity, strength, Young's modulus etc. (e.g. Schon 1998) [1]. Therefore, measurements of density and porosity using drilling core samples retrieved from depths and rock samples taken from geological outcrops are the most popular and important investigation items in both geo- science and geo-engineering fields. Several measurement techniques of bulk density and porosity are available and currently applied in the fields (Franclin, 1979) [2]. To ensure the data quality, comparison and examination of the measurement results by the different measurement techniques are necessary since the techniques are based on different principles and test procedures. In this experimental study, for comparison of measurement results we collected eight types of rock samples including a gabbro, a granite, four sandstones, a welded tuff and a mudstone as test materials, and also several metal samples and then measured the density and porosity of the rock samples by different techniques. The porosities of the eight rocks covered a very wide range from 0.3% to 50% approximately. We employed three methods (caliper, buoyancy and helium-displacement pycnometer also called gas pycnometer) to measure volume of regularly-shaped specimens and to determine the bulk densities and porosities of the rock samples (Franclin, 1979 and Blum, 1997) [2] [3]. In addition, we also measured the porosities and determined pore size distributions by mercury intrusion porosimetry (American Society for Testing and Materials, 1999) [4].

Citation: Lin, W. , Tadai, O. , Takahashi, M. , Sato, D. , Hirose, T. , Tanikawa, W. , Hamada, Y. and Hatakeda, K. (2015) An Experimental Study on Measurement Methods of Bulk Density and Porosity of Rock Samples. Journal of Geoscience and Environment Protection, 3, 72-79. doi: 10.4236/gep.2015.35009.

ROCK SAMPLES AND TEST METHODS

Samples

As test materials for the planned experiments of different methods, eight types of rock samples which cover a very wide porosity range were collected. The rock types of the samples are middle-grained Belfast gabbro (Symbol used in this study: BG) retrieved from South Africa, fine-grained Aji Granite (AG) from Kagawa Pref., Japan, an yellow sandstone (AAS) from Australia, Rajasthan sandstone (RS) from India, Shirahama sandstone (SS) from Wakayama Pref., Japan, Berea sandstone (BS) from Ohio State, USA, Tage welded tuff (TWT) from Tochigi Pref., Japan and Nankai mudstone (NM) from an off shore ocean drilling site in southwest Japan. The Nankai mudstone was obtained by a scientific deep ocean drilling project from a depth of 476 meters be-low sea floor with about 3800 m water depth at IODP (Integrated Ocean Drilling Program) drilling site C0006F in west Pacific Ocean (Expedition 314 Scientists, 2009) [5]. The other rocks were taken from quarries on land. All of the rock samples are relative homogenous and fresh, e.g., not weathered and visual crack free. The rock types and number of specimens used for various experiments were shown in **Table 1**. Addition to the rock samples, we also prepared five metal specimens (two aluminum, two brass and one stainless specimens) for the volume measurements because they do not have pore and are regularly shaped in a better quality than rock samples.

Table 1. Rock types and number of test specimens for the various experiments

Sample Type	Specimen number for individual method			
	Caliper method	Buoyancy method	Pycnometer	Porosimetry
Belfast gabbro [BG]	6	6	3	2
Aji granite [AG]	6	6	3	2
Australia A Sandstone [AAS]	3	3	3	2
Rajasthan Sandstone [RS]	3	3	3	2
Shirahama Sandstone [SS]	3	3	3	2
Berea Sandstone [BS]	3	3	3	2
Tage Welded Tuff [TWT]	3	3	3	2
Nankai Mudstone [NM]	—	3	3	2
Metals	5	5	5	—

We prepared rock specimens for the measurements in a shape of regular cylinder for seven rock types except the NM. Sizes of all the cylinder specimens are of the same diameter (approximately 25 mm), but of different length of 20 - 35 mm dependent on the original rock sample size. Because the dimension of NM drilling core sample was not enough for recoring to make up the 25 mm-diameter cylindrical specimens, we used irregular specimens for NM.

Measurement Methods and Procedures

For determining bulk density of rock samples, the simplest method is to measure their volume and mass (or weight) as showed in Equation (1);

$$\rho_B = M / V_B \qquad (1)$$

where ρ_B is bulk density (g/cm^3), M is mass (g), and V_B is the bulk volume including both solid's and pore's volume (cm^3). If mass M was measured using dry rock specimen, ρ_B is the bulk dry density. On the other hand, bulk wet density is determined by the mass of the rock specimen at water saturated state. In this study, we deal with the bulk dry density. Porosity (n in %) is defined as the ratio of accumulated (total) pore volume (V_P in cm^3) included in a rock specimen to the bulk volume V_B of the specimen as follow:

$$n = 100 \times V_P / V_B \qquad (2)$$

Based on these equations, it is clear that the measurements of bulk volume and pore volume control measurement accuracy of density and porosity of rock specimens rather than mass (or weight) measurements which can be performed by an electrical balance relatively easily and accurately. Therefore, to examine results of volume measurements by different methods is the key issue.

For bulk volume measurements, three methods *i.e.* caliper (Franclin, 1979; resolution of the caliper used: 0.05 mm) [2], buoyancy based on Archimedes' principle (Franclin, 1979; US620H and SMK-102 of Shimadzu Cor-poration, Japan) [2], and helium-displacement Penta-Pycnometer which is according to Boyle's Law (Blum, 1997; Pentapycnometer of Quantachrome Corporation, USA) [3], were employed in this study.

First, the rock specimens were soaked in a vacuumed desiccator for about three days for water saturation. Then, their bulk volumes were independently measured by the three methods; and their weights at wet state by an electrical balance with 0.001 g resolution. After these measurements, the specimens were dried in an oven at 110 °C for more than 24 hours. After cooling the specimens to the room temperature in a dried desiccator, their dry weights were measured by the same electrical balance and dry volumes (volumes of solid portion only) were determined by pycnometer. We use the following equation to determine pore volume V_P (cm^3):

$$V_P = (M_{wet} - M_{dry}) / \rho_{water} \qquad (3)$$

where M_{wet} (g) is the mass of the specimen at water saturated state, M_{dry} (g) is the mass at dry state and ρ_{water} (g/cm^3) is the water density at the room temperature.

Here, we use the same rock specimens for the different measurement methods, and believe it is important for such experimental comparison studies of different measurement methods. After the measurements of caliper, buoyancy and pycnometer methods; the specimens used were finally resized (cut) to fit sample holder of the mercury intrusion porosimetry apparatus (Mercury Porosimeter Auto Pore IV 9500 of Micromeritics, USA). Then, the pore volumes and pore size distributions in dry specimens were determined by the porosimetry method (American Society for Testing and Materials, 1999) [4].

MEASUREMENT RESULTS AND DISCUSSIONS

Bulk Volume Measured by Different Methods

The all bulk volume data of the eight rock types and metal specimens measured independently by the caliper, buoyancy and pycnometer methods were listed in **Table 2**. Pycnometer measurements were carried out at both wet and dry states of the rock specimens, respectively. Because the measured volume at dry state is corresponding the volume of solid portion only, the bulk volume (including both solid and pore volumes) can be calculated by adding the measured solid volume to the pore water volume obtained from the difference between wet and dry specimen weights.

Table 2. A comparison of bulk volumes measured by different methods

Sample Type	Specimen ID	Bulk volume (cm^3) Caliper method	Buoyancy method	Pycnometer (wet)	Solid volume + Pore volume	Pore volume (cm^3)	Solid volume (Pycnometer, dry) (cm^3)
Belfast gabbro [BG]	BG-1	17.51	17.45	17.45	17.46	0.03	17.42
	BG-2	17.08	16.99	16.97	17.03	0.03	17.00
	BG-3	16.74	16.63	16.54	16.62	0.03	16.59
	BG-4	11.25	11.27			0.02	
	BG-5	9.63	9.62			0.02	
	BG-6	8.90	8.90			0.02	
Aji granite [AG]	AG-1	18.59	18.52	17.96	18.55	0.14	18.41
	AG-2	18.21	18.14	17.50	18.20	0.17	18.03
	AG-3	17.77	17.69	17.02	17.72	0.15	17.57
	AG-4	10.05	10.03			0.07	
	AG-5	9.58	9.50			0.06	
	AG-6	8.93	8.93			0.06	
Australia A Sandstone [AAS]	AAS-1	17.83	17.73	17.08	17.26	1.49	15.77
	AAS-2	17.66	17.51	16.86	17.00	1.70	15.30
	AAS-3	15.85	15.74	15.02	15.18	1.18	14.00
Rajasthan Sandstone [RS]	RS-1	17.67	17.52	16.80	17.32	1.85	15.48
	RS-2	17.63	17.51	16.76	17.30	1.86	15.44
	RS-3	17.48	17.36	16.63	17.15	1.87	15.28
Shirahama Sandstone [SS]	SS-1	17.17	17.09	16.04	16.78	2.27	14.51
	SS-2	16.31	16.24	15.06	15.99	2.16	13.83
	SS-3	14.97	14.87	14.11	14.59	2.08	12.51
Berea Sandstone [BS]	BS-1	16.25	16.01	15.43	15.93	3.09	12.84
	BS-2	15.80	15.53	14.79	15.39	3.09	12.30
	BS-3	14.78	14.57	13.83	14.41	2.91	11.50
Tage Welded Tuff [TWT]	TWT-1	15.46	15.47	14.86	15.48	4.87	10.60
	TWT-2	15.17	15.17	14.44	15.13	4.81	10.32
	TWT-3	14.09	14.11	13.37	14.18	4.60	9.58
Nankai Mudstone [NM]	NM-1		9.09	8.19	9.06	4.46	4.60
	NM-2	—	8.79	7.80	8.79	4.32	4.48
	NM-3		8.30	7.45	8.27	4.06	4.22
Metals	Aluminum 1	14.63	14.50	14.59			
	Aluminum 2	25.34	25.23	25.43			
	Brass 1	14.37	14.29	14.35	—	—	—
	Brass 2	25.06	24.98	25.22			
	Stainless	14.98	14.98	14.96			

Figure 1 is cross plots of the bulk volumes measured by buoyancy and the other methods, showing that the bulk volume values located in the vicinity of 1:1 line approximately. In detail, the values by pycnometer at wet state (blue circles) look like to be slightly smaller than the buoyancy method.

To make a quantitative comparison among the bulk volumes obtained by different methods, we defined a parameter "error" E (%) as follow:

$$E = 100\times (V_B - V_{B-B}) / V_{B-B} \qquad (4)$$

where V_B: the bulk volume obtained by individual method; V_{B-B}: the bulk volume by the buoyancy method. This parameter in percentage more clearly shows how much different between the individual and buoyancy measurements (**Figure 2**).

Basically, the bulk volumes of metal specimens (Aluminum, Brass and Stainless) by the three methods are well consistent with each other; the differences were less than 1%. For the rock specimens, if the specimen surface is not smooth caliper may measure "convex" parts, then showed slightly larger values than the buoyancy method. Within the two pycnometer measurements, the dry one shows a better agreement with the buoyancy and caliper methods. The wet one, however, had approximately 5% errors for six rocks except BG and NM. From this result, it can be considered that the pycnometer measurement using the dry specimen seems to be better than using the wet one. In addition, a trend showing the error level of the wet pycnometer data increases with the water content should be noted.

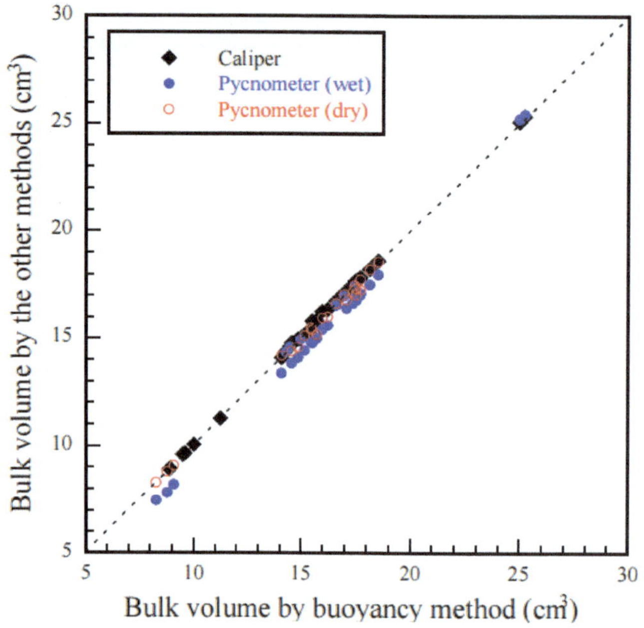

Figure 1. A comparison between bulk volumes measured by buoyancy and the other methods.

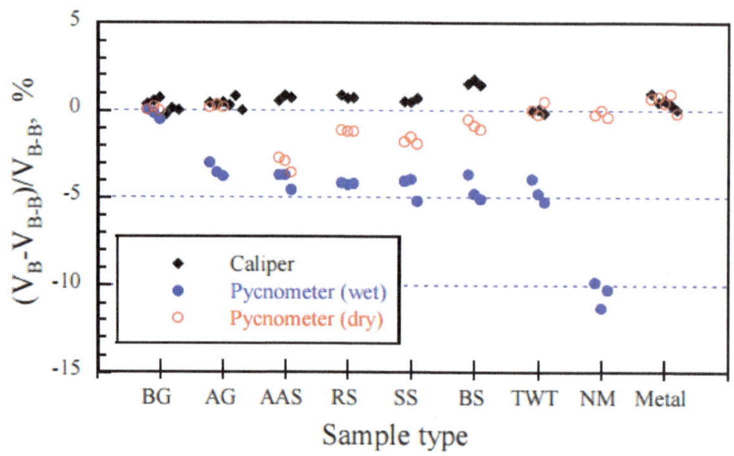

Figure 2. A quantitative comparison among the bulk volumes obtained by different methods.

Bulk Density and Porosity Measured by Different Methods

We determined bulk dry densities of all the rock and metal specimens and porosities of the rock specimens by the different methods (**Table 3**). The densities were calculated based on the bulk volumes (**Table 2**) and specimen masses (weights) at dry state measured by the electrical balance according to Equation (1); and the porosities based on the mass (weight) difference between wet and dry states and the same bulk volumes by Equation (2). Mercury intrusion porosimetry also yields the porosity by measuring mercury volume injected into the rock specimens at high pressures; and bulk dry density using bulk volumes measured by the apparatus based on mercury displacement principle.

Table 3. A comparison of bulk dry densities and porosities measured by various techniques

Sample Type	Specimen ID	Bulk dry density (g/cm^3)				Porosity (%)		
		Caliper method	Buoyancy method	Pycnometer (dry)	Porosimetry	Buoyancy method	Pycnometer (dry)	Porosimetry
Belfast gabbro [BG]	BG-1	2.942	2.954	2.952		0.20	0.20	
	BG-2	2.942	2.958	2.952		0.18	0.18	
	BG-3	2.921	2.941	2.942		0.19	0.19	
	BG-4	2.952	2.945		2.940	0.21		0.35
	BG-5	2.942	2.945		2.935	0.24		0.35
	BG-6	2.940	2.945			0.24		
	Average	2.940	2.948	2.949	2.937	0.21	0.19	0.35
Aji granite [AG]	AG-1	2.635	2.647	2.643		0.76	0.75	
	AG-2	2.636	2.647	2.638		0.95	0.95	
	AG-3	2.635	2.648	2.643		0.85	0.85	
	AG-4	2.639	2.646		2.633	0.70		0.66
	AG-5	2.625	2.647		2.633	0.63		0.66
	AG-6	2.641	2.642			0.63		
	Average	2.635	2.646	2.641	2.633	0.75	0.85	0.66
Australia A Sandstone [AAS]	AAS-1	2.406	2.419	2.485	2.398	8.39	8.62	10.87
	AAS-2	2.360	2.380	2.451	2.409	9.72	10.01	10.51
	AAS-3	2.405	2.422	2.510		7.48	7.76	
	Average	2.390	2.407	2.482	2.403	8.53	8.80	10.69
Rajasthan Sandstone [RS]	RS-1	2.325	2.345	2.372	2.349	10.55	10.67	10.54
	RS-2	2.327	2.343	2.371	2.337	10.60	10.72	10.50
	RS-3	2.323	2.340	2.368		10.77	10.90	
	Average	2.325	2.342	2.370	2.343	10.64	10.76	10.52
Shirahama Sandstone [SS]	SS-1	2.275	2.286	2.328	2.291	13.28	13.52	14.15
	SS-2	2.273	2.283	2.319	2.285	13.30	13.50	13.78
	SS-3	2.268	2.283	2.327		13.98	14.25	
	Average	2.272	2.284	2.325	2.288	13.52	13.76	13.96
Berea Sandstone [BS]	BS-1	2.107	2.140	2.151	2.138	19.29	19.39	19.41
	BS-2	2.084	2.122	2.141	2.125	19.92	20.10	20.62
	BS-3	2.086	2.117	2.140		19.96	20.17	
	Average	2.093	2.126	2.144	2.131	19.72	19.89	20.02
Tage Welded Tuff [TWT]	TWT-1	1.713	1.710	1.710	1.727	31.50	31.49	27.71
	TWT-2	1.713	1.712	1.717	1.747	31.71	31.79	28.59
	TWT-3	1,707	1.704	1.695		32.59	32.42	
	Average	1.711	1.709	1.707	1.737	31.93	31.90	28.15
Nankai Mudstone [NM]	NM-1	—	1.373	1.392	1.454	49.14	49.26	43.56
	NM-2		1.372	1.387	1.409	49.09	49.09	44.39
	NM-3		1.381	1.401		48.83	49.03	
	Average		1.376	1.393	1.431	49.02	49.13	43.98
Metals	Aluminum 1	2.655	2.680	2.663	—	—	—	—
	Aluminum 2	2.790	2.802	2.781				
	Brass 1	8.441	8.488	8.453				
	Brass 2	8.459	8.485	8.405				
	Stainless	7.867	7.870	7.882				

In a general sense, the bulk dry densities determined by the four methods were consistent well with each other (Table 3). For the porosities, the buoyancy method and pycnometer (dry) method showed almost the same values for all the rocks. The porosities by porosimetry, however, were larger than those of the other two methods for BG with very low porosity; and smaller for NM with very large porosity (Table 3). The difference for BG might be caused by the measurement accuracy of injected mercury volumes which are approximately 0.03 cc.

On the other hand, the reason for the smaller porosity result for NM might be that high pressure applied for mercury injection causes deformation of the soft rock specimen and decrease its pore volume. **Figure 3** shows a very good general relationship between the bulk dry density and the porosity of the eight rocks, *i.e.* the bulk dry density increases with the porosity decrease.

Although porosities obtained by the mercury porosimetry showed some uncertainty, it has a great advantage on measuring pore size distribution which could not be obtained by the other porosity measurement methods employed in this study (Lin *et al.*, 2011) [6]. For example, the peak of pore size distribution of BS is around 10 μm being more than 100 times of that of NM although the porosity (43.56%) of NM is much larger than that (19.41%) of BS (**Figure 4**).

SUMMARY

Several different measurement techniques of bulk density and porosity of rocks are available and being applied in the geoscience and geo-engineering fields. To ensure the data quality and to make its quality assessment, we carried out an experimental comparison study in which different measurement techniques (caliper, buoyancy, helium-displacement pycnometer and mercury intrusion porosimetry) were employed for the same rock specimens. We collected eight types of rock samples including a gabbro, a granite, four sandstones, a welded tuff and a mudstone as test materials. The porosities of the eight rocks covered a very wide range from 0.3% to 50% approximately. Therefore, it can be said that the collected rock samples are proper for such comparison study of different measurement techniques. As a result, the techniques using caliper, buoyancy and pycnometer yielded the almost same bulk dry density and porosity results. In detail, the helium-displacement pycnometer measurement using the dry specimens seems to be better than using the wet ones. However, the porosimetry underestimated porosity values in the case of very high-porosity rock (*i.e.* soft rock) probably due to the specimen deformation under intrusion pressure; and showed some errors for the very low-porosity rock samples probably due to insufficient accuracy of mercury intrusion volume measurement. Ability to determine pore size distribution is an important advantage of the porosimetry method.

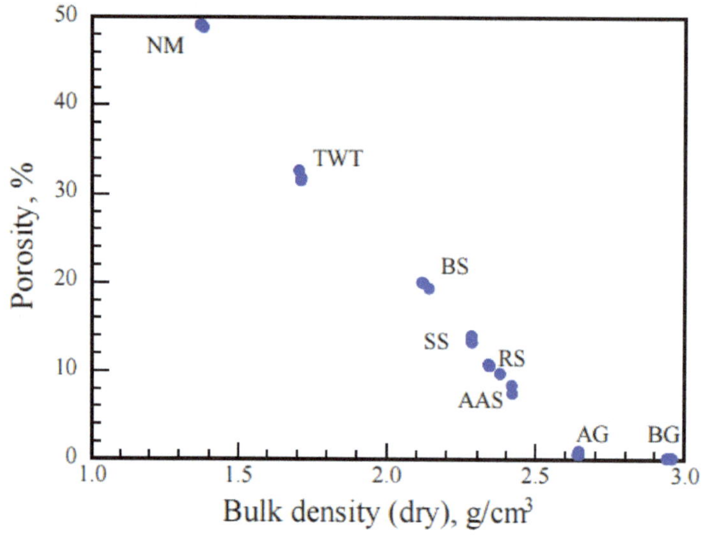

Figure 3. Relationship between porosity and bulk dry density measured by buoyancy method.

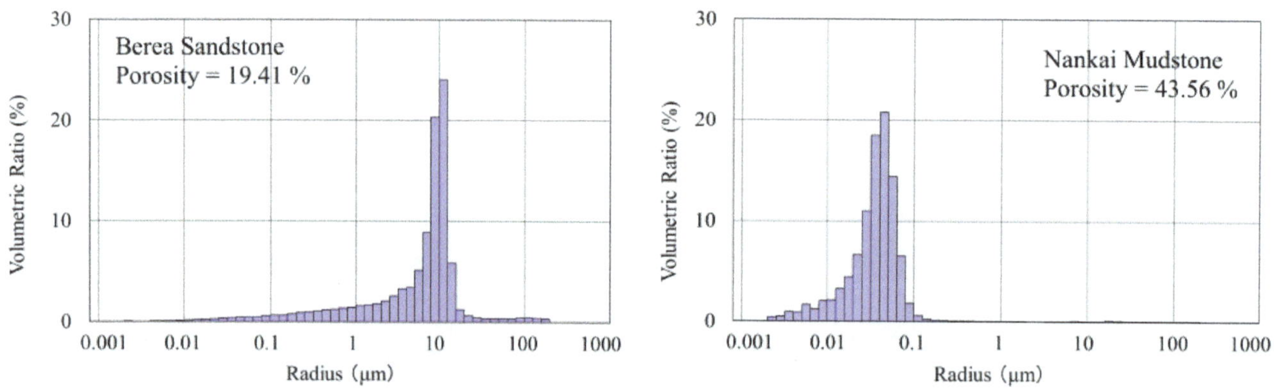

Figure 4. Two examples of pore size distribution obtained by mercury intrusion porosimetry.

ACKNOWLEDGEMENTS

The Nankai Mudstone core samples were provided by the Integrated Ocean Drilling Program (IODP). Part of these works were supported by Grants-in-Aid for Scientific Research 25287134 (JSPS), Japan.

REFERENCES

1. Schon, J.H. (1998) Ch. 2, Pore Space Properties: Porosity, Specific Internal Surface, and Permeability, 2nd Edition. Handbook of Geophysical Exploration, Seismic Exploration, Vol. 18, Pergamon, Netherlands, 23-58.
2. Franklin, J.A. (1979) Suggest Methods for Determining Water Content, Porosity, Density, Absorption and Related Properties and Swelling and Slake-Durability Index Properties. International Journal of Rock Mechanics and Mining Science & Geomechanics Abstracts, 16, 141-156.
3. Blum, P. (1997) Ch. 2, Moisture and Density (by Mass and Volume), in Physical Properties Handbook: A Guide to the Shipboard Measurement of Physical Properties of Deep-Sea Cores. ODP Technical Notes, 26, 2-1-2-15.
4. American Society for Testing and Materials (1999) Standard Test Method for Determination of Pore Volume and Pore Volume Distribution of Soil and Rock by Mercury Intrusion Porosimetry, Designation D 4404-84 (Reapproved 1992). Annual Book of ASTM Standards, 04.08, 588-592.
5. Expedition 314 Scientists (2009) Expedition 314 Site C0006. In: Kinoshita, M., Tobin, H., Ashi, J., Kimura, G., Lallemant, S., Screaton, E.J., Curewitz, D., Masago, H., Moe, K.T., and the Expedition 314/315/316 Scientists, Proc. IODP, 314/315/316, Integrated Ocean Drilling Program Management International, Inc., Washington DC.
6. Lin, W., Tadai, O., Hirose, T., Tanikawa, W., Takahashi, M., Mukoyoshi, H. and Kinoshita, M. (2011) Thermal Conductivities under High Pressure in Core Samples from IODP NanTroSEIZE Drilling Site C0001. Geochemistry, Geophysics, Geosystems, 12, Q0AD14.

CHAPTER 5

GEOMORPHOMETRY AND EARTH SURFACE DYNAMICS

Giulia Sofia, John K. Hillier, and Susan J. Conway

INTRODUCTION

Elucidating the dynamics of Earth surface processes through analysis of digital elevation models (DEMs), or "geomorphometry" (Evans et al., 2003; Hengl and Reuter, 2008; Pike, 1995, 2000), has become a flourishing interdisciplinary subject, with applications in numerous fields (e.g. geomorphology, hydrology, planetary science, archaeology, geo-biology, natural hazards, and computer science). The Earth's morphology can be measured at all scales, from macro (e.g. globally via space missions), to micro (e.g. using laser scanners and most recently structure-from-motion techniques). These data sets have been widely used to analyze both natural (Tarolli, 2014) and anthropogenic (Tarolli and Sofia, 2016) landscapes, and they underpin much modern geomorphological research.

Conceptually any analysis in geomorphometry is a two-step process. Firstly, data must be obtained and their accuracy assessed. Errors are propagated and amplified in surface derivatives (Burrough and McDonnell, 1998; Heuvelink, 1998), thus the usefulness and validity of the results obtained in geomorphometry are intimately associated with the quality of the original data (Felicísimo, 1994). Secondly, these data are used to derive simplified indices, or are integrated into modelling to portray and understand the specific process of interest. Initially, geomorphometry was used mainly for drainage basin analysis from topographic maps (Miliaresis, 2008), but with time, quantitative techniques and a range of geomorphometric parameters have been developed and applied in an attempt to characterize the landscape and identify processes (Evans, 2012). There are advantages and dis-advantages of each method, technique, parameter and topographic data type, which vary depending on the objectives of the analysis; often significant weaknesses or methodological limitations exist, which prevent us from gaining the insights into processes that we otherwise might. The interdisciplinarity of geomorphometry is its greatest strength and also one of its major challenges. Specifically, process-focused fields (e.g. soil science, glaciology, volcanology, hydrology) use their own set of established geomorphometric approaches, and geomorphological specialists often play a key role in developing these. However, these specialists in turn struggle to incorporate the most innovative approaches and theory being developed in the associated technical domains (such as, computer vision, machine learning, and statistics), or even approaches being used in neighboring disciplines. So, interesting geomorphometric applications struggle to jump be-tween process-specific disciplines, but also struggle to take advantage of advances in computer science and technology.

Citation: Giulia Sofia, John K. Hillier, and Susan J. Conway, "Geomorphometry and Earth Surface Dynamics," doi:10.5194/esurf-4-721-2016.

Table 1. Main themes covered by the research in the Special Issue

Perspectives	Research and innovative techniques	Research and innovative applications
Synthetic DEMs: Hillier et al., 2015	*Sediment dynamics/Fluvial incision*: Hergarten et al., 2016	*Past tectonic history*: Andreani and Gloaguen, 2016
Structure-From-Motion (SfM) Photogrammetry: Eltner et al., 2016	*Stage dependent patterns in rivers*: Brown et al., 2016	*Past interaction between ice sheets and glacial systems*: Wickert, 2016
Learning Algorithms: Valentine and Kalnins, 2016	*Erosion and connectivity at the hillslope or catchment scale*: Sklar et al., 2016; Trevisani and Cavalli, 2016; Bigelow et al., 2016; Grieve et al., 2016	*Multitemporal data set to evaluate erosion patterns*: Loye et al., 2016; Bechet et al., 2015;
		SfM for glacial processes: Westoby et al., 2016; Piermattei et al., 2016

If we are to best exploit the wealth of information held within DEMs it is important to

- gather knowledge about the current technical state-of-the-art in order to consolidate and disseminate established advances;
- evaluate stubbornly unproductive areas to identify key future challenges and opportunities;
- provide specific and innovative case studies to assist in cross-disciplinary communication;
- provide clear and understandable translations from the technical domains where algorithms and techniques find their basis.

In light of the challenges set out above, this special issue in Earth Surface Dynamics highlights current frontiers in geomorphometry. In order to collect recent research advancements and motivate further research in this direction, we organized a "Frontiers in geomorphometry" session at the European Geosciences Union General Assembly in 2015, and it has continued successfully since then. The session was focused on facilitating cross-fertilization of best practice across disciplines, highlighting novel technical developments, and showcasing innovative applications of geomorphometry to various Earth-surface processes. The issue collects a variety

of contributions, which fall into two main categories: Perspectives and Research, where Research is further divided into "Research and innovative techniques' and 'Research and innovative applications" (Table 1).

The collected Perspective works are reviews of state-of-the-art developments as applied to geomorphometry, with a forward-looking component seeking to identify opportunities and challenges. They are intended to stimulate discussion and new experimental approaches, and they offer a general framework for scientists in different disciplines, dealing with geomorphometry. The papers in the Research section present developments of novel techniques, or showcase in-novative application(s) of existing methods; the novel techniques are applicable to a variety of dominant geomorphic features, whilst the applications cover different spatial and temporal scales (Fig. 1). The works display how geomorphometry can provide sets of useful techniques and tools for research in different geomorphic and spatiotemporal con-texts, given that sufficient data, in sufficient quality, are avail-able.

FRONTIERS

Perspectives

The collected perspectives investigate three major questions.

- Physical processes, including anthropogenic feedbacks sculpt planetary surfaces (e.g., Earth's). A fundamental tenet of geomorphology is that mapping and, increasingly, quantifying landform features produced can yield insights into the processes. However, the precision and accuracy of mapped data are not well understood. So, how good are these geo-morphological data that underpin analyses, and how can we more objectively investigate this?
- The human brain has a remarkable capability for identifying patterns in complex, noisy data sets, and then applying this knowledge to problem solving. Can we transfer and replicate this ability via computational means, to advance geosciences?
- One of the most recent revolution in geomorphology is the multiview photogrammetry, or Structure-from-Motion (SfM) technique (Fonstad et al., 2013; Micheletti et al., 2015; Smith et al., 2015; Westoby et al., 2012). What are the key developments and potential future avenues for research in this field, and how do they relate to geomorphometry?

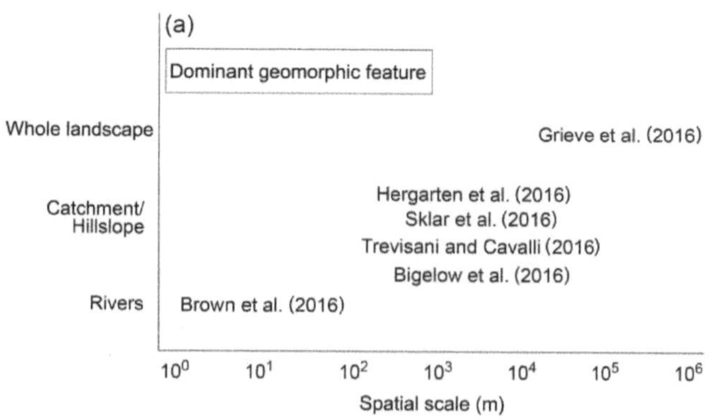

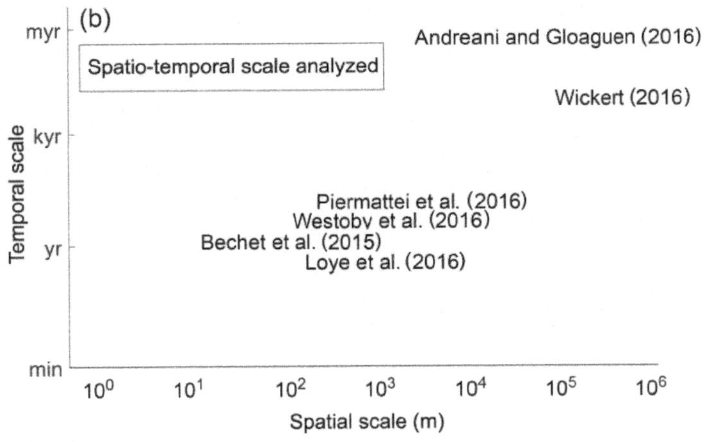

Figure 1. Dominant geomorphic feature(s) and spatial and temporal scales investigated by the research papers in this special issue: (a) dominant geomorphic feature(s) and spatial extent of the suggested applicability of the innovative techniques (Sect. 2.2); (b) dominant temporal scale and spatial extent covered in upon which the innovative applications are demonstrated (Sect. 2.3).

To respond to the first point, Hillier et al. (2015) introduce synthetic DEMs. This perspective reviews the possible approaches to the generation of artificial DEMs. highlighting their limitations, potential, and the opportunities for application. Realistic synthetic DEMs offer a way to assess and understand geomorphological data, allowing users to proceed with uncertainty-aware landscape analysis to examine physical processes.

Valentine and Kalnins (2016) offer an overview about ma-chine learning and its potential in geosciences. Learning al-gorithms come from the computer science world, and they are designed to replicate the human approach of inferring in-formation from a data set, and then apply that information predictively. In this work, the authors provide a review of the existing applications in geosciences, and discuss some of the factors that determine whether a learning algorithm approach is suited to geomorphological problems.

Eltner et al. (2016) provide a summary for researchers wanting to apply the SfM method. They summarize the state of the art of published research on SfM photogrammetry applications in geomorphometry. In addition, they give an overview of terms and fields of application, and they identify key future challenges, with a specific focus also on the errors associated with such a technique.

Research and Innovative Techniques

A fundamental operation in geomorphometry is the extraction of parameters from DEMs to understand the underlying process. How these parameters or objects are evaluated and identified still presents a challenge, and there is still room for improvement. Papers included here extend our knowledge about sediment dynamics and fluvial incision, or stage-dependent patterns in rivers. A further collection of work focuses on sediment, erosion and connectivity at the hillslope or watershed scale.

Hergarten et al. (2016) develop and explore an extension of the chi-transformation ( ) to small catchment sizes. They solve the limitation of the technique for different water-shed sizes, extending the stream power equation to headwater areas dominated by debris flows. In addition, the authors introduce an alternative optimization scheme to linearize the chi-elevation relation.

Brown and Pasternack (2016) demonstrate a relatively new method of analysis for stage-dependent patterns in rivers named geomorphic covariance structures (GCSs). Using meter-scale resolution DEMs, their approach aims to understand if and how the covariance of bed elevation and flow-dependent channel top width are organized in a partially con-fined, incising gravel-cobble bed river with multiple spatial scales of anthropogenic and natural landform heterogeneity across a range of discharges.

Trevisani and Cavalli (2016) propose a flow-oriented directional measure of surface roughness based on geostatistics that takes into account surface gravity-driven flow directions. Their approach shows the potential impact of considering directionality in the calculation of roughness indices. In addition, they demonstrate how the use of flow-directional rough-ness can improve the geomorphometric modelling of sediment connectivity, and the interpretation of landscape morphology.

Sklar et al. (2016) propose a novel way to quantify the three-dimensional geometry of catchments. The authors develop an empirical algorithm for generating synthetic source-area power distributions, parameterized with data from natural catchments. Their model can be used to explore the effects of topography on the distribution on fluxes of water, sediment, isotopes and other landscape products passing through catchment outlets.

Bigelow et al. (2016) focus on erosion and sedimentation, and the identification of sediment sources and sinks across landscapes from a practitioners' point of view. Their approach demonstrates a modern analysis of important geomorphic processes affected by land use that can be easily applied by agencies to solve common problems in watersheds, improving the integration between science and environmental management. Grieve et al. (2016) present software for the automatic ex-traction and processing of relevant topographic parameters to rapidly generate non-dimensional erosion rate and relief data. This application allows identification of whether landscapes are in topographic steady state, and to identify clear signals of an erosional gradient, or evidence for a landscape decaying following uplift.

Research and Innovative Applications

In this section, the collected papers expand the applications of geomorphometry to a larger spatial and temporal domain, investigating past tectonic history, or past interactions between ice sheets and climate in glacial systems. Other researchers show the effectiveness of multitemporal data sets at the hillslope or catchment scale to give new insights into sediment dynamics and the seasonal pattern of erosion processes. Finally, two more papers push the frontier of which processes can be examined using SfM for quantitative analysis in the glaciological field.

Andreani and Gloaguen (2016) present a study that uses geomorphic indices to classify the landscape into different regions in order to unravel its tectonic history. These observations and interpretations allow for a better understanding of the recent evolution of the diffuse triple junction between the North American, Caribbean, and Cocos plates in northern Central America.

Wickert (2016) offers a general method to compute past river flow paths, drainage basin geometries, and river discharges at the continental-scale. By integrating numerical modelling (i.e. ice sheet, isostatic adjustment and climate) with field data including geomorphology, his work builds new insights into past glacial systems and climate–ice-sheet interactions.

In Loye et al. (2016), terrestrial Laser Scanning (TLS) is used as a monitoring tool at the catchment scale to analyze the coupling between sediment dynamics and torrent responses in terms of debris flow events. Similarly, Bechet et al. (2015) provide a novel example of how high-resolution time-lapse DEM collection can give insights into processes, in particular for understanding the seasonal pattern of erosion processes for black marls badland-type slopes.

The work by Piermattei et al. (2016) demonstrates the ad-vantages and potential of SfM to calculate the geodetic mass balance of glacier in the Ortles-Cevedale Group, Eastern Italian Alps. In addition, they investigated the feasibility of using the image-based approach for the detection of the surface displacement rate of an active rock glacier. Westoby et al. (2016) analyze the surface evolution of an Antarctic blue-ice moraine using multi-temporal DEMs from TLS and SfM.

The authors' results provide an additional understanding of inter-annual development of moraine systems.

CLOSING REMARKS

The availability of DEMs at multiple scales in terms of resolution and spatial and temporal coverage offers great opportunities for the investigation of Earth-surface processes. Geomorphometry has become inter-disciplinary, with focus on new techniques in digital terrain production but also analyses, independent of the subject, and/or field. This special issue showcases exciting developments and tools (e.g. synthetic DEMs, neural networks, Structure-From-Motion) that are the building blocks for the next step-change in the field. Research continues to evolve as computing power increases, and new instrumentation is developed to observe and analyze the Earth and its interacting processes. Geomorphome-try is becoming essential to the understanding of global is-sues, such as natural hazards, sediment production and anthropogenic changes to the Earth system, among others. Such multidisciplinary analytical tools will only become more effective in improving our knowledge of the Earth at a variety of spatio-temporal scales. In reading and compiling the contributions in this Special Issue, we hope that you, the scientific community, will be inspired to seek out collaborations and share your ideas across subject-boundaries, be-tween technique-developers and users, enabling us as a com-munity to fully exploit the wealth of knowledge inherent in our increasingly digital landscape.

ACKNOWLEDGEMENTS

We thank the two anonymous reviewers whose comments helped improve this manuscript, and Tom Coulthard and ESurf's editorial team for their support throughout producing this Special Issue.

REFERENCES

1. Andreani, L. and Gloaguen, R.: Geomorphic analysis of transient landscapes in the Sierra Madre de Chiapas and Maya Moun-tains (northern Central America): implications for the North American–Caribbean–Cocos plate boundary, Earth Surf. Dy-nam., 4, 71–102, doi:10.5194/esurf-4-71-2016, 2016.
2. Bechet, J., Duc, J., Loye, A., Jaboyedoff, M., Mathys, N., Malet, J.-P., Klotz, S., Le Bouteiller, C., Rudaz, B., and Travelletti, J.: Detection of seasonal erosion processes at the scale of an elemen-tary black marl gully from time series of Hi-Resolution DEMs, Earth Surf. Dynam. Discuss., 3, 1555–1586, doi:10.5194/esurfd-3-1555-2015, 2015.
3. Bigelow, P., Benda, L., and Pearce, S.: Delineating incised stream sediment sources within a San Francisco Bay tributary basin, Earth Surf. Dynam., 4, 531–547, doi:10.5194/esurf-4-531-2016, 2016.
4. Brown, R. A. and Pasternack, G. B.: Analyzing bed and width os-cillations in a self-maintained gravel-cobble bedded river using geomorphic covariance structures, Earth Surf. Dynam. Discuss., doi:10.5194/esurf-2015-49, in review, 2016.
5. Burrough, P. A. and McDonnell, R. A.: Principles of Geographical Information Systems, in: Economic Geography, edited by: De By, R. A., Oxford University Press, Vol. 75, 422–423, 1998.
6. Eltner, A., Kaiser, A., Castillo, C., Rock, G., Neugirg, F., and Abel-lán, A.: Image-based surface reconstruction in geomorphometry– merits, limits and developments, Earth Surf. Dynam., 4, 359– 389, doi:10.5194/esurf-4-359-2016, 2016.
7. Evans, I. S.: Geomorphometry and landform mapping: What is a landform?, Geomorphology, 137, 94–106, doi:10.1016/j.geomorph.2010.09.029, 2012.
8. Evans, I. S., Dikau, R., Tokunaga, E., Ohmori, H. and Hirano, M. (Eds.): Concepts and Modelling in Geomorphology: Interna-tional perspectives, TERRAPUB, Tokyo, 2003.
9. Felicísimo, A. M.: Parametric statistical method for error detection in digital elevation models, ISPRS J. Photogramm., 49, 29–33, doi:10.1016/0924-

2716(94)90044-2, 1994.

10. Fonstad, M. A., Dietrich, J. T., Courville, B. C., Jensen, J. L., and Carbonneau, P. E.: Topographic structure from motion: a new development in photogrammetric measurement, Earth Surf. Proc. Land., 38, 421–430, doi:10.1002/esp.3366, 2013.
11. Grieve, S. W. D., Mudd, S. M., Hurst, M. D., and Milodowski, D. T.: A nondimensional framework for exploring the relief structure of landscapes, Earth Surf. Dynam., 4, 309–325, doi:10.5194/esurf-4-309-2016, 2016.
12. Hengl, T. and Reuter, H. I. (Eds.): Geomorphometry: Concepts, Software, Applications, Elsevier, Amsterdam, 2008.
13. Hergarten, S., Robl, J., and Stüwe, K.: Tectonic geomorphol-ogy at small catchment sizes – extensions of the stream-power approach and the method, Earth Surf. Dynam., 4, 1–9, doi:10.5194/esurf-4-1-2016, 2016.
14. Heuvelink, G. B. M.: Error propagation in environmental modelling with GIS, Taylor & Francis, London, UK, 1998.
15. Hillier, J. K., Sofia, G., and Conway, S. J.: Perspective – syn-thetic DEMs: A vital underpinning for the quantitative fu-ture of landform analysis?, Earth Surf. Dynam., 3, 587–598, doi:10.5194/esurf-3-587-2015, 2015.
16. Loye, A., Jaboyedoff, M., Theule, J. I., and Liébault, F.: Headwa-ter sediment dynamics in a debris flow catchment constrained by high-resolution topographic surveys, Earth Surf. Dynam., 4, 489–513, doi:10.5194/esurf-4-489-2016, 2016.
17. Micheletti, N., Chandler, J. H., and Lane, S. N.: Investigating the geomorphological potential of freely available and accessi-ble structure-from-motion photogrammetry using a smartphone, Earth Surf. Proc. Land., 40, 473–486, doi:10.1002/esp.3648, 2015.
18. Miliaresis, G. C.: Quantification of Terrain Processes, in Advances in Digital Terrain Analysis, Springer Berlin Heidelberg, Berlin, Heidelberg, 13–28, 2008.
19. Piermattei, L., Carturan, L., de Blasi, F., Tarolli, P., Dalla Fontana, G., Vettore, A., and Pfeifer, N.: Suitability of ground-based SfM-MVS for monitoring glacial and periglacial processes, Earth Surf. Dynam., 4, 425–443, doi:10.5194/esurf-4-425-2016, 2016.
20. Pike, R. J.: Geomorphometry – progress, practice, and prospect, Z. Geomorphol. Suppl., 101, 221–238, 1995.
21. Pike, R. J.: Geomorphometry – diversity in quantita-tive surface analysis, Prog. Phys. Geogr., 24, 1–20, doi:10.1177/030913330002400101, 2000.
22. Sklar, L. S., Riebe, C. S., Lukens, C. E., and Bellugi, D.: Catchment power and the joint distribution of elevation and travel distance to the outlet, Earth Surf. Dynam. Discuss., doi:10.5194/esurf-2016-9, in review, 2016.
23. Smith, M. W., Carrivick, J. L., and Quincey, D. J.: Structure from motion photogrammetry in physical geography, Prog. Phys. Ge-ogr., 1–29, doi:10.1177/0309133315615805, 2015.
24. Tarolli, P.: High-resolution topography for understanding Earth sur-face processes: Opportunities and challenges, Geomorphology, 216, 295–312, 2014.
25. Tarolli, P. and Sofia, G.: Human topographic signatures and derived geomorphic processes across landscapes, Geomorphology, 255, 140–161, doi:10.1016/j.geomorph.2015.12.007, 2016.
26. Trevisani, S. and Cavalli, M.: Topography-based flow-directional roughness: potential and challenges, Earth Surf. Dynam., 4, 343– 358, doi:10.5194/esurf-4-343-2016, 2016.
27. Valentine, A. and Kalnins, L.: An introduction to learning al-gorithms and potential applications in geomorphometry and Earth surface dynamics, Earth Surf. Dynam., 4, 445–460, doi:10.5194/esurf-4-445-2016, 2016.
28. Westoby, M. J., Brasington, J., Glasser, N. F., Hambrey, M. J., and Reynolds, J. M.: "Structure-from-Motion" photogrammetry: A low-cost, effective tool for geoscience applications, Geomor-phology, 179, 300–314, doi:10.1016/j.geomorph.2012.08.021, 2012.
29. Westoby, M. J., Dunning, S. A., Woodward, J., Hein, A. S., Mar-rero, S. M., Winter, K., and Sugden, D. E.: Interannual surface evolution of an Antarctic blue-ice moraine using multi-temporal DEMs, Earth Surf. Dynam., 4, 515–529, doi:10.5194/esurf-4-515-2016, 2016.
30. Wickert, A. D.: Reconstruction of North American drainage basins and river discharge since the Last Glacial Maximum, Earth Surf. Dynam. Discuss., doi:10.5194/esurf-2016-8, in review, 2016.

CHAPTER 6 GEOMORPHOMETRY AND EARTH SURFACE DYNAMICS

Andrew Valentine and Lara Kalnins

INTRODUCTION

The human brain has a remarkable capability for identifying patterns in complex, noisy data sets, and then applying this knowledge to solve problems or negotiate new situations. The research field of "learning algorithms" (or "machine learning") centers around attempts to replicate this ability via computational means, and is a cornerstone of efforts to create "artificial intelligence". The fruits of this work may be seen in many different spheres – learning algorithms feature in everything from smartphones to financial trading. As we shall discuss in this paper, they can also prove useful in scientific research, providing a route to tackling problems that are not readily solved by conventional approaches. We will focus particularly on applications falling within geomorphometry and Earth surface dynamics, although the fundamental concepts are applicable throughout the geosciences, and beyond.

This paper does not attempt to be comprehensive. It is impossible to list every problem that could potentially be tackled using a learning algorithm, or to describe every technique that might somehow be said to involve "learning". Instead, we aim to provide a broad overview of the possibilities and limitations associated with these approaches. We also hope to highlight some of the issues that ought to be considered when deciding whether to approach a particular research question by exploring the use of learning algorithms.

The artificial intelligence literature is vast, and can be con-fusing. The field sits at the interface of computer science, engineering, and statistics: each brings its own perspective, and sometimes uses different language to describe essentially the same concept. A good starting point is the book by Mackay (2003), although this assumes a certain level of mathematical fluency; Bishop (2006) is drier and has a somewhat narrower focus, but is otherwise aimed at a similar readership. Unfortunately, the nature of the field is such that there are few good-quality reviews targeted to the less mathematically inclined, although one example can be found in Olden et al. (2008). There is also a wealth of tutorials and other course material available online, varying in scope and quality. Finally, we draw readers' attention to a recent review by Jordan and Mitchell (2015), offering a broad survey of machine learning and its potential applications.

Learning algorithms are computational tools, and a number of software libraries are available which provide users with a relatively straightforward route to solving practical problems. Notable examples include pybrain and scikitlearnfor Python (Schaul et al., 2010; Pedregosa et al., 2011), and the commercially available "Statistics and Machine Learning" and "Neural Network" toolboxes for Matlab. Most major techniques are also available as packages within the statistical programming language R. Nevertheless, we encourage readers with appropriate interest and skills to spend some time writing their own implementations for ba-sic algorithms: our experience suggests that this is a powerful aid to understanding how methods behave, and provides an appreciation for potential obstacles or pitfalls. In principle, this can be achieved using any mainstream programming language, although it is likely to be easiest in a high-level language with built-in linear algebra support (e.g. Matlab, Python with NumPy). Efficient "production" applications are likely to benefit from use of the feature-intensive and highly optimized tools available within the specialist software libraries mentioned above.

This paper begins with a brief overview of the general framework within which learning algorithms operate. We then introduce three fundamental classes of problem that are often encountered in geomorphological research, and which seem

Citation: Valentine, A. and Kalnins, L.: geomorphometry and Earth surface dynamics, Earth Surf. Dynam., 4, 445-460, doi:10.5194/esurf-4-445-2016, 2016.

particularly suited to machine learning solutions. Motivated by this, we survey some of the major techniques within the field, highlighting some existing applications in geomorphology and related fields. Finally, some of the practical considerations that affect implementation of these techniques are discussed, and we highlight some issues that should be noted when considering exploring learning algorithms further.

Learning Algorithms: A General Overview

Fundamentally, a learning algorithm is a set of rules that are designed to find and exploit patterns in a data set. This is a familiar process when patterns are known (or assumed) to take a certain form – consider fitting a straight line to a set of data points, for example – but the power of most learning algorithm approaches lies in their ability to handle complex, arbitrary structures in data. Traditionally, a distinction is drawn between "supervised" learning algorithms – which are aimed at training the system to recognize known patterns, features, or classes of object – and "unsupervised" learning, aimed finding patterns in the data that have not previously been identified or that are not well defined. Supervised learning typically involves optimizing some pre-defined measure of the algorithm's performance – perhaps minimizing the difference between observed values of a quantity and those predicted by the algorithm – while in unsupervised learning, the goal is usually for the system to reach a mathematically stable state.

At a basic level, most learning algorithms can be regarded as "black boxes": they take data in, and then output some quantity based upon that data. The detail of the relationship between inputs and outputs is governed by a number of adjustable parameters, and the "learning" process involves tuning these to yield the desired performance. Thus, a learning algorithm typically operates in two modes: a learning or "training" phase, where internal parameters are iteratively updated based on some "training data", and an "operational" mode in which these parameters are held constant, and the algorithm outputs results based on whatever it has learned. Depending on the application, and the type of learning algorithm, training may operate as "batch learning" – where the entire data set is assimilated in a single operation – or as "online learning", where the algorithm is shown individual data examples sequentially and updates its model parameters each time. This may be particularly useful in situations where data collection is ongoing, and it is therefore desirable to be able to refine the operation of the system based on this new information.

In the context of learning algorithms, a "data set" is generally said to consist of numerous "data vectors". For our purposes, each data vector within the data set will usually correspond to the same set of physical observations made at different places in space or time. Thus, a data set might consist of numerous stream profiles, or different regions extracted from a lidar-derived digital elevation model (DEM). It is possible to combine multiple, diverse physical observations into a single data vector: for example, it might be desirable to measure both the cross section and variations in flow rate across streams, and regard both as part of the same data vector. It is important to ensure that all data vectors constituting a given data set are obtained and processed in a similar manner, so that any difference between examples can be attributed solely to physical factors. In practice, pre-processing and "standardizing" data to enhance features that are likely to prove "useful" for the desired task can also significantly impact performance: we return to this in Sect. 4.1.

SOME GENERAL CLASSES OF GEOMORPHOLOGICAL PROBLEM

Broadly speaking, we see three classes of problem where learning algorithms can be particularly useful in geomorphology: classification and cataloguing; cluster analysis and dimension reduction; and regression and interpolation. All represent tasks that can be difficult to implement effectively via conventional means, and which are fundamentally data-driven. However, there can be considerable overlap between all three, and many applications will not fit neatly into one category.

Classification and Cataloguing

Classification problems are commonplace in observational science, and provide the canonical application for supervised learning algorithms. In the simplest case, we have a large col-lection of observations of the same type – perhaps cross sections across valleys – and we wish to assign each to one of a small number of categories (for example, as being of glacial or riverine form). In general, this kind of task is straightforward to an experienced human eye. However, it may be difficult to codify the precise factors that the human takes into account, preventing their implementation as computer code: simple rules break down in the face of the complexities inherent to real data from the natural world. With a learning algorithm approach, the user typically classifies a representative set of examples by hand, so that each data vector is associated with a "flag" denoting the desired classification. The learning algorithm then assimilates information about the connection between observations and classification, seeking to replicate the user's choices as closely as possible. Once this training procedure has been completed, the system can be used operationally to classify new examples, in principle without the need for further human involvement.

Beyond an obvious role as a labor-saving device, automated systems may enable users to explore how particular factors affect classification. It is straightforward to alter aspects of data processing, or the labelling of training examples, and then re-run the classification across a large data set. Another advantage lies in the repeatable nature of the classification: observations for a new area, or obtained at a later date, can be processed in exactly the same manner as the original data set, even if personnel differ. It is also possible to use multiple data sets simultaneously when performing the classification – for example, identification of certain topographic features may be aided by utilizing high-resolution local imagery, plus lower-resolution data showing the surrounding region, or land use classification may benefit from using topography together with satellite imagery.

It is sometimes claimed that it is possible to somehow "interrogate" the learning algorithm so as to discover its internal state and understand which aspects of the data are used to make a particular classification. This information could offer new insights into the physical processes underpinning a given problem. For the simplest classifiers, this may be possible, but in general we believe it ought to be approached with some scepticism. Classification systems are complex, and subtle interactions between their constituent parts can prove important. Thus, simplistic analysis may prove misleading. A more robust approach, where feasible, would involve classifying synthetic (artificial) data and exploring how parameters con-trolling the generation of this affect results (see also Hillier et al., 2015). It may also be instructive to explore how performance varies when different subsets of observables are used.

Conventionally, classification problems assume that all examples presented to the system can be assigned to one category or another. A closely related problem, which we choose to call "cataloguing", involves searching a large data set for examples of a particular feature – for example, locating moraines or faults in regional-scale topographic data. This introduces additional challenges: each occurrence of the feature should be detected only once, and areas of the data set that do not contain the desired feature may nevertheless vary considerably in their characteristics. As a result, cataloguing problems may require an approach that differs from other classification schemes.

Examples of classification problems in geomorphology where machine learning techniques have been applied include classifying elements of urban environments (Miliaresis and Kokkas, 2007), river channel morphologies (Beechie and Imaki, 2014), and landslide susceptibility levels (e.g. Brenning, 2005). An example of a cataloguing problem is given in Valentine et al. (2013), aimed at identifying seamounts in a range of tectonic settings.

Cluster Analysis and Dimension Reduction

Classification problems arise when the user has prior knowledge of the features they wish to identify within a given data set. However, in many cases we may not fully understand how a given process manifests itself in observable phenomena. Cluster analysis and dimension reduction techniques provide tools for "discovering" structure within data sets, by identifying features that frequently occur, and by finding ways to partition a data set into two or more parts, each with a particular character. An accessible overview of the conceptual basis for cluster analysis, as well as a survey of available approaches, can be found in Jain (2010).

In many applications, data vectors are over parameterized: the representations used for observable phenomena have more degrees of freedom than the underlying physical sys-tem. For example, local topography might be represented as a grid of terrain heights. If samples are taken every 10 m, then 11 samples span a distance of 100 m, and a 100 m 100 m area is represented by 121 distinct measurements. A single 121-dimensional data vector is obtained by "unwrapping" the grid according to some well-defined scheme – perhaps by traversing lines of latitude. Nevertheless, the underlying process of interest might be largely governed by a handful of parameters. This implies that there is a high level of redundancy within the data vector, and it could therefore be transformed to a lower-dimensional state without significant loss of information. This may be beneficial, either in its own right or as an adjunct to other operations: low-dimensional systems tend to be easier to handle computationally, and similarities or differences between examples may be clearer after transformation. Dimension reduction algorithms aim to find the optimal low-dimension representation of a given data set.

One particularly important application of dimension reduction lies in visualization. Where individual data vectors are high-dimensional, it may be difficult to devise effective means of plotting them in (typically) two dimensions. This makes it difficult to appreciate the structure of a data set, and how different examples relate to one another. In order to tackle this problem, learning algorithms may be used to identify a two-dimensional representation of the data set that somehow preserves the higher-dimensional relation-ships between examples. This process involves identifying correlations and similarities between individual data vectors, and generally does not explicitly incorporate knowledge of the underlying physical processes. Thus, the coordinates of each example within the low-dimensional space may not have any particular physical significance, but examples that share common characteristics will yield closely spaced points. Thus, it may be possible to identify visual patterns, and hence discover relationships within the high-dimensional data.

Geomorphological applications of cluster analysis include identifying flow directions from glacial landscapes (Smith et al., 2016), identifying different types of vegetation (Belluco et al., 2006), or extracting common structural orientations from a laser scan of a landslide scarp (Dunning et al., 2009). An example of dimension reduction comes again from landslide susceptibility studies, this time aimed at identifying the most influential observables within a suite of possibilities (Baeza and Corominas, 2001).

Regression and Interpolation

The third class of problem involves learning relationships be-tween physical parameters, in order to make predictions or to infer properties. Very often, it is known that one set of observable phenomena are closely related to a different set – but the details of that relationship may be unknown, or it may be too complex to model directly. However, if it is possible to obtain sufficient observations where both sets of phenomena have been measured, a learning algorithm can be used to rep-resent the link, and predict one given the other – for example, an algorithm might take measurements of soil properties and local topography and then output information about expected surface run-off rates. Alternatively, the same training data could be used to construct a system that infers the soil parameters given topography and run-off measurements. This may be useful when there are fewer measurements avail-able for one of the physical parameters, perhaps because it is harder or more expensive to measure: once trained on examples where this parameter has been measured, the algorithm can be used to estimate its value in other locations based on the more widely available parameters.

Questions of this sort may be framed deterministically – so that the system provides a single prediction – or statistically, where the solution is presented as a probability distribution describing the range of possible outcomes. The choice of approach will depend upon the nature of the underlying problem, and upon the desired use of the results. In general, probabilistic approaches are desirable, since they provide a more realistic characterization of the system under consideration – deterministic approaches can be misleading when more than one solution is compatible with available data, or where uncertainties are large. However, in some cases it may be difficult to interpret and use information presented as a probability distribution. For completeness, we observe that most learning algorithms have their roots in statistical theory, and even when used "deterministically", the result is formally defined within a statistical framework.

In geomorphology, a common application of machine learning for regression and interpolation is to link widely available remote sensing measurements with underlying parameters of interest that cannot be measured directly: for example, sediment and chlorophyll content of water from color measurements (Krasnopolsky and Schiller, 2003) or marine sediment properties from bathymetry and proximity to the coast (Li et al., 2011; Martin et al., 2015).

SOME POPULAR TECHNIQUES

The aforementioned problems can be tackled in almost any number of ways: there is rarely a single "correct" approach to applying learning algorithms to any given question. As will become clear, once a general technique has been selected, there remains a considerable array of choices to be made regarding its precise implementation. Usually, there is no clear reason to make one decision instead of another – often, the literature describes some "rule of thumb", but its underlying rationale may not always be obvious. A certain amount of trial and error is generally required to obtain optimal results with a learning algorithm. This should perhaps be borne in mind when comparisons are drawn between different approaches: although many studies can be found in the literature that conclude that one method outperforms an-other for a given problem, it is unlikely that this has been demonstrated to hold for all possible implementations of the two methods. It is also worth noting that the relationship be-tween performance and computational demands may differ between algorithms: a method that gave inadequate performance on a desktop computer a decade ago may nevertheless excel given the vastly increased resources of a modern, high-performance machine.

In what follows, we outline a selection of common methods, with an emphasis on conveying general principles rather than providing precise formal definitions. There is no particular rationale underpinning the methods we choose to include here, beyond a desire to cover a spectrum of important approaches. Other authors would undoubtedly make a different selection (for example, see Wu et al., 2008, although this has a narrower scope than the present work). In an effort to promote readability, we order our discussion roughly according to complexity, although this is not an objective assessment. A brief summary of the methods discussed can be found in Table 1.

Table 1. Summary of methods discussed in this paper. For each of the "popular techniques" discussed in Sect. 3, we indicate the classes of problem for which they are generally used (as described in Sect. 2); whether the method generally operates as "supervised" learning (based on optimizing a pre-determined performance measure), or unsupervised (attempting to reach a stable state); and whether the method is typically "deterministic", so that it is guaranteed to yield identical results each time it is applied to a given data set. Note that Bayesian inference is itself deterministic, but it is most often encountered in contexts where it is applied to randomly chosen observations. The indications given here are not intended to be exhaustive: it may be possible to adapt each technique to suit the full range of applications

	Generally used for:				Learning mode		
Method	Classification	Clustering	Dimension reduction	Regression	Supervised	Unsupervised	Deterministic
Decision trees	✓				✓		
K-means		✓				✓	
PCA			✓			✓	✓
Neural networks	✓		✓	✓	✓	✓	
SVMs	✓					✓	
SOMs		✓	✓			✓	
Bayesian inference	✓	✓		✓	✓	✓	(✓)

Decision Trees and Random Forests

A decision tree is a system that takes a data vector, and processes it via a sequence of if-then-else constructs ("branch points") until an output state can be determined (see Fig. 1). This is clearly well suited to addressing simple classification problems, and to predicting the states of certain physical systems: essentially, the system resembles a flowchart. In this context, "learning" involves choosing how the data set should be partitioned at each branch.

Typically, each data vector contains a number of "elements" – distinct observations, perhaps made at different points in space or time, or of different quantities relevant to the phenomenon of interest. Each vector is also associated with a particular "desired outcome" – the classification or state that the tree should output when given that example. Basic decision tree generation algorithms aim to identify a test that can be applied to any one element, which separates desired outcomes as cleanly as possible (e.g. Fig. 1a, b). This is typically quantified using a measure such as "information entropy", which assesses the degree to which a system behaves predictably. Once the training data have been partitioned into two sets, the algorithm can be applied recursively on each, until a complete tree has been constructed. Commonly en-countered tree generation schemes include ID3 (Quinlan, 1986) and its successor C4.5 (Quinlan, 1993).

Tree generation assumes that the data are perfect, and will therefore continue adding branch points until all training data can be classified as desired. When real-world data sets – which invariably contain errors and "noise" – are used, this tends to result in overly complex trees, with many branches. This phenomenon is known as "over fitting", and tends to result in a tree with poor generalization performance: when used to process previously unseen examples, the system does not give the desired outcome as often as one might hope. It is therefore usual to adopt some sort of "pruning" strategy, by which certain branches are merged or removed. Essentially, this entails prioritizing simple trees over perfect performance for training data; a variety of techniques exist, and the choice will probably be application-specific.

Another approach to this issue, and to the fact that in many problems the number of data elements vastly exceeds the number of possible outcomes, lies in the use of "random forests" (Breiman, 2001). By selecting data vectors from the training set at random (with replacement), and discarding some randomly chosen elements from these, we can construct a number of new data sets. It is then straightforward to build a decision tree for each of these. Typically, this results in trees that perform well for some – but not all – examples. However, if we use each tree to predict an outcome for a given data vector, and then somehow compute the average of these predictions, performance is usually significantly better than can be achieved with any one tree. This strategy, where a number of similar, randomized systems are constructed and then used simultaneously, is sometimes referred to as an "ensemble" method. Again, when treated in more detail, a variety of approaches are possible, and the intended application may help dictate which should be used.

A recent example of an application of random forests to Earth surface data can be found in Martin et al. (2015). Here, the goal is to predict the porosity of sediments lying on the ocean floor from a range of different data, including measures such as water depth or distance from sediment-producing features. Such observations are much easier to obtain than direct measurement of the seafloor, which obviously requires ocean-bottom sampling. Using training data from locations where such samples have been collected, a random forest is constructed that enables porosity predictions to be made throughout the oceans. This concept could readily be adapted to a variety of other situations where local physical properties must be estimated from remotely sensed data. Further discussion of the use of random forests for interpolation between samples can be found in, for example, Li et al. (2011), who test a variety of techniques including random forests and support vector machines in order to produce a regional map of mud content in sediments based on discrete sampling. Another example comes from Bhattacharya et al. (2007), who construct models of sediment transport using a variant of decision trees where each "leaf" is a linear regression function, rather than a single classification; the decision tree is essentially used to choose which mathematical model to apply in each particular combination of circumstances.

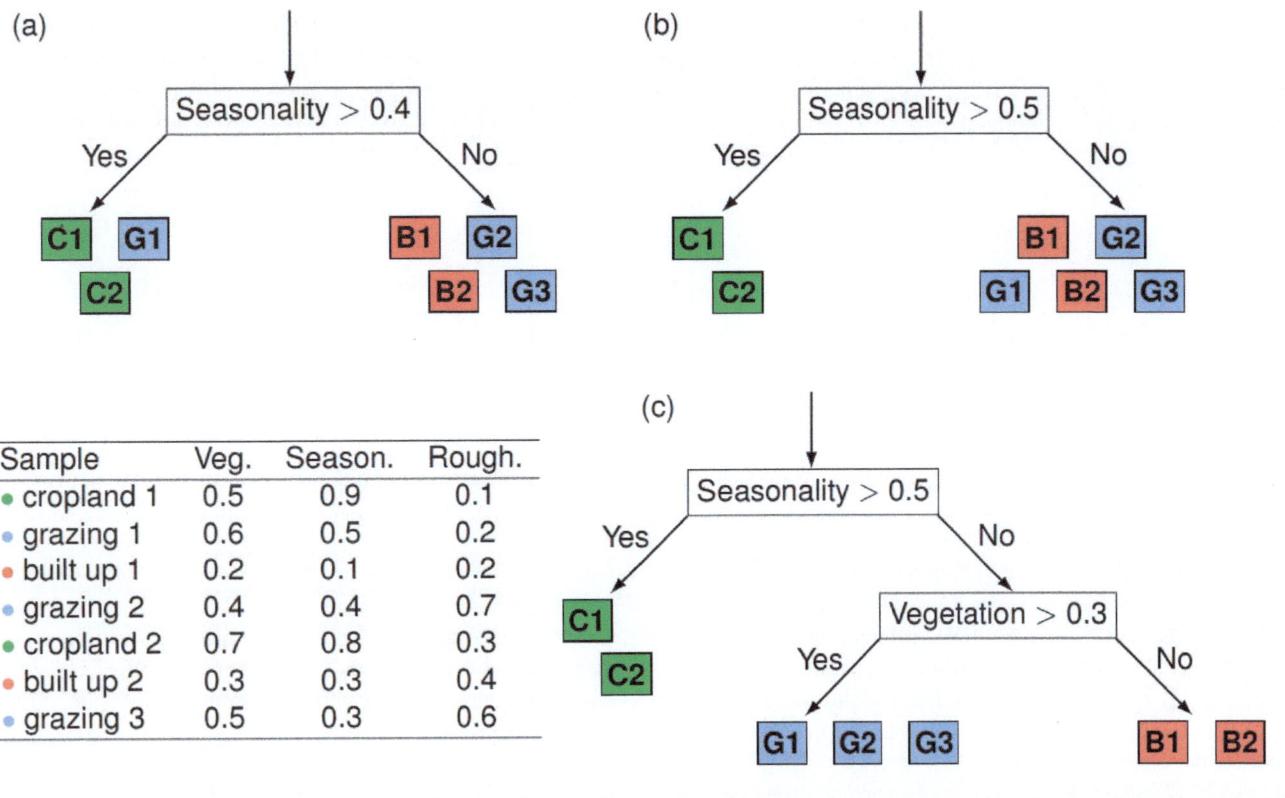

Sample	Veg.	Season.	Rough.
• cropland 1	0.5	0.9	0.1
• grazing 1	0.6	0.5	0.2
• built up 1	0.2	0.1	0.2
• grazing 2	0.4	0.4	0.7
• cropland 2	0.7	0.8	0.3
• built up 2	0.3	0.3	0.4
• grazing 3	0.5	0.3	0.6

Figure 1. Evolution of a decision tree for land use data based on normalized parameters for vegetation index, seasonal color variability, and topographic roughness. (a) The algorithm tests a variety of conditions for the first branch point, looking for a condition such as (b) that cleanly separates different classes. (c) A second branch point is added to differentiate between built-up and grazing land.

The *k*-Means Algorithm

By far the most well-known technique for cluster analysis, [1] ***k***-means is usually said to have its origins in work eventually published as Lloyd (1982), but disseminated earlier (thus, for example, Hartigan and Wong, 1979, set out a specific implementation). The algorithm is designed to divide a set of N data vectors, each consisting of $\boldsymbol{M}$ elements, into $\boldsymbol{K}$ clusters. These clusters are defined so that the distance of each point in the cluster from its center is as small as possible.

The algorithm is readily understood, and is illustrated in Fig. 2. We begin by generating $\boldsymbol{K}$ points, which represent our initial guesses for the location of the center of each cluster – these may be chosen completely at random, or based on various heuristics which attempt to identify a "sensible" initial configuration. We then assign each element of our training data to the cluster with the nearest center (Fig. 2b). Once this has been done, we recompute the position of the central point by averaging the locations of all points in the cluster (Fig. 2c). This process of assignment and averaging is repeated (Fig. 2d) until a stable configuration is obtained (Fig. 2e). The resulting clusters may then be inspected to as-certain their similarities and differences, and new data can be classified by computing its distance from each cluster center.

In order to implement this, it is necessary to define what the word "distance" means in the context of comparing any two data vectors. There are a number of possible definitions, but it is most common to use the "Euclidean" distance: the sum of the squared difference between each element of the two vectors. Thus, if x is a vector with $\boldsymbol{M}$ elements $(x_1; x_2; ......; x_M)$ and y a second vector $(y_1, y_2, ..., y_M)$, then the Euclidean distance between them is defined

$$d = \sum_{i=1}^{M} (x_i - y_i)^2. \tag{1}$$

This definition is a natural extension of our everyday understanding of the concept of "distance". However, where data vectors are comprised of more than one class of observation

- perhaps combining topographic heights with soil properties
- problems can arise if the measurements differ considerably in typical scale. The Euclidean distance between two two-element data vectors $(1;10^{9})$ and $(1;10^{5})$ is very small, despite the second elements differing by 4 orders of magnitude, because both are negligible in comparison to the first element. It may be necessary to rescale the various measurements that make up a data vector to ensure that they have a similar magnitude and dynamic range. Alternatively, and equivalently, the definition of "distance" can be adapted to assign different weights to various data types.

In its basic form, the ***k***-means algorithm requires the user to specify the number of clusters to be sought as a priori information. In many cases, this may not be known, and a range of different solutions have been proposed – see, for example, Jain (2010). Fundamentally, these entail balancing increased complexity (i.e., an increased number of clusters) against any resulting reduction in the average distance between samples and the cluster center. In general, this reduction becomes insignificant once we pass a certain number of clusters, and this is taken to provide an appropriate description of the data.

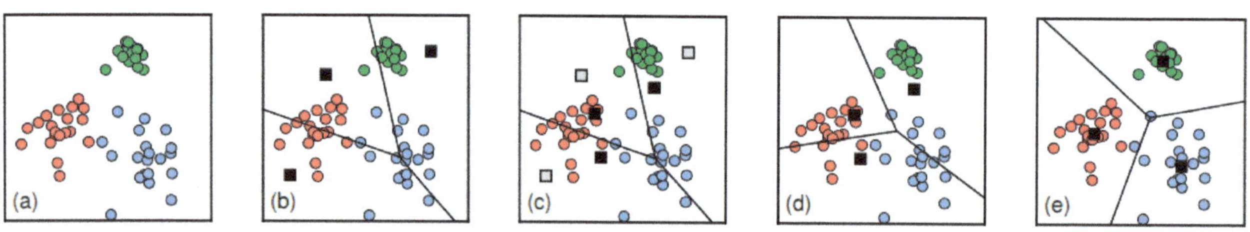

Figure 2. The ***k***-means algorithm. (a) A data set contains points clustered around three distinct locations (colour-coded for ease of reference). (b) We first guess locations for these centers at random (black squares) and assign each datum to the nearest cluster (shown divided by black lines). (c) We then re-compute the location of each cluster center by averaging all points within the cluster (old centers, grey squares; new centers, black squares), and (d) update the cluster assignment of each point to reflect this. It may be necessary to repeat steps (c) and (d) for several iterations until a stable partitioning of the data set is found (e).

Clustering algorithms such as ***k***-means have seen use in geomorphology as an aid to analysis or interpretation of a variety of data sets. It is common to first apply some trans-formation or pre-processing to raw data so as to improve sensitivity to particular classes of feature. For example, Miliaresis and Kokkas (2007) take lidar DEMs, apply various filters designed to enhance the visibility of the built environment within the image, and then use ***k***-means to distinguish different areas of urban environment (such as differentiating between vegetation, buildings, and roads/pavements). Similarly, Belluco et al. (2006) use the technique to assist in map-ping vegetation based on remotely sensed spectral imaging, although they find that which method performs best depends on the type of imaging and field data. On a much smaller scale, Dunning et al. (2009) use ***k***-means and other clustering algorithms to extract discontinuity orientations from laser-derived observations of landslide scarps, which help constrain the mechanism behind the slope failure.

Principal Component Analysis

Often, the different observations comprising a given data vector are correlated, and thus not fully independent: for example, topographic heights at adjacent sites are likely to be reasonably similar to one another, and an imprint of topography will often be found in other data sets such as soil thickness or temperature. For analysis purposes, it is often desirable to identify the patterns common to multiple data elements, and to transform observations into a form where each parameter, or component, is uncorrelated from the others. Principal component analysis (PCA) provides one of the most common techniques for doing so, and has its roots in the work of Pearson (1901). Essentially the same mathematical operation arises in a variety of other contexts, and has acquired a different name in each: for example, "singular value decomposition", "eigenvector analysis", and the concept of "empirical orthogonal functions" are all closely related to PCA.

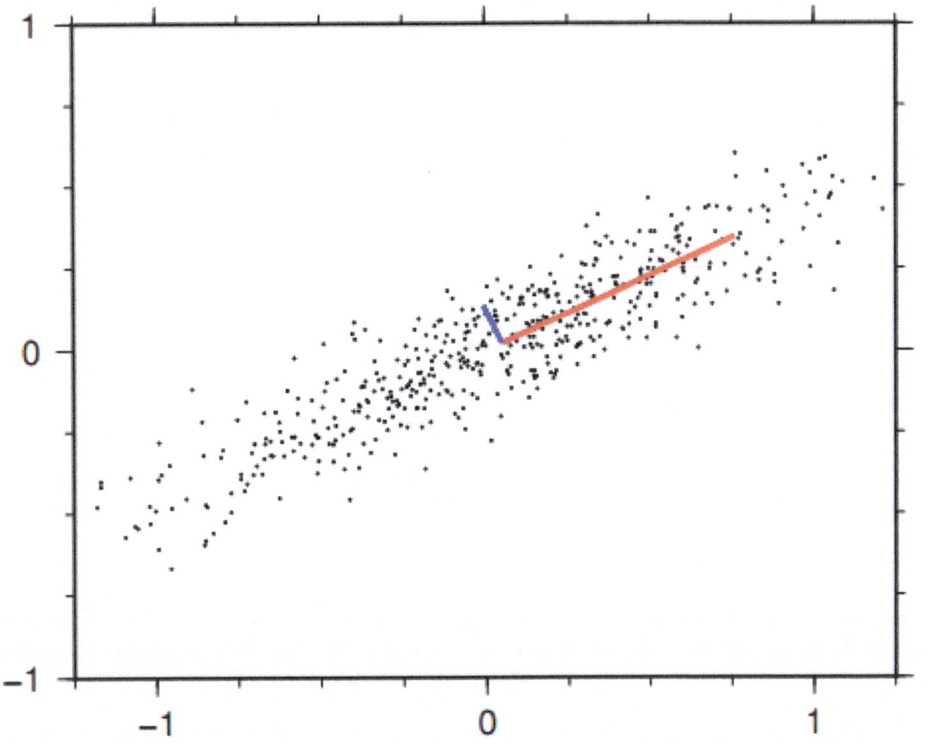

Figure 3. Principal component analysis of a simple data set. Dominant principal component shown in red; secondary principal component shown in blue. Line lengths are proportional to the weight attached to each component. It is apparent that the principal components align with the directions in which the data set shows most variance.

Numerical algorithms for performing PCA are complex, and there is usually little need for the end user to understand their intricacies. In general terms, PCA involves finding the "direction" in which the elements of a data set exhibit the greatest variation, and then repeating this with the constraint that each successive direction considered must be at right angles (orthogonal) to those already found. Although easiest to visualize in two or three dimensions (see Fig. 3), the principle works in the same way for data vectors with more elements: for a set of data vectors with M elements, it is possible to construct up to M perpendicular directions.

Thus, the outcome of PCA is an set of orthogonal directions (referred to as principal components) ordered by their importance in explaining a given data set: in a certain sense, this can be regarded as a new set of co-ordinate axes against which data examples may be measured. The principal components may be regarded as a set of weighted averages of different combinations of the original parameters, chosen to best describe the data. Often, much of the structure of a data set can be expressed using only the first few principal components, and PCA can therefore be used as a form of dimensionality-reduction (see Fig. 4). In a similar vein, it can form a useful precursor to other forms of analysis, such as clustering, by virtue of its role in unpicking the various signals contributing to a data set: often, each component will exhibit sensitivity to particular physical parameters. However, particularly where the original data are composed of a variety of different physical data types, the results of PCA may not be straightforward to interpret: each principal component may be influenced by a number of disparate measurements.

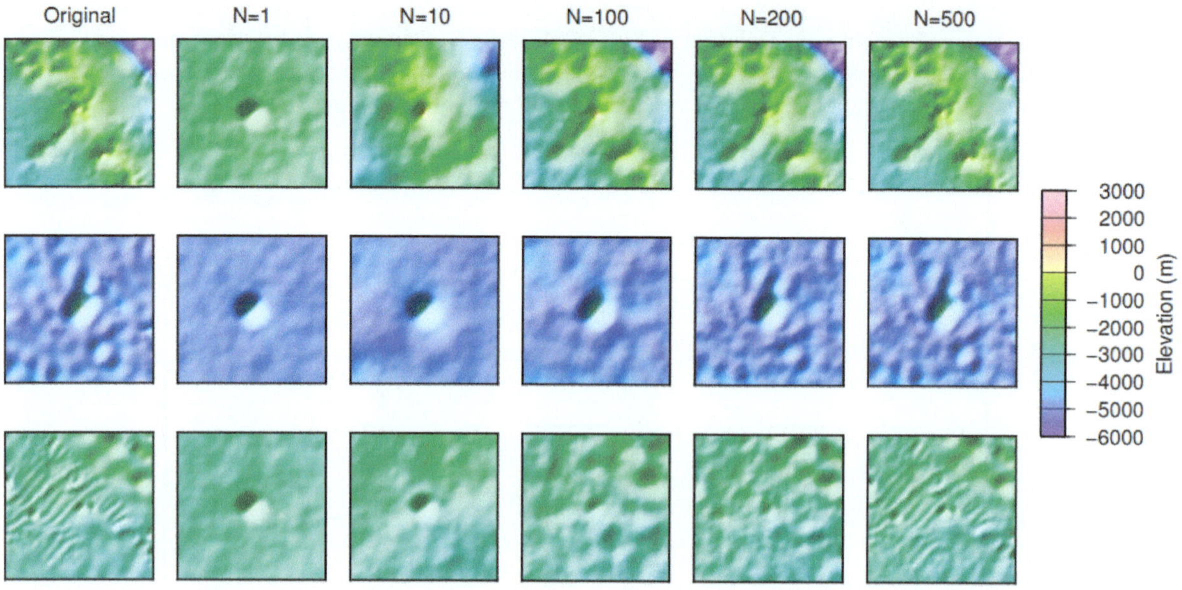

Figure 4. Image reconstruction using principal components. PCA has been performed on a data set containing 1000 square "patches" of bathymetric data, each representing an area of dimension 150 km 150 km, center upon a seamount (the "training set" used in Valentine et al., 2013). Each patch is comprised of 64 64 samples – thus, each can be seen as a 4096-dimensional object. Three examples from this set are shown here (one per row). In the left-most column we show the original bathymetry for each; then, we present reconstructions of this bathymetry using only the N most significant principal components, for N D 1, 10, 100, 200, and 500. It is apparent that the large-scale structures within this data set can be represented using only 100–200 dimensions, while around 500 dimensions are required to allow some of the fine-scale structure to be represented, particularly in the third example. This still represents almost an order of magnitude reduction, in comparison to the original, 4096-dimensional, data.

One example of this can be found in Cuadrado and Perillo (1997), where PCA is performed on a data set consisting of bathymetric measurements for a given region repeated over several months. The first principal component is then found to describe the mean bathymetry of the period, while the second provides information about the general trend of change in bathymetry over time, highlighting areas of de-position and erosion. Another typical application occurs in Baeza and Corominas (2001), where PCA is used to identify the observable parameters that best serve as predictors of landslide hazard. A more recent example is Tamene et al. (2006), who use PCA to identify which observable parameters best explain variability in sediment yield between different river catchments in Ethiopia. This example also shows how a single principal component may be a combination of observables: their first component, which explains about 50 % of the variability, is dominated by a combination of topographic variables, such as height difference, elongation ratio, and catchment area. The second component, which ex-plains approximately 20 % of the variability, is dominated by variables associated with lithology and land use/vegetation cover.

Neural Networks

Perhaps the most varied and versatile class of algorithm discussed in this paper is the neural network. As the name suggests, these were originally developed as a model for the workings of the brain, and they can be applied to tackling a wide range of problems. It has been shown (e.g. Hornik, 1991) that neural networks can, in principle, be used to rep-resent arbitrarily complex mathematical functions, and their use is widespread in modern technology – with applications including tasks such as voice recognition or automatic language translation. They may be used to tackle problems falling in all three of the classes discussed in Sect. 2. A comprehensive introduction to neural networks may be found in Bishop (1995); the aforementioned book by Mackay (2003) also discusses them at some length. Some readers may also be interested in a review by Mas and Flores (2008), targeted at the remote sensing community.

A neural network is constructed from a large number of interconnected "neurons". Each neuron is a processing unit that takes a number of numerical inputs, computes a weighted sum of these, and then uses this result to compute an out-put value. The behavior of the neuron can therefore be con-trolled by altering the weights used in this computation. By connecting together many neurons, with the outputs from some being used as inputs to others, a complex system – or network – can be created, as in Fig. 5. The weights associated with each neuron are unique, so that the behavior of the net-work as a whole is controlled by a large number of adjustable parameters. Typically, these are initially randomized; then, a "training" procedure is used to iteratively update the weights until the network exhibits the desired behavior for a given training set.

Such a brief description glosses over the richness of approaches within "neural networks": choices must be made regarding how individual neurons behave, how they are connected together (the "network architecture"), and how the training algorithm operates. These are generally somewhat application-dependent, and may dictate how effectively the network performs. The simplest form of neural network, sometimes called a "multi-layer perceptron" (MLP), consists of neurons arranged in two or three "layers", with the outputs from neurons in one layer being used as the inputs for the next layer (see Fig. 5). Traditionally, these are trained by "back-propagation" (Rumelhart et al., 1986), which involves calculating how network outputs are influenced by each individual weight. The canonical use for such a network is as a classifier (e.g. Lippmann, 1989; Bischof et al., 1992), although they are exceptionally versatile: for example, Ermini et al. (2005) demonstrate that MLPs can be used for landslide susceptibility prediction. They may also be used to model a wide variety of physical relationships between observables, for example, in interpolation problems such as estimating difficult-to-measure surface properties from satellite observations (e.g. Krasnopolsky and Schiller, 2003), and to detect unusual or unexpected features within data sets (e.g. Markou and Singh, 2003).

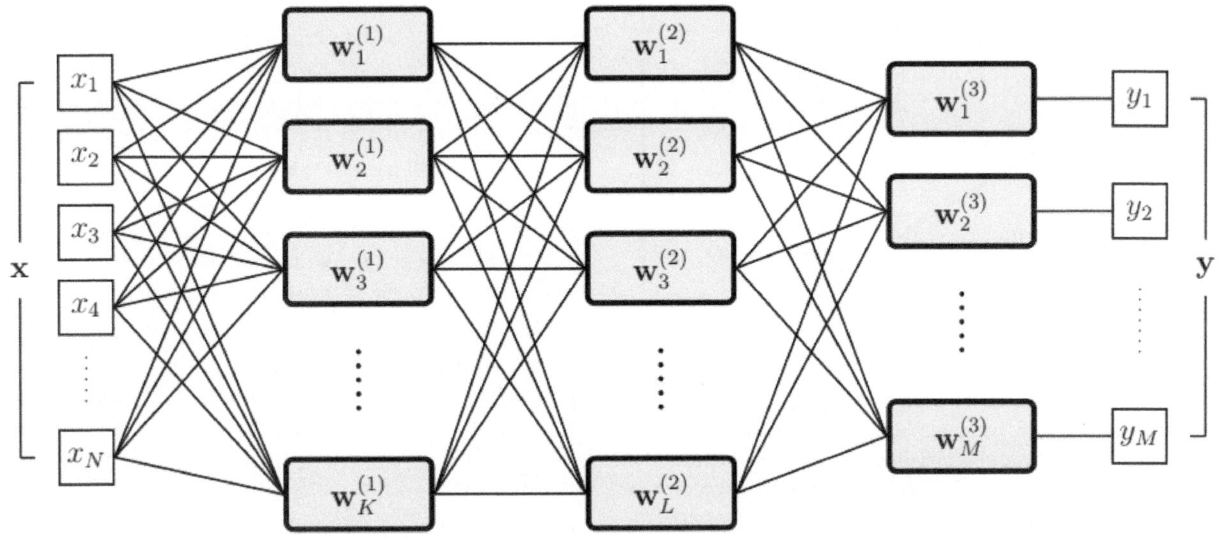

Figure 5. Schematic of a simple neural network (a "multi-layer perceptron"). The network takes an N-dimensional vector, x, on the left side and transforms it into an M-dimensional vector y on the right side. Each grey box represents a "neuron", and is a simple mathematical operation which takes many inputs (lines coming in from the left) and returns a single output value which is sent to every neuron in the next "layer" (lines coming out from the right). The neuron's behavior is governed by a unique set of "weights" (w): one common mathematical relation has a neuron in a layer with K inputs return $y = \tanh(w_0 + \sum_{i=1}^{K} w_i x_i)$, where the single output element y will become part of the inputs for the next layer.

In recent years, attention has increasingly focused on "deep learning", where many more layers of neurons are used. This has proven effective as a means to "discover" structure in large, complex data sets and represent these in a low-dimensional form (Hinton and Salakhutdinov, 2006). Typically, these require specialized training algorithms, since the number of free parameters in the network is exceptionally large. Such systems are particularly useful for applications in cluster analysis and dimension reduction, and these proper-ties can be exploited to enable cataloguing of geomorphological features in large data sets, as in Valentine et al. (2013).

Support Vector Machines

The modern concept of the support vector machine (SVM) stems from the work of Cortes and Vapnik (1995), although this builds on earlier ideas. The approach is targeted towards classification problems, framed in terms of finding, and then utilizing, "decision boundaries" that separate one class from another. In the simplest case, linear decision boundaries can be found – that is, any two classes can always be separated by a straight line when two-dimensional slices through the data set are plotted on a graph. The SVM method provides an algorithm for constructing linear decision boundaries that maximize the "margin" between boundary and adjacent data points, as shown in Fig. 6.

However, in most realistic cases, the data set cannot be cleanly categorized using linear boundaries: all possible linear decision boundaries will misclassify some data points. To handle this scenario, the SVM approach uses a mathematical trick to create nonlinear decision boundaries in a way that is computationally tractable. Data are first mapped into a higher-dimensional "feature space" (the opposite of dimensionality reduction); in this space, the data can then be separated using linear boundaries (see Fig. 6). This mapping, or transformation, may be nonlinear, so that lines in the feature space may correspond to curves in the original data space. Various extensions to this approach exist that allow for a less-than-perfect division of the data set, to reflect the presence of classification errors or observational noise. Once boundaries have been determined, the SVM may then be used for classification of new examples. SVMs have some similarities in structure to simple neural networks, although the "training" or optimization procedure is quite distinct.

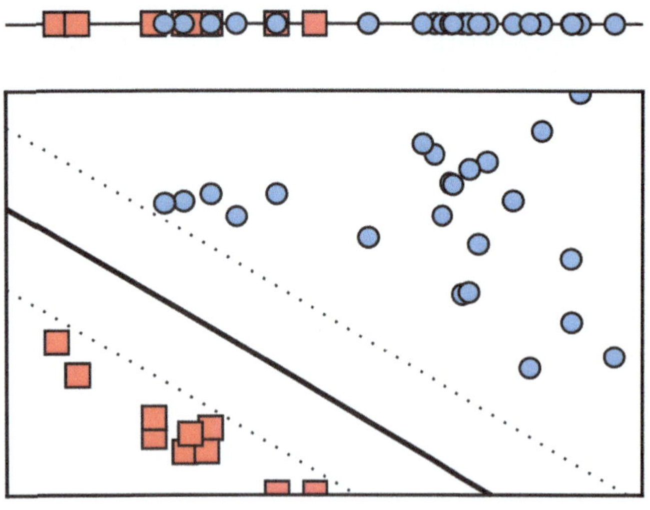

Figure 6. Segmenting data sets with linear decision boundaries. In the original one dimension (top), it is not possible to separate red squares from blue circles. However, by mapping this data set into an artificial second dimension, it becomes possible to draw a linear "decision boundary" that distinguishes the two classes. The support vector machine provides a technique for achieving this in such a way that the "margin" between the boundary and the nearest data points is maximized (as shown by the dotted lines).

Again, landslide susceptibility assessment offers one aspect of geomorphology where SVMs have found significant application (e.g. Brenning, 2005; Yao et al., 2008; Marjanović et al., 2011; Peng et al., 2014). They have also been used to differentiate between fluvial and glacial river valleys (Matías et al., 2009). Another similar use can be found in Beechie and Imaki (2014): there, the authors use an SVM to classify river channel morphologies based on geospatial data including DEMs, precipitation, and geology. This can then be used in river conservation and restoration to infer the natural patterns that existed in areas that have been subject to extensive human intervention.

Self-Organizing Maps

The concept of the self-organizing map (SOM) stems from the work of Kohonen (1990), and it can be viewed as a particular class of neural network. The SOM implements a form of dimensionality reduction, and is generally used to help identify clusters and categories within a data set. The basic premise is to take a (usually) two-dimensional grid and, through training, create a mapping between that 2-D space and a higher-dimensional data set (see Fig. 7). The 2-D representation can then be used to help visualize the structure of the data. The SOM is also typically designed to be significantly smaller than the training set, spanning the same data space with fewer points, and is thus easier to analyze.

To create an SOM, we start with a map consisting of a number of "nodes", often arranged as a regular grid, so that it is possible to define a "distance" between any two nodes, and hence identify the set of nodes that lie within a certain radius of a given node, known as its "neighborhood". Each node is associated with a random "codebook vector" with the same dimensionality as the data (Fig. 7a). During training, we iteratively select a data example at random from the training set, identify the node with the closest-matching codebook vector, and then adjust this vector, and those of neighboring nodes, to better match the training example (Fig. 7b). Given sufficient training, the codebook vectors come to mirror the distribution of data in the training set (Fig. 7c). Typically, both the radius used to define the "neighborhood" of a given node and the extent to which codebook vectors are allowed to change are reduced as training proceeds in order to promote fine tuning of performance.

Once the SOM is trained, various approaches exist to en-able visualization of the codebook vectors, with the goal of highlighting any underlying structure within the data set. One common approach is to try to identify clusters in the data set by examining the distances between the codebook vectors for neighboring nodes, often by plotting the SOM grid colored according to these distances (sometimes described as depicting the "U-matrix"). Alternatively, a method can be used to distort the grid in such a way that when plotted in 2-D, the distance between nodes is proportional to the distance between their codebook vectors; one common technique for this is "Sammon's mapping" (Sammon, 1969). An-other visualization approach, sometimes called "component plane analysis", looks for correlations between input parameters, e.g. between rainfall and elevation or slope orientation and landslide risk. Here, the SOM grid is colored according to the values of particular elements of the codebook vectors, with each element corresponding to an input parameter. Correlated parameters can then be identified by their similar color patterns.

Potential applications in geomorphology are numerous. Marsh and Brown (2009) use an SOM-based method to analyze and classify bathymetry and backscatter data, al-lowing near-real-time identification of regions with particular seafloor characteristics, e.g. for benthic habitat map-ping. Similarly, Ehsani and Quiel (2008) demonstrate that SOMs can be used to classify topographic features, identifying characteristic morphologies contained within DEMs such as channels, ridge crests, and valley floors. In a third example, Friedel (2011) uses SOMs to assess post-fire hazards in recently burned landscapes. Both PCA and ***k***-means clustering are then used to divide the 540 areas studied into 8 distinct groups, which the author suggests could be used for focusing future field research and development of empirical models. As illustrated by these examples, one of the key strengths of the SOM method is that learning is unsupervised, and does not rely on the user having any prior knowledge of the "important" structures contained within the data set.

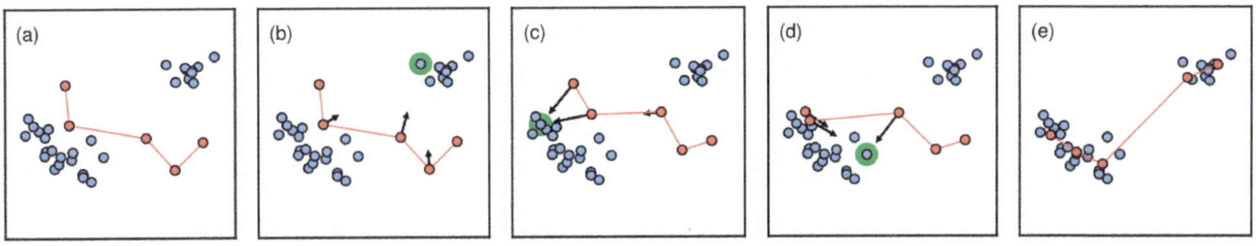

Figure 7. The self-organizing map (SOM). A data set consists of numerous data points (blue), and is to be represented by a one-dimensional SOM (red). The SOM contains of a number of nodes, with a well-defined spatial relationship (here, depicted by lines connecting points). Initially, all SOM nodes are associated with random locations in data space (a). During training (b–d) a datum is selected at random (high-lighted in green). The closest SOM node is identified, and this and its nearest neighbors are adjusted so as to be closer to the chosen point (arrows), with the scale of adjustment proportional to the separation. After many iterations of this procedure, the distribution of SOM nodes mirrors that of the data set (e). For ease of illustration, this figure depicts a two-dimensional data set and a one-dimensional SOM; in practice, the data set will usually have much higher dimensionality, and the SOM is usually organized as a two-dimensional grid.

Bayesian Inference

To conclude this section, we mention the concept of Bayesian inference. This is a much broader topic than any of the methods discussed so far; indeed, in many cases these methods are themselves formally derived from Bayesian concepts. Bayes' theorem (Bayes, 1763) underpins a significant fraction of modern statistical techniques, and explains how new observations can be used to refine our knowledge of a system. It tells us that the probability that the system is in a certain state, given that we have made a particular observation ("obs."), ***P*** (state jobs*:*), can be expressed in the form

$$P(\text{state}\mid\text{obs.}) = \frac{P(\text{obs.}\mid\text{state})P(\text{state})}{P(\text{obs.})}. \qquad (2)$$

Here, ***P*** (state) represents our prior knowledge – that is, our assessment of the probability that the system is in the given state without making any measurements – while ***P*** (obs*:*) represents the probability with which we expect to obtain the exact measurements we did, in the absence of any knowledge about the state of the system. Finally, ***P*** (obs*:* jstate) expresses the probability that we would get those measurements for a system known to be in the relevant state.

In many cases, it is possible to estimate or compute the various probabilities required to implement Bayes' theorem, and thus it is possible to make probabilistic assessments. This is often useful: for example, hazard assessment is generally better framed in terms of the chance, or risk, of an event, rather than attempting to provide deterministic predictions. An extensive discussion of Bayesian analysis can be found in Mackay (2003). Sivia (1996) is another useful reference, and a paper by Griffiths (1982) provides some geomorphological context. A wide range of computational techniques have been developed to enable Bayesian calculations in various settings, for which a review by Sambridge and Mosegaard (2002) may be a good starting point.

As a simple example, suppose we are interested in classifying land use from satellite imagery. Grassland will appear as a green pixel 80 % of the time, although it may also be brown: thus, ***P*** (greenjgrass) D 0*:*8. On the other hand, desert environments appear as brown in 99 % of cases. A particular region is known to be mainly desert, with only 10 % grassland – so an image of the area will consist of 8.9 % green pixels (D ((0*:*01 0*:*9) C (0*:*8 0*:*1))). If we look at one specific pixel, and observe that it is green, Bayes' theorem tells us there is a 90 % chance that the location is grass-land: ***P*** (grassjgreen) D ***P*** (greenjgrass)***P*** (grass)=***P*** (green) D 0*:*8 0*:*1=0*:*089 0*:*899. On the other hand, if a pixel is seen to be brown, we can classify it as desert with 98 % certainty. This illustrates a property that emerges from the Bayesian formalism: unusual observations convey more information than routine ones. Before we obtained any satellite imagery, we could have guessed with 90 % accuracy that a particular location in the region was desert; observing a brown pixel leads to only a relatively modest increase in the certainty of this determination. However, if we were to observe a green pixel, the chance that the location is a desert drops markedly, to only 10 %.

Most examples in geomorphology again come from land-slide susceptibility: Lee et al. (2002) and Das et al. (2012) use Bayesian techniques to assess landslide susceptibility in Korea and the Himalayas, respectively. In a similar application, Mondini et al. (2013) map and classify landslides in southern Taiwan using a Bayesian framework. Examples from other areas of geomorphology include Gutierrez et al. (2011), who use a Bayesian network to predict shoreline evolution in response to sea-level change, and Schmelter et al. (2011), who use Bayesian techniques for sediment transport modelling.

PRACTICAL CONSIDERATIONS

Each of the techniques discussed in the previous section – and the wide variety of alternatives not mentioned here – has its strengths and weaknesses, and a number of practical is-sues may need to be considered when implementing a solution to a particular problem. Here, we discuss some topics that may be relevant across a range of different approaches, and which may affect the viability of a learning algorithm solution in any case.

Constructing a Training Set

Unsurprisingly, the training data used when implementing a learning algorithm can have a major impact upon results: everything the system "knows" is derived entirely from these examples. Any biases or deficiencies within the training data will therefore manifest themselves in the performance of the trained system. This is not, in itself, necessarily problematic – indeed, the landform cataloguing system introduced by Valentine et al. (2013) relies upon this property – but it must be borne in mind when tools are used and results interpreted. If training data are low-quality or contain artefacts, glitches and other observational "noise", results will suffer. In general, time invested in properly selecting and processing data sets will be well spent. However, it is also important that training data remain representative of the data that will be used during operation.

One particular issue that can arise stems from the fact that the learning algorithm lacks the trained researcher's sense of context: it has no preconception that certain structures in the data are more or less significant than others. For example, suppose a system is developed to classify valley profiles as being formed by either a river or a glacier. The training data for this system would consist of a sequence of hand-classified valley profiles, each expressed as a vector of topographic measurements. If, for the sake of example, all glacial examples happen to be drawn from low-elevation regions, and all riverine examples from high-elevation regions, it is likely that the system would learn to treat elevation as an important factor during classification.

A second potential pitfall arises from the statistical basis underpinning most learning algorithms. Typically, the extent to which a particular feature or facet of the data set will be "learnt" depends on its prevalence within the training examples as a whole. This can make it difficult for a system to recognize and use information that occurs infrequently in the training data, since it gets "drowned out" by more common features. Again, considering the problem of valley classification, if 99 % of training examples are glacial, the system is likely to learn to disregard its inputs and simply classify everything as glacial, since this results in a very low error rate; for best results, both types should occur in roughly equal pro-portions within the training set. As before, this should be regarded as a natural property of the learning process, rather than as being inherently problematic; indeed, it can be exploited as a tool for "novelty detection", allowing unusual features within a data set to be identified (for a review, see e.g. Marsland, 2002). Nevertheless, it is a factor that ought to be borne in mind when a system is designed and used.

To avoid problems, it is important to choose training data with care, and to develop strategies for evaluating and monitoring the performance of the trained system. It is often beneficial to invest time in finding the best way to represent a given data type, so as to accentuate the features of interest, and remove irrelevant variables. This process is sometimes referred to as "feature selection" (e.g. Guyon and Elisseeff, 2003). Thus, in the situation mentioned above, the valley pro-files might be more usefully represented as variations relative to their highest (or lowest) points, rather than as a sequence of absolute elevations. In line with our comments in Sect. 3.2, it is often helpful to "de-mean" and normalize the data: using the training set, it is straightforward to compute the mean input vector, as well as the standard deviation associated with each component of this. Then, all examples – during training, and during operation – can be standardized by subtraction of this mean and rescaling to give unit standard deviation. Clearly, the inverse of these transformations may need to be applied before results are interpreted. For more complex applications, specialist representations such as the "geomorphons" developed by Jasiewicz and Stepinski (2013) may be beneficial, and provide a targeted route to encoding geomorphological information. Training examples should be chosen to cover a spread of relevant cases, perhaps including different regions or measurements made at different times of year, to avoid introducing unintended biases into results.

Assessing Performance

For "supervised" learning algorithms, where a pre-defined relationship or structure is to be learnt, it is possible to assess performance using a second set of user-selected data – often referred to as a "test" or "monitoring" data set. These test data intended to provide an independent set of examples of the phenomena of interest, allowing quantitative measures of effectiveness to be evaluated; this strategy is sometimes referred to as "cross-validation". It is important to do this using examples separate from the training set in order to ensure that the system's "generalization performance" is measured: we want to be sure that the algorithm has learned proper-ties that can be applied to new cases, rather than learning features specific to the particular examples used during training. As an analogy: a dog may learn to respond to particular commands, but only when issued in a very specific manner; it cannot then be said to properly understand a spoken language.

The metric by which performance is assessed is likely to be situation-dependent. In general, a supervised learning algorithm will be designed to optimize certain "error measures", and these are likely to provide a good starting point. Nevertheless, other statistics may also prove useful. For classification systems, analysis of "receiver operating characteristics" (ROCs) such as hit and false positive rates may be instructive (e.g. Fawcett, 2006). More difficult to quantify, but still potentially valuable, is the experienced researcher's sense of the plausibility of a system's predictions: do results exhibit geomorphologically reasonable patterns? For example, does a prediction of landslide risk seem plausible in its relationship with the topography and underlying bedrock?

Assessing performance in unsupervised learning is more challenging, as we may not have prior expectations against which to measure results: fundamentally, it may be challenging to define what "good performance" should signify. In many cases, application-specific statistics may be helpful – for example, in cluster analysis it is possible to calculate the standard deviation of each cluster, quantifying how tightly each is defined – and again, the researcher's sense of plausibility may provide some insight. It may also prove instructive to repeat training using a different subset of examples, to assess how stable results are with respect to variations in the training data: a structure or grouping that appears consistently is more likely to be real and significant than one that is very dependent on the precise examples included in the training set.

Overtraining and Random Noise

The phenomenon of "overtraining" or "overfitting", which typically arises in the context of supervised learning algorithms, has already been alluded to. It occurs when an iterative training procedure is carried out for too many iterations: at some point, the system usually begins to learn information that is specific to the training examples, rather than being general. This tends to reduce the performance of the system when subsequently applied to unseen data. It can of-ten be detected by monitoring the algorithm's generalization performance using an independent set of examples as training proceeds: this enables the training procedure to be terminated once generalization performance begins to decrease. In certain cases, post-training strategies can be used to re-duce the degree of over-fitting: the example of "pruning" decision trees has already been mentioned. It has also been shown that "ensemble methods" may be useful (indeed, "random forests" provide one example of an ensemble method), whereby multiple instances of the same learning algorithm are (over-) trained, from different randomized starting points, and their outputs are then somehow averaged (e.g. Dietterich, 2000). Because each instance becomes sensitive to distinct aspects of the training data, due to their different initialization conditions, each performs well on a subset of examples, and poorly on the remainder. However, the overall performance of the ensemble (or "committee") as a whole is typically better than that of any individual member.

Another strategy that is adopted is to add random noise to the training data. The rationale here is that training examples typically exhibit an underlying pattern or signal of interest, overprinted by other processes and observational errors. In order to learn the structure of interest, we wish to desensitize our training procedure to the effects of this overprinting. If we can define a "noise model" that approximates the statistical features of the unwanted signal, adding random realizations of this to each training example allows us to limit the extent to which the algorithm learns to rely on such features: by making their appearance random, we make them less "useful" to the algorithm. Returning again to the example of valley classification, local variations in erosion and human intervention might be modelled as correlated Gaussian noise on each topographic measurement. During training, each example is used multiple times, with different noise on each occasion; in theory, this results in only the gross features of the valley profile being taken into account for classification purposes. However, it may be challenging to identify and construct appropriate noise models in many realistic cases.

It is worth noting here that a similar strategy may prove useful in other cases where it is desirable to desensitize systems to particular aspects of the data. For example, spatial observations are typically reported on a grid, aligned with geographic coordinates. However, natural phenomena typically do not display any such alignment, and orientation in-formation may be irrelevant in many cases. If 2-D spatial in-formation is used in a particular case, it may be desirable to make use of multiple, randomly rotated copies of each training set example. This allows an effectively larger training set to be created, and reduces the chance that features are treated differently depending on their alignment.

Operational Considerations

As with any analysis technique, results obtained using learning algorithms can be misleading if not treated carefully. This is especially true where a technique invites treatment as a "black box", and where the mechanism by which it operates is not always easily understood. The great strength of artificial intelligence is that it enables a computer to mimic the experienced researcher – but this is also a potential draw-back, tending to distance the researcher from their data. In some sense, this is an inevitable consequence of the ever-increasing quantity of data available to researchers – but there is a risk that it leads to subtleties being missed, or results interpreted wrongly due to misapprehensions surrounding computational processing. To minimize the risk of these issues arising, it is important that researchers develop heuristics and procedures that enable "intelligent" systems to be monitored. For example, users of automated data classification systems should monitor the statistical distributions of classification outputs, and investigate any deviations from the norm; it is also desirable to spot-check classifications. This is particularly true in settings where there is a risk that new examples lie outside the scope of the training set – perhaps where data are drawn from new geographic regions, or were collected at a different time of year.

Learning algorithms have immense potential in enabling exploration of large, complex data sets and in automating complicated tasks that would otherwise have to be done manually. However, developing a learning algorithm system – especially those targeted at classification, regression, or interpolation – can also be time-consuming and resource-intensive. Computational demands may also be significant: although many applications require no more than a standard laptop or desktop computer, some may only be viable with access to large-scale parallel computing resources. By way of an illustration for the more computationally intensive end of the spectrum: training the learning algorithm used to catalogue seamounts in Valentine et al. (2013) currently requires a few hundred CPU hours on a 1.9 GHz machine, in addition to considerable computational demands for processing the raw bathymetric data. Thus, learning algorithms do not currently present an economic solution to all problems – al-though this balance will undoubtedly change as technology and algorithms continue to evolve.

OUTLOOK

In this paper, we have attempted to provide a general survey of the field of learning algorithms, and how they might be useful to researchers working in the fields of geomorphometry and Earth surface dynamics. These fields benefit from extensive, large-scale, feature-rich data sets, bringing opportunities and challenges in equal measure. Although currently dominated by a few specific topics in the field, such as land-slide hazard assessment, the use of artificial intelligence to help explore, process, and interpret geomorphological data is almost certain to be an increasingly significant aspect of research in coming years.

An increased use of learning algorithms in geomorphological communities is likely to require developments in computational infrastructure. There are obvious requirements for access to appropriate hardware and software resources, as well as skill development. In particular, larger problems or more complex algorithms may make use of parallel computing and other high-performance computing techniques essential. In addition, in light of the potentially substantial development cost of implementing some of the more complex learning algorithms, it is worth trying to plan for a flexible implementation. Most of these approaches can, in principle, be implemented in a fairly general framework, allowing the same underlying algorithm to be applied to many problems.

The computational implications of both parallelization and generalization are beyond the scope of this review, but one area of particular relevance to the geomorphology community concerns data input and output, and hence data format: the ability to reuse an algorithm across multiple data types is a key element of flexibility in a field with such a diversity of measurements. The ability to handle large file sizes and support efficient data access, potentially in parallel, is also an important consideration. As these techniques develop, this places increasing importance on the development and use of robust community standards for data formats. File frame-works, which allow the development of multiple specialist file formats all adhering to a common set of rules, may be particularly valuable in combining consistency from an algorithmic point of view with flexibility to accommodate varied data.

However, learning algorithms are not a panacea. "Traditional" approaches to data analysis will remain important for the foreseeable future, and are well suited to tackling many problems; we do not advocate their wholesale replacement. In addition, it is important that the use of more advanced computational techniques is not allowed to become a barrier between researchers and data; it is almost certain that the nature and practice of "research" will need to evolve to accommodate increasing use of these technologies. Some interesting perspectives on these issues may be found in – for example – an issue of Nature dealing with "2020 computing" (e.g. Muggleton, 2006; Szalay and Gray, 2006), and in work aimed at developing a "robot scientist" (e.g. King et al., 2009). Nevertheless, artificial intelligence opens up many new possibilities within the fields of geomorphology and Earth surface processes – particularly given the common scenarios of large data sets; complex, interacting processes; and/or natural variability that make many geomorphological situations difficult to classify or model using conventional techniques. Learning algorithms such as those discussed here thus represent a powerful set of tools with the potential to significantly expand our capabilities across many branches of the field.

ACKNOWLEDGEMENTS

We are grateful to the associate editor, John Hillier, and to Niels Anders, J. J. Becker, Ian Evans, and Evan Goldstein for reviews and comments on the initial draft of this manuscript. We also thank Jeannot Trampert for numerous useful discussions. A. P. Valentine is supported by the European Research Council under the European Union's Seventh Framework Programme (FP/2007-2013)/ERC grant agreement no. 320639.

REFERENCES

1. Baeza, C. and Corominas, J.: Assessment of shallow landslide sus-ceptibility by means of multivariate statistical techniques, Earth Surf. Proc. Land., 26, 1251–1263, 2001.
2. Bayes, T.: An essay towards solving a problem in the doctrine of chances, Philos. T. R. Soc. A, 53, 370–418, 1763.
3. Beechie, T. and Imaki, H.: Predicting natural channel patterns based on landscape and geomorphic controls in the Columbia River basin, USA, Water Resour. Res., 50, 39–57, 2014.
4. Belluco, E., Camuffo, M., Ferrari, S., Modenese, L., Silvestri, S., Marani, A., and Marani, M.: Mapping salt-marsh vegetation by multispectral and hyperspectral remote sensing, Remote Sens. Environ., 105, 54–67, 2006.
5. Bhattacharya, B., Price, R., and Solomatine, D.: Machine learn-ing approach to modeling sediment transport, J.Hydraul. Eng.-ASCE, 133, 440–450, 2007.
6. Bischof, H., Schneider, W., and Pinz, A.: Multispectral classifica-tion of Landsat-images using neural networks, IEEE T. Geosci. Remote S., 30, 482–490, 1992.
7. Bishop, C.: Neural Networks for Pattern Recognition, Oxford Uni-versity Press, Oxford, UK, 1995.
8. Bishop, C.: Pattern Recognition and Machine Learning, Springer, New York, USA, 2006.
9. Breiman, L.: Random Forests, Mach. Learn., 45, 5–32, 2001. Brenning, A.: Spatial prediction models for landslide hazards: review, comparison and evaluation, Nat. Hazards Earth Syst. Sci., 5, 853–862, doi:10.5194/nhess-5-853-2005, 2005.
10. Cortes, C. and Vapnik, V.: Support-Vector Networks, Mach. Learn., 20, 273–297, 1995.
11. Cuadrado, D. and Perillo, G.: Principal component analysis applied to geomorpholigic evolution, Estuar. Coast. Shelf S., 44, 411– 419, 1997.
12. Das, I., Stein, A., Kerle, N., and Dadhwal, V. K.: Landslide sus-ceptibility mapping along road corridors in the Indian Himalayas using Bayesian logistic regression models, Geomorphology, 179, 116–125, 2012.
13. Dietterich, T.: Ensemble methods in machine learning, in: Multiple classifier systems, in: Lecture Notes in Computer Science, edited by: Kittler, J. and Roli, F., 1857, 1–15, Springer-Verlag, Berlin, Germany, 2000.
14. Dunning, S., Massey, C., and Rosser, N.: Structural and geomor-phological features of landslides in the Bhutan Himalaya de-rived from terrestrial laser scanning, Geomorphology, 103, 17– 29, 2009.
15. Ehsani, A. and Quiel, F.: Geomorphometric feature analysis using morphometric parameterization and artificial neural networks, Geomorphology, 99, 1–12, 2008.
16. Ermini, L., Catani, F., and Casagli, N.: Artificial neural networks applied to landslide susceptibility assessment, Geomorphology, 66, 327–343, 2005.
17. Fawcett, T.: An introduction to ROC analysis, Pattern Recogn. Lett., 27, 861–874, 2006.
18. Friedel, M. J.: Modeling hydrologic and geomorphic hazards across post-fire landscapes using a self-organizing map approach, Env-iron. Modell. Soft., 26, 1660–1674, 2011.
19. Griffiths, G.: Stochastic Prediction in Geomorphology Using Bayesian Inference Models, Math. Geol., 14, 65–75, 1982.
20. Gutierrez, B. T., Plant, N. G., and Thieler, E. R.: A Bayesian net-work to predict coastal vulnerability to sea level rise, J. Geophys. Res.-Earth, 116, F02009, doi:10.1029/2010JF001891, 2011.
21. Guyon, I. and Elisseeff, A.: An introduction to variable and feature selection, J. Mach. Learn. Res., 3, 1157–

1182, 2003.

22. Hartigan, J. and Wong, M.: A K-Means Clustering Algorithm, J. Roy. Stat. Soc. C-App., 28, 100–108, 1979.
23. Hillier, J., Conway, S., and Sofia, G.: Perspective – Synthetic DEMs: A vital underpinning for the quantitative future of land-form analysis?, Earth Surface Dynamics, 3, 587–598, 2015.
24. Hinton, G. and Salakhutdinov, R.: Reducing the Dimensionality of Data with Neural Networks, Science, 313, 504–507, 2006.
25. Hornik, K.: Approximation capabilities of multilayer feedforward networks, Neural Networks, 4, 251–257, 1991.
26. Jain, A.: Data clustering: 50 years beyond K-means, Pattern Recogn. Lett., 31, 651–666, 2010.
27. Jasiewicz, J. and Stepinski, T.: Geomorphons – a pattern recogni-tion approach to classification and mapping of landforms, Geo-morphology, 182, 147–156, 2013.
28. Jordan, M. and Mitchell, T.: Machine learning: Trends, perspec-tives, and prospects, Science, 349, 255–260, 2015.
29. King, R., Rowland, J., Aubrey, W., Liakata, M., Markham, M., Soldatova, L., Whelan, K., Clare, A., Young, M., Sparkes, A., Oliver, S., and Pir, P.: The robot scientist Adam, Computer, 42, 46–54, 2009.
30. Kohonen, T.: The Self-Organizing Map, Proceedings of the IEEE, 78, 1464–1480, 1990.
31. Krasnopolsky, V. and Schiller, H.: Some neural network applica-tions in environmental sciences. Part I: forward and inverse prob-lems in geophysical remote measurements, Neural Networks, 16, 321–334, 2003.
32. Lee, S., Choi, J., and Min, K.: Landslide susceptibility analysis and verification using the Bayesian probability model, Environ. Geol., 43, 120–131, 2002.
33. Li, J., Heap, A., Potter, A., and Daniell, J.: Application of machine learning methods to spatial interpolation of environmental vari-ables, Environ. Modell. Soft., 26, 1647–1659, 2011.
34. Lippmann, R.: Pattern classification using neural networks, IEEE Commun. Mag., 27, 47–50, 1989.
35. Lloyd, S.: Least squares quantization in PCM, IEEE T. Inform. The-ory, 28, 129–137, 1982.
36. Mackay, D.: Information Theory, Inference and Learning Algo-rithms, Cambridge University Press, Cambridge, UK, 2003.
37. Marjanovic,′ M., Kovacevi˘c,′ M., Bajat, B., and Voženílek, V.: Land-slide susceptibility assessment using SVM machine learning al-gorithm, Eng. Geol., 123, 225–234, 2011.
38. Markou, M. and Singh, S.: Novelty detection: a review – part 2: neu-ral network based approaches, Signal Process., 83, 2499–2521, 2003.
39. Marsh, I. and Brown, C.: Neural network classification of multi-beam backscatter and bathymetry data from Stanton Bank (Area IV), Appl. Acoust., 70, 1269–1276, 2009.
40. Marsland, S.: Novelty detection in learning systems, Neural Com-puting Surveys, 3, 1–39, 2002.
41. Martin, K., Wood, W., and Becker, J.: A global prediction of seafloor sediment porosity using machine learning, Geophys. Res. Lett., 42, 10640–10646, 2015.
42. Mas, J. and Flores, J.: The application of artificial neural networks to the analysis of remotely sensed data, Int. J. of Sens., 29, 617– 664, 2008.
43. Matías, K., Ordóñez, C., Taboada, J., and Rivas, T.: Functional sup-port vector machines and generalized linear models for glacier geomorphology analysis, Int. J. Comput. Math., 86, 275–285, 2009.
44. Miliaresis, G. and Kokkas, N.: Segmentation and object-based clas-sification for the extraction of the building class from LIDAR DEMs, Comput. Geosci., 33, 1076–1087, 2007.
45. Mondini, A. C., Marchesini, I., Rossi, M., Chang, K.-T., Pasquar-iello, G., and Guzzetti, F.: Bayesian framework for mapping and classifying shallow landslides exploiting remote sensing and to-pographic data, Geomorphology, 201, 135–147, 2013.
46. Muggleton, S.: Exceeding human limits, Nature, 440, 409–410, 2006.
47. Olden, J., Lawler, J., and Poff, N.: Machine Learning Methods Without Tears: A Primer for Ecologists, Q. Rev. Biol., 83, 171– 193, 2008.
48. Pearson, K.: On Lines and Planes of Closest Fit to Systems of Points in Space, Philos. Mag., 2, 559–572, 1901.
49. Pedregosa, F., Varoquaux, G., Gramfort, A., Michel, V., Thirion, B., Grisel, O., Blondel, M., Prettenhofer, P., Weiss, R., Dubourg, V., Vanderplas, J., Passos, A., Cournapeau, D., Brucher, M., Per-rot, M., and Duchesnay, E.: Scikit-learn: Machine Learning in Python, J. Mach. Learn. Res., 12, 2825–2830, 2011.
50. Peng, L., Niu, R., Huang, B., Wu, X., Zhao, Y., and Ye, R.: Land-slide susceptibility mapping based on rough set theory and sup-port vector machines: A case of the Three Gorges area, China, Geomorphology, 204, 287–301, 2014.
51. Quinlan, J.: Induction of Decision Trees, Mach. Learn., 1, 81–106, 1986.
52. Quinlan, J.: C4.5: Programs for Machine Learning, Morgan Kauf-mann, San Francisco, CA, USA, 1993.
53. Rumelhart, D., Hinton, G., and Williams, R.: Learning representa-tions by back-propagating errors, Nature, 323, 533–536, 1986.
54. Sambridge, M. and Mosegaard, K.: Monte Carlo methods in geo-physical inverse problems, Rev. Geophys., 40, 1–29, 2002.
55. Sammon, J.: A Nonlinear Mapping for Data Structure Analysis, IEEE T. Comput., C-18, 401–409, 1969.
56. Schaul, T., Bayer, J., Wierstra, D., Sun, Y., Felder, M., Sehnke, F., Rückstieß, T., and Schmidhuber, J.: PyBrain, J. Mach. Learn. Res., 11, 743–746, 2010.
57. Schmelter, M., Hooten, M., and Stevens, D.: Bayesian sediment transport model for unisize bed load, Water

Resour. Res., 47, W11514, doi:10.1029/2011WR010754, 2011.

58. Sivia, D.: Data analysis: A Bayesian tutorial, Oxford University Press, Oxford, UK, 1996.

59. Smith, M., Anders, N., and Keesstra, S.: CLustre: semi-automated lineament clustering for paleo-glacial reconstruction, Earth Surf. Proc. Land., 41, 364–377, 2016.

60. Szalay, A. and Gray, J.: Science in an exponential world, Nature, 440, 413–414, 2006.

61. Tamene, L., Park, S., Dikau, R., and Vlek, P.: Analysis of factors de-termining sediment yield variability in the highlands of northern Ethiopia, Geomorphology, 76, 76–91, 2006.

62. Valentine, A., Kalnins, L., and Trampert, J.: Discovery and analysis of topographic features using learning algorithms: A seamount case-study, Geophys. Res. Lett., 40, 3048–3054, 2013.

63. Wu, X., Kumar, V., Quinlan, J., Ghosh, J., Yang, Q., Motoda, H., McLachlan, G., Ng, A., Liu, B., Yu, P., Zhou, Z.-H., Steinbach, M., Hand, D., and Steinberg, D.: Top 10 algorithms in data min-ing, Knowl. Inf. Syst., 14, 1–37, 2008.

64. Yao, X., Tham, L., and Dai, F.: Landslide susceptibility mapping based on Support Vector Machine: A case study on natural slopes of Hong Kong, China, Geomorphology, 101, 572–582, 2008

CHAPTER 7

SEISMIC AND MECHANICAL STUDIES OF THE ARTIFICIALLY TRIGGERED ROCKFALL

P. Bottelin, D. Jongmans, D. Daudon, A. Mathy, A. Helmstetter, V. Bonilla-Sierra, H. Cadet, D. Amitrano, V. Richefeu, L. Lorier, L. Baillet, P. Villard, and F. Donzé

INTRODUCTION

Rockfalls are sudden events which are able to cause considerable loss of life and property, depending on their location and characteristics. As their occurrence is still difficult to predict, field measurements during rockfalls are scarce, which makes it difficult to gain information on their dynamics. In recent years, increasing attention has been paid to rockfall-induced seismic signals recorded by permanent arrays, which often constitute the only information available during the events (Dammeier et al., 2011). In parallel to seismic record processing, numerical modeling of rockfalls has developed considerably with the aim of gaining knowledge on the rock mass propagation and making the link be-tween landslide dynamics and generated seismic waves (e.g., Favreau et al., 2010). In this paper, the term "rockfall" de-scribes all events with a rock mass detaching and propagating on a steep slope, regardless of volume, dynamics or failure mode. When a specific rockfall type will be discussed the terms "rock avalanches", "mid-size rockfalls" and "block falls" will be employed for large (> 10^6 m^3), intermediate (10^3–10^6 m^3) and small volumes (< 10^3 m^3), respectively. Block falls are characterized by the independent movement of individual particles, in contrast to rock avalanches that generate extremely rapid flows of debris (Evans and Hungr, 1993).

Seismic records induced by rockfalls have specific characteristics that can be used for event detection, identification and location. Such signals generally last about tens to hundreds of seconds with an emergent onset and progressive decaying tail (Dammeier et al., 2011). The seismogram du-ration is linked to the duration of the propagation, whereas its shape reflects the dynamics (sometimes in several stages) of the event (Deparis et al., 2008; Hibert et al., 2011). Hibert et al. (2011) studied the rockfall activity that occurred in the Dolomieu crater (Piton de la Fournaise volcano, Reunion) during some periods in 2006 and 2007. These granular flow like events occurred in very destructured rocks and the signal envelope showed a smooth cigar shape similarly to snow avalanches (Biescas et al., 2003; Surinach et al., 2005) or lahars (Cole et al., 2009), reflecting progressive initiation, propagation and deposition phases. In contrast, other studies have shown that rockfalls have more irregular envelopes with some high-amplitude peaks, probably generated by blocks impacts (Deparis et al., 2008; Vilajosana et al., 2008; Helmstetter and Garambois, 2010; Dammeier et al., 2011). In most cases, no clear ***P*** and ***S*** wave arrivals are spotted and the seismogram is dominated by surface waves (Rousseau, 1999; Deparis et al., 2008; Helmstetter and Garambois, 2010). For all the above-mentioned events, most of the energy lies be-tween tenths of Hz and dozens of Hz, with predominant frequency about a few Hz (Dammeier et al., 2011; Hibert et al., 2011). The low-frequency limit is controlled by the event size whereas the high-frequency corner depends also on the distance to the source.

Rockfall-induced signals show characteristic time-frequency distributions, hence spectrograms are frequently used to gain insights into rockfall dynamics and to separate phases which occur at close times but with different frequency content. In their studies, Surinach et al. (2005) and Dammeier et al. (2011) reported that rockfall spectrograms show a typical triangular shape that could be related to source effects, material entrainment during propagation and/or progressive failure and individual block impacts (Norris, 1994; McSaveney and Downes, 2002). All these characteristics can be used for event detection at different scales (Helmstetter and Garambois, 2010; Yamada et al., 2012; Kao et al., 2012).

Citation: P. Bottelin, D. Jongmans, D. Daudon, A. Mathy, A. Helmstetter, V. Bonilla-Sierra, H. Cadet, D. Amitrano, V. Richefeu, L. Lorier, L. Baillet, P. Villard, and F. Donzé ,"Seismic and mechanical studies of the artificially triggered rockfall," Nat. Hazards Earth Syst. Sci., 14, 3175–3193, 2014, doi:10.5194/nhess-14-3175-2014.

Several studies have shown that rockfalls can be located using records from multiple seismic stations. When computed in near-realtime, this information can be of critical importance for rescue organizations. As no clear ***P*** and ***S*** waves are visible on seismograms, conventional earthquake location techniques cannot be applied. Specific methods have been developed, based on the seismogram envelope (Yamada et al., 2012, 2013), amplitudes (Battaglia et al., 2003), wave polarization (Vilajosana et al., 2008), signal cross-correlation (Lacroix and Helmstetter, 2011) and arrival times (Hibert, 2012). Due to the emergent seismogram onset, this last technique requires advanced picking methods (e.g., Baillard et al., 2013). In addition to rockfall detection, identification and location, numerous studies have focused on determining the fallen volume using seismic signals. Norris (1994) and McSaveney and Downes (2002) found a linear relation-ship between volume and seismogram amplitude for rock avalanches in USA and New Zealand, although the relation-ship appeared site-specific. In the same volume range, Yamada et al. (2012) observed a linear relationship between signal energy (square of the amplitude) and the square of the landslide volume. Dammeier et al. (2011) studied a set of 20 mid-size rockfalls in the Alps. The fallen volume was best correlated with signal duration, in good agreement with De-paris et al. (2008) that related signal duration with potential energy and run-out. However, Hibert et al. (2011) showed that the seismogram duration is controlled by numerous parameters, reflecting complex and simultaneous phenomena. They proposed a method for granular flows which uses the seismic energy ($\boldsymbol{E_s}$) dissipated over the entire signal to de-rive the potential energy and the fallen volume.

The conversion between initial potential energy ($\boldsymbol{E_p}$) and seismic energy ($\boldsymbol{E_s}$) has been studied by many authors. Weichert (1994) computed conversion ratios of about 10 ^{3}– 10 ^{7} for four rock avalanches and suggested that $\boldsymbol{E_s}$=$\boldsymbol{E_p}$ de-pends on the slope angle. For mid-size rockfalls, Deparis et al. (2008) observed ratios in the range 10 ^{3} to 10 ^{6}, in good agreement with Hibert et al. (2011), who found $\boldsymbol{E_s}$=$\boldsymbol{E_p}$ 10 ^{4}. In contrast, Vilajosana et al. (2008) determined a high conversion ratio (0.25) for the impact of a mass (75 m^3) on a terrace after a 120 m free fall. The $\boldsymbol{E_s}$=$\boldsymbol{E_p}$ conversion ratio appears then highly variable, and is probably a function of the fallen volume, block unitary size, fall type, slope angle and geotechnical ground conditions (Hibert, 2012).

Besides the experimental approach, numerical modeling techniques have been increasingly used to reproduce the behavior of rockfalls, pursuing the ultimate goal of propagation prediction (Hungr et al., 2005). Rockfall numerical simulations can be divided into two main families. First, granular numerical methods – which simulate a macroscopic behavior using microscopic interacting particles – were developed and have been applied to rockfalls with promising results (Dorren, 2003; Cleary and Prakash, 2004). These methods have been successfully compared with laboratory experimental tests (Banton et al., 2009; Richefeu et al., 2012) but still need validation on real cases. These methods are suitable to model a large range of rockfall volumes, including moderate volumes such as mid-size rockfalls. Another approach consists of simulating the fallen mass as a depth-averaged flow motion (Iverson, 1997; Mangeney Castelnau, 2003), which provides accurate results for large volumes of granular material such as rock avalanches (Pirulli and Mangeney, 2008; Moretti et al., 2012). For most of the studies, the efficiency of simulation models is assessed by comparison between the experimental and numerical deposit geometries.

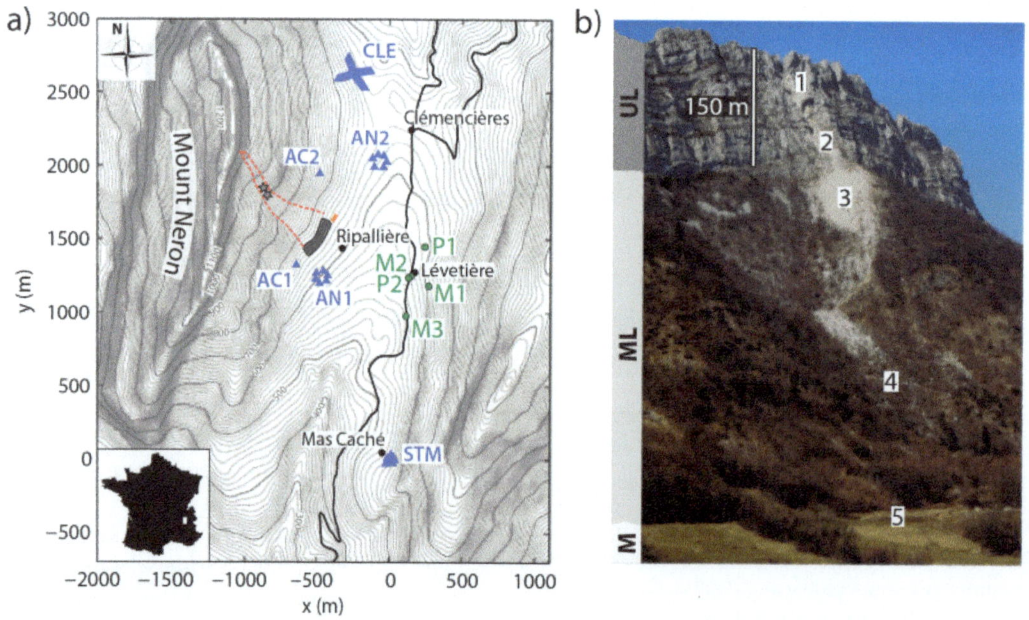

Figure 1. Panel **(a)**: map of the settings, the topography is shown by grey contour lines and the major roads are marked with black thick lines. Black dots point out the hamlets. The red dashed line demarcates the rockfall propagation zone for the provoked event. The footprint of the ditch–earthen barrier is marked out with the dark grey polygon, and the flexible rockfall barrier is shown by the thick orange line. Blue triangles point out the seismic sensors. The green dots indicate the location of cameras used for photogrammetric analysis (P1 and P2) or event shooting (M1 to M3). The two grey stars show the location of the block bounces studied in Sect. 3.4. Panel **(b)**: picture of Mount Néron taken from the southeast (Mas Caché) after the provoked event, with Urgonian limestone (UL), marly limestone (ML) and marls (M) indicated. (1) rockfall scar (light grey), (2) zone of mass impact in the lower part of the cliff, (3) scree deposits, (4) forest and (5) ditch and earthen barrier system.

Recent numerical models allow computing synthetic seismograms induced by large, granular rockfalls (i.e., rock avalanches), which allows a physical, quantitative and time-dependent comparison with the experimentally recorded ground motion (Favreau et al., 2010). Seismic signal generation processes are related to flow dynamics and trajectory changes (Surinach et al., 2001; Brodsky et al., 2003; Favreau et al., 2010; Schneider et al., 2010; Allstadt, 2013) as well as changes in coupling with topography (Caplan-Auerbach and Huggel, 2007; Mangeney et al., 2007). Thanks to the low-frequency content of the seismograms, the waveform is preserved even for large source–sensor distances allowing the signals to be inverted. The seismic source can be modeled as a single horizontal force (Kanamori and Given, 1982; Kanamori et al., 1984; Dahlen, 1993) in which magnitude and direction vary over time and depend on the fallen volume, basal friction and dynamic effects (La Rocca et al., 2004). Sliding mass models (Brodsky et al., 2003; Allstadt, 2013; Yamada et al., 2013) or numerical depth-averaged flows models (Favreau et al., 2010; Schneider et al., 2010; Moretti et al., 2012) showed their ability to reproduce the seismic source characteristics for volcanic, rocky or rock–ice debris flows.

The case study of the Néron site, where two mid-size rockfalls of similar volumes (~2 ′ 10^3 m^3) occurred at the same place, offers an opportunity to improve our understanding of rockfall kinematics and its link with induced seismic signals. Video clips and seismic signals recorded during the provoked rockfall were jointly interpreted to determine the chronology of the successive rockfall phases and to characterize them in terms of energy and propagation velocities. 3-D modeling of the rockfall propagation was performed and calibrated on the deposit observations to compare numerical propagation velocities with experimental data.

SITE DESCRIPTION

Mount Néron is located 5 km NW of Grenoble (Isère, France), at the southern end of the Chartreuse carbonate massif (Fig. 1a). The upper layer of its eastern slope is 150 m high (between 1150 and 1300 m in altitude) and made up of lower-Urgonian massive limestone (Fig. 1b). Below the cliff, a 40 inclined slope (1150–600 m) is first covered by scree deposits overlying marly limestone, as its lower part shows a gentle slope with marl outcrops (Gidon and Arnaud, 1978). In summer 2003, the *SAGE* engineering office detected a 4700 m^3 unstable rock column in the limestone cliff, between 1210 m and 1260 m altitude.

A field structural analysis of the cliff revealed that the back of the unstable column was bounded by a potential sliding plane (F1, Fig. 2a and b, Fabre et al., 2013) striking 10 N, 53–61 SE, while the bedding (S0) strikes 15 N and dips 45 W inwards (Table 1). Two families of fractures (F2 and F3) striking 70 N and 125 N were identified in the rock mass, with dips ranging from 80 S to vertical and 70 N to vertical, respectively. These discontinuities divide the column into many compartments, with the five larger of them reaching 200 to 2000 m^3.

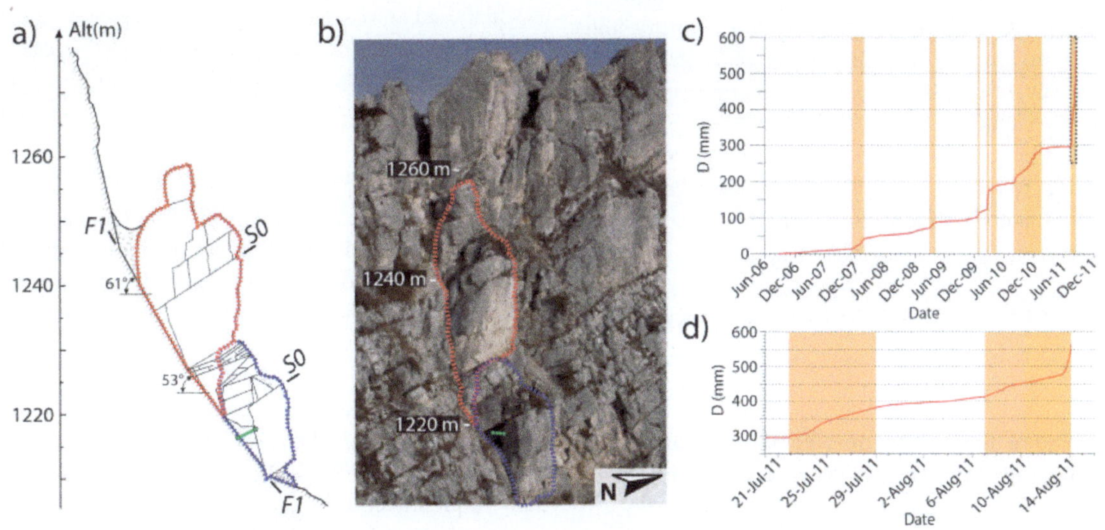

Figure 2. Panel **(a)**: 2-D schematic cross-section of the unstable column. The cliff is indicated with the hatched line, discontinuities are shown with black lines. Panel **(b)**: picture of the unstable rock column before the two rockfalls, taken from the southeast. Altitude is given with the white marks. For **(a)** and **(b)**, blue and red dashed lines demarcate the natural and blasted compartments, respectively. The location of the extensometer E1 installed between the rock mass and the lower part of the unstable column is shown by the green line. Panel **(c)**: displacement versus time measured by E1, from the setup in summer 2006 until the natural rockfall on 14 August 2011 (red curve). The black dashed rectangle shows the extent of zoom **(d)**. Panel **(d)**: zoom of **(c)** between 20 July and 14 August 2011. For **(c)** and **(d)**, the crises periods are outlined in orange.

A geotechnical study conducted in 2004 showed that the Ripaillère hamlet (70 inhabitants) was directly threatened by potential rockfalls (Fig. 1a). Consequently, four extensometers and one tiltmeter were installed in summer 2006 to monitor the movements of the rock column. The measurements were continuously recorded and transmitted to an alert center at the *SAGE* engineering office, which was in charge of the monitoring. In 2007, a ditch and a 9 m high by 300 m long earthen barrier were built up at the toe of the slope to protect the hamlet (grey patch in Fig. 1a). From 2006 to 2011, the extensometer installed in the lower part of the unstable column (E1, Fig. 2c) exhibited a steady displacement trend disrupted by some sharp rises, which were some-times associated with small rockfall events. Following a period of heavy rainfalls

(130 mm over 9 days), a strong acceleration in displacement occurred between 22 and 28 July 2011 (Fig. 2c and d) with a velocity reaching 12 mm day $^{-1}$ before decreasing to 3 mm day $^{-1}$. However, between 7 and 13 August 2011, centimetric daily opening rates were once again reached with peaks close to 50 mm day [1] on 13 August 2011. In the evening of the same day, authorities decided to preventively evacuate the hamlet. After an additional 30 mm displacement, a rockfall occurred at 02:30 UTC on 14 August.

Table 1. Mean discontinuity orientation revealed by field (left column) and stereoscopic (right column) stereographic analyses

	Field analysis		Stereoscopic analysis	
	Strike (°)	Dip (°)	Strike (°)	Dip (°)
S0	15 N	45 W	05 N	30 W
F1	10 N	53–61 SE	0 N	60 E
F2	70 N	80–90 S	87 N	80 S
F3	125 N	70–90 N	–	–

Only the 2000 m^3 lower part of the column (demarcated with the dashed blue line, Fig. 2a and b) detached from the cliff and fell freely about 100 m. Although most of the fallen mass was quickly deposited on the upper part of the slope, some approximately 15 m size blocks (3 to 10 m^3) propagated down the slope and were trapped in the ditch (Fig. 3d) after a run-out of 800 m. The 2700 m^3 upper compartment of the rock column (circled with the red dashed line in Fig. 2a and b) remained attached to the cliff. Owing to its low residual stability, the decision was taken to blast it. An additional 60 m long, flexible rockfall barrier was built at the northern extremity of the earthen barrier to extend the protection area (orange dash in Fig. 1a). Seismometers, video and photo-graphic cameras were deployed in the area to record the artificially triggered rockfall. The blasting, undertaken by *Hydrokarst* company, occurred on 13 December 2011.

THE ARTIFICIALLY TRIGGERED ROCKFALL

Description

The remaining 2700 m^3 unstable compartment (delineated with the red dashed line, Fig. 2a and b) was blasted on 13 December 2011 at 02:01:55 UTC, using 400 kg of ex-plosives. As seen in Fig. 3a, the size of the blasted blocks strongly varies, from centimetric to metric scale. The largest blocks are located at the bottom of the compartment owing to a larger spacing between the blasting boreholes. From video clips, several phases of the rockfall could be identified and time-referenced (Fig. 3e). The blasted volume (Fig. 1b, spot 1, Fig. 3a) free-fell for 3 to 6 s before impacting the ground at the toe of the cliff (spot 2) and the top of the rocky scree (spot 3) about 1 s later. Smaller blocks dropped continuously from the cliff for about 30 s. Simultaneously, most of the fallen rock mass settled rapidly on the upper part of the slope (rocky scree), generating a dense aerosol. Only a few large blocks propagated downwards across the forest (spot 4 in Fig. 1b), bouncing on the ground and felling trees. Seven meter-size blocks (0.8 to 12 m^3*)* reached the toe of the slope and were stopped by the ditch–earthen barrier system (spot 5 in Figs. 1b and 3d). One of them (block 6 in Fig. 3d) broke into two pieces when impacting the upper rock wall of the earthen barrier (Fig. 3c), about 45 s after the initial blast. Another block propagated further north in the slope (Fig. 3b), and ended in the flexible rockfall barrier (orange line in Fig. 1a). There was no sign of movement on the videos later than 90 s after the blast.

Instrumentation

A permanent seismic array (STM, Fig. 1a) located about 2.5 km southeast from the site recorded both the natural and the provoked rockfalls. This array was made up of six vertical short-period (2 Hz) sensors with one three-component (3C) seismometer at the center. Two additional short-period 3C sensors (AC1 and AC2); two 6-channel, vertical sensors arrays (AN1 and AN2) and one 48-channel, 4.5 Hz vertical geophone array (CLE) were also deployed at the toe of the slope, to record the provoked rockfall (see location in Fig. 1a).

Photographs of the cliff were taken both before and after the provoked event from spots P1 and P2 (Fig. 1a), in or-der to carry out photogrammetric analysis (Poropat, 2001). A *Canon* EOS 5D Mark II digital camera with 21.1 megapixel resolution was used in combination with an Electro Focus 135 mm f/2L Ultra Sonic Motor lens. Despite the large distance to the rock cliff (more than 1000 m), it provided good sharpness, contrast and color rendition. A laser range finder was used to pin control points on the rock cliff for georeferencing the photographs.

A *Canon* EOS 500D digital camera with a 58 mm lens was installed at M1 (Fig. 1a) to shoot the whole propagation of the blasting event, providing 25 frames per second, 1920 1080 pixel videos. In addition, one pair of similar digital cameras was installed about 40 m lower in altitude to get the trajectories of the fallen blocks (M2 and M3, Fig. 1a). The positions of all the instruments were determined by GPS.

Aerial lidar acquisitions were conducted both before and after the provoked event to obtain a high-resolution DSM (digital surface model) of the site. The first scan was conducted by *Sintegra* company on 29 September 2011 with a *Riegl* 2-D laser scan LMS-Q560 brought together with an inertial measurement unit (IMU) *IXSEA* and a dual-frequency GNSS receiver. The scan after the blast was acquired on 27 April 2012, using *Hélimap System* composed of a *Riegl* 2-D laser scan, a *Hasselblad* digital camera, an *iMAR* IMU and a dual-frequency GNSS receiver.

Volume Estimation

The volume of the artificially triggered rock compartment was estimated using the aerial lidar scans and photographic images. lidar point clouds of the site area (0.7 km^2) acquired before and after the provoked rockfall were treated using *3DReshaper* software (www.technodigit.com). The points corresponding to the ground and cliff surface were kept, filtering points with low reflectivity coefficient (reflections on the vegetation) and manually rejecting aberrant points. The 2.4 million point large cloud was meshed with 469 611 tri-angles. The blasted volume was determined by subtracting the DSM before (dark grey) and after the blast (light grey, Fig. 4a), leading to a 2570 m^3 estimation. Two close views of the blasted column are shown in Fig. 4b and c.

Detailed 3-D images were also obtained by processing stereographic pairs of photographs using *Sirovision* software (CSIRO, 2010a), and were georeferenced within the local co-ordinate system using control points on the rock slope. Two photographs of the same slope were taken at a distance close to 1000 m. The cameras were positioned to obtain a baseline ratio between camera positions and the distance to the face of 1 V 7 or 1 V 6. Adjusting the 3-D images collected before (dark grey) and after (light grey) the blast, the volume difference and the displacement variation into the selected area are computed (Fig. 4d and e). The potential accuracy of the method given the equipment used in this study equates to the resolution of the range measurement at 150 m, i.e., 5 mm for a displacement within the 2-D image of 1 pixel if the baseline is 1 V 7 (e.g., Poropat, 2006, Wolter et al., 2013). In optimum conditions (i.e., plane accuracy), the corresponding precision of the range measurement is approximately 3 cm. However, the final model precision is more controlled by the accuracy of the georeferencing procedure than by the predicted photogrammetric precision stated above.

In the present case, with a distance of 1000 m, the accuracy of the laser range was found to be close to 1 m, leading to a precision of the 3-D georeferenced image close to the meter as well. The blasted volume was estimated at 2380 m^3, which is consistent with the volumes estimated by the consulting company *SAGE* (2700 m^3) and determined from lidar data (2570 m^3). In this case, the aerial lidar provided a more complete topography than the photogrammetric technique because it could have access to the backside of the hanging block (see Fig. 2), which was not visible from the places where the pictures were taken. However, the photogrammetric technique offers the advantage of being much less expensive and more flexible.

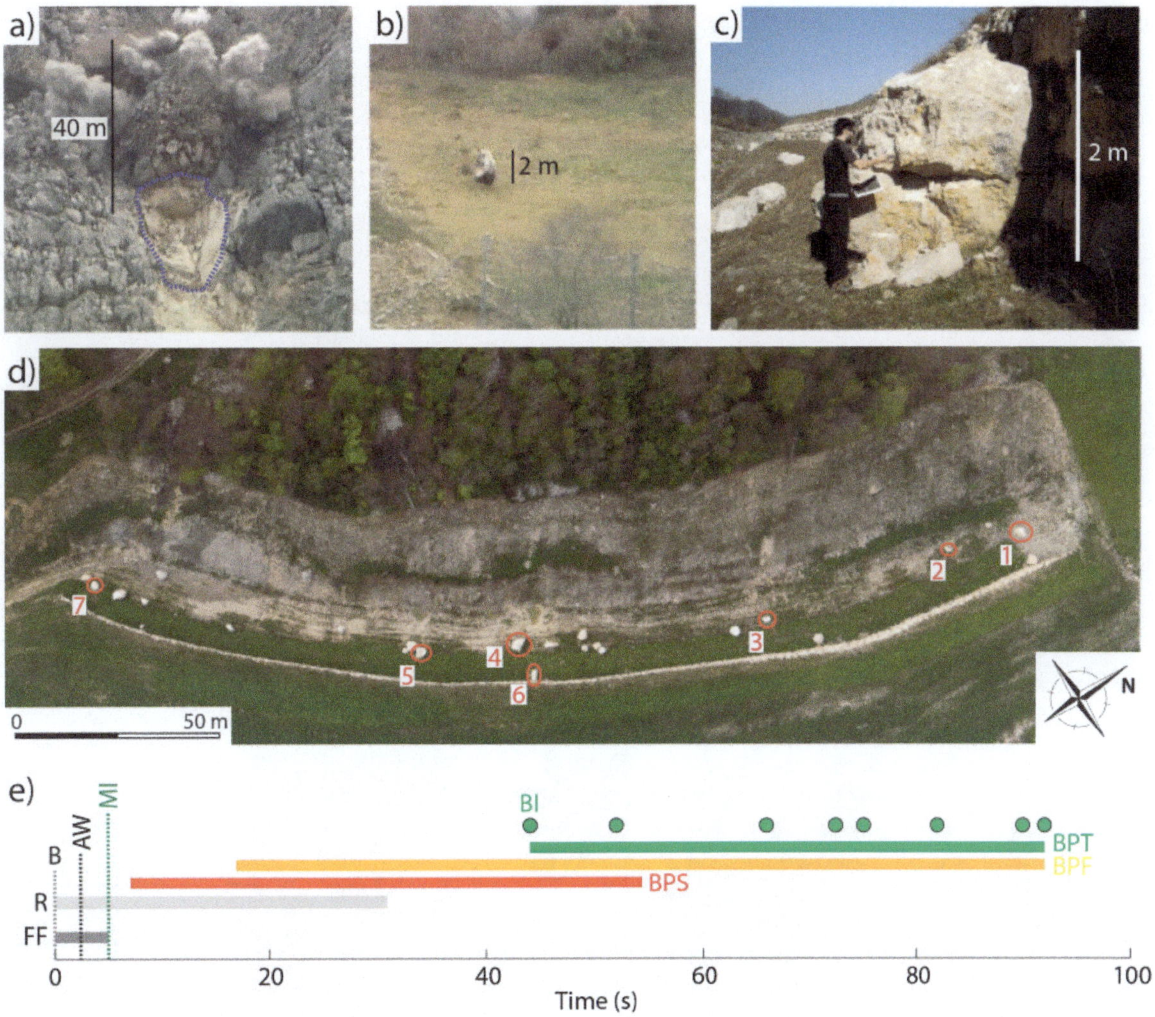

Figure 3. Pictures of the different phases of the blasting event. Panel **(a)**: initiation phase. Picture of the unstable remaining 2600 m^3 just after its blasting. The blue dashed line shows the footprint of the previous natural rockfall. Panel **(b)**: propagation phase. Picture of a 3 m^3 isolated block that propagated down the slope into the flexible rockfall barrier, visible at the bottom right. Panel **(c)**: deposit phase. Picture of block 6 (cf. Fig. 3d) that split into two parts when impacting the rock-reinforced uphill side of the earthen barrier. Panel **(d)**: deposit phase. Picture taken on 10 May 2012, after the natural and provoked rockfalls. The 15 blocks from the natural event are not numbered, with volumes ranging from 3 to 10 m^3. The blocks generated by the provoked event are circled in red and numbered from 1 to 7, with volumes of 3, 0.8, 4, 12, 9, 9 and 4 m^3, respectively. Panel **(e)**: chronology of the provoked rockfall derived from the video clips. From the left to the right: blast (B), free fall (FF), air wave (AW), main mass impact (MI), rocks falling from the cliff (R), block propagation in the scree (BPS), in the forest (BPF), at the toe of the slope (BPT) and block impacts in the ditch–earthen barrier (BI).

Discontinuity Pattern

Structural features of the cliff were extracted from the 3-D digital image taken after the provoked rockfall, using the *Sirojoint* software (CSIRO, 2010b). The planes to process for discontinuity orientation characterization are manually selected (colored areas in Fig. 5a) and the corresponding poles of each facet are shown by colored triangles on a stereographic diagram using a Wulff lower hemisphere projection (Fig. 5b). Three sets of discontinuities were identified, and the average pole (square marker) and average plane (circle line) for each family are drawn in Fig. 5b and listed in Table 1. The bedding (S0, in green) strikes 353–11 N with 19–53 W inward dipping (mean 05 N, 30 W), while the two fracture sets F1 (blue) and F2 are oriented 322–24 N, with a dip of 53–73 E (mean orientation: 0 N, 60 E) and 84–93 N, with a dip of 77–86 S (mean orientation: 87 N, 80 S), respectively. Despite the metric resolution of these measurements, they are in good accordance with the structural analysis from the consulting company *SAGE* (see Table 1 for comparison), which supports the reliability of the photogrammetric data. However, the fracture set F3 observed on field on the outcrops was not identified by the photogram-metric analysis. The limiting factors appear to be the resolution of the 3-D images and their manual processing, potentially leading to less accurate or fragmentary structural information. Nevertheless, the photogrammetric technique showed its ability of retrieving the main discontinuity pattern of the cliff even from such remote viewpoint, over 1 km in this case (see also Sturzenegger and Stead, 2009).

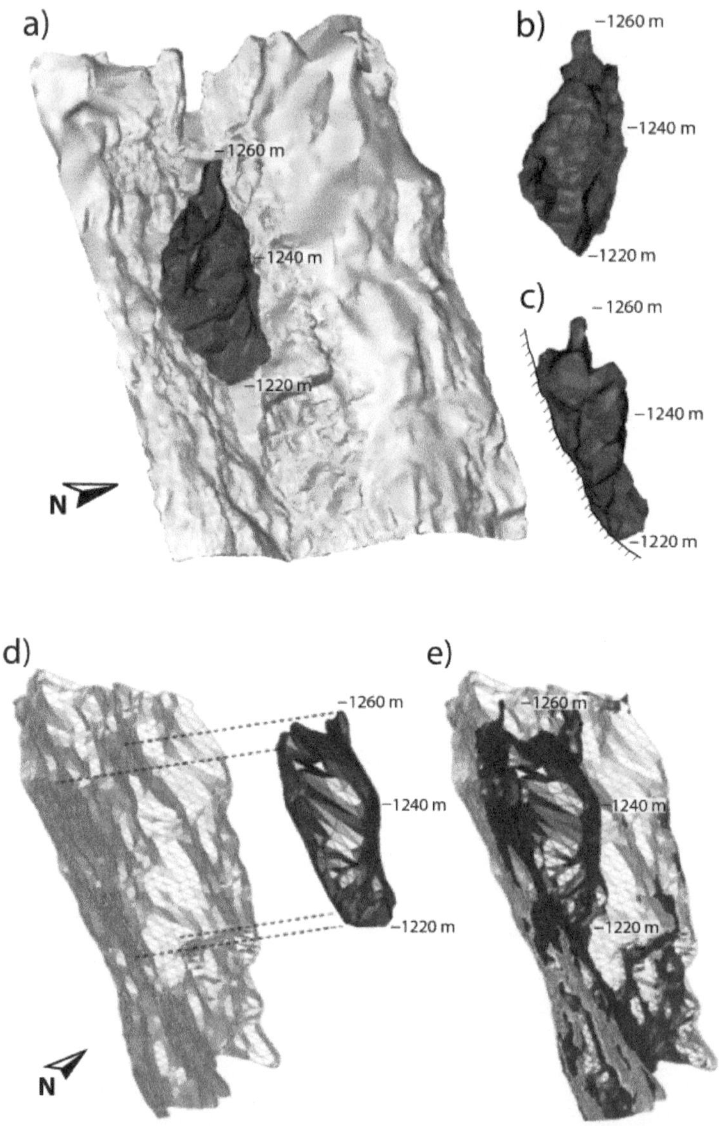

Figure 4. Panel **(a)**: Discrete element method (DEM) before and after the blasting (in dark and light grey, respectively). Views of the unstable compartment **(b)** from east to west and **(c)** from south to north with the elevation shown by the black marks. The hatched line draws the cliff section in **(c)**. Panel **(d)**: 3-D images collected before (dark grey) and after (light grey) the blasting event. Panel **(e)**: Sirovision model with superimposition of the 3-D images for volume determination.

SEISMIC SIGNAL ANALYSIS AND EVENT IDENTIFICATION

Figure 6 shows the seismograms and spectrograms of the vertical component recorded by seismic station STM (2.5 km away) for both the natural (a) and artificially triggered (b) rockfalls. While the volume estimated for the second event (2570 m^3) is slightly larger than for the first one (2000 m^3), the peak amplitude of the seismograms and the signal duration (

100 s) are similar for the two events. The envelope of the first signal is smoother (Fig. 6a), with a progressive rise in amplitude. The highest amplitude is observed about 10 s after the first wave arrival, and then the amplitude decreases and remains roughly constant for 50 s with some seismic pulses. For the triggered event (Fig. 6b), there is an impulsive start followed by a spike 7 s later owing to the acoustic waves generated by the explosion. For the time period between 15 and 40 s, the signal amplitude is slightly smaller than for the natural event. There is a clear peak at 45 s after the blast, synchronously with the first blocks impacting the toe of the slope. Both natural and artificially triggered event spectrograms show an irregular envelope (Fig. 6c and d), with some energetic seismic pulses that probably result from multiple sources and the propagation of different waves (Deparis et al., 2008; Dammeier, 2011). Most of the energy lies below 20 Hz, in good agreement with the range tenths of Hz to dozens of Hz observed for other rockfalls (Dammeier et al., 2011; Hibert et al., 2011). Excluding the acoustic waves, the main energetic phases in the signals are the ground impact following the free fall and later events which could correspond to impacts of rock blocks at the slope toe. This issue is discussed below by comparing close seismograms and the videos. Figure 7 shows the seismic signals and spectrograms of the vertical component recorded for the artificially triggered rockfall by stations AC1 (a, c) and AC2 (b, d). The signal amplitude is generally larger for station AC2, which is located closer to the blast. The seismic phases are interpreted in light of the event sequence determined from video clips (Fig. 3e). Both signals show an impulsive start and high amplitudes for the first 2 s (arrow 1 in Fig. 7) associated with the blasting. The high-frequency acoustic wave arrives about 3 s after the blast (arrow 2), followed by the large impact of the fallen mass at the base of the cliff beginning about 3 s later (arrow 3). There is a clear decrease in signal amplitude for the time period between 15 and 30 s (arrow 4), when most of the rock mass gradually settles in the rocky scree, the latter acting probably as a mat-tress and attenuating the seismic waves transmitted into the ground. The amplitude then rises gradually from 30 to 45 s (arrow 5), especially on AC1 located at the toe of the slope. This probably results from seismic source movement, as the blocks propagate downwards and approach the sensors.

At station AC1, the largest amplitude in the seismogram is observed 45 s after the blast (arrow 6). It corresponds to block 6 that hit the uphill rock reinforced wall of the earthen barrier (Fig. 3d and c). This seismic peak is also visible at AC2, but is weaker due to greater source–sensor distance. Several pulses with large amplitudes (notably for AC2) occur from 60 to 90 s (arrow 7) and are associated with the late propagation of blocks along the slope, predominantly in its northern part. Finally, seismogram amplitudes for both sensors decrease rapidly for the time period after 90 s, in good accordance with the videos. For the period after than 15 s, spectrograms (Fig. 7c and d) show an overall triangular shape that has been commonly observed in rockfalls and snow avalanches (Suriñach et al., 2005; Vilajosana et al., 2008) and related to material propagation effects. The energy lies mainly in a frequency range between 1 and 50 Hz, which is higher than that recorded by STM because of less attenuation at stations closer to the source. For the two main dominant phases, the energy mainly lies in the 1–20 Hz frequency range, with spectral peaks between 2 and 6 Hz (Fig. 7c and d).

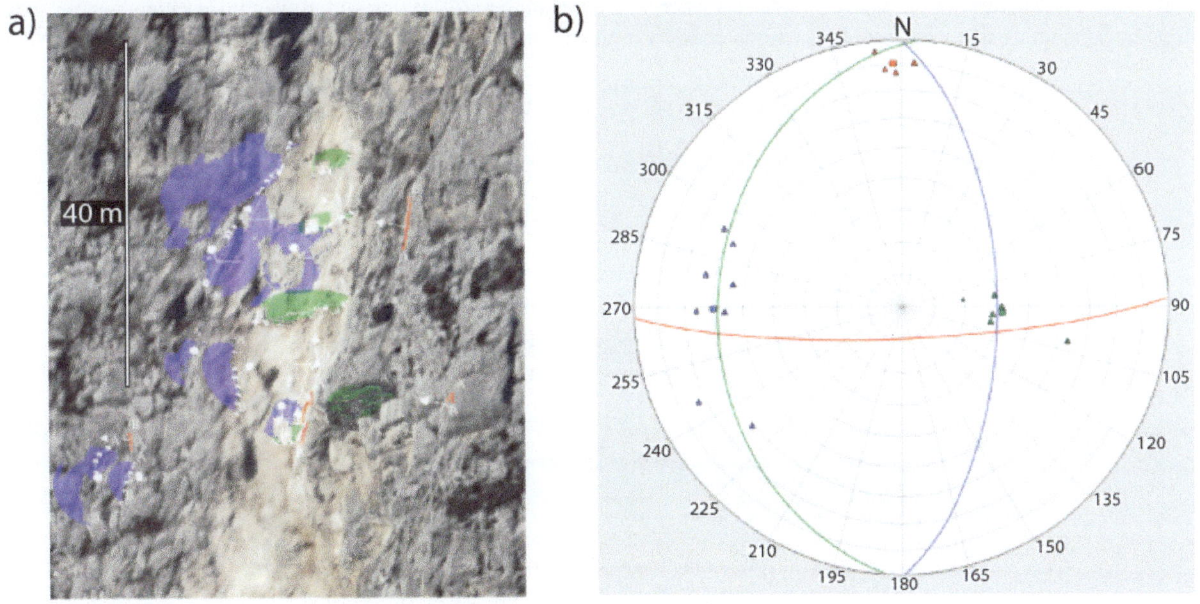

Figure 5. Panel **(a)**: 3-D picture of the cliff after the provoked rockfall. The planes manually selected for discontinuity orientation characterization are shown with the colored areas, along with their respective normal vector (white arrow). The mean cliff plane strikes 15 N. Panel **(b)**: stereographic diagram of **(a)** using Wulff lower hemisphere projection. The pole of each facet is shown by a triangle, whereas the average pole is pointed out with a square marker. The average plane is drawn with a circle line. The bedding (S0, in green) as well as F1 (blue) and F2 (red) fracture sets are retrieved.

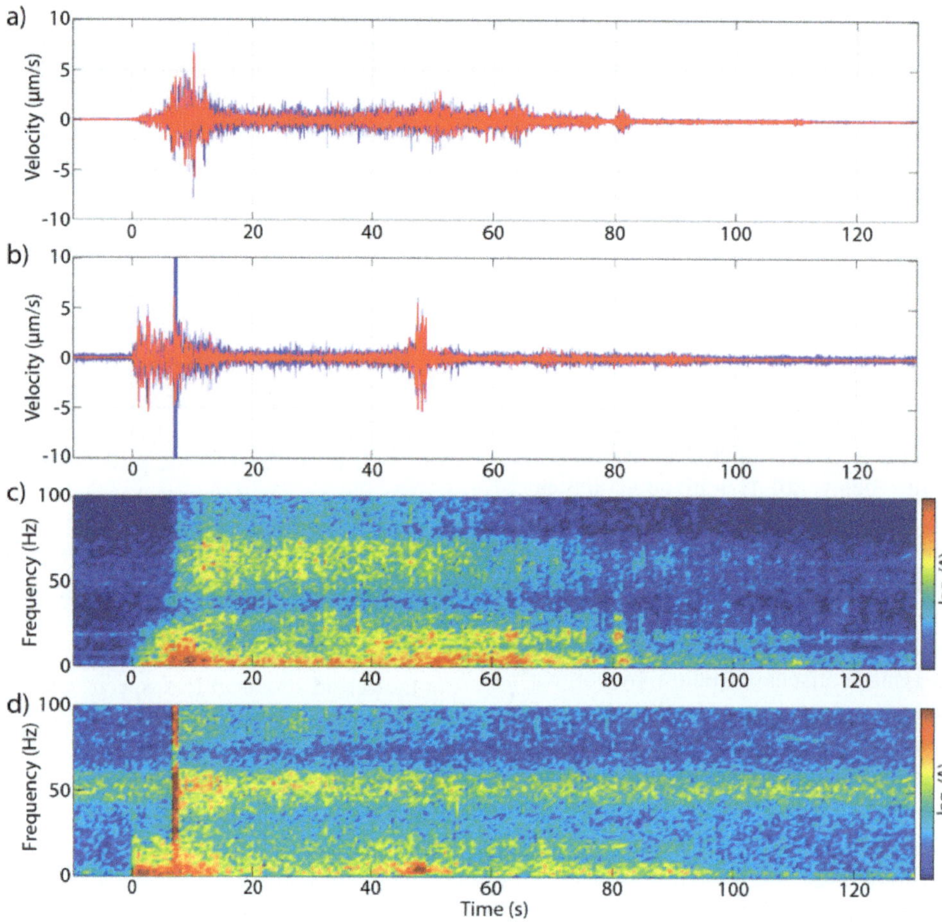

Figure 6. Seismograms (**a**, **b**, in blue) and spectrograms (**c, d**) of the vertical component seismic signals recorded by station STM for the natural (**a, c**) and artificially triggered (**b, d**) rockfalls. The seismograms filtered in the 1–20 Hz range for magnitude estimation are shown in red. Spectrograms are computed using a fast Fourier transform of 256 samples and are shown using the same logarithmic color scale for both events.

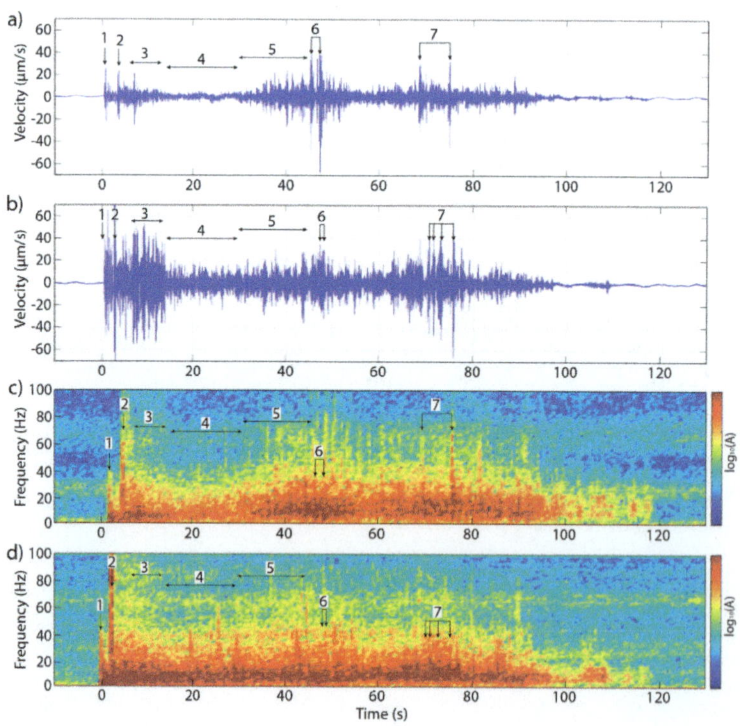

Figure 7. Seismograms (**a, b**) and spectrograms (**c, d**) of the vertical component seismic signals recorded by station AC1 (**a, c**) and AC2 (**b, d**) for the artificially triggered rockfall. Spectrograms are computed using a fast Fourier transform of 256 samples and are shown using the same logarithmic color scale for both stations.

EVENT CHARACTERIZATION AND LOCALIZATION

Magnitude

Natural and Provoked Rockfalls

The magnitude of the natural and provoked rockfalls was estimated from the peak amplitude of the seismic signals detected at seismic station STM, located about 2.5 km south from the blast. The relationship between distance, peak amplitude and magnitude proposed by Lacroix and Helmstetter (2011) for the Séchilienne rockslide (located 20 km away) was employed. This relation was calibrated using magnitudes computed by SISMALP seismic network for regional earthquakes (http://sismalp.obs.ujf-grenoble.fr). The shots performed at Séchilienne for tomographic purposes were also used to better constrain this relationship for short distances. The magnitude is computed from the peak amplitude A (in µm s^{-1}) of the signal filtered between 1 and 20 Hz (Fig. 6) and from the distance d (in km) to the source using the relationship $M = \log^{10} [A \exp(d/d_0)/d^n]$, with $n = 0.95$ and $d_0 = 151$ km. The magnitude is 1.14 for the natural event and 1.05 for the triggered one, considering the peak amplitude generated by the impact of the mass on the ground ($t \approx 6$ to 10 s, Fig. 6). These values are in good agreement with values of magnitude and volumes reported by Deparis et al. (2008) for other rockfalls in the French Alps. Applying the magnitude (M) – seismic energy (E_s) relationship $\log^{10}E_s = 1.5M + 4.8$ (Kanamori, 1977) gives $E_s = 3.2 \times 10^6$ and 2.4×10^6 J for the natural and provoked rockfall, respectively. Assuming that the potential energy E_p of the fallen mass is fully converted into kinetic energy E_k during free fall, the ratios between the seismic energy released by the impact and the kinetic energy are $8\times10{-}4$ and 4.7×10^{-4} for the natural and provoked events, respectively. These values lie in the conversion ratio range commonly found in the literature (see introduction). Magnitude was also computed for the impact of block 6 into the earthen barrier ($t \approx 45$ s, Fig. 6) from the provoked rockfall seismogram recorded at station STM, leading to $M = 0.98$. Applying the same magnitude – seismic energy relationship yields $E_s = 1.9\times10^6$ J.

Subsidiary Rockfall

The seismic network remained in operation for 3 days after the blast. The continuous seismic signal recorded by stations AN1 and AN2 were analyzed using the automatic detection method of Helmstetter and Garambois (2010) which is based on the spectrogram of the signal. During this time, one subsidiary rockfall was detected which occurred on 13 December 2011 at 23:50 UTC, i.e., 9 h after the shot.

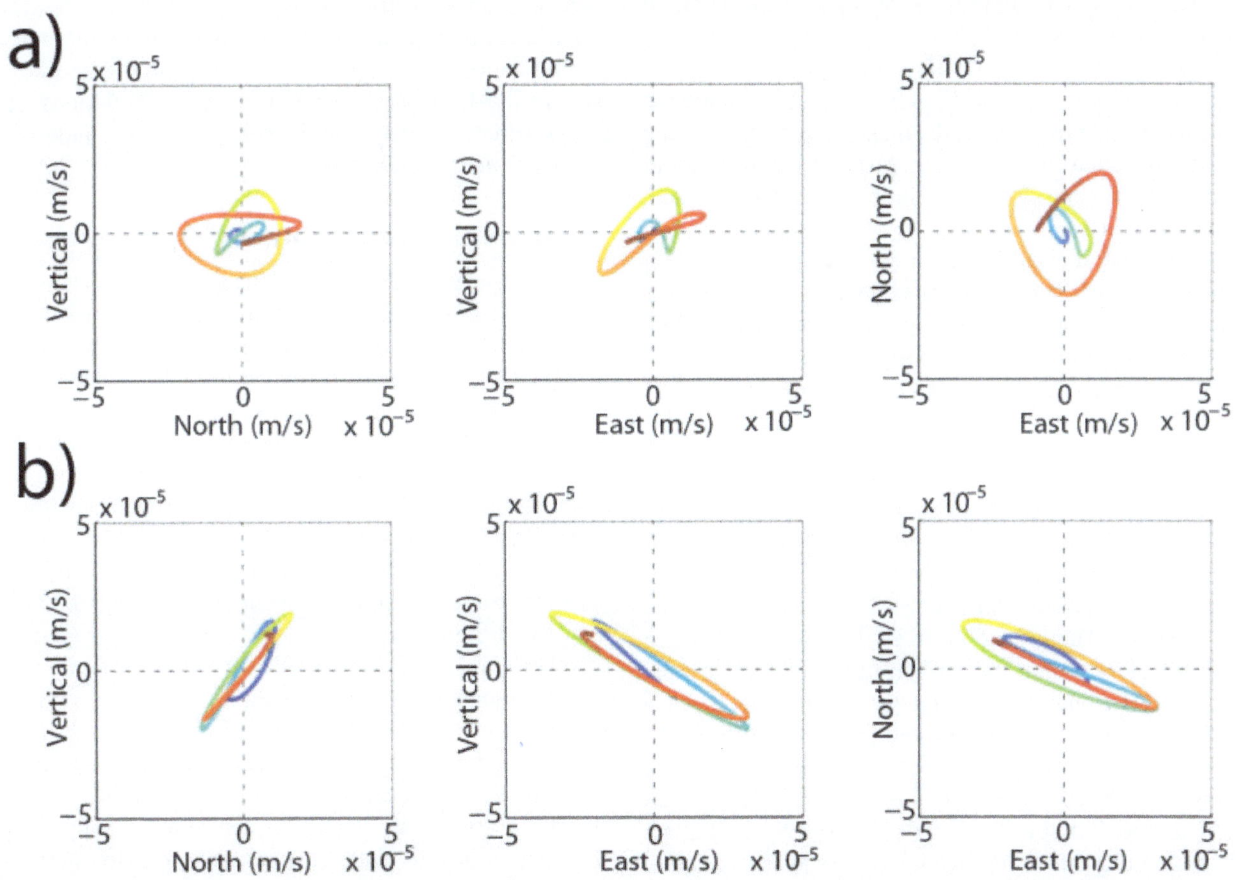

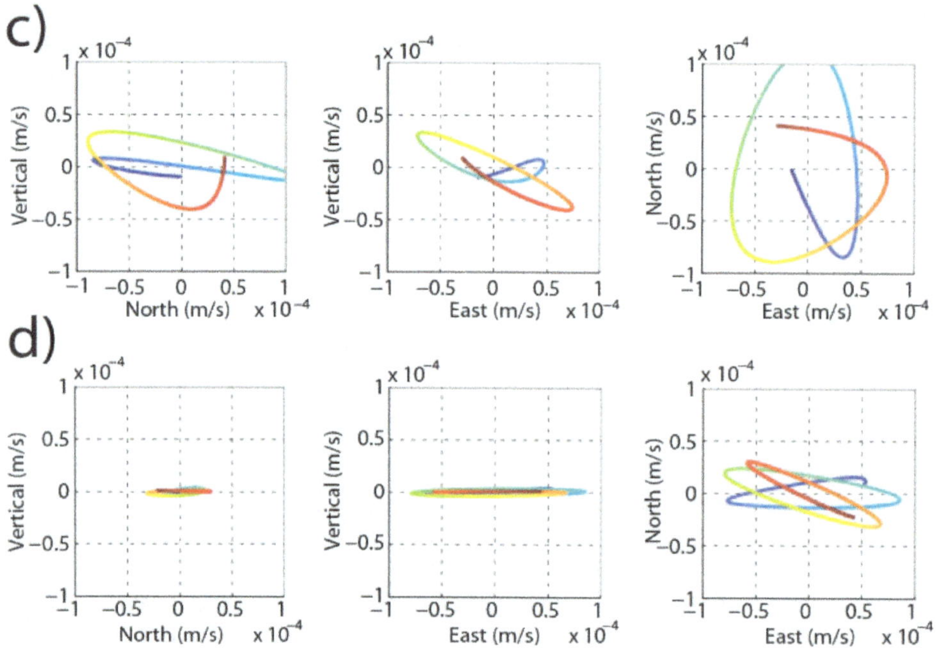

Figure 8. Polarization analysis using the 3C sensors AC1 (**a** and **b**) and AC2 (**c** and **d**) for both the impact at the toe of the cliff (**a** and **c**) and the impact of block 6 into the earthen barrier (**b** and **d**). The particle motion is shown by the continuous line, whose color indicates the time (from blue to red). The axes show the motion velocity in the vertical, north or east direction.

The seismic signal amplitude of this event (ground impact) at station AN1 was 140 times smaller than the provoked rockfall. Under the hypothesis that these two events originate from the same area of the cliff, we can estimate the volume of the second event from the magnitude of the seismic signal. Assuming that the seismic moment is proportional to the rockfall volume and that the seismic moment scales with magnitude M as $10^{1.5\,M}$ (Hanks and Kanamori, 1979), the volume of the second event is estimated to about 1.5 m^3. This value agrees with the field observations made on the cliff scar by the consulting company (*SAGE*, personal communication, 2012). These results indicate the ability of seismic arrays for monitoring rockfalls and estimating the fallen volumes.

Polarization Study for the Main Identified Phases

For the provoked rockfall, polarization analysis was conducted on the seismograms in order to identify the nature of the waves (compressional, shear or surface waves) recorded by the 3C sensors. The ground impact following free fall and the impact of block 6 into the earthen barrier, which are the two main energetic phases in the signal, were studied. Signals from AC1 and AC2 were windowed and band-pass filtered in the 2–6 Hz range, where most of the energy lies (see Fig. 7c and d). Figure 8 shows the particle motion at the two stations AC1 (a, b) and AC2 (b, c). The signal generated by the impact of the mass on the ground after free fall (a, c) shows a complex pattern with no specific polarization regarding the visualization plane, probably due to multiple impacts and complex seismic paths. In contrast, the signals generated by the impact into the earthen barrier (b, d) exhibit strong linear ground motion at both sensors.

At AC2, the movement is purely horizontal and predominantly oriented along the east–west direction, whereas there is a slight vertical component of motion at AC1 with vibration striking ESE–WNW. These linear, mainly horizontal ground motions appear perpendicular to the source–sensor path in the horizontal plane (Fig. 9a). Simulations of wave propagation are carried out in Sect. 6 to better understand the origin of such polarization pattern.

Location of Seismic Sources

The beam-forming method of Lacroix and Helmstetter (2011) is applied to locate the seismic source. This method has been developed for locating seismic sources when the first arrival is difficult to pick with precision and allows monitoring the position of a moving source over time. In this method, the position of the source and the seismic wave velocities are inverted by maximizing the temporal correlation between all traces after shifting each trace by the travel time. The inversion starts with a 100 m cell size grid search and an initial seismic wave velocity. For each grid point, the travel times between this point and all sensors are computed. The traces are then shifted by this travel time and the average weighted correlation between all couples of traces is computed. The weight associated with each couple of traces decreases with the distance between sensors to give more weight to nearby sensors with a better intertrace correlation. The weights were also tuned so that each seismic antenna has the same weight, otherwise the results would be controlled mainly by station CLE which has 48 channels compared to 6 for stations AN1 and AN2. The grid point with the largest correlation is then used as the initial point for a simplex optimization of intertrace correlation. Although the seismic velocity is likely very heterogeneous, a uniform velocity is assumed for simplicity and fixed to 2000 m s^{-1} for seismic waves and 340 m s^{-1} for acoustic waves.

This method was applied to different time windows, starting with using the first 2.5 s of the seismic signals (explosive blast) filtered in the range 3–20 Hz. The inverted apparent velocity is 1750 m s^{-1}, corresponding probably to ***S*** waves or surface waves, which are the most energetic. The estimated source is located 110 m to the west of the actual location. The location uncertainty can be estimated from the area where the correlation is within 5 % of the peak correlation (Lacroix and Helmstetter, 2011). In this case, the uncertainty is 555 m (blue contour in Fig. 9a). This limited resolution is due to the weak correlation between signals and to heterogeneities of seismic wave velocities. The same method was then applied to acoustic waves produced by the explosion. The signal was selected in the time window between 2.65 and 3.85 s after the blast and filtered in the range 20–80 Hz to remove lower-frequency ***P*** and ***S*** waves. We assume an initial velocity of 340 m s^{-1} and obtain an inverted value of 337 m s^{-1}. In this case, the estimated location is very close to the actual one, the location error is of 17 m and the estimated uncertainty is 32 m (magenta contour in Fig. 9a). The accuracy is much better for acoustic waves because the air velocity is much more uniform, even if the correlation quality between traces is much smaller for acoustic waves than for seismic waves (the weighted average inter-trace correlation is 0.8 for seismic waves and 0.37 for acoustic waves).

We then used the beam-forming method to monitor the source position over the full duration of the seismic signal, using a sliding time window of 1 s. The velocity was fixed to 1750 m s^{-1}, and the signal was filtered between 3 and 20 Hz. Because of the limited resolution of the method, the source location was imposed along the mean path of the rockfall (dashed black line in Fig. 9a). For each time window, the point that maximizes the inter-trace correlation is computed and the estimated sources are shown as colored points in Fig. 9a. Figure 9b presents the distance ***d*** along the path from the blast point as a function of time (top), with a simplified chronology of the provoked rockfall phases from videos (bot-tom). During the first 30 s of signal, several sources were located very close to the explosion (***d*** ~ 0 m). They were probably generated by the blast (B, Fig. 9b) and by the blocks continuously falling from the cliff (R). Then (15–55 s), the blast point–source distance linearly increases with time from ***d*** = 0 to approximately 1000 m, with an average velocity of 22 m s^{-1}. This phase corresponds to the block propagation in the scree (BPS) and in the forest (BPF) spotted on the videos. After this ~55 s time limit, the source location is relatively stable with some scattering. It matches the time when the 7 m size blocks arrived at the toe of the slope (BPT) and were trapped in the ditch–earthen barrier. Although there were still blocks propagating on the upper part of the slope, the seismic signal is likely dominated by the propagation front which is closer to the seismic sensors.

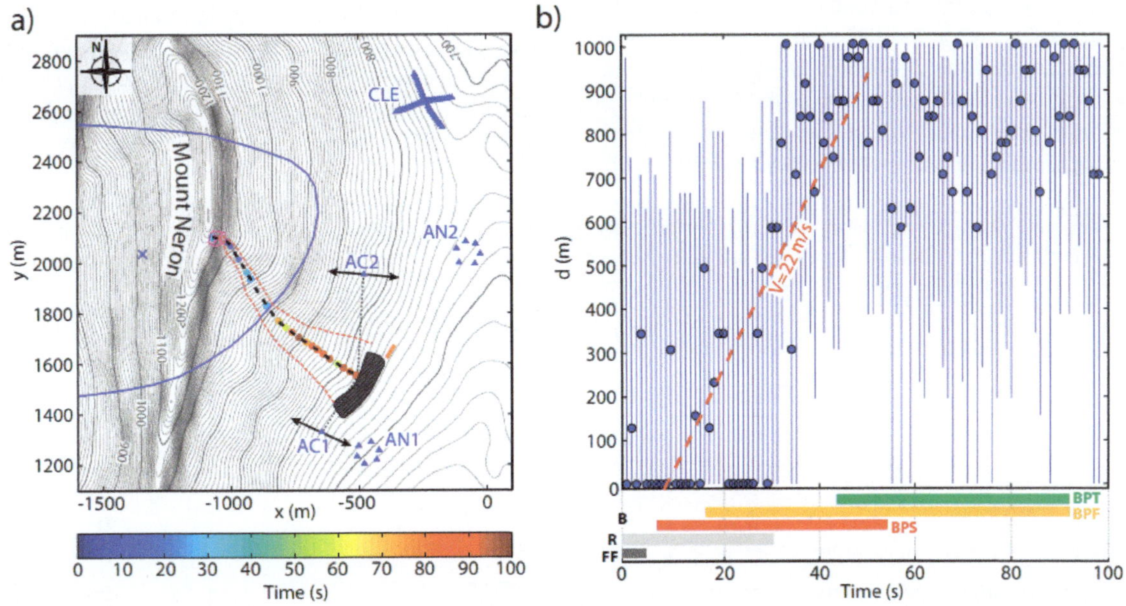

Figure 9. Panel **(a)**: seismic source location estimated using different waves and time windows. Source location with the beam-forming method applied either to the first 2.5 s of the signal (blue cross) or to the acoustic waves produced by the explosion (magenta cross). The location uncertainties for the beam-forming method are estimated by the area where the mean inter-trace correlation is larger than 95 % of its peak value (blue and magenta contours). The colored circles illustrate the propagation of the source using a 1 s long sliding time window, constrained on the mean rockfall path (black dashed line). Time ranges from the blast (in blue) until 100 s later (dark red). Similarly to Fig. 1a, blue triangles show the seismometers, the hatched grey contour indicates the earthen barrier and the red dashed line represents the area of provoked rockfall propagation. The black double arrows show the direction of ground motion polarization for the impact of block 6 into the earthen barrier (see Fig. 8b and d). The grey dotted line points out the source–sensor path. Panel **(b)**: top, estimated location of the seismic signal source (blue points) computed every second since the blast. The source is imposed to be along the main propagation path shown as a black dashed line in **(a)**. The vertical blue lines are error bars (points along path with correlation larger than 95 % of the maximum value). The distance (vertical axis) is measured from the blast location, along the main propagation path. The red line is a linear fit for the first 50 s of the signal giving an average propagation velocity of 22 m s^{-1}. Bottom: simplified chronology of the provoked rockfall from the video clips (see Fig. 3e). From the left to the right: blast (B), free fall (FF), rocks falling from the cliff (R), block propagation in the scree (BPS), in the forest (BPF) and at the toe of the slope (BPT).

Estimation of the Block and Mass Velocities from the Videos

The video clips were processed in order to assess the block velocities during the fall and propagation phases using two different approaches. The image processing software 7-D (Vacher et al., 1999) was employed to calculate the displacement field in a given window (Fig. 10) from one frame to the next, for the first 17 s of the video.

The software applies a pattern recognition algorithm based on image correlation technics. Knowing the real pixel size and the sampling frequency of the frames, the displacement is transformed into velocity. The block velocities are spatially averaged, as the block density in each pixel is not taken into account. The intrinsic processing error was estimated by analyzing the histogram of the displacements calculated on zones that were supposedly stable. The histogram (not shown here) displays a log-normal shape with a mean of 0.5 pixels and a standard deviation of 0.5 pixels. Only displacements larger than one pixel between two successive frames are considered in the following. The displacement field (magnified 10 times in pixel size) at three different times (1, 3 and 11 s) is superimposed with the corresponding photographs in Fig. 10a. The mean velocity amplitude in the vertical direction (V_y, solid red line) and the mean velocity norm (V_n, solid blue line) in the photograph plane are displayed in Fig. 10b. After the initial sharp acceleration related to the blast, the two curves tend to remain parallel, indicating that V_x (in the horizontal direction) slightly oscillates between 10 and 15 m s^{-1}. In contrast, V_y drops for the period between 0.5 and 1 s before increasing again rapidly for 2 s (Fig. 10a, left), following a constant acceleration. The velocity drop probably results from the contribution of the blocks projected up-ward, while the rise fits well with the free fall hypothesis ($V_y = g\,t$, black dashed line in Fig. 10b) until the main mass impacts the toe of the cliff for the period between 3 and 6 s after the blast (Fig. 10a, center). Afterwards, the mean vertical velocity in the given window settles at a constant value of about 12 m s^{-1}. As visible in the right picture in Fig. 10a, the dust generated by the rockfall, however, produces artifacts in the image correlation, affecting the later time periods and x direction velocities as the wind laterally pushed the smoke.

Stereographic videos of the blasting were also used to determine the 3-D trajectories of the blocks individually propagating down the slope, with the aim of quantifying their translational and rotational velocities. The planar coordinates of specific blocks were extracted manually with *ImageJ* software (http://rsb.info.nih.gov) from each frame of the video clips simultaneously by the M2 and M3 cameras (Fig. 1a). The use of 12 reference points pinned in each frame allowed us to reconstruct the 3-D trajectory of particular blocks. The conversion was optimized by minimizing the error between the measured and calculated point positions. However, the poor definition of the videos which cover the entire slope and the dense vegetation prevented a detailed tracking of most of the blocks (Fig. 10c). Consequently, velocity values were only determined for block 6, the trajectory of which is visible in both videos (Fig. 10d). Two specific bounces occurring at mid-slope (940 and 900 m in altitude, grey stars in Fig. 1a) and showing long free-flight phases were studied in more detail. The block translational speeds before and after the bounces lie in the range 12–14.5 and 22–28 m s^{-1}, respectively (Table 2). At the toe of the slope, the translational and rotational velocities of block 6 were estimated to 20 to 25 m s^{-1} and to about 10 rad s^{-1}, respectively, before the impact into the earthen barrier. Considering the 9 m^3 block as a homogeneous sphere with a density of 2500 kg m^{-3}, the total kinetic energy (Ec) is about 6 × 10^6 J prior to the impact.

Table 2. Comparison of average propagation velocities for different types of rock instabilities, determined from video and/or seismic analysis. The mean slope angle and the difference in altitude are given between the initial position of the unstable mass and the toe of the deposit area

Type	Name	Rock type	Mean slope angle (°)	Difference in altitude (m)	Max punctual speed (m s^{-1})	Mean velocity (m s^{-1})	Reference
Rock avalanches ($>10^5$ m^3)	Thurwieser	Limestone and dolostone	17	~ 1300	60–65	36–38	Sosio et al. (2008)
	Mount Fletcher	Sandstone and siltstone	22	~ 1300	–	80	McSaveney (2002)
	Mount Cook	Graywacke and argillite	32	~ 2600	–	55–60	McSaveney (2002)
Rockfalls (10^3–10^5 m^3)	Mount Néron	Limestone	40	550			This work
Free fall					–	30 [1]	
					–	20 [2]	
Block propagation					12–28 [1,3]	–	
					20–25 [1,4]	–	
Isolated block falls ($<10^3$ m^3)	–	–	34	30	~ 15	–	Ma et al. (2011)
	–	–	38	200	24–31	12–15	Dorren et al. (2006)

[1] Derived from video processing (Sect. 3.4); [2] obtained by seismic source location technique (Sect. 3.3); [3] at mid-slope; [4] at the slope toe (see Fig. 1a).

The ratio between the seismic energy released by the impact (E_s, Sect. 3.1) and the above-computed kinetic energy before impact (E_k) is about 0.3, 3 orders of magnitude larger than the ratios (5 10 ^{4} and 8 10 ^{4}) obtained for the two Néron events. This strong difference could result from (1) the strong nonlinear behavior in the soil deposits (scree overlying marly layer) at the cliff toe, (2) the free fall hypothesis of a rigid mass, which is probably incorrect because of disintegration and impacts of blocks on the cliff face or (3) the progressive fall of the mass for a few seconds, which makes the seismic energy transfer not instantaneous. These three effects tend to underestimate the ratio E_s=E_k. In contrast, the effect of the rigid rock-reinforced barrier at the slope toe, on which the block 6 broke, could explain the high conversion ratio from mechanical to seismic energy.

Translational velocities at mid-slope derived from video analysis (12–28 m s^{-1}) are of the same order of magnitude as those derived from seismic source location (22 m s^{-1}, Sect. 3.3). Even so, the seismic source location technique provides an average location of all the seismic sources over a given time window which may be related to the dynamics of the propagation front along the slope, whereas the video analysis focuses on the dynamics of one single fast propagating block. These propagation velocities are compared with the results of other studies using image analysis and seismic signal processing (Table 2). Such data are quite rare for rock-falls of medium size like the one studied and the comparison was extended to rock avalanches (volume > 10^{5} m^{3}) and isolated blocks. With the largest fallen volumes and the greatest difference in altitude, rock avalanches show the greatest propagation velocities (36–80 m s^{-1} mean velocity), even if the mean slope angle over the entire path is moderate. On the contrary, the two studies concerning isolated rock blocks deal with human-triggered, block drop field experiments on steep slopes, but with small differences in elevation. Thus, maximum block velocity reaches about 30 m s^{-1}, and the mean velocity ranges around 12–15 m s^{-1}. Mount Néron mid-size rockfall is characterized by a relatively steep slope (about 40 on average) with an intermediate difference in altitude (550 m). Three different stages of propagation can be defined during the rockfall. At the early stage, the falling rock mass followed the free-fall equation with maxi-mal mass velocity of □30 m s^{-1}. After the impact at the toe of the cliff, the fallen volume behaved as a rock mass with moderate propagation speed (□20 m s^{-1} on average over the whole event) and settles rapidly. Several isolated blocks emerge from this mass, and propagate down the slope. Velocities of 12–28 m s^{-1} at mid-slope are reported for one single block, which reaches a maximal velocity of 20–25 m s^{-1} at the toe of the slope. These values measured during the Mount Néron rockfall are bracketed by the values found in the literature (15–80 m s^{-1}). The interesting feature is that one block (9 m^{3}) generated a signal as energetic as the mass impacting the ground after the free-fall sequence: this is probably due to the collision at high speed (translational and rotational) in the highly compacted, rock-reinforced earthen barrier.

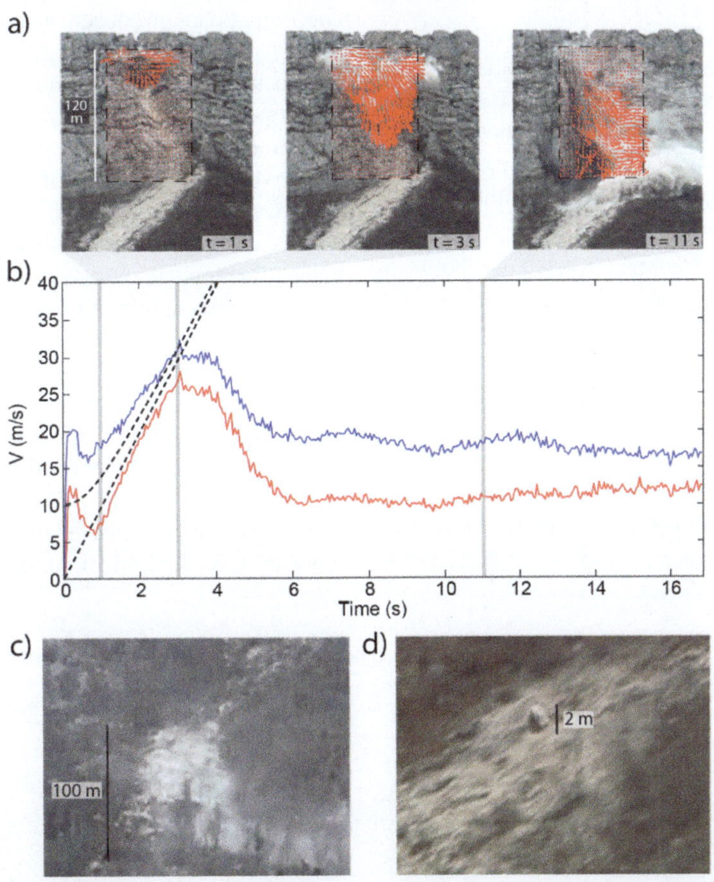

Figure 10. Panel **(a)**: images of the artificially triggered rockfall taken at three different times (t_1, t_2 and t_3) with displacement field (magnified 10 times in pixel size, red arrows) superimposed. Panel **(b)**: mean vertical (V_y , in red) and norm (V_n, in blue) velocities of the blocks retrieved from image correlation analysis (solid lines) and from numerical discrete element simulation (dashed lines). The free fall equations along the vertical direction ($V_y = g\,t$) and norm $(V_n = [(g\,t)^2 + V_0^2]^{1/2})$ are shown by the black dashed

line, where g = 9.81 m s^{-2} is the gravitational acceleration, V0 = 10 m s^{-1} is the initial horizontal velocity due to the blast, and t is the time. The times corresponding to the pictures in (a) are marked with the vertical grey lines. Panel (c): TIFF image extracted from the 1920× 1080 pixels videos showing the toe of the scree deposits and the upper part of the slope (labeled 3 and 4, respectively, in Fig. 1b). Panel (d): zoom-in of block 6 during propagation.

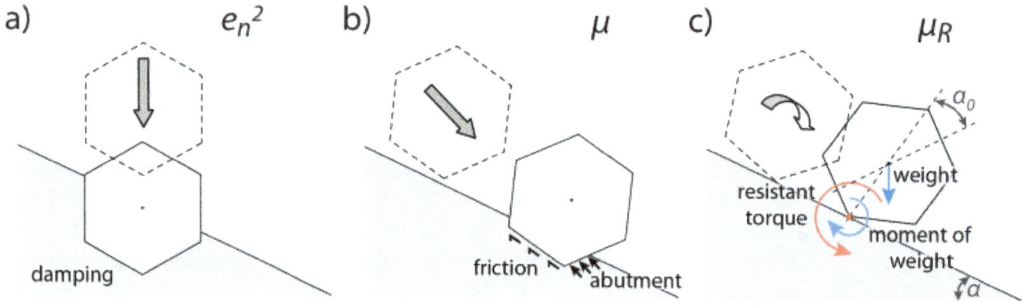

Figure 11. Schematic sketch illustrating the normal (e_n^2, **a**) and tangential (μ, **b**) restitution coefficients, as well as the resistive rolling coefficient (μ_R, **c**). The grey arrow shows the direction of motion. The resistant torque introduced in the contact law at the contact point is shown in red in **(c)**.

ROCKFALL PROPAGATION MODELING

The artificially triggered rockfall propagation was simulated using the discrete element method (DEM) based on rigid body dynamics. The objective of this simulation was to compare numerical propagation velocities with experimental data. Within the simulations, the motion of each block is powered by gravity and the model accounts for both translation and rotation.

This numerical model takes account of realistic block shapes and specific contact laws which are able to represent the main dissipation phenomena at the contact–rebound point. It was validated by comparison with experimental results from Manzella and Labiouse (2009) involving a granular flow of small bricks on an inclined plane.

The normal and tangential contact force laws between discrete elements (Richefeu et al., 2012) are controlled by two dissipation coefficients: e_n^2 and m. The parameter e_n^2 reflects the amount of energy restored in the perpendicular direction to the impact plane (Fig. 11a). In the case of a vertical drop of a block on a horizontal plane, this coefficient is the ratio between the height of rebound of the particle and the height of the initial drop. A value of 1 corresponds to a perfectly elastic collision, and a value close to 0 refers a strongly dissipative contact. The parameter m is a coefficient which allows one to introduce the energy dissipation in the tangential direction to the contact plane. For a contact between rocks and soft soil, this coefficient incorporates abutment force and shear strength of the soil of the impacted area in addition to frictional forces (Fig. 11b).

Different values and definitions of the restitution coefficients are proposed in the literature, leading to a wide range of possible parameters (Heidenreich, 2004). It was shown that on the experiment scale, the type of experiments performed and the nature of the soil have a great influence on the values of the restitution parameters obtained. According to these works, high values of the normal restitution coefficient (e_n^2) need to be considered for block impacts involving hard surface, paving or bedrock (0.6–0.15 respectively) and small values for impacts on soft soil slopes (0.2–0.1). The abutment forces integrated into the tangential restitution co-efficient (m) strongly depend on the blocks' shape, the nature of the soil, the impact intensity and angle. For impacts involving hard surface and bricks, the tangential restitution co-efficient corresponds to the dynamic friction angle (Richefeu et al., 2012), whereas values greater than the friction angle need to be considered for impacts on soft soil. The tangential restitution coefficient remains poorly constrained owing to a lack of an experimental, representative rock block rebound data set.

Figure 12. Panel **(a)**: the digital surface model is built by joining basic forms (top right corner) on the mesh of the lidar DSM. A detailed view is shown for the earthen barrier mesh (bottom right corner). Panel **(b)**: spatial zoning of the site, depending on the type of terrain: solid rock (green), rocky scree (orange), soft ground with trees (red) and earthen barrier (light grey). The initial velocity V_0 given to the blocks is shown by the black arrow.

To adapt the numerical model to the resistive phenomenon occurring during the rotation of angular blocks on a soft soil, a resistive moment was introduced by means of a resistive rolling coefficient (m_R). Without rolling resistance, the moment of the weight force with respect to the contact point on the slope (function of the block shape) acts as a source of rotational motion and cannot be compensated by the contact force itself, since the latter generates no moment at the con-tact point (Fig. 11c). The rolling resistant torque C, inversely proportional to the angular velocity of the block w, was there-fore defined in the model and scaled with the normal force f_n (i.e., with the ground indentation):

$$C = -\min\left(\gamma_R \|\omega\|,\ \mu_R \ell f_n\right) \frac{\omega}{\|\omega\|}, \tag{1}$$

where g_R is a parameter of regularization and m_R is the co-efficient of rolling friction. The latter dimensionally scales with a length and plays a role similar to the friction coefficient in classical Coulomb law. It is physically interpreted as the amount of soil punching that happens during a collision expressed as a portion of the mean size of the blocks. The minimum value of μ_R necessary to stop the rolling movement of a block under its own weight is a function of the slope angle (a) and the block geometry (a $_0$), and can be ex-pressed by the following expression (see also Fig. 11c):

$$\mu_{R0} = \sin(\alpha - \alpha_0)/\cos(\alpha). \tag{2}$$

Values greater than μ_{R0} slow down the propagation of a single block on a slope.

Table 3. Dissipation coefficients along the normal direction (e^2_n), tangential direction (μ) and rolling friction (μ_R) defined for (i) block–solid rock (including another block), (ii) block–rocky scree, (iii) block–soft ground with trees and (iv) block–earthen barrier contacts

	Solid rock	Rocky scree	Soft ground with trees	Earthen barrier
e_n^2	0.30	0.05	0.01	0.001
μ	0.50	1.00	0.20	1.20
μ_R	0.00	0.20	0.10	0.90

The unstable rock column is modeled by an amount of randomly arranged blocks with an apparent volume of about 2500 m^3 (Fig. 13b). Each rock block is considered as a discrete element of regular polyhedral shape with rounded edges and vertices. Block unitary volumes are uniformly distributed between 1 and 9 m^3, matching the size of blocks that propagated down to the slope toe in the real event. Smaller blocks were not included in the simulations for saving computation time as the video clips suggest that they settled rapidly.

The topography of the slope corresponds to a coarse version of the lidar DSM (triangular elements of 10 m in characteristic size) to speed up the computation. The edges and vertices of each facet have also been rounded (Fig. 12a). This allows unambiguous definition of the contact directions be-tween the blocks and the terrain or in-between the blocks. Due to the great influence of the soil nature on the restitution coefficients, the DSM was zoned into four different areas according to the type of terrain: solid rock, rocky scree, soft ground with trees and earthen barrier material (Fig. 12b). For each type of zone, the dissipation coefficients (e_n^2; μ and μ_R, Table 3) were defined from a set of values from the literature (Heidenreich, 2004). A value of m= 1 was retained for the impact between blocks and rocky scree in order to take into account the tangential interaction and the pushing of the rocky blocks during impact. This set of parameters was improved by trial and error until the simulated rockfall time propagation behaved closely to the real one, i.e., with most of the mass (~80 %) being deposited on the upper half of the slope (rocky scree) and presence of blocks stopped by the ditch–earthen barrier (Figs. 13c and 14c).

The triggering of the rockfall consisted of setting an initial horizontal velocity (V_0) to the blocks to mimic the initial blasting energy (Fig. 13a and b). V_0 was set at 10 m s^{-1} out-wards and normal to the cliff (Fig. 12b), which is the value measured by digital image correlation (Sect. 3.4). Right after the triggering, the blocks remained together as a whole during free fall until they impacted the toe of the cliff (Fig. 13c).

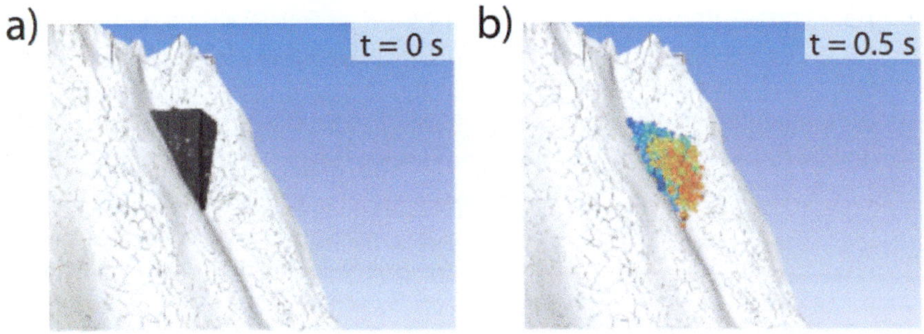

c)

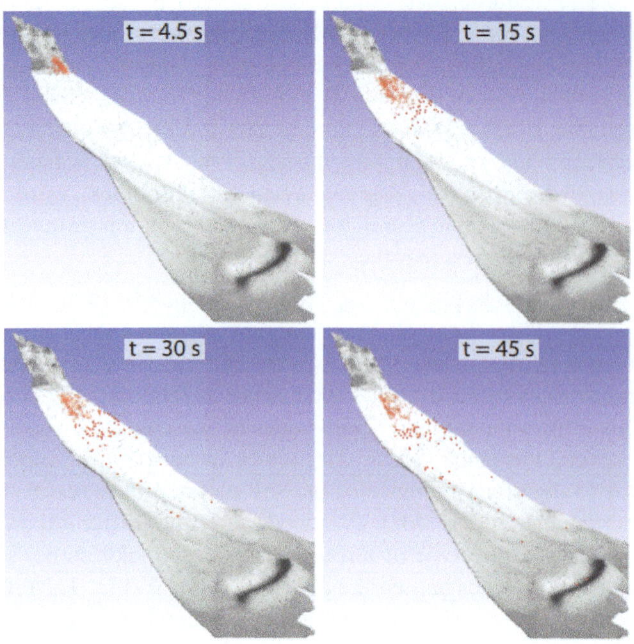

Figure 13. Panel **(a)**: picture of the zone with the unstable mass confined in an apparent 2500 m^3 volume ($t = 0$ s). Panel **(b)**: artificial triggering of the rockfall simulation after wall removing and initial velocity push ($t = 0.5$ s). Panel **(c)**:screenshots of a simulation taken at four different times. Rocks blocks are shown in red on the light grey DSM.

The numerical translational (Vt) and rotational (ω) block velocities are shown as a function of the distance to the blast d in Fig. 14a and b, as well as the final mass distribution on the slope in Fig. 14c. The 30–35 m s^{-1} maximum mean block velocity is reached at the toe of the cliff (d ≈ 100 m) during the free-fall phase, very similarly to what observed at t ≈ 4 s with image correlation (Fig. 10b). At the same distance, the mean rotational speed peaks at 10 rad s^{-1} . Most of the blocks then slow down rapidly, due to ground impact. Between 150 and 450 m (rocky scree), the mean translational and rotational speeds remain low (about 5 m s^{-1} and 5 rad s^{-1}) because most of the blocks stop there (see final mass distribution in Fig. 14c). The velocity range is yet wide, with some blocks speeding up to Vt ≈ 60 m s^{-1} and ω ≥ 50 rad s^{-1} , as observed in the real case. Most of the blocks with large unitary volume propagate further down and stop in the soft ground area with trees, between d = 450 m and 900 m (Fig. 14c). The mean translational and rotational speeds increase slowly and irregularly, as the few blocks still propagating speed up and bounce on the ground. The translational velocities studied in Sect. 4.4 (black stars at d = 380 and 410 m, Fig. 14a) lie in the fastest 25 % numerical blocks range, probably because the video analysis focuses on large, fast-propagating blocks.

The average seismic source velocity of 22 m s^{-1} determined using seismic source location (see Sect. 4.3) is compatible with the numerical block propagation (thick grey dashed line in Fig. 14a) and reflects the global, averaged kinematics of the seismogenic processes occurring in the rockfall. At the toe of the slope (d ≥ 900 m), translational and rotational speeds lie in the 10–35 m s−1 (mean 20 m s^{-1}) and 7–23 rad s^{-1} ranges (mean 14 rad $^{s-1}$), respectively. These values are in good accordance with the video analysis of block 6 before its collision with the earthen barrier (see Sect. 4.4), which yielded Vt ≈ 20–25 m $^{s-1}$ and ω ≈ 10 rad $^{s-1}$ (black stars, Fig. 14a and b). At the end of the simulation (about 80 s), six blocks have reached the ditch–earthen barrier, with the first impact into the earthen barrier occurring at t ≈ 40 s, which is similar to the real case.

The modeling of the Néron artificially triggered rockfall showed that the discrete element technique can provide valuable insights into rockfall kinematics once appropriately calibrated. In particular, numerical translational and rotational block velocities agreed well with seismic and video observations, with mean values ranging from 5 to 35 m s^{-1} and 0 to 14 rad s^{-1}, respectively. Such simulations, which provide full information about the volume, velocity and trajectory of each individual block, could be of critical help to design protection structures. They need, however, to be applied to other case studies and gain forecasting ability through reliable, a priori determination of the restitution coefficients.

SEISMIC MODELING OF A BLOCK IMPACT

The seismic wave field generated by impact of block 6 (Fig. 3d) into the ditch–earthen barrier was numerically simulated to explain the discriminative feature pattern of polarization seen in Sect. 3.2. The 3-D finite element model (FEM) Comsol software (www.comsol.com) was used for dynamic elastic simulations. A two-layer model was designed (Fig. 15a) according to the results from a refraction seismic survey. The top soft layer was found to be 5 m thick, with P and S wave velocities of 700 and 400 m s^{-1} , respectively. A higher velocity layer (VP = 2450 m s^{-1} and VS = 800 m s^{-1}) was detected below, probably corresponding to the marly bedrock. Densities in the two layers were fixed to 1900 and 2500 kg m^{-3} , from top to bottom. The impact was modeled by a unitary load applied to the model surface for 0.2 s and striking perpendicularly to the source– sensor direction with 45° inclination. Four 3C sensors (blue triangles in Fig. 15a) were setup along this same axis to study the wave field propagation, with source–sensor distance ranging from 0 to 250 m. The medium was meshed with second-order tetrahedrons, with smaller elements setup along the source–sensor path to fulfill the FEM requirements for adequate spatial resolution of propagating waves (Bazant et al., 1978; Moser et al., 1999). The generalized-alpha, implicit time integration algorithm was used (Chung and Hulbert, 1993). The 5×10^{-3} s time step provides accurate resolution for frequencies up to 10 Hz (ANSYS, 1992), and could be related to the time of propagation between two successive nodes in the mesh for the fastest wave of interest (Moser et al., 1999).

The radial (red), tangential (blue) and vertical (black) seismic signals computed at the four sensors, from the load point location ($x = 0$ m) to the last sensor ($x = 250$ m) over the first 1.5 s of simulation are presented in Fig. 15b. The corresponding particle motions are presented in the radial–vertical (***X–Z***, top), tangential–vertical (***Y –Z***, center) and radial– tangential (***X–Y***, bottom) planes in Fig. 15c. The seismograms exhibit waves propagating at different velocities. The first arrival (marked with grey dashed line in Fig. 15b) shows high velocity and low amplitude, and corresponds to the head wave refracted on the top of the marly bedrock. Then, the signal is dominated by slow and large amplitude waves corresponding to surface waves. The wavefield is dominated by the tangential component (Fig. 15b), with a maximal amplitude about 2 to 3 times greater than the others. The polarization in the tangential–vertical plane (Fig. 15c) exposes a 45 dipping particle motion at load point location which rapidly flattens with distance, indicating the predominance of Love waves.

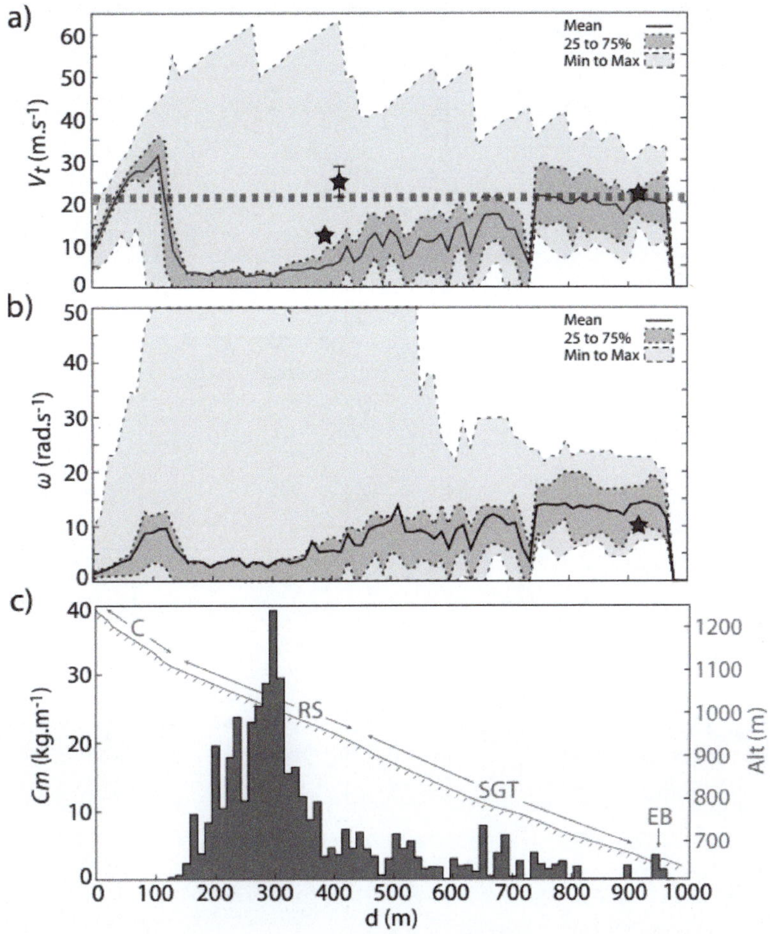

Figure 14. Numerical modeling of the rockfall propagation. Panel (a): translational velocity Vt as a function of the propagation distance d from the blast. The mean value is shown as a black continuous line, the dark grey and light grey show the 25–75 % and min– max intervals, respectively. The black stars point out the velocities derived from video analysis. The thick grey dashed line shows the seismic source mean velocity. Panel (b): rotational velocity ω as a function of the propagation distance d from the blast. Same legend as (a). Panel (c): bar graph of the final mass distribution over distance d from the blast. The cliff (C), rocky scree (RS), soft ground with trees (SGT) and earthen barrier (EB) are shown along the slope section (grey).

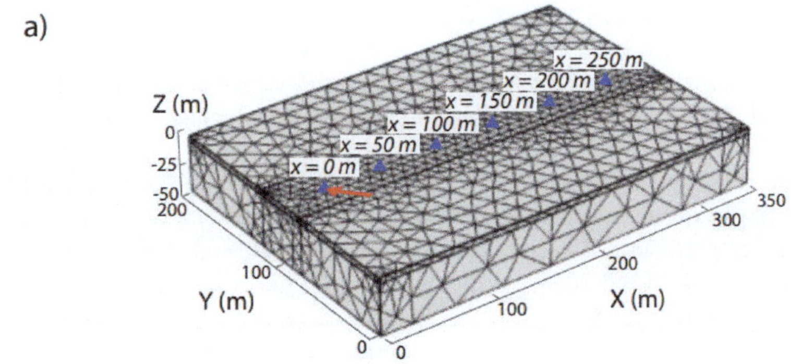

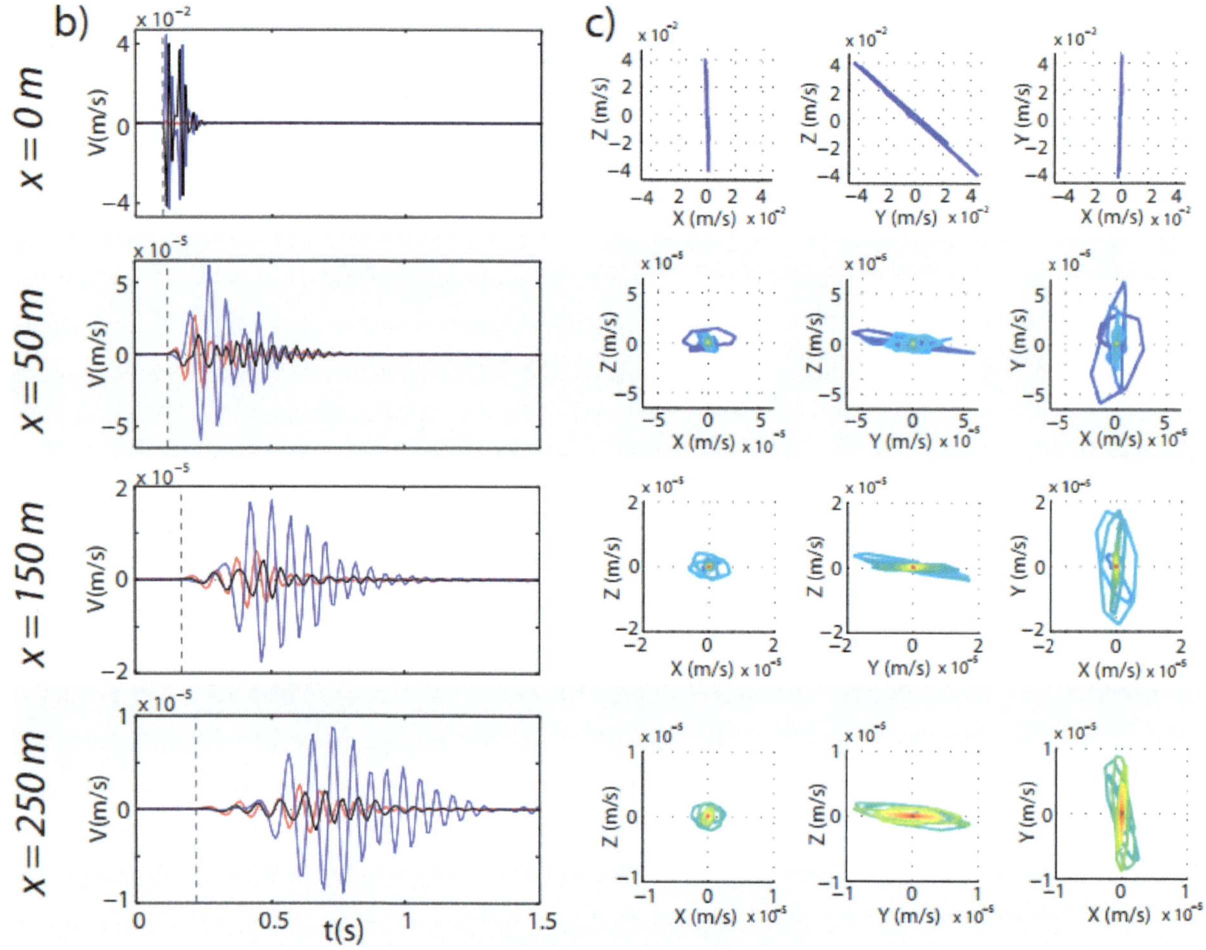

Figure 15. Numerical modeling of the impact of block 6 on the earthen barrier. Panel (a): tetrahedral mesh of the two-layer finite element model. The red arrow symbolizes the point load inclined at 45° in the Z–Y plane and blue triangles show the four velocity sensors setup along the x axis. Panel (b): synthetic seismograms computed during 1.5 s at the four 3C sensors ranging from 0 to 250 m in distance x to the source. Vertical (Z, black), radial (X, red) and tangential (Y , blue) component are presented. Panel (c): particle motions of the signals displayed in (b). The radial– vertical (XZ, top), tangential–vertical (Y − −Z, center) and radial– tangential (X − −Y , bottom) planes are shown. Color scales with time, from blue (t = 0 s) to red (t = 1.5 s). Note that the amplitude scale depends on source–sensor distance in (b) and (c).

In contrast, Rayleigh waves which appear in the radial–vertical plane are only little excited. Numerical particle motions can be compared with experimental motions related to the impact of block 6 into the earthen barrier. Both the numerical diagram at x= 250 m (Fig. 15b) and the experimental diagram at sensor AC1 (Fig. 8b), that have similar source–sensor distance, demonstrate clear, horizontal, trans-verse particle motions related to Love waves. Although simulations were not carried out for greater source–sensor distances for computation time reasons, the same numerical particle motion is likely to be observed at AC2 ($x \approx 400$ m), in good accordance with the recorded, horizontal, transverse motion (Fig. 8d).

CONCLUSIONS

The medium-size artificially triggered rockfall of December 2011 at Mount Néron provided a unique opportunity to compare artificial to natural rockfalls regarding their respective propagation and seismic characteristics, using various ground and remote techniques. The following conclusions can be drawn from the present study. First, field survey, photogrammetric and aerial lidar surveys yielded consistent volume estimations for this event (from 2380 to 2700 m^3). Although its resolution is lower, the low-cost ground photogrammetry technique has shown its potential for evaluating the volume and performing a structural analysis of the cliff, even with a ranging distance of 1 km. Secondly, joint video and seismic data interpretation enabled the establishment of the relationship between provoked rockfall stages and seismic phases, particularly, signals related to the initial blast, to the ground impact of the mass, to its settlement on the upper part of the scree slope and to isolated block impacts were clearly identified. The two main energetic phases correspond respectively to the ground impact following the free fall and a 9 m^3 block impacting the earthen barrier at the final slope toe. In contrast, the mass propagation along the scree slope generated little seismic energy. A simultaneous use of the image correlation and the seismic source location techniques provided the rockfall kinematics characterization. It was observed that maximal propagation speed was reached at the end of the free fall (~30 m s^{-1}) and the mass velocity abruptly dropped to about 20 m s^{-1} after the ground impact. From stereoscopic videos, the velocity of one isolated block ranged from 12 to 28 m s^{-1} at mid-slope, depending on the kinetic energy lost and gained during the propagation. Velocities ranging from 20 to 25 m s^{-1} for the translational mode and 10 rad s^{-1} for rotational mode were recorded at the slope toe. For the blocks impacting the protective earthen barrier, strong associated seismic signals ($M \approx 98$) were recorded, resulting from a high conversion ratio from kinetic energy to induced seismic waves. These observations are of primary importance for correctly interpreting the seismic phases. Moreover, polarization analysis and modeling of the impact sequences showed surface waves dominating the wavefield.

The main features of the rockfall (free fall phase, major block settlement on the rocky scree area and propagation of isolated blocks down the slope) were numerically reproduced with the aim of calibrating a discrete element model. The free fall simulation agreed well with image correlation analysis and theoretical free fall equations, assuming initial horizontal block velocities induced by the blast. The propagation path of isolated blocks and the event duration were consistently reproduced, as well as the order of magnitude of translational and rotational velocities. The model predicted impact speeds comparable to the ones estimated from the videos, i.e., ranging from 10 to 35 m s^{-1} and 7 to 23 rad s^{-1}. This case study provided a straightforward way to calibrate a set of dissipation coefficients which could be used for future works. These values yet need to be further constrained, for example by processing very high resolution videos of real events, tracking numerous block trajectories.

This study provided an exceptional opportunity to compare natural and provoked rockfalls of similar volumes (2000 and 2570 m^3, respectively) occurring at the same location. Neglecting the initial blast and the associated acoustic wave, natural and provoked rockfalls generated comparable seismic signals (duration, peak amplitude, spectrograms), yielding close magnitude estimations ($M \approx 1$). This supports that the source mechanism has little control on the global mass propagation. In contrast, the number of blocks reaching the ditch was higher for the natural event (15) than for the provoked one (7), suggesting a dramatic effect of blasting on the block size.

ACKNOWLEDGEMENTS

This work was partially funded by the Interreg project MASSA, the federal body VOR (Vulnérabilité des Ouvrages aux Risques) and the laboratory ISTerre. The used seismic instruments belong to the French national collection Sismob-RESIF.

REFERENCES

1. Allstadt, K.: Extracting source characteristics and dynamics of the August 2010 Mount Meager Landslide from broad-band seismograms, J. Geophys. Res.-Earth, 118, 1472–1490, doi:10.1002/jgrf.20110, 2013.
2. ANSYS user's manual for revision 5.0, Houston, TX, Swanson Analysis Systems, 1992.
3. Baillard, C., Crawford, W. C., Ballu, V., Hibert, C., and Mangeney, A.: An automatic Kurtosis-based P and S phase picker designed for local and regional seismic networks, B. Seism. Soc. Am., 104, 394–409, doi:10.1785/0120120347, 2013.
4. Banton, J., Villard, P., Jongmans, D., and Scavia, C.: Two-dimensional discrete element models of debris avalanches: Pa-rameterization and the reproducibility of experimental results, J. Geophys. Res., 114, F04013, doi:10.1029/2008JF001161, 2009.
5. Battaglia, J., Got, J.-L., and Okubo, P.: Location of long-period events below Kilauea Volcano using seismic amplitudes and ac-curate relative relocation, J. Geophys. Res., 108, ESE 2-1–2-16, doi:10.1029/2003JB002517, 2003.
6. Bazant, Z. P.: Spurious reflection of elastic waves in non-uniform fi-nite element grids, Comput. Method. Appl. M., 16, 91–100, 1978 Biescas, B., Dufour, F., Furdada, G., Khazaradze, G., and Surinach, E.: Frequency content evolution of snow avalanche seismic signals, Surv. Geophys., 24, 447–464, doi:10.1023/B:GEOP.0000006076.38174.31, 2003.
7. Brodsky, E. E., Gordeev, E., and Kanamori H.: Landslide basal friction as measured by seismic waves, Geophys. Res. Lett., 30, 2236, doi:10.1029/2003GL018485, 2003.
8. Caplan-Auerbach, J. and Huggel, C.: Precursory seismicity asso-ciated with frequent, large ice avalanches on Iliamna volcano, Alaska, USA, J. Glaciol., 53, 128–140, 2007.
9. Chung, J. and Hulbert, G. M.: A time integration algorithm for structural dynamics with improved numerical dissipation: the generalized-alpha method, J. Appl. Mech., 60, 371–375, 1993.
10. Cleary, P. W. and Prakash, M.: Discrete–element modelling and smoothed particle hydrodynamics: potential in the envi-ronmental sciences, Philos. T. R. Soc. A, 362, 2003–2030, doi:10.1098/rsta.2004.1428, 2004.
11. Cole, S., Cronin, S., Sherburn, S., and Manville, V.: Seismic signals of snow-slurry lahars in motion: 25 September 2007, Mt Ruapehu, New Zealand, Geophys. Res. Lett., 36, L09405, doi:10.1029/2009GL038030, 2009.
12. Commonwealth Scientific and Industrial Research Organization (CSIRO): Siro 3-D, Sirovision 3-D Imaging Mapping System, User Manual, Version 4.1, 2010a.
13. Commonwealth Scientific and Industrial Research Organization (CSIRO), Sirojoint, Sirovision structural mapping and analysis system, User manual, Version 4.1, 2010b.
14. Dahlen, F.: Single-force representation of shallow landslide sources, B. Seismol. Soc. Am., 83, 130–143, 1993.
15. Dammeier, F., Moore, J. R., Haslinger, F., and Loew, S.: Characterization of alpine rockslides using statistical anal-ysis of seismic signals, J. Geophys. Res., 116, F04024, doi:10.1029/2011JF002037, 2011.
16. Deparis, J., Jongmans, D., Cotton, F., Baillet, L., Thouvenot, F., and Hantz, D.: Analysis of rock-fall and rock-fall avalanche seismo-grams in the French Alps, B. Seismol. Soc. Am., 98, 1781–1796, 2008.
17. Dorren, L. K. A.: A review of rockfall mechanics and modelling approaches, Prog. Phys. Geogr., 27, 69–87, 2003.
18. Dorren, L. K. A., Berger, F., and Putters, U. S.: Real-size ex-periments and 3-D simulation of rockfall on forested and non-forested slopes, Nat. Hazards Earth Syst. Sci., 6,

145–153, doi:10.5194/nhess-6-145-2006, 2006.

19. Evans, S. G. and Hungr, O.: The assessment of rockfall hazard at the base of talus slopes, Can. Geotech. J., 30, 620–636, 1993.
20. Fabre, D., Lorier, L., Mathy, A., and Hantz, D.: Gestion d'un risque rocheux en zone urbanisée: le cas des rochers du Néron près de Grenoble (Isère – France), Proceedings of the 1st International Conference on Landslides risk, 14–16 March 2013, Tabarka, Tunisia, 149–162, 2013.
21. Favreau, P., Mangeney, A., Lucas, A., Crosta, G., and Bouchut, F.: Numerical modeling of landquakes, Geophys. Res. Lett., 37, L15305, doi:10.1029/2010GL043512, 2010.
22. Gidon, M. and Arnaud, H.: Carte géologique détaillée de la France à 1/50 000ème, feuille Grenoble, 2 Edn., Bureau de Recherches Géologiques et Minières, Orléans, 1978.
23. Hanks, T. C. and Kanamori, H.: A moment magnitude scale, J. Geo-phys. Res., 84, 2348–2350, 1979.
24. Heidenreich, B.: Small- and half-scale experimental studies of rock-fall impacts on sandy slopes, PhD thesis, EPFL, Lausanne, 2004.
25. Helmstetter, A. and Garambois, S.: Seismic monitoring of Séchili-enne Rockslide (French Alps): analysis of seismic signals and their correlation with rainfalls, J. Geophys. Res., 115, F03016, doi:10.1029/2009JF001532, 2010.
26. Hibert, C., Mangeney, A., Grandjean, G., and Shapiro, N.: Slopes instabilities in the Dolomieu crater, la Réunion island : from the seismic signal to the rockfalls characteristics, J. Geophys. Res.-Earth, 116, F04032, doi:10.1029/2011JF002038, 2011.
27. Hibert, C.: Apport de l'écoute sismique pour l'étude des éboule-ments du cratère Dolomieu, Piton de la Fournaise (Ile de la Réu-nion), PhD thesis, Institut de Physique du Globe de Paris, Paris, France, 2012.
28. Hungr, O., Corominas, J., and Eherhardt, E.: Estimating landslide motion mechanism, travel distance and velocity, in: Landslide Risk Management, edited by: Hungr, O., Fell, R., Couture, R., and Eherhardt, E., Taylor and Francis Group, London, ISBN 041538043X, 2005.
29. Iverson, R. M.: The physics of debris flows, Rev. Geophys., 35, 245–296, doi:10.1029/97RG00426, 1997.
30. Kanamori, H.: The energy release in great earthquakes, J. Geophys. Res., 82, 2981–2987, 1977.
31. Kanamori, H. and Given, J. W.: Analysis of long-period seismic waves excited by the May 18, 1980, eruption of Mount St. Helens-A terrestrial monopole?, J. Geophys. Res.-Sol. Ea., 87, 5422–5432, 1982.
32. Kanamori, H., Given, J. W., and Lay, T.: Analysis of seismic body waves excited by the Mount St. Helens eruption of May 18, 1980, J. Geophys. Res.-Sol. Ea., 89, 1856–1866, 1984.
33. Kao, H., Kan, C.-W., Chen, R.-Y., Chang, C.-H., Rosenberger, A., Shin, T.-C., Leu, P.-L., Kuo, K.-W., and Liang, W.-T.: Locating, monitoring, and characterizing typhoon-linduced landslides with real-time seismic signals, Landslides, 9, 557–563, 2012.
34. Lacroix, P. and Helmstetter, A.: Location of seismic signals asso-ciated with microearthquakes and rockfalls on Séchilienne land-slide, B. Seism. Soc. Am., 101, 341–353, 2011.
35. La Rocca, M., Galluzzo, D., Saccorotti, G., Tinti, S., Cimini, G. B., and Del Pezzo, E.: Seismic signals associated with landslides and with a tsunami at Stromboli volcano, Italy, B. Seis. Soc. Am., 94, 1850–1867, 2004.
36. Ma, G., Matsuyama, H., Nishiyama, S., and Ohnishi, Y.: Practical studies on rockfall simulation by DDA, J. Rock Mech. Geotech. Eng., 3, 57–63, 2011.
37. Mangeney-Castelnau, A., Vilotte, J.-P., Bristeau, M. O., Perthame, B., Bouchut, F., Simeoni, C., and Yerneni, S.: Numeri-cal modeling of avalanches based on Saint Venant equa-tions using a kinetic scheme, J. Geophys. Res., 108, 2527, doi:10.1029/2002JB002024, 2003.
38. Mangeney, A., Tsimring, L. S., Volfson, D., Aranson, I. S., and Bouchut, F.: Avalanche mobility induced by the presence of an erodible bed and associated entrainment, Geophys. Res. Lett., 34, L22401, doi:10.1029/2007GL031348, 2007.
39. Manzella, I. and Labiouse, V.: Flow experiments with gravel and blocks at small scale to investigate parameters and mecha-nisms involved in rock avalanches, Eng. Geol., 109, 146–158, doi:10.1016/j.enggeo.2008.11.006, 2009.
40. McSaveney, M. J.: Recent rockfalls and rock avalanches in Mount Cook national park, New Zealand, in: Catastrophic landslides: Effects, occurence and mechanisms: Boulder, Colorado, edited by: Evans, S. G. and DeGraff, J. V., Geological Society of Amer-ica Reviews in Engineering Geology, 15, 35–70, 2002.
41. McSaveney, M. J. and Downes, G. L.: Application of landslide seis-mology to some New Zealand rock avalanches, in: Landslides: Proceedings of the First European Conference on Landslides, edited by: Rybar, J., Stemberk, J., and Wagner, P., 24–26 June 2002, Lisse, the Netherlands, 649–654, 2002.
42. Moretti, L., Mangeney, A., Capdeville, Y., Stutzmann, E., Huggel, C., Schneider, D., and Bouchut, F.: Numerical modeling of the Mount Steller landslide flow history and of the generated long period seismic waves, Geophys. Res. Lett., 39, L16402, doi:10.1029/2012GL052511., 2012.
43. Moser, F., Jacobs, L. J., and Qu, J.: Modeling elastic wave propa-gation in waveguides with the finite element method, NDT & E International, 32, 225–234, 1999.
44. Norris, R. D.: Seismicity of rockfalls and avalanches at three Cas-cade Range volcanoes: Implications for seismic detection of haz-ardous mass movements, B. Seis. Soc. Am., 84, 1925–1939, 1994.
45. Pirulli, M. and Mangeney, A.: Results of Back-Analysis of the Prop-agation of Rock Avalanches as a Function of the Assumed Rhe-ology, Rock Mech. Rock Eng., 41, 59–84 doi:10.1007/s00603-007-0143-x, 2008.
46. Poropat, G. V.: New methods for mapping the structure of rock masses, in: Proceedings, Explo 2001, Hunter

Valley, New South Wales, 28–31 October 2001, 253–260, 2001.

47. Poropat, G. V.: Remote 3-D Mapping of rock mass structure, Laser and photogrammetric methods for rock face characterization, American Rock Mechanics Association, Golden, Colorado, 17– 18 June 2006.
48. Richefeu, V., Mollon, G., Daudon, D., and Villard, P.: Dis-sipative contacts and realistic block shapes for modelling rock avalanches, Eng. Geol., 149–150, 78–92, doi:10.1016/j.enggeo.2012.07.021, 2012.
49. Rousseau, N.: Study of seismic signal associated with rockfalls at 2 sites on the Réunion Island (Indian Ocean): Mahavel Cascade and Soufrière cavity, PhD Thesis of the Institut de Physique du Globe de Paris, Paris, France, 134 pp., 1999.
50. Schneider, D., Bartelt, P., Caplan-Auerbach, J., Christen, M., Huggel, C., and McArdell, B. W.: Insights into rock-ice avalanche dynamics by combined analysis of seismic recordings and a numerical avalanche model, J. Geophys. Res.-Earth, 115, F04026, doi:10.1029/2010JF001734, 2010.
51. Sosio, R., Crosta, G. B., and Hungr, O.: Complete dynamics modeling calibration for the Thurwieser rock avalanche (Italian Central Alps), Eng. Geol., 100, 11–26, doi:10.1016/j.enggeo.2008.02.012, 2008.
52. Sturzenegger, M. and Stead, D.: Close-range terrestrial digital pho-togrammetry and terrestrial laser scanning for discontinuity char-acterization on rock cuts, Eng. Geol., 106, 163–182, 2009.
53. Surinach, E., Furdada, G., Sabot, F., Biescas, B., and Vilaplana, J.: On the characterization of seismic signals generated by snow avalanches for monitoring purposes, Ann. Glaciol., 32, 268–274, 2001.
54. Suriñach, E., Vilajosana, I., Khazaradze, G., Biescas, B., Furdada, G., and Vilaplana, J. M.: Seismic detection and characterization of landslides and other mass movements, Nat. Hazards Earth Syst. Sci., 5, 791–798, doi:10.5194/nhess-5-791-2005, 2005
55. Vacher, P., Dumoulin, S., Morestin, F., and Mguil-Touchal, S.: Bidi-mensional strain measurement using digital images, Proceedings of the Institution of Mechanical Engineers, J. Mech. Eng. Sci., 213, 811–817, 1999.
56. Vilajosana, I., Suriñach, E., Abellán, A., Khazaradze, G., Garcia, D., and Llosa, J.: Rockfall induced seismic signals: case study in Montserrat, Catalonia, Nat. Hazards Earth Syst. Sci., 8, 805–812, doi:10.5194/nhess-8-805-2008, 2008.
57. Weichert, D., Horner, R. B., and Evans, S. G.: Seismic signature of landslides: the 1990 Brenda mine collapse and the 1965 Hope rockslides, B. Seismol. Soc. Am., 84, 1523–1532, 1994.
58. Wolter, A., Stead, D., and Clague, J.: A morphologic characterisa-tion of the 1963 Vajont Slide, Italy, using long-range terrestrial photogrammetry, Geomorphology, 206, 147–164, 2013.
59. Yamada, M., Matsushi, Y., Chigira, M., and Mori, J.: Seismic recordings of landslides caused by Typhoon Talas (2011), Japan, Geophys. Res. Lett., 39, L13301, doi:10.1029/2012GL052174, 2012.
60. Yamada, M., Kumagai, H., Matsushi, Y., and Matsuzawa, T.: Dynamic landslide processes revealed by broadband seismic records, Geophys. Res. Lett., 40, 2998–3002, doi:10.1002/grl.50437, 2013.

CHAPTER 8

ENERGY DISSIPATION OF ROCKFALLS BY COPPICE STRUCTURES

G. Ciabocco, L. Boccia, and M. N. Ripa

INTRODUCTION

Forests play an undeniably important role in mitigating the risk of rockfall, even though it is difficult to quantify this role, as forests often absorb the smaller fallen boulders. (Leibundgut et al., 1986; Lafortune et al., 1997; Hetu´ and Gray, 2000; Stoffel et al., 2006).

Italian legislation has officially recognized this role since 1923 (Royal Decree Law num. 3267 – 30.12.1923).

However, although much of the Italian territory is characterized by geomorphologic features that are very precarious in nature, it is also densely populated and rich in infrastructures. Therefore, continuous maintenance and protection of the infrastructures is required, generating high costs, technical difficulties and possible penal responsibilities for the technicians involved in the management of mountain regions.

These costs are often unsustainable and therefore, there is a great need for research into methodologies to improve the cost/benefit ratio of invested funds. To overcome deterministic factors (e.g., concepts of vulnerability, hazard and risk), it is imperative to quantify the contribution of forests in mitigating the likelihood of damage.

With regard to rockfalls, forest management models should take the multifunctional role of forests into consideration, and in particular, their role in mitigating risks from fallen boulders.

Moreover, planners and foresters should take into account the natural protective function of forests when defining measures to protect forests from rockfall events and optimize forest planning. On a practical level, foresters can optimize the natural protective function of forests by applying target values to structure parameters such as tree density, tree spatial distribution, species composition, tree conditions, diameter distribution and the basal area. (Chauvin et al., 1994; Wasser and Frehner, 1996; Frehner et al., 2005; Stoffel et al., 2006) How-ever, the development of technical specifications is required for coppice structures.

Four different phenomena have been identified in the physical effect of forest structures on the dynamics of fallen boulders: (a) kinetic energy absorption through direct impact between a boulder and a trunk, Gsteiger (1989), Zinggeler (1989), Stokes et al. (2005), Brauner (2005), Dorren et al. (2005, 2006), Lundstrom¨ et al. (2007, 2009), (b) kinetic energy absorption through interaction between a rock and shrub vegetation, (c) increasing rugosity of slope and a consequent reduction of coefficient restitution, Jones et al. (2000), (d) the positive effect of forest vegetation on geotechnical soil characteristics, Pfeiffer (1989).

The theoretical energy dissipated by forest structures depends on the number of tree/boulder impacts and on the kinetic energy dissipated by a single impact. Gsteiger (1989) proposed an index to describe the number of tree/boulder impacts in a given forest structure. This index is defined as the "Average Distance between Contacts" (ADC). Since the ADC index describes the average distance between two impacts, if the distance travelled inside the forest structure by a given boulder is known, then the hypothetical number of impacts can be estimated. In the ADC index, a random distribution of trees in the examined area is assumed, using the number of trees per hectare.

More recently, Brauner et al. (2005) and Dorren et al. (2007) resumed that in a tree/boulder impact, the tree dissipates the boulder's energy through the rotation and shifting of parts

Citation: G. Ciabocco, L. Boccia, and M. N. Ripa, "Energy Dissipation of Rockfalls by Coppice Structures," Nat. Hazards Earth Syst. Sci., 9, 993–1001, 2009. http://www.nat-hazards-earth-syst-sci.net/9/993/2009/nhess-9-993-2009.pdf.

of the root system, the deformation and oscillation of the tree stem, and local penetration of the rock at the point of impact.

Dorren et al. (2007) summarized as follows the methods proposed in the literature to quantify energy absorption during tree/boulder impact:

- Couvreur (1982) and Zinggeler (1989), this method derives the fracture energy from standard dynamics tests on a small sample;
- Stokes et al. (2005), this method is based on static winching experiments that calculate the energy required to cause a failure in the root system,
- Brauner et al. (2005), this method assumes that the banding energy is fully dissipated by the work expended to bend the tree stem to the point at which it breaks,
- Dorren et al. (2006), derived the tree/boulder energy absorption capacity through real-size rockfall experiments on living trees, by measuring energy absorption as a function of tree diameter at breast height (DBH) and of rockfall impact eccentricity,
- Lundstrom¨ et al. (2007, 2009), derived the energy absorption of trees subject to rockfall and root sys-tem anchorage mechanics, for many Alpine Conifer species, with a new approach and an experimental method,
- Jonsson (2007), used full-scale impact test data to calibrating a numerical single tree impact model, using Finite Element Method (FEM), in this way the author can consider energy absorption of tree dependent on: DBH, tree/boulder impact height, angle and eccentricity, level of tree damage and material properties.

• Many of these studies were predominantly conducted in the Alpine area using characteristic tree species, Berger et al. (2001, 2004), Dorren et al. (2004, 2005, 2006), Brauner et al. (2005), Stokes et al. (2005), Lundstrom¨ et al. (2007, 2009). Through these studies, numerous important steps have been made towards understanding the mitigation of rockfall risks. They have enabled a shift from a descriptive ap-proach to silvicultural systems to a more pragmatic method.

With reference to fallen boulders, the behavior of coppice is very different from that of high forest, depending on the probability of tree/boulder impacts, and the energy dissipation per impact. Since the spatial distribution of plants differs, the same approach cannot be employed to determine the probability of an impact. Moreover, the effect of an impact upon the group of stump shoots of a coppiced tree is not necessarily the same as that of a similar impact on a non-coppiced plant of equivalent section.

The aim of this research was to make an initial contribution to estimating the energy absorbed by a coppice. In particular, it is focused on evaluating the energy dissipated by a specific Appennine forest structure, by comparing the kinetic energy dissipation calculated by means of equations proposed in the literature with the actual energy measured in a full-scale test, in order to adapt the ADC index equation.

Assuming that the spatial distribution of trees in a coppice differs from that of high forest, a modification to the original equation published by Gsteiger (1989) to calculate the ADC index is proposed, introducing what the authors term the "*Structure of Interception*".

Considering the absence of specific studies on the mechanical behavior of tree stems in a coppice subject to rockfall events, a simplified approach was adopted to determining the kinetic energy absorption per impact. Recent studies (Dorren et al., 2006, Lundstrom¨ et al., 2007, 2009) have made considerable progress towards understanding the mechanism of kinetic energy absorption during tree/boulder impact, through real-size tests. However, the results of these studies are not easily applicable to coppices, due to the differences in structure between the trees in a coppice and those in high for-est areas (e.g., size of stems, root system/stem ratio, type of wood).

This study also aimed to modify the equation for calculating the number of impacts using the ADC index proposed by Gsteiger (1989), and to investigate the possibility of pro-viding forest managers with some parameters and technical aids in order to increase the protective function of Apennine coppice against falling rocks by optimizing traditional forestry management techniques.

MATERIALS AND METHOD

The experimental site is located in central Italy (42°56'14' N, 13°07›03›› E), 700 m above mean sea level (a.m.s.l.); the rocks are characterized by a compact lithology of limestone, while the forests are predominantly of flowering ash (Fraxinus ornus) and hop hornbeam (Ostrya carpinifolia) forest, once managed with a simple coppice system, but now entirely abandoned. The dendrometric data characterizing the forest area studied in this work are reported in Table 1.

Table 1. Data acquired in two forest sample plots. The data are input to calculate the average distance between contacts with Gsteiger 1989 equation and with modified equation. The two forestry sampling characterize the real structure forest crossed by analyzed boulders

Parameter	ADS_1	ADS_2
Geographic sample area (m^2)	400	400
Slope (°)	40	34
Topographic sample area (m^2)	306	332
Forest type	flowering ash and hop hornbeam forest	flowering ash and hop hornbeam forest
Silvicultural system	simple coppice system	young degradation forest
Metres Above Sea Level (m)	709	683
Exposure	SE	SE
Skeletal (%)	60	90
Fragmental (%)	3	0
Basal area (m^2)	25.929	8.11
Mean basal area (m^2)	0.0029	0.0017
Mean diameter (m)	0.061	0.047
Stand density trees/hectare (num.)	9076	4641
Stump/hectare (num.)	3688	2807
N of ADC (num.) Gsteiger, 1989	4438	3349
A (m^2)	10 000	10 000
Ns (num.) modif. Gsteiger, 1989	750	542
Nc (num.) modif. Gsteiger, 1989	3688	2807
dbh_m (m)	0.061	0.047
dbh_{int} (m)	0.5	0.5
dbh_{si} (m)	0.561	0.547
h_m (m)	6	4
V_m (m^3)	0.0107	0.0059

The first step in producing the model was to simulate rockfall dynamics using PC-Massi software© Bruschi Alberto Ver. 3.2.0, by means of which the experimental site was reproduced using the recorded data to recreate the conditions (i.e., rock mass, slope, track length and morphometric data), while considering the trajectory without forest cover. This software, concerning rockfall dynamic trajectory simulation, utilized a model based on stochastic analysis. The software was derived from calculations based on the Colorado Rockfall Simulation Program (C.R.S.P) model, developed by Pfeiffer et al. (1988, 1990, 1993). Input data and geotechnical coefficients were derived from the examination of a digital video analysis of the trajectory of fallen boulders at the experimental site, and are reported in Tables 2 and 3.

Once the boulder dynamics were known, it was possible to estimate the kinetic energy of the boulder dissipated during its track through the forest structure. The parameters used for the simulation were those of a real Apennine coppice case, reported in Table 1.

Table 2. Rock geometry and material properties and some input data for rockfall dynamics simulation. The parameters were acquired in the starting zone of rockfalls, before the controlled removing, and in the deposition zone.

Parameter	
Shape of boulder	Cylinder
Height of boulder (m)	0.8
Diameter of boulder (m)	0.56
Specific gravity of rock (kg/m^3)	2600
Simulation (num.)	999
Start horizontal speed v_x (m/s)	0.3
Start vertical speed v_y (m/s)	−0.3
Control point (horizontal distance) (m)	135

The number of impacts was calculated using Eq. (1), Gsteiger (1989), and Eq. (2) proposed in the literature. By means of Eq. (1) it is possible determine Average Distance between Contacts (ADC) index that depends on: (a) the diameter of the boulder (B), (b) the horizontal projection of the sample area (A), (c) the mean trunk diameter (dbh_m) in the sample area (A), (d) the number of trees (N) in the sample area (A).

$$\text{ADC} = \frac{A}{N \cdot B + N \cdot dbh_m} = \frac{\text{m}^2}{\text{m}} \tag{1}$$

Since the ADC index describes the average distance between two impacts, if the course length (L_{tra}) that is travelled by the boulder inside the forest structure is known, then the hypothetical number of impacts (N_{ip}) can be estimated by applying Eq. (2):

$$N_{ip} = \frac{L_{\text{tra}}}{\text{ADC}} \tag{2}$$

Table 3. Slope geometry and geotechnical properties that we have used as input data for rockfall dynamics simulation with PCMassi© (by Alberto Bruschi) simulation software. The data were acquired through a slope analyzed survey in real size and through Geographical Information Systems G.I.S. software

Segment	X_i	Y_i	X_f	Y_f	r_n	r_t	θ
1	0	126	10	106	0.31	0.87	0.25
2	10	106	13	86	0.4	0.9	0.001
3	13	86	26.95	76	0.31	0.83	0.15
4	26.95	76	42.8	66	0.31	0.83	0.15
5	42.8	66	48.11	61	0.31	0.83	0.15
6	48.11	61	56.36	56	0.31	0.83	0.15
7	56.36	56	70.69	46	0.31	0.83	0.15
8	70.69	46	84.66	36	0.31	0.83	0.15
9	84.66	36	97.43	26	0.31	0.83	0.15
10	97.43	26	115.42	16	0.31	0.83	0.15
11	115.42	16	134.59	6	0.31	0.83	0.15
12	134.59	6	144.33	2	0.31	0.83	0.15
13	144.33	2	152.68	0	0.31	0.83	0.15

In this work, the kinetic energy dissipated by a single impact (E_{diss}), was estimated adopting the approach proposed by Brauner (2005), and using Eq. (3). This approach is based on the assumption that all the kinetic energy of the boulder is required to break the trunk, taking into account the volume of the trunk itself. Therefore it can be assumed that all of the energy is dissipated virtually in the work required to reach the breaking point of the trunk (W_{dbhm}) (Brauner, 2005).

Thus, the work involved in breaking the trunk concerns only the area affected by the deformation of wood fibers. In order to utilize the recorded data relative to tree size, a magnitude which represents the work involved in breaking the trunk as a unit of volume must be introduced, this is termed $U_{fr-Wood}$ and is expressed in J/m (Brauner, 2005). $U_{fr-Wood}$ represents a relationship between the energy absorbed during impact, which produces the trunk fracture, and the volume of wood fractured the volume of the trunk. In other words, the work required to break the trunk for a given impact, or the energy dissipated (W_{dbhm}) is calculated by considering the average trunk volume (V_m), calculated using real data from forest measurements, the average diameter at breast height (d_{bhm}) and the average height (h_m):

$$W_{dbhm} = V_m \cdot U_{\text{fr-Wood}} \tag{3}$$

In this test, the work involved in breaking the trunk per volume unit ($U_{fr-Wood}$), was derived from literature data proposed by Niemz, 1993; Berger et al., 2001; Brauner 2005, reported in Table 4.

Figure 1. Aerial photograph of forest sample plots, located in two varieties of forest structure, with trajectory of boulder highlighted.

Total dissipated kinetic energy (E_{diss}), along course length into the forest structure, is calculated according to Eq. (4)

$$E_{\text{diss}} = \frac{L_{\text{tra}}}{\text{ADC}} \cdot V_m \cdot U_{\text{fr-Wood}} = J \tag{4}$$

To carry out the analysis, it was necessary to determine the parameters required to calculate the ADC index. Two sample plots along the trajectory of the boulder's fall were taken into consideration. The two sample plots represent the variety of typologies in forest structure crossed by the boulder, see Fig. 1 and Table 1.

The residual kinetic energy of the boulder obtained by simulation, derived from the difference between the kinetic energy and the energy dissipated by the forest (estimated from the control section), was compared to the authentic residual kinetic energy, estimated using the digital video images of the experimental tests.

The real speed of the boulder leaving the forest structure, and, therefore, its acquired kinetic energy, were determined by a digital video analysis of the experimental test, using the software AviStep (Michel Delabaere) (Version 2.1.1). This software allows the speed and acceleration of an object to be calculated through film, taking into account the camera data acquisition parameters (i.e., number of frames per second) and references in space relative to the object (i.e., a known distance). To make this analysis, we used a 10 frames video captured when the boulder, leaving the forest structure, has gone through the first 3 m.

In examining data from the experimental video, the kinetic energy, $E_{c-(real)}$, leaving the forest after impact with the forest structure was obtained. This energy was compared with the difference between $E_{c-(simulate)}$, (the kinetic energy calculated at the same point with simulated rockfall dynamics, which was then subtracted) and $E_{diss-(calculate)}$ (the energy dissipated by the forest, using the Gsteiger approach).

$$E_{C-(\text{simulate})} - E_{\text{diss}-(\text{calculate})} \text{ compare to } E_{C-(\text{real})} \quad (5)$$

Table 4. Literature data reported in Brauner 2005, that we used to calculate dissipated kinetic energy from a single impact. In this table we can see work of breaking as a unit of volume ($U_{fr-Wood}$)

Trees species	E_{-mod} (N/mm^2)	$\varepsilon - fr$	U_{frWood} (J/m^3)
Conifers *(Picea abies)*	5281	49.98	92 213
Forest broadleaves *(Fagus selvatica)*	7501	104.00	102 800

Table 5. Average output values on boulder speed and kinetic energy carried out with dynamic simulation model of rockfalls, hypothesized without tree/boulder impacts

Parameter	
Speed (m/s)	20.94
Kinetic energy (kj)	122

Using the simulation of rockfall dynamics, hypothesized without the interaction between boulder dynamics and the tree trunk, $E_{c-(simulate)}$ was determined. Simulations began with the data input reported in Tables 1 and 2, while the average output values are reported in Table 5.

To achieve $E_{diss-(calculate)}$, input data obtained by survey-ing were used, as well as the data relative to the forest sample plots reported in Table 6. Equation (4) was applied for each different forest structure, and a value of 34 kJ of total energy absorption was obtained.

In order to obtain $E_{c-(real)}$, the digital video data of the experimental test was analyzed. The camera was placed at the end of the forest structure and boulder trajectory. The boulder speed v_u=15,6 (m/s) leaving the forest structure was determined with AviStep software. $E_{c-(real)}$ was measured taking into account both the specific gravity of the rock, assumed at 2600 kg/m^3, Geological Survey of Italy (1976) and the volume of the boulder. The boulder's volume has been calculated using its cylindrical shape. Its geometric measures, have been taken at the end of the test, when the boulder has arrived at the depositional zone. By applying the universal kinetic energy equation to these input data and a value of 63kJ of real kinetic energy leaving the forest structure was obtained.

RESULTS

As stated above, one aim of this study was to evaluate the energy dissipated by a specific Apennine forest structure, comparing the kinetic energy dissipation calculated by means of equations proposed in the literature with the actual energy measured in a full-scale test, in order to adapt the ADC index equation.

Table 6. Resume of data input to calculate kinetic energy dissipated with Gsteiger (1989) Eq. (4). These data are divided into two sections: ADS 1 and ADS 2, that representing respectively two forests structures crossed by boulder in real size test

Parameter	ADS_1	ADS_2
L_{tra} (m)	67.5	67.5
B (m)	0.68	0.68
U_{frWood} (J/m^3)	102 800	102 800
W_{dbhm} (J)	1099.96	606.52

Using the values obtained, by comparing $E_{c-(simulate)} - E_{c-(real)}$ (122 kJ–63 kJ) with E_{diss}–(calculate) (valued at 34 kJ), it emerged clearly that the kinetic energy dissipation calculated using Gsteiger's approach was slightly underestimated (34 kJ compared to 59 kJ). This suggests strongly that is not possible to apply the ADC index proposed for high forests to coppice.

In fact, coppice structures differ from those of high forest. As a number of coppice shoots grow from each stump, it is possible to assume that they occupy approximately the same position in the area, see Fig. 2. Initially, this study used as number of trees (*N*) in its sample area (*A*), the sum of healthy trees and the number of stumps, assuming that the boulder could strike only one sample per stump of average diameter. In doing this, the actual protection value was underestimated.

In fact, considering both the fact that a number of coppice shoots grow from each stump and the spatial distribution of these coppice shoots, it is clear that the ADC index has to be increased, with a consequent reduction in dissipated energy.

However, it should also be taken into consideration that a coppice can have a "sail" effect, increasing the interceptive surface for each stump. Therefore, the W_{dbhm} increases as well as the energy dissipation per impact see Fig. 3.

To improve understanding of the action of coppice structures with regard to rockfall events, it is necessary to take into consideration the average diameter of a given falling boulder. If the boulder is small, it will have more opportunity than a large boulder to pass through the coppice shoots of the same stump. This phenomenon is further amplified when boulders bounce higher, so that they cross the coppice shoots at a greater height, where the distance between trunks naturally increases see Fig. 3.

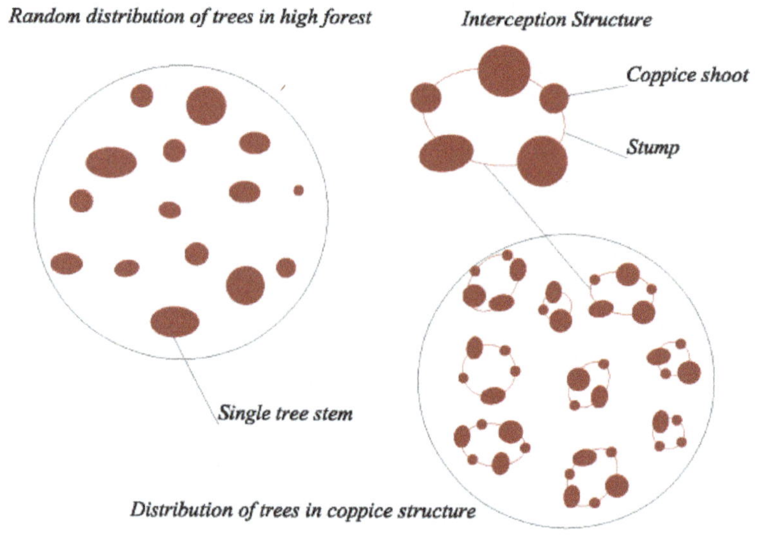

Figure 2. Graphical representation of random distribution of trees in high forest, and difference with cluster distribution (Interception Structure) in the coppice structure.

Therefore, if the average diameter of a boulder (B) is smaller than the average distance between coppice shoots (μd), we assume that forest structure is similar to a high forest with random distribution of trees per hectare. On the contrary, if (B) is greater than (μd), an "Interception Structure" in the ADC index calculations, has to be considered. Thus, if (•) (B)(μd) the authors propose the following Eq. (6) to calculate the ADC index:

$$\text{ADC} = \frac{A}{(N_c + N_s) \cdot B + N_c \cdot dbh_{si} + N_s \cdot dbh_m} = $$

$$= \frac{A}{(N_c + N_s) \cdot B + N_c \cdot \left(dbh_m + \overline{dbh}_{\text{int}}\right) + N_s \cdot dbh_m} \quad (6)$$

where the mean diameter of the stump is associated with the number of stumps (N_c) and is defined as the mean diameter of the "Interception Structure", proposed as (dbh_{si}). dbh_{si} is the sum of the mean diameter at breast height (dbh_m) plus the mean distance between the coppice shoots of the stump (dbh_{int}). In this case, the (dbh_{int}) value can be calculated arithmetically as the mean of distances ($d_1 \div d_n$) see Fig. 3, based on a significant sample of stumps. In this study a mean distance between coppice shoots equal to $\overline{dbh_{int}} = 0.5$ m was used.

In this work, mean dbh_{int}, has been obtained through the enumeration's measures of a coppice shoots sample. Changing dbh_{int} value in Eq. (6) into a rank of ±10 cm, is possible to see that there are not big changes of ADC index, ±0.03 m. At the same time, it is necessary to consider that, a silvicultural treatment with the cut of one or more coppice shoots, involves a big increase of dbh_{int}. So, if dbh_{int} is bigger than the boulder diameter, we should use Eq. (4) to calculate ADC index, that in this way, will have changed significantly its values.

Figure 3. Graphical representation of distances between coppice shoots in the "Interception Structure".

In recalculating $E_{diss-(calculate)}$ through Eqs. (6) and (4), a resulting dissipated kinetic energy value of 53.831 kJ was obtained. This value is closer to the value determined by using data from the experimental site, i.e., 59 kJ.

DISCUSSION

Discussion of the Approach to Measuring/ Evaluating Energy Absorption by Coppice

The results obtained in this study allowed a new approach to calculating the ADC index to be elaborated, requiring a knowledge of the number of trees per hectare so that what the authors term as "*Interception Structures*" (SI) can be considered.

The concept of the ADC is based on the assumption that plants are locally randomly distributed. However, since a coppice tends to have a more clustered distribution, this assumption is not satisfactory. For example, the trajectory of a given boulder might pass through a cluster, in which case it would be likely to hit several stems, on the other hand, its trajectory might be through a relatively empty area, in which case it might hit none at all. Therefore the ADC index cannot be applied in the same way to coppice structures. In the ADC index, using the number of trees per hectare, a random distribution of trees in the examined area is assumed.

Table 7. Silvicultural system in Marche region, data from Regional Forest Inventory (Sistema Informativo Forestale Regionale – SIFR 2001)

Forest structure types	Area (ha)	Area (%)
Coppice with reserves	87 050	43.3
Natural high forest	28 550	14.2
Coppice with a lot of reserves	30 275	15.0
Young forest	13 400	6.6
Coppice with standards system	13 250	6.6
Forest without management	7900	3.9
High forest (forestation)	16 775	8.3
Coppice in conversion	3150	1.5
Coppice selection system	350	0.1
Total	200 700	100

However, it should be borne in mind that it is difficult to compare the results obtained because the values compared are all relatively small, even when taking into consideration the statistical variability of dynamic rockfall simulations, in terms of kinetic energy, to the controlled section.

Discussion of Coppice Management

According to the Forest Inventory data System of the Marche Region (VV.AA., 2001), coppice is the most widespread (66.5%) silvicultural system of forest, see Table 7. The majority of regional forest areas (49%) exert a commercial-protective function, while (11%) of these forests perform only protective roles, see Table 8. Therefore coppice structures play a potentially major role in protection from rockfall.

A number of features of coppice influence the quantity of dissipated energy in a rockfall event: specific composition, spatial distribution and stand growth parameters. Therefore, it is necessary to obtain data on the quantity of energy dissipated by the application of various management techniques (which vary in relation to alternate phases of growth and re-generation) when applied to forests that perform protective functions.

In the Marche region, as occurs throughout the Italian Apennines, the large majority of coppices are managed as simple coppice structures, with harvest falling and reserve systems with a minimum number of 60 plants per hectare (Ministry Decree Law, 1964). This type of management, considering for example, a boulder similar to that in the ex-perimental test, with an average reserve diameter at breast height of 20 cm, would obtain ADC values of around 190 m. It is generally known that a boulder reaches maximum speed, and thus, maximum kinetic energy, in about 30 m (Bozzolo, 1987). For forest structures to play a fundamental role in rockfall protection, it is necessary for the ADC index to be less than 30 m. Therefore, the value estimated above for simple coppice structures (i.e., around 190 m), can be considered comparable to the absence of any forest structure, and offers poor protection.

Table 8. Forest function in the Marche region, data from Regional Forest Inventory (Sistema Informativo Forestale Regionale – SIFR 2001)

Forest function	Area (ha)	Area (%)
Commercial- protective forest	99 100	49.3
Naturalistic	59 850	29.8
Protective forest	22 875	11.4
Commercial forest	11 775	5.8
Public use	1250	0.6
Free growth	5850	2.9
Total	200 700	100

Among the types of management for coppices, the most pertinent to rockfall protection are those closest to high forests. For example, the coppice selection system is characterized by ideal features, even though this system is not often practiced in the region (0.1% of regional forest). In this sys-tem, two or more coppice shoots, having an age equal to, or a multiple of, the rotation (i.e., the time that elapses between one harvest and the following one), are always left growing on the stump. The traditionally silvicultural systems recognized that the benefit of such silvicultural systems lies in the possibility of obtaining fuel wood for the local population. This study suggests that a further benefit is a reduction in the

Average Distance between Contacts and an improvement in the *Structure of Interception* function, which could lead to an increase in rockfall protection.

Another positive aspect of such silvicultural systems is the roughness of the ground which characterizes them and the normal restitution coefficient. These parameters are improved by the effect of the stumps, which tend to be particularly large and are often above ground. However, it is quite clear that this type of management requires specialized skills in cutting, which are much more complex than those required for simple coppice management.

Other types of management which can improve rockfall protection include coppices with large reserves and coppices with a standards system. Among the types of coppice management closer to high forest, these are the most frequently used in the region, representing respectively 15.0% and 6.6% of regional forest.

Coppices with large reserves have structures similar to even-aged forests in their layers of reserves. Coppices with large reserves have the advantage of having a lower ADC than simple coppice systems, although they do not have the reserves of uneven-aged structures which would allow such a form of management not to increase the ADC during cutting operations.

Coppice with standards systems, instead, are characterized by the simultaneous presence of two structures: coppice and high forest. High forest can be pure or mixed, consisting of reserves with different classes of age, which are equal to, or multiples of, the rotation age. In this type of structure, two components that favor rockfall protection can be found: firstly, reserves which have a structure similar to that of high forest are preserved, where the ground is partially covered. Secondly, the presence of coppice guarantees the *Structure of Interception* effect.

The advantages of these types of forestry treatments can be attributed to the flexibility they give to forest management, whereby a coppice can be managed so that it produces fuel-wood, without relinquishing the permanent protection guaranteed by the uneven-aged structure of the reserves.

CONCLUSIONS

The probabilistic approach to estimating the energy absorbed by a coppice based on ADC is effectively valid, but when a forest structure is not formed by single plants, as in the case of Apennine coppice, it is necessary to use a modified model.

The experimental test to assess coppice structure energy dissipation rate (selected among others performed during the initial stages of this study) suggests that the energy dissipation flow is higher than that predicted by Gsteiger's approach, Gsteiger (1989). It also indicates that corrections taking into account the actual spatial distribution of coppice shoots are needed. Consequently the proposed equation, derived from the *Interception Structure* concept, appears to be more suit-able for estimating the ADC in the case of coppice.

According to the proposed methodology a coppice structure energy dissipation rate of 40 kJ along 100 m each has been assessed. To uphold the result a new mechanical model to describe the shots/boulder impact based on data by full scale test will be needed.

In should be borne in mind that the type of protection that forest or coppice can offer is probabilistic, unlike that offered by traditional rockfall defense systems, which is deterministic. In other words, a coppice cannot guarantee that any given boulder will be stopped. The risk of a rockfall event can be considered as equal to the probability of that event happening as a function of the vulnerability of the structures which could be damaged by it. Therefore, where the vulnerability is low, i.e. where little damage could occur to people or structures, then forest protection could be taken into consideration as a viable, ecological and cheap alternative. Where the risk is high, on the other hand, alternative protection should be used.

As regards the best practices for improving protection against rockfall by coppice structures, the results of this study suggest that the management system should be based on either the coppice selection system, or coppices with large re-serves, or coppices with a standard system. In particular, the coppice selection system, once considered anachronistic, appears to be the best choice in order to provide strong protective effects against rockfall risks. Therefore, this practice should be revalued and reproposed.

Taking into account both the costs of installing traditional rockfall defense systems (steel grid, dynamics and static barriers, etc.), and their notable impact on the visual landscape, good practices of forest management could represent a suitable and effective alternative in the mitigation of rockfall hazards.

REFERENCES

1. Berger, F., Corominas, J., Lopez-Carreras, C., Brauner, M., Kien-holz, H., and Grassl, H.: ROCKFOR Project, Rockfall – Forest Interrelation, Efficiency of the Protective Function of Mountain Forest against Rockfall, First Annual Report, 2001.
2. Berger, F., Corominas, J., Lopez-Carreras, C., Brauner, M., Kien-holz, H., and Perry, B.: ROCKFOR Project, Rockfall–Forest Interrelation, Efficiency of the Protective Function of Mountain Forest against Rockfall, First Annual Report, 2004.
3. Brauner, M., Weinmeistera, W., Agnera, P., Vospernikb, S., and Hoesle, B.: Forest management decision support for evaluating forest protection effects against rockfall, Forest Ecol. Manag., 207, 75–85, 2005.
4. Bruschi, A.: Meccanica delle rocce. Nella pratica geologica ed in-gegneristica. Ed. Flaccovio Dario, 2004.
5. Chauvin, C., Renaud, J. P., and Rupe,´ C.: Stabilite´ et gestion des foretsˆ de protection, ONF Bull. Techn., 27, 37–52, 1994.
6. Ciancio, O. and Nocentini, N.: Il bosco ceduo. Selvicoltura Assesta-mento Gestione, Accademia Italiana di Scienze Forestali, 2004.
7. Couvreur, S.: Les foretsˆ de protection contre les risques naturels.
8. Ecole Nationale du Genie´ Rural des Eaux et Forets,ˆ 89 p., 1982. Dorren, L. K. A., Berger, F., and Putters, U. S.: Real-size ex-periments and 3-D simulation of rockfall on forested and non-forested slopes, Nat. Hazards Earth Syst. Sci., 6, 145–153, 2006, http://www.nat-hazards-earth-syst-sci.net/6/145/2006/.
9. Dorren, L. K. A., Berger, F., Imeson, A. C., Maier, B., and Rey, F.: Integrity, stability and management of protection forests in the European Alps, Forest Ecol. Manag., 195, 165–176, 2004.
10. Dorren, L. K. A. and Berger, F.: Stem breakage of trees and en-ergy dissipation during rockfall impacts, Tree Physiol., 26, 63– 71, 2005.
11. Dorren, L. K. A., Berger, F., Le Hir, C., Mermin, E., and Tardif P.: Mechanisms, effects and management implications of rockfall in forests, Forest Ecol. Manag., 215, 183–195, 2005.
12. Dorren, L., Berger, F., Jonsson, M., Krautblatter, M., Molk,¨ M., Stoffel, M., and Wehrli, A.: State of the art in rockfall – forest interactions, Schweiz Z Forstwes, 158-6, 128–141, 2007.
13. Frehner, M., Wasser, B., and Schwitter, R.: Nachhaltigkeit und Erfolgskontrolle im Schutzwald. Wegleitung fur¨

Pflegemassnah- men in Waldern¨ mit Schutzfunktion. Bundesamt fur¨ Umwelt, Wald und Landschaft, Bern, 2005.

14. Geological Survey of Italy – Italian Cartographic Boby: Geological Map of Italy at the scale 1:100 000, 1976.
15. Gsteiger, P.: Steinschlag Wald Relief Empirische Grundlagen zur Steinschlagmodellierung. Diploma Thesis Geography, Depart-ment of Geography, University of Berne, 1989.
16. Hetu,´ B. and Gray, J. T.: Effects of environmental change on scree slope, development throughout the postglacial period in the Chic-Choc Mountains in the northern Gaspe´ Peninsula, Quebec,´ Geo-morphology, 32, 335–355, 2000.
17. Hoesle, B.: Rock Fall Problems and its Protection, Course script, Department of Civil and Structural Engineering, The Hong Kong Polytechnic University, 2001.
18. Jahn, J.: Proceedings, Entwaldung und Steinschlag. International Congress INTERPRAEVENT, Graz, 1, 185–198, 1988.
19. Jones, C. L., Higgins, J. D., and Andrew, R. D.: Colorado Rockfall Simulation Program Version 4.0. Colorado Department of Trans-portation, 2000.
20. Jonsson, M. J. O.: Energy absorption of trees in a rockfall protection forest. DISS. ETH No. 17214, 2007.
21. Lafortune, M., Filion, L., and Hetu,´ B.: Dynamique d'un front forestier sur un talus d'eboulis´ actif en climat temper´e´ froid (Gaspesie,´ Quebec),´ Geogr. Phys. Quat., 51 , 1–15, 1997.
22. Leibundgut, H.: Unsere Gebirgswa¨ lder. Paul Haupt Verlag, Bern, Stuttgart, 1986.
23. Lundstrom,¨ T., Jonas, T., Stockli,¨ V., and Ammann, W.: Anchor-age of mature conifers: resistive turning moment, root–soil plate geometry and root growth orientation, Tree Physiol., 27, 1217– 1227, 2007.
24. Lundstrom,¨ T., Jonsson, M. J., Volkwein, A., and Stoffel, M.: Reac-tions and energy absorption of trees subject to rockfall: a detailed assessment using a new experimental method, Tree Physiol., 29, 345–359, 2009.
25. Meissl, G.: Modellierung der Reichweite von Felsstuerzen, Innsbr. Geogr. Studien, 28, Inst. fur¨ Geogr., University of Innsbruck, 1998.
26. Ministry Decree Law 11/11/1964: Summary of Requirements of Forest Police. 1964.
27. Pfeiffer, T. J.: Rockfall hazard analysis using computer simula-tion of rockfalls, Master of Science Thesis, Colorado School of Mines, 1989.
28. Pfeiffer, T. J., Higgins, J. A., Andrew, R. D., Barrett, R. K., and Beck, R. B.: Colorado Rockfall Simulation Program Version 3.0, Users Manual, Colorado Transportation Institute, 1993.
29. Royal Decree Law num. 3267 – 30.12.1923, Gazz. Uff., num. 117– 17 May, 1924.
30. Stoffel, M, Wehrli, A., Kuhne,¨ R., Dorren, L. K. A., Perret, S., and Kienholz, H.: Assessing the protective effect of mountain forests against rockfall using a 3-D simulation model, Forest Ecol. Manag., 225, 113–122, 2006.
31. Stokes, A., Salin, F., Kokutse, A. D., Berthier, S., Jeannin, H., Mochan, S., Dorren, L., Kokutse, N., Ghani, M. A., and Four-caud, T.: Mechanical resistance of different tree species to rock-fall in the French Alps, Plant and Soil, 278, 107–117, 2005.
32. Various authors: I tipi forestali delle Marche, Monografia dell'Istituto per le Piante da Legno e L'Ambiente IPLA e Servizio Valorizzazione Terreni Agricoli e Forestali Regione Marche, 252 pp., 2001.
33. Wasser, B. and Frehner, M.: Minimale Pflegemassnahmen f ur¨ Walder¨ mit Schutzfunktionen, Wegleitung. Bundesamt fur¨ Umwelt, Wald und Landschaft (BUWAL), Berne, 1996.
34. Zinggeler, A.: Steinschlagsimulation in Gebirgswaldern:¨ Model-lierung der relevanten Teilprozesse. Diploma Thesis, Department of Geography, University of Berne, 1989.

CHAPTER 9

SHORT-TERM VELOCITY VARIATIONS AT THREE ROCK GLACIERS AND THEIR RELATIONSHIP WITH METEOROLOGICAL CONDITIONS

V. Wirz, S. Gruber, R. S. Purves, J. Beutel, I. Gärtner-Roer, S. Gubler, and A. Vieli

INTRODUCTION AND RELATED WORK

Rock glaciers are a widespread landform in alpine permafrost terrain related to the creep of frozen debris or sediment. They shape and alter the landscape and contribute to the transport of sediment. In recent years, strong variations in the behavior of rock glaciers have been detected, with a general trend towards increasing velocities over the past decade (see, e.g., Delaloye et al., 2008a), raising questions about the glaciers' stability in a changing climate.

We define rock glaciers by morphology, irrespective of their genesis, as lobate or tongue-shaped debris–ice mixtures, frequently with steep sides and snouts and longitudinal and/or transverse ridges and furrows, that are slowly creeping downslope (cf. Vitek and Giardino, 1987). Ice within rock glaciers may include buried surface ice (including glacier ice; see, e.g., Potter, 1972) or ice formed in situ (see, e.g., Wahrhaftig and Cox, 1959; Humlum et al., 2007). Rock glaciers with periglacial origins typically have an ice content of 50–90 % (Potter, 1972; Barsch, 1977; Haeberli et al., 1998). In addition, the permafrost body is mostly permeable and air voids and liquid water can occur throughout the rock glacier body (Johnson and Nickling, 1979; Vonder Mühll, 1992; Wagner, 1992; Arenson et al., 2002; Buchli et al., 2013).

Horizontal displacements of alpine rock glaciers typically range between several centimetres to around 1 m per year (Haeberli et al., 2006), and maximum velocities occur along the central flow line and decrease towards the margins (see, e.g., Barsch, 1992; Haeberli, 1985; Roer et al., 2005).

In areas of compressive flow, surface structures such as transverse ridges and furrows are typical. However, related variations in flow speed are very small (Kääb et al., 2003; Kääb and Weber, 2004) and these surface structures are advected downstream with the speed of the rock glacier body (Kääb and Weber, 2004).

Typically, rock glacier flow velocities correlate with slope angle (Francou and Reynaud, 1992; Sloan and Dyke, 1998; Konrad et al., 1999). Movement observed at the rock glacier surface is in general due to deformation of the frozen rock glacier material; however, the magnitude of deformation is not necessarily uniform with depth (Fig. 1). The few existing studies on borehole-deformation measurements in rock glaciers all show that within the permafrost body, thin layers with distinctly different rheological properties (e.g. lower viscosity) exist, in which 50–97 % of the horizontal deformation occurs (Wagner, 1992; Haeberli et al., 1998; Hoelzle et al., 1998; Arenson et al., 2002; Buchli et al., 2013, Fig. 1). Typically, no internal deformation occurs below such shear horizons, and only limited internal deformation occurs above them. Hence, surface displacement measurements appear to be a good approximation of the slope movement of the entire rock glacier.

In recent years an increasing number of studies have detected temporal variations in rock glacier movement distinguishing three timescales: decennial, inter-annual, and intra-annual (Delaloye et al., 2010).

In the Alps, long-term variations have been observed since the 1950s, with significant acceleration in the 1980s (see, e.g., Roer et al., 2005; Kääb et al., 2007; Kauf-mann and Ladstädter, 2007; Lugon et al., 2008; De-laloye et al., 2008a) which has been related to longer-term variations in climatic conditions. The majority of studies are, however, on the interannual (year-to-year) timescale (e.g. Krummenacher et al., 2008; Perruchoud

---

Citation: Wirz, V., Gruber, S., Purves, R. S., Beutel, J., Gärtner-Roer, I., Gubler, S., and Vieli, A.: Short-term velocity variations at three rock glaciers and their relationship with meteorological conditions, Earth Surf. Dynam., 4, 103-123, doi:10.5194/esurf-4-103-2016, 2016.

and De-laloye, 2007). Annually repeated measurements at 16 rock glaciers across the European Alps, suggest that most have strong, generally synchronized, interannual velocity variations (Delaloye et al., 2008a, 2010; PERMOS, 2013), which have been related to external climatic factors. The proposed main mechanism linking long-term and interannual variations to climate is through heat conduction and the related change in temperature and hence deformation of the rock glacier body (see, e.g., Hoelzle et al., 1998; Delaloye et al., 2008a, 2010; Schoeneich et al., 2015). Furthermore, enhanced pore water pressure caused by snow meltwater in- filtration has also been suggested as a potentially influential factor for interannual variations in rock glacier movement (Ikeda et al., 2008; Krainer and He, 2006; Delaloye et al., 2010).

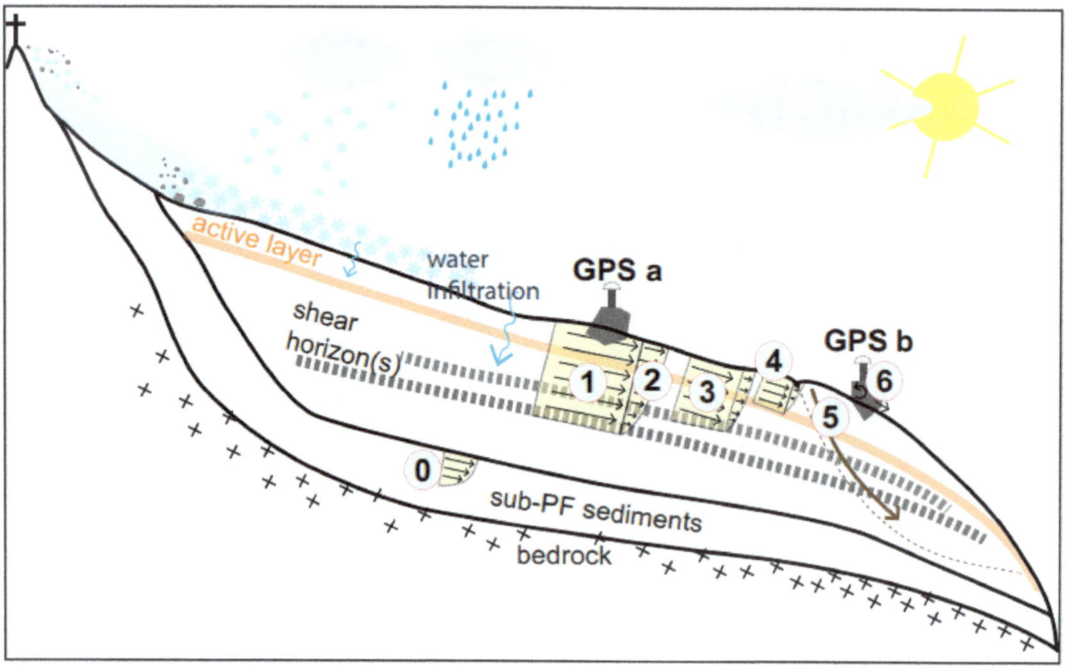

Figure 1. Schematic profile of a rock glacier. The observed movement at the surface might result from movement at different depths of the rock glacier (Roer et al., 2008): (0) deformation of subpermafrost sediments (Roer et al., 2008); (1) displacement within main shear horizon (see, e.g., Arenson et al., 2002); (2) (a little) internal deformation above shear horizon (see, e.g., Arenson et al., 2002); (3) displacement within additional shear horizon(s) (see, e.g., Delaloye et al., 2013); (4) displacement at base or within the active layer (AL) (see, e.g., Buchli et al., 2013); (5) displacement due to rotational slide (see, e.g., Roer et al., 2008); and/or (6) displacement or rotation of boulder at surface (Lambiel and Delaloye, 2004). Thick dashed lines within the rock glacier body indicate potential shear horizons.

Observations of short-term, intra-annual variations in velocities, which are the main focus of this study, are less common, mainly because of limitations in measurement techniques. Most studies on these timescales are either based on measurements with low temporal resolutions (~ 1–4 months, e.g. Haeberli, 1985; Kääb et al., 2003; Roer, 2005; Perruchoud and Delaloye, 2007; Krummenacher et al., 2008) or had relatively short total observation periods typically of a few months (Roer, 2005; Krainer and He, 2006; Buchli et al., 2013, Table 1). An overview of previous studies on intra-annual variability in rock glacier movements and the related timescales is given in Table 1. For various rock glaciers, seasonal variations in velocity up to 50 % of the mean annual velocity were observed (see, e.g., Haeberli, 1985; Hausmann et al., 2007; Perruchoud and Delaloye, 2007), but the magnitude often varied significantly from year to year (Delaloye et al., 2010). Such seasonal fluctuations occur mostly at the same time of the year (Delaloye et al., 2010), with highest velocities in autumn or late summer and lowest velocities in winter (Kääb et al., 2005; Perruchoud and Delaloye, 2007; Buchli et al., 2013).

Table 1. Overview of previous studies on the intra-annual variability in rock glacier movement found in literature. For each study (if available) the following information is listed: name and location of investigated rock glacier(s) (location), the number of measurement points (*N*), the surveying technology (technology), the observation period (period), and the temporal resolution (resolution). In addition, it is indicated for each study whether the following observations have been made (yes: observed; no: not observed; ?: no statement): a seasonal cycle of the movement (seasonal cycle), a smoother seasonal deceleration than acceleration (smoother deceleration), a seasonal acceleration beginning during the snowmelt period (acceleration during snowmelt)

Location	*N*	Technology	Period	Resolution	Seasonal cycle	Smoother deceleration	Acceleration during snowmelt	Reference
Murtel/Muragl (Grison, Switzerland)	45/14	photogrammetry	1971–1973		yes	?	?	Barsch and Hell (1975) Wagner (1994)
Gruben (Valais, Switzerland)	8	electro-optical distance measurement	August 1979–August 1982	~ 3 months	yes	?	?	Haeberli (1985)
Furggentaelti (Bernese Alps, Switzerland)	32	tachymetry	autumn 1998–1999 (1 year)	monthly	yes	yes	yes	Mihajlovic et al. (2003) Krummenacher et al. (2008)
Muragl (Grison, Switzerland)	20	total station	1998–2001	~ 4 months	yes	no	no	Kääb et al. (2003) Kääb et al. (2005)
Muragl (Grison, Switzerland)	2	borehole inclinometer	1999–2000	~ monthly	yes	no	no	Arenson et al. (2002)
Reichenkar (Stubai Alps, Austria)	3	GPS	July–September 2003 and June–October 2002	10 s (87 d)	no	no	no	Krainer and He (2006)
Huhh1 in Hungerlitälli (Valais, Switzerland)	25	terrestrial geodetic survey	July–August 2003	36 days	yes	?	?	Roer (2005)
Oelgrube (Stubai Alps, Austria)	21	GPS	July–September 2003	55 days	yes	?	?	Krainer and He (2006) Hausmann et al. (2007)
Tsarmine (Valais, Switzerland)		GPS	October 2004–August 2005	≥ 8 months	yes	?	?	Lambiel and Delaloye (2004) Lambiel et al. (2005)
Becs-de-Bosson (Valais, Switzerland)	6(85[1])	GPS	2004–2006	30–60 days	yes	yes	yes	Perruchoud and Delaloye (2007) Delaloye et al. (2010)
Mount Gibbs (Sierra Nevada, USA)	1	InSAR	April 2007–May 2008	48–138 days	yes	?	?	Liu et al. (2013)
Furggwanghorn (Valais, Switzerland)	1	borehole inclinometer	October 2010–May 2011	daily	yes	yes	yes	Buchli et al. (2013)
Büz North (Grison, Switzerland)	1	borehole inclinometer	2000–2003/05	3 h	yes	yes	yes	Ikeda et al. (2008)

Typically, such seasonal velocity fluctuations have been observed to lag behind ground surface temperature (GST) by a few weeks to months (Kääb, 2005; Delaloye et al., 2010; Buchli et al., 2013), suggesting that changes in the temperature of the rock glacier body through vertical heat conduction are an important influence on seasonal velocity variations (timing and amplitude; Kääb et al., 2005; Lambiel et al., 2005; Delaloye et al., 2010). Furthermore, the observed decrease in rock glacier velocities in winter was typically smooth and gradual and started a few weeks to months after the initiation of the seasonal cooling of GSTs (Delaloye et al., 2010; Wirz et al., 2015). The observed acceleration either occurred progressively and lagging behind the temperature increase (3–4 months after surface melt started; Kääb et al., 2005), or it occurred rapidly during the snowmelt period (Krummenacher et al., 2008; Perruchoud and Delaloye, 2007; Buchli et al., 2013; Wirz et al., 2015), indicating a potentially strong influence of meltwater infiltration.

Similarly, in addition to seasonal cycles, velocity observations with high temporal resolutions on the Furggwanghorn rock glacier (Buchli et al., 2013) and on the rock glaciers in the Mattertal discussed in this study (Wirz et al., 2015) detected strong peaks with very short durations of a few days during the snowmelt period. These observations suggest a very rapid and direct response of flow speeds to meteorological forcing, again likely involving infiltration of meltwater. However, these studies have only made qualitative links between short-term variations in flow speed and external environmental forcing.

In this paper we therefore extend our previous work which focused on extracting and describing the velocities of these three rock glaciers (see Wirz et al., 2015) by quantitatively exploring the link between surface velocity, and in particular short-term peaks, and meteorological and snowpack factors. We analyze a continuous time series of surface velocities from six GPS stations located on three rock glaciers in the Mattertal, Switzerland, with the factors analyzed including GST, air temperature, precipitation, snow cover duration, and snowmelt period. Velocities are derived from GPS measurements over 3 years with a temporal resolution of 1 day and a spatial accuracy in the sub-centimeter range (Wirz et al., 2014).

Thus, we focus here on addressing the following research questions:

- Are continuous GPS measurements with high temporal resolutions consistent with previous observations of intra-annual variability in rock glacier movement?
- Can meteorological and snowpack factors influencing intra-annual variability in rock glacier velocity be quantitatively analyzed, and how can these factors be related to a better process understanding?

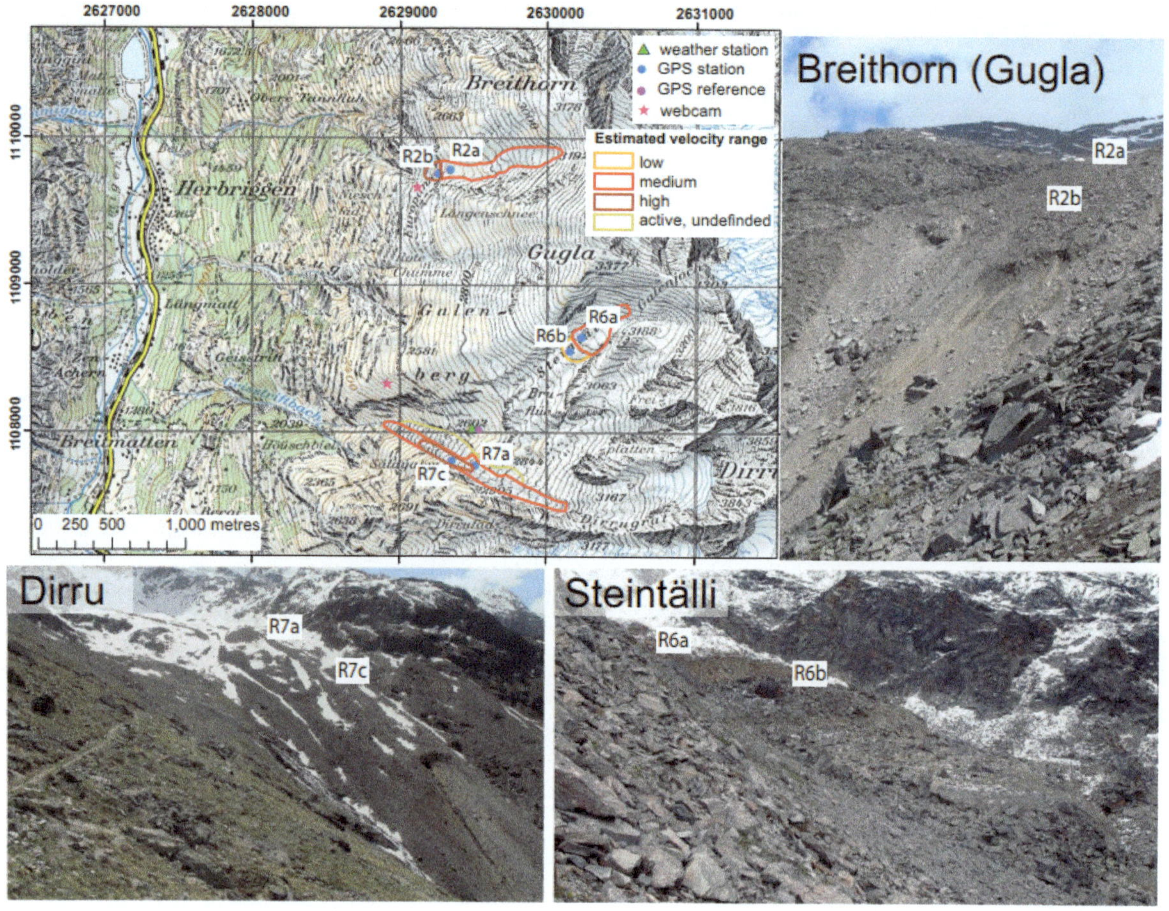

Figure 2. Overview of study site and main instrumentation. Estimated velocity range are a few decimetres per year for the category "low", several decimetres per year for the category "medium", and more than 1 m per year for the category "high". This was estimated based on field observations, expert knowledge (Hugo Raetzo, Federal Office for the Environment), and published observations (Delaloye et al., 2008b; Strozzi et al., 2009a, b; Delaloye et al., 2010, 2013; Wirz et al., 2015). Right and bottom: field impressions of rock glacier Breithorn (R2), Steintälli (R6), and Dirru (R7). Background: topographic map LK25 from the year 2013 reproduced by permission of Swisstopo BA15027.

STUDY SITE

The study site is located above the villages of Herbriggen and Randa on the orographic right side of the Mattertal in the Swiss Alps (see Fig. 2). Permafrost is likely to exist beneath much of the study area, especially above 2850 m a.s.l. (Boeckli et al., 2012). The main lithology is gneiss belonging to the crystalline Mischabel unit (Labhart, 1995).

Climate in the study area can be described as inner-alpine continental, with an average annual precipitation of 523 mm and mean annual air temperature of 5.2 C (long-term climatic mean (1961–1990) obtained from the closest MeteoSwiss station, located in Grächen, at 1550 m a.s.l., 9 km away from the study site). Over the experimental period of this study (20 August 2011–19 August 2014), the mean annual air temperature (MAAT) in Grächen was approximately 1 C warmer and the mean annual precipitation was approximately 100 mm more than the long-term climatic mean. During this period a local weather station installed at the study site at 2697 m a:s:l: measured an MAAT of about 0.5 C and an average annual liquid precipitation of about 630 mm. Due to a limitation of the instrument used (Vaisala WXT520), only liquid and not solid precipitation was measured.

Within the project PermaSense (http://www.permasense.ch), a total of 18 GPS stations were installed on various land-forms within the study site (Wirz et al., 2015). Six of these GPS stations are located on three active rock glaciers within the boundaries of the study site. All three rock glaciers have a distinct morphology, and for two, exceptionally high movement rates at their tongues were detected by InSAR (Delaloye et al., 2008b; Strozzi et al., 2009a) and later confirmed by GPS measurements (Delaloye et al., 2013; Wirz et al., 2015). These active rock glaciers and their associated measurements constitute the core subjects studied within this pa-per. The GPS stations are installed on large well-grounded boulders in such a way that the displacement is as representative as possible for this rock glacier lobe (e.g. in the middle of the rock glacier lobe, not at its margins; further details on the strategy for the selection of GPS locations are given in Wirz et al., 2013). Table 2 gives an overview of the topographic, geomorphologic, and kinematic situation of each rock glacier.

Table 2. Main characteristics regarding topography, morphology, and kinematics of the three investigated rock glaciers (Wirz et al., 2013; Delaloye et al., 2010, 2013; Wirz et al., 2015). Description of kinematics is based on expert knowledge (Dr. Hugo Raetzo, Federal Office for the Environment) and published observations (Delaloye et al., 2008b; Strozzi et al., 2009a, b; Delaloye et al., 2010, 2013; Wirz et al., 2015). In addition, for each GPS station the average velocities from summer 2011 to summer 2015 are given

Rock glacier unit	Topography	Morphology	Kinematics
Breithorn R2	Situated in valley between the slopes of Breithorn and Gugla; convex topography, where ~100 m above the front, the slope angle rapidly increases from about 20 to 35°.	The unit consists of various lobes. The main lobe can be divided into two main zones: upper, flatter zone (R2a) – transverse furrows and ridges indicating compressive flow; lower zone (R2b) – destabilized tongue with open cracks.	Upper zone with average velocities of several decimetres (R2a: 1.7 m a^{-1}); lower steep zone (tongue) with velocities of several metres per year (R2b: 6.8 m a^{-1}).
Steintälli R6	Situated in a small valley below the peak Chli Dirruhorn; the upper lobe (R6a) is slightly steeper (20 °) than the lower lobe (11 °, R6b).	Two main lobes of different lithologies, one above the other; steep fronts and transversal furrow and ridges indicating compressive flow on both lobes. Both lobes may originate from ice-cored moraines.	The upper lobe, moving at several decimetres per year (R6a: 0.6 m a^{-1}), currently overrides the lower lobe, moving at a few decimetres per year (R6b: 0.2 m a^{-1}).
Dirru R7	Below Dirru glacier, on the orographic right side of Dirrugrat; convex topography, where ~400 m above the front, the slope angle rapidly increases from about 15 to 35°.	The unit includes various tongues and lobes from different generations with different degrees of activity (inactive–destabilized); the currently active lobe can be divided into an upper, flatter zone (R7a) with several ridges and furrows, partly nested (both transversal and longitudinal), and various depressions; and a lower steep zone (R7c) with a few longitudinal ridges but no transversal ridge–furrow structure.	Upper, flatter part moving at velocities of several decimetres per year (R7a: 1.2 m a^{-1}); the steep fast-moving tongue advancing with velocities of several metres per year (R7c: 4.9 m a^{-1}).

The rock glacier Breithorn (R2) is situated between the mountains Breithorn and Gugla and extends into the Bielzigji torrent (Fig. 2). It stretches from approximately 2550 to 2850 m a.s.l., is about 600 m long, 130 wide, and about 40 m thick (Delaloye et al., 2013). It has a convex profile curvature: approximately 100 m above the front the slope angle rapidly increases from about 20 to 35°. 1 GPS station R2a2 is located on the flatter part of the rock glacier at 2704 m a.s.l. that also moves more slowly than the steeper frontal part (tongue) where GPS R2b is located at 2650 m a.s.l. (Delaloye et al., 2013; Wirz et al., 2015). Since about 1995 this tongue has accelerated to about 2–3 m a^{-1}. The cause of this acceleration is not well known but has been attributed to a combination of the effects of topography, geometry of the landform, and increased permafrost temperatures over the last 2 decades (Delaloye et al., 2013).

The rock glacier Steintälli (R6) consists of two superimposed lobes of different sediment lithologies and moving at different velocities (Fig. 2). The rock glacier is approximately 550 m long and 160 m wide, and each lobe is about 25 m thick (derived from analysis of a digital elevation model with a resolution of 2 m). The upper lobe currently overrides the lower lobe. The GPS station at position R6a is located at 3020 m a.s.l., with a slope angle of 20° on the upper lobe, while position R6b is located on the lower lobe at 2991 m a.s.l. and has a slope angle of 15°.

The rock glacier Dirru (R7) is composed of various lobes and fronts (Fig. 2), originating from different rock glacier generations (Delaloye et al., 2013). The currently active lobe, which is located on the orographic right side of Dirrugrat, has a total length of more than 1 km, is about 60 to 120 m wide, and is approximately 20 m thick (Delaloye et al., 2013). It has a convex profile and slope angles increase from about 15° in the upper part to more than 30° in the lower part towards its front. Since the 1970/80s this frontal steep part (tongue) has progressively accelerated and reached surface velocities above 5 m a^{-1}, potentially indicating a phase of destabilization (Delaloye et al., 2013). From 2008 to 2010 velocities measured at the front of R7 using manually executed annual GPS surveys have decreased slightly (Delaloye et al., 2013) with recent increases observed since 2011 (Wirz et al., 2015). The GPS station at position R7a is located on the upper flatter part at 2772 m a.s.l. The GPS at position R7c was installed on a small longitudinal ridge at 2673 m a.s.l. more or less in the middle of the steep fast-moving tongue.

Further details on the description of the individual rock glaciers as well as a detailed description of the field instrumentation can be found in Wirz et al. (2015).

DATA AND METHODS

GPS and Inclinometer Data

Each GPS station consists of a low-cost single-frequency GPS receiver (u-blox LEA-6T) with an active antenna (Trimble Bullet III) and a two-axis inclinometer (VTI SCA830- D07) mounted on a mast 1–1.5 m above ground (Beutel et al., 2011; Buchli et al., 2012; Wirz et al., 2015). GPS sensors as well as inclinometers are sampled continuously at 30 s and 2 min intervals respectively. From this primary raw data, daily position solutions and inclinometer values are computed. A static approach based on a single-frequency differential carrier-phase technique with short baselines (0.2 to 1.5 km, using the Bernese software) was used for the postprocessing of the raw GPS to positioning data (Limpach and Grimm, 2009; Dach et al., 2007). The standard deviation (SD) of the GPS position solutions derived is typically about 1.5 mm in the horizontal and about 3.5 mm in the vertical direction, and related covariances are small (< 0.01 mm, Wirz et al., 2014). The inclinometer measurements mounted directly on the GPS mast allow correction of GPS positions for mast tilt (Wirz et al., 2014). SDs in inclination (θ) are typically around 0.4°, and they are around 2.5° for the orientation (az) of the mast tilt. It has been found that the correction for mast tilt is only appropriate where tilting is greater than the accuracy of the inclinometer measurements (see Supplement and Wirz et al., 2014).

GPS position data were collected continuously between summer 2011 and summer 2014. However, all stations have some data gaps complicating the data analysis and interpretation. During the first winter (2011/12) a 30-day gap occurred for all positions, due to a power outage at the GPS reference location, which did not allow a differential processing of the GPS positions. Position R7a has a large data gap from 17 July 2012 to 23 April 2013 caused by a missing instrument. For this study we structure the total study period from 20 August 2011 to 19 August 2014 into three periods of 1 year each: the study years 2011/12, 2012/13 and 2013/14 lasting from 20 August of the first year to 19 August of the following year; for example, the study year 2011/12 lasts from 20 Au-gust 2011 to 19 August 2012.

Meteorological and Auxiliary Data

GST is measured next to each GPS station within a radius of about 7 m using five miniature temperature loggers (Maxim iButton DS1922L; see Gubler et al., 2011). These temperature loggers were distributed in small spaces between the surrounding boulders at depths of approximately 5 to 30 cm to prevent direct solar exposure. Those GST measurements have an accuracy of ±0.5°C (near 0°C; Gubler et al., 2011) and a temporal resolution of 3 h. For each GPS station, a daily mean GST was calculated by averaging across all measurements of the respective temperature data loggers deployed at the site in question. Applying the approach of Schmid et al.

(2012), the duration of the winter snow cover and zero curtain period were derived from GST. The zero curtain period is defined as the duration of the zero curtain, the effect of latent heat in maintaining temperatures near 0 Cover extended periods in freezing or thawing soils (see, e.g., Outcalt et al., 1990). The zero curtain period is defined as period during which at least for one iButton a zero curtain was detected. Winter snow cover and the zero curtain period were calculated for each individual station as well as jointly across all GST data points at the study site (Wirz et al., 2015).

Air temperature and liquid precipitation, measured at the local weather station (Vaisala WXT520, Fig. 2), are available with a temporal resolution of 2 min and an accuracy of ±0.3°C for air temperature and ±5 % for liquid precipitation measurements. Air temperature was aggregated to a daily mean value. The measured amount of liquid precipitation was aggregated for daily mean and daily maximum.

Three webcams distributed within the study site provide additional information about (actual) surface characteristics (Fig. 10). Based on these daily webcam images taken at noon, for each day it was subjectively classified whether new snow (true/false), strong melt (true/false), a wet front (values: 1 – clearly absent; 2 – possibly absent; 0 – no statement possible; 3 – visible to some extent; 4 – clearly visible), or water outflow were visible at the active front of each rock glacier (see Fig. 10).

Velocity Estimations using SNRT

For the calculation of the magnitude of the horizontal velocity (*v*; in this paper, referred to as velocity) and the direction of movement (azi_v), the SNRT (signal-to-noise thresholding) method was applied. Detailed information and a discussion of limitations and advantages of this method are given in Wirz et al. (2014). When using SNRT, temporal resolution (duration of applied smoothing window, here called velocity period) directly depends on the signal-to-noise ratio (SNR) of the position data and therefore varies with time.

For each velocity period the SNR must be higher than a predefined threshold (SNR-t). A suitable SNR-t is found empirically (Wirz et al., 2015), using a similar approach as Laube and Purves (2011). In this study, a threshold of 40 seems optimal because (a) the influence of noise on estimated velocities is strongly reduced and (b) the same SNR-t can be applied for comparatively slow (≤ 0.5 m a^{-1}) and fast (≥ 4 m a^{-1}) stations, facilitating the comparison of individual stations (Wirz et al., 2015). The sensitivity of the timing of velocity changes to SNR-t is reported in Sect. 4.1.

Mean annual horizontal velocities of each study year (MAV_{12}, MAV_{13}, MAV_{14}) are given as total horizontal displacements divided by time and were only calculated if at least one data point existed within the first and last 10 days of the period. MAV_{12-14} refers to the mean velocity over the entire measurement period. Interannual variability between the study years (2011/12, 2012/13, and 2013/14) was calculated with respect to the MAV of the previous year. The intra-annual variability in velocity of each GPS station is given as relative deviation to the MAV_{12-14}. The SNR of velocities in the vertical direction is much lower and therefore does not allow the investigation of the intra-annual variability.

Table 3. Overview of potential explanatory variables used. As potentially important time periods, we used days during the peak, also including the total of days during the peak and 7 days before the peak. For precipitation for the logistic regression models, a log transformation was applied (log)

Name	Description	Data used	Log
Prec	mean daily rainfall	weather station at the study site	yes
mPrec	maximum daily rainfall	weather station at the study site	yes
PDD_{air}	mean number of positive degree days derived from daily mean air temperature	weather station at the study site	no
PDD_{gst}	mean number of positive degree days derived from GST	GST of iButtons next to the GPS	no
n.snow	mean number of days with fresh snow	Webcam images	no
zc.ind	index for zero curtain derived from GST (yes/no)	GST of iButtons next to the GPS	no
snow.ind	index for insulating snow cover derived from GST (yes/no)	GST of iButtons next to the GPS	no
With adjunct	**Time period**	**Example**	
0	Days during velocity period	Prec.0	
7	Days during velocity period as well as 7 days before	Prec.7	

Detection of Peaks in Velocity Data

Especially at two GPS stations (R2b, R7c) sudden strong peaks in velocities were observed (Wirz et al., 2015). We here define a peak as a velocity period where (a) the velocity is distinctly higher, both compared to the velocity before and after the peak, and (b) the strength of the peak ($diff_p$) must be greater than a threshold tp. $Diff_p$ is the percentage velocity ratio between the peak velocity and the mean (μ) of the previous (v_i−1) and following (v_i+1) velocity periods ($diff_p = \frac{v_i}{\mu(v_{i-1}, v_{i+1})}$). The detection of peaks depends on (a) the SNR-t used to calculate velocities (here SNR-t = 40) and (b) the threshold t_p (here t_p = 6) used for the peak detection (Fig. 9, Sect. 4.1).

Analysis of Velocity Peaks

To investigate the relationship between individual meteorological factors (see Sect. 3.2) and the occurrence of a peak in velocity, we applied both qualitative and quantitative analyses. All analyses are performed with the software R using the packages MASS (Venables and Ripley, 2002) and pROC (Robin et al., 2011). As potential explanatory variables we used the following: cumulative amount and maximum of daily amount of liquid precipitation (Prec. and mPrec.), cumulative amount of positive degree days of air temperature (PDD_{air}) and GST (PDD_{gst}), the number of days with visible fresh snow (n.snow), and the number of days during the zero curtain period (zc; Table 3). All variables were calculated once for the duration of the peak only and once for the days during and 7 days before peak. Because the duration of the peaks varies, the cumulative amount is normalized by the duration of the peak (number of days).

To visualize potential relationships between individual explanatory variables and the occurrence of peaks, we used spineplots (Friendly, 1994), showing the frequencies of the occurrence of velocity peaks as opposed to the absence of velocity peaks (later called no-peak periods) in relation to potential meteorological variables. The width of the individual bars relate to the 25 % quantiles of the explanatory variable. Since a range of processes and meteorological variables may have been relevant for the occurrence of peaks during snow cover and snow-free conditions, we distinguished between periods with an insulating snow cover (snow cover period) and without (snow-free period). The snow cover period includes the days where an insulating snow cover can be derived from GST data (Sect. 3.2). In addition, binary values (0/1) are applied to distinguish between periods during and outside the zero curtain period, which was also derived from GST measurements.

In order to quantify the importance of the individual variables, we applied the Wilcoxon rank-sum test (Baur, 1972), with the alternative hypothesis that the distribution of explanatory variables differs for peak periods and no-peak periods. Based on this preliminary analysis, relevant explanatory variables could be preselected.

Logistic regressions were used to study the relationship between a set of explanatory variables and a binary response variable, such as avalanche release (Jomelli et al., 2007) or the occurrence of debris flows (Xu et al., 2012), landslides (Bernknopf et al., 1988; Ohlmacher and Davis, 2003; Can et al., 2005), or rock glaciers (Brenning and Trombotto, 2006). In this study, the occurrence of a peak is the binary variable, and the logistic model has the form of

$$\text{logit}(x) = \ln\frac{\Pi(x)}{1-\Pi(x)} = \beta^{T}x, \quad (1)$$

where P(x) is the conditional probability of the occurrence of a peak given a set of explanatory variables x, and β is a coefficient vector determining the contribution of each explanatory variable to a prediction (Hosmer et al., 2013). The odds of P (x) are given by the ratio of P (x)/(1 − P (x)). If the variable xi with coefficient β_i is increased by one unit, the odds of P (x) change by a factor of $e^{\beta i}$.

Table 4. Characteristics of individual GPS stations. N: number of daily GPS solutions obtained; from and to: measurement period; disptot: total displacement; σ_{disp}: standard deviation of total displacement (estimated based on 2000 Monte Carlo simulations); disp-ele$_{tot}$: total vertical displacement; disp-eleα: vertical displacement related to surface slope angle with related uncertainties from 5° variations in surface slope ; MAV_{12-14}: mean velocity for the entire measurement period; $MAV1_2$, MAV_{13}, and MAV_{14}: mean annual velocities. Note that for R2a only GPS measurements since 23 February 2012 exist and, thus, results for MAV_{12-14} and especially R2a results for MAV_{12} may not be fully representative of the "true" displacement for the period 2011/12. For R7a, there is a data gap from summer 2012 to spring 2013

Pos.	*N*	From	To	$disp_{tot}$ (m)	σ_{disp} (mm)	disp-ele$_{tot}$ (m)	disp-ele$_{\alpha}$ (m)	MAV_{12-14} ($m a^{-1}$)	MAV_{12} ($m a^{-1}$)	MAV_{13} ($m a^{-1}$)	MAV_{14} ($m a^{-1}$)	$MAGST_{12}$ (°C)	$MAGST_{13}$ (°C)	$MAGST_{14}$ (°C)
R2a	846	25 February 2012	19 August 2014	4.29[1]	0.006	1.59[1]	2.0[1] (0.5)	1.7[1]	1.4[2]	1.6	2.0	NA	0.93	0.59
R2b-f	981	20 August 2011	19 August 2014	20.32	0.01	14.97	13.7 (2.7)	6.8	5.9	7.1	7.3	1.38	1.09	1.67
R6a	958	20 August 2011	19 August 2014	1.70	0.006	0.25	0.62 (0.2)	0.6	0.4	0.6	0.7	0.26	0.31	0.02
R6b	1003	20 August 2011	19 August 2014	0.65	0.005	0.15	0.17 (0.1)	0.2	0.2	0.2	0.2	0.53	0.37	−0.11
R7a	700	20 August 2011	19 August 2014	5.00[3]	0.005	1.60[3]	1.1 (0.3)	1.6[3]	1.4[4]	NA	1.7	0.31	−0.45	0.03
R7c	784	20 August 2011	19 August 2014	14.59	0.006	9.62	9.5 (1.9)	4.9	4.2	4.3	6.1	0.38	0.15	0.38

[1] Refers to measurements from 23 February 2012 to 19 August 2014.
[2] Refers to measurements from 23 February 2012 to 19 August 2012.
[3] Has data gap from 18 July 2012 to 22 April 2013.
[4] Refers to measurements from 20 August 2011 to 17 July 2012.

All periods of both stations (R2b and R7c) were included in the logistic regression models (number of periods used for logistic regression models: N = 276; with N peaks = 37 and a median duration of the velocity periods of 6 days). As potential variables for the logistic regression, all variables with a significant P value from the Wilcoxon rank-sum test (pw ≤ 0.05) were selected. In addition, interaction terms with the indices for snow cover (snow.ind), for zero curtain (zc.ind), and for position (R2b and R7c) were included in the models to account for seasonal effects and differences between the individual rock glaciers. As precipitation is not normally distributed, we applied a log transformation for Prec. and mPrec. To obtain the most appropriate models (later called final models), an iterative stepwise model selection by the Akaike information criterion was applied (AIC; Venables and Ripley, 2002). Final model performance was evaluated using the area under the receiver-operating characteristics curve (AUROC; Hosmer et al., 2013).

RESULTS AND INTERPRETATION

Sensitivity Tests for Peak Detection

Applying different SNR-t values has a negligible influence on the observed seasonal variability for all stations (e.g., Fig. 9b). The detection of peaks, however, is sensitive to the applied thresholds. The higher the SNR-t, the fewer peaks are detected and the start of a peak is often delayed (Fig. 9a and b). Hence, for SNR-t a balance between the reliability of a peak and the detection of all peaks needs to be considered.

For small SNR-t (< 20), velocities are influenced by noise and, hence, velocity peaks are potentially caused by noise in the GPS positions (Wirz et al., 2014). For higher SNR_t (≥ 30), variability is strongly reduced, velocities become more stable, and differences between the variability in estimated velocities (and detected peaks) for different SNR-t become smaller. In addition, the number of peaks during winter (December–March) is strongly reduced. However, with an SNR-t of 40, peaks were almost solely detected for stations with high velocities (R2b and R7c). Furthermore, the number of detected peaks also depends slightly on the applied threshold t_p (Fig. 9c). The influence of t_p is larger for R2b than for R7c. For R7c, for $t_p \geq 3$ % the number of peaks is almost stable. For R2b, with $t_p \geq 6$ % the number of peaks becomes quite stable and detected peaks correspond well with manually identified peaks.

We studied the sensitivity of our statistical analysis of peaks to a change in SNR-t (20, 40) and t_p (6, 10) and found that a change in the SNR-t and t_p has little influence (Tables S3, S4, and S6 in the Supplement): for example, independent of the applied thresholds, the estimates for precipitation, temperature, and zero curtain are similar in all final models and the highest AUROC ("best" model) was always obtained if precipitation and PDD_{gst} during the peak were included.

Mean Velocities and their Interannual Variability

The average mean annual horizontal velocities (MAV_{12-14}) in the study site over the 3 study years ranged from 0.2 to 6.5 m a^{-1} (Table 4) and confirmed earlier observations based on 2 years of data (Wirz et al., 2015). Highest velocities were measured at the steep fronts of the rock glaciers Breithorn (R2b: 6.5 m a^{-1}) and Dirru (R7c: 4.9 m a^{-1}). The average vertical displacement ranges from 5 cm to more than 5 m per year. A comparison with expected vertical movement from horizontal displacement and slope angle indicates a slight thinning for stations R2b and R7a and a slight thickening for the stations R2a and R6a, but uncertainties are high and are not conclusive regarding mass change in these rock glaciers. On both the Dirru and Breithorn rock glacier, the 3–4 times faster flow of the lower compared to the upper stations indicated extensional flow. By contrast, the much slower Steintälli rock glacier instead shows compressional flow.

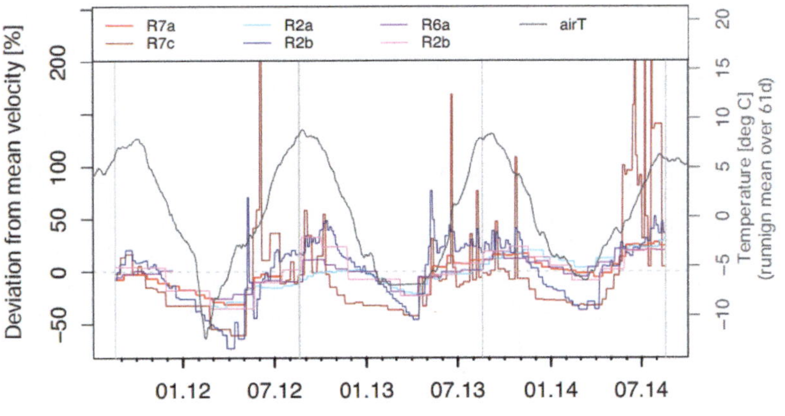

Figure 3. Intra-annual variability in horizontal velocities. Left y axis: intra-annual variability, expressed as deviation from MAV12–14. Right y axis: filtered air temperature (running mean over 61 days). Vertical grey lines separate the 3 different years of observation (2011/12, 2012/13, 2013/14).

At all stations, mean annual velocities (MAV) increased over the entire observation period (2–42 %, Table 4), and this is most pronounced for R2a (Fig. 4). However, the percentage increase from year to year is not simultaneous between stations. The winter minima in velocity increased for all stations from the study year 2011/12 to 2013/14.

At the study site and at the MeteoSwiss weather station in Grächen (1550 m a.s.l.), both MAAT and average mean annual ground temperatures (MAGST) over all iButtons were highest in the study year 2011/12 and lowest in 2012/13 Table 4). In contrast, in winter (mean of January and February), both air temperature and GST were lowest in 2011/12 and highest in 2013/14 (Fig. 3), both at the study site and in Grächen.

The annual amount of precipitation at the study site and at the MeteoSwiss station in Zermatt (1638 m a.s.l.) was highest in the period 2013/14 and lowest in 2012/13, whereas in Grächen the annual amount of precipitation was about +10 % higher in 2012/13 than in 2011/12 and 2012/13. The snow cover duration at the study site derived from GST was longest during the period 2012/13 (226 days, median of GST measurements) and shortest in 2013/14 (195 days; 2011/12: 218 days).

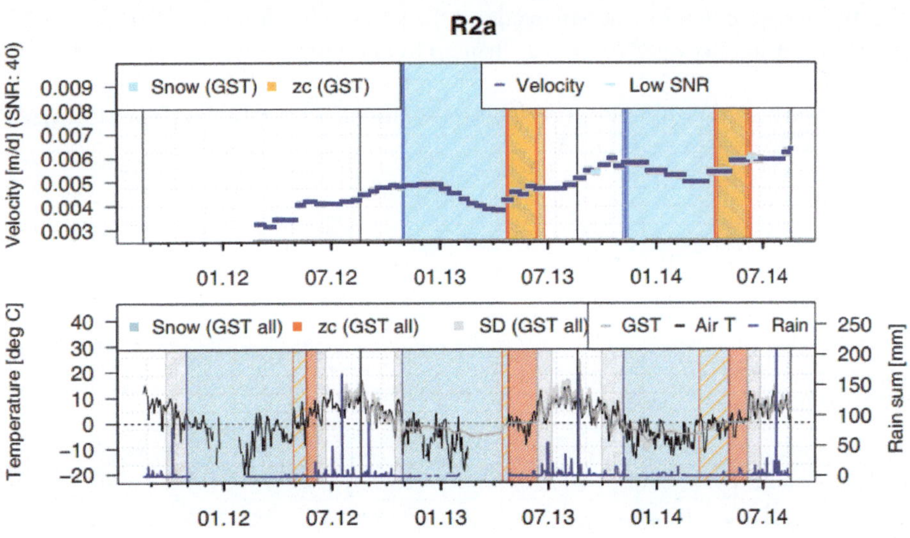

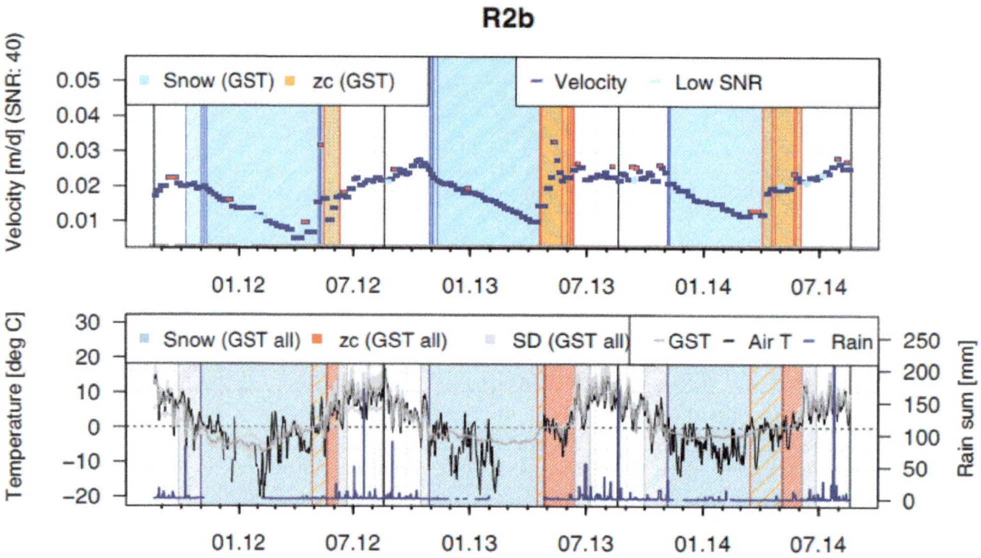

Figure 4. Comparison of the horizontal velocities to meteorological factors for Breithorn rock glacier (R2a and b). Vertical black lines separate the 3 different years of observation (2011/12, 2012/13, 2013/14). Top: horizontal velocity (dark blue), its standard deviation (grey), and velocity periods detected as peaks (red dots), periods with an SNR below SNR-t are indicated in light blue. Grey dots at the bottom indicate days with available GPS positions. Zero curtain period (orange) and insulating snow cover (light blue) are both estimated from individual iButtons next to the GPS station (the vertical blue lines indicate the start of insulating snow cover at the individual iButtons and the vertical orange line represents the end of the zero curtain period). Bottom: daily amount of liquid precipitation (dark blue bars), air temperature (black line), and GST (grey line, iButtons next to GPS station; bold line refers to the median over all iButtons). Insulating snow cover (blue, earliest in grey) and zero curtain periods (dark orange, earliest shaded in orange, latest in grey) as derived from all iButtons in the study area.

Intra-Annual Variability: Seasonal

At all stations, horizontal velocities have a quasi-sinusoidal form over each year, with a maximum in autumn (often around October, if neglecting the short velocity peaks at R2b and R7c) and a minimum in late winter or early spring (often around March, Figs. 3, 4, 5, and 6). Briefly, the minimum in velocity occurred the latest in the study year 2012/13 and earliest in 2013/14. The amplitudes of the intra-annual variability range from less than 30 to more than 2000 %, with greatest amplitudes at R7c ($\geq$ 240 %), especially in the period 2013/14 (2800 %). Also at R2b, the intra-annual variability was large and above 80 %. For the other stations (R2a, R6a, R6b, R7a) the yearly amplitude in velocity was around 30 % (from −15 to +15 %). At individual stations the amplitudes were similar in the different years (Fig. 3).

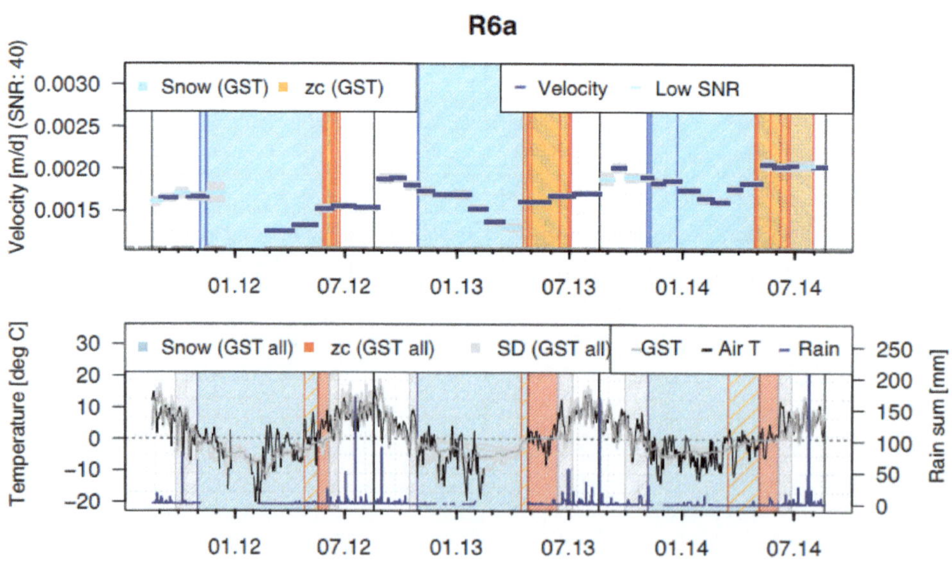

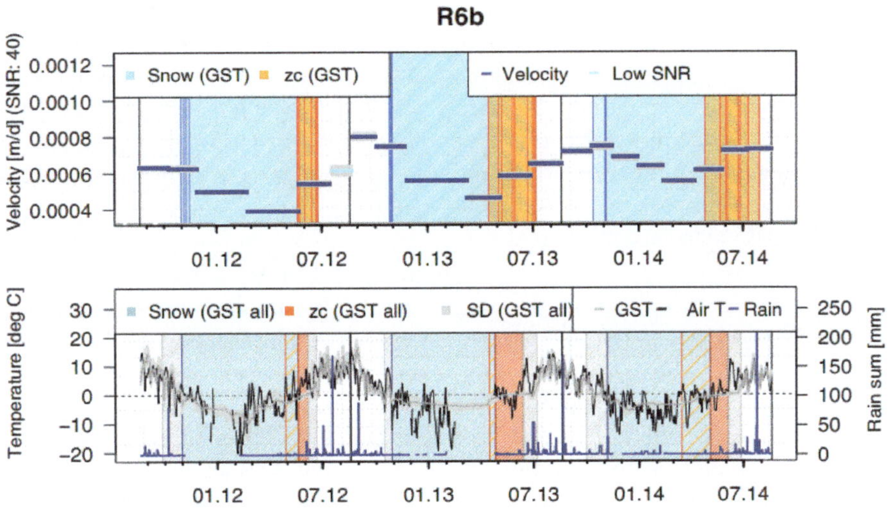

Figure 5. As Fig. 4 but for rock glacier Steintälli (R6a and b).

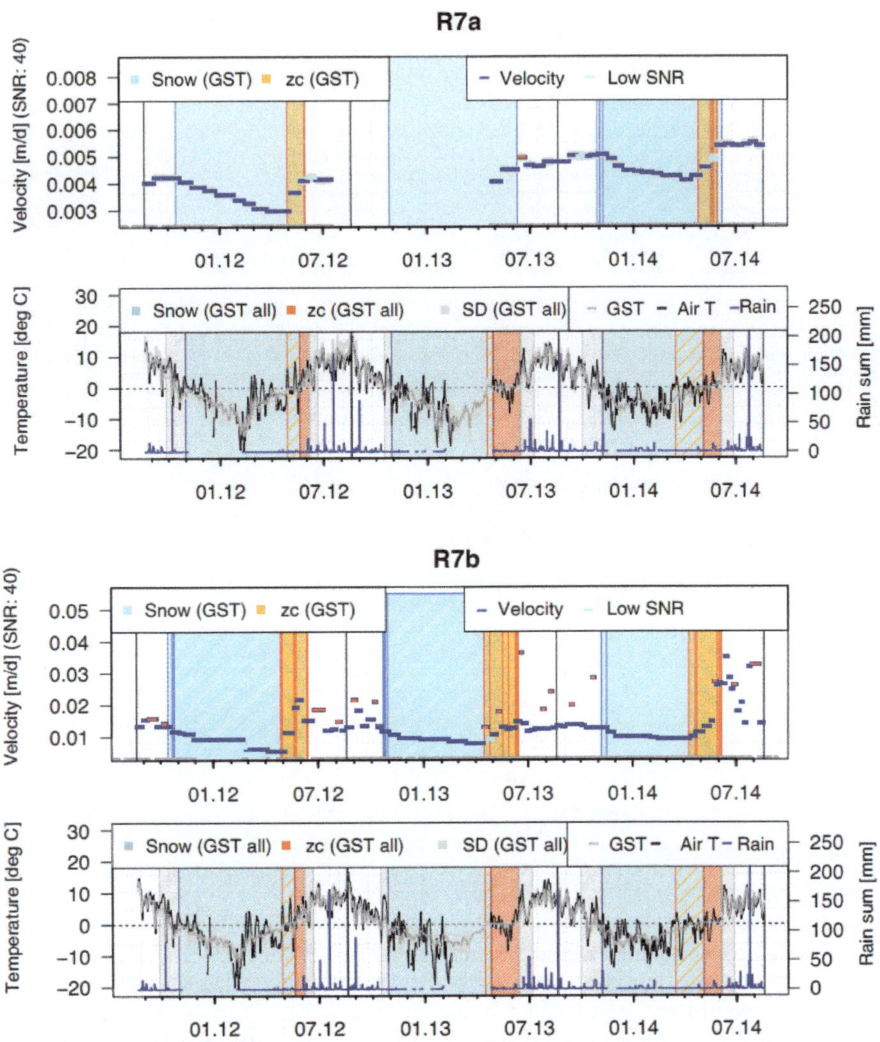

Figure 6. As Fig. 4 but for rock glacier Dirru (R7a and b).

The exact timing of acceleration and deceleration is often difficult to clearly determine because either additional short peaks in velocity occurred (R2b and R7c; see next section) or the temporal resolution differs between individual stations or is too coarse (e.g. > 1 months for R6b). Nevertheless, some observations can be made: spring acceleration of different stations and for different years started between March and May, frequently in April. The winter deceleration started between September and December, mostly in October. Differences in the start of winter deceleration between individual stations were larger (~ 3 months) compared to the differences in the start of spring acceleration (~ 3 weeks). For all rock glaciers, the deceleration started earlier at lower (R2b, R6b, R7c) than at upper stations (R2a, R6a, R7a). Where data are available, the deceleration started latest at R2a (December). Importantly, the acceleration in spring always started during the zero curtain period (Figs. 4, 5, and 6). Furthermore, the timing of acceleration frequently occurred nearly simultaneously for the individual stations located upon the same rock glacier (e.g. R7a and R7c in 2011/12 and 2013/14 (R7a has no data in 2012/13); Figs. 3 and 6).

For the stations on R2 and R7, spring acceleration occurred more abruptly than the continuous and smooth deceleration in winter (Fig. 3). For R6a, and especially R6b, this is less clear because of lower flow speeds and the resulting coarser resolution in velocity periods (R6b: 27 d; R6b: 69 d). Nevertheless, also for R6a, a strong increase in velocity is visible during zero curtain in all study years with available data. Furthermore, at four stations (R2b, R6a, R6b, R7a), the movement direction during the period of acceleration differs slightly from that during the period of deceleration. At all these stations, the movement direction pointed slightly more (12–15°) towards the north during the period of acceleration and towards the south during the phase of deceleration.

Over the entire observation period, the yearly maxima in air temperature and GST (running mean over 60 days) were in all 3 study years rather similar and typically occurred in late August or September (Fig. 3). By contrast, minimum values for GST and air temperature in winter increased over the 3 study years. Lowest air temperatures and GST were observed in February and March. The phase lag between temperature (air and GST) and velocity differs among the individual stations and years and is often difficult to determine (Fig. 3). It ranges from less than 1 month (R6a; minimum in 2011/12) to more than 3 months (R2a; maximum in 2012/13). As both the depth where the main movement occurs and the thermal conditions of the rock glacier body are unknown, we did not perform any statistical analyses on the observed seasonal variability in the rock glacier velocities.

Table 5. Peaks observed at R2b and R7a during the 3 different study years. For each year and for both stations, the duration (days) of the snow-free, the snow cover, and the zero curtain period, the number of peaks per period (snow-free, snow cover, zero curtain; the number outside the brackets refers to the total number during this study year), and summary statistics of the strength of the peaks (diffp) for the different periods are given. The first number in brackets refers to the snow-free, the second to the snow cover (excluding days with zero curtain), and the third to the zero curtain period. Both winter snow cover and the zero curtain period are derived from GST measurements next to the GPS station (Sect. 3.2)

Study year	Duration:		N Peaks:		Min ($diff_p$):		Max ($diff_p$):		Mean ($diff_p$):		Median ($diff_p$):	
	R7c	R2b	R7c	R2b	R7c	R2b	R7c	R2b	R7c	R2b	R7c	R2b
2011/12	(143, 197, 26)	(126, 220, 20)	5 (4, 0, 1)	5 (2, 3, 0)	(7, NA, 314)	(7, 9, NA)	(37, NA, 314)	(11, 98, NA)	(19, NA, 314)	(9, 57, NA)	(15, NA, 314)	(9, 56, NA)
2012/13	(130, 176, 59)	(142, 172, 51)	7 (5, 0, 2)	5 (3, 1, 1)	(37, NA, 50)	(6, 7, 30)	(155, NA, 52)	(14, 7, 30)	(75, NA, 51)	(10, 7, 29)	(49, NA, 51)	(11, 7, 30)
2013/14	(154, 154, 57)	(157, 148, 60)	8 (7, 0, 1)	7 (5, 1, 1)	(22, NA, 33)	(7, 12, 11)	(2424, NA, 34)	(16, 12, 12)	(548, NA, 33)	(10, 12, 12)	(102, NA, 33)	(10, 12, 12)
Total			20 (16 / 0 / 4)	17 (10 / 5 / 2)	(7, NA, 33)	(6, 7, 12)	(2424, NA, 314)	(17, 98, 30)	(268, NA, 213)	(10, 38, 21)	(46, NA, 51)	(10, 12, 21)

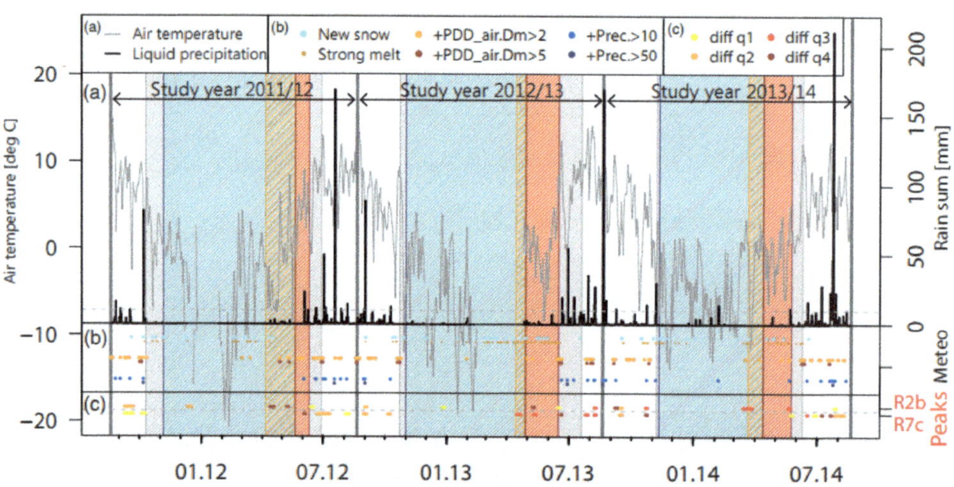

Figure 7. Comparison of velocity peaks at R2b and R7c to meteorological factors. Panel (a): air temperature and liquid precipitation, both measured at the weather station within the study. Vertical grey lines separate the 3 different years of observation (2011/12, 2012/13, 2013/14). Panel (b): in the middle of the plots, the colored dots refer to the occurrence of exceptional values of meteorological factors: new snow – on this day fresh snow was clearly visible on webcam images; strong melt – on this day strong melt was clearly visible on webcam images; $+PDD_{air}$. $Dm \geq 2$ – the difference of PDD_{air} on that day compared to the monthly mean of PDD_{air} was above 2 °C; $+PDD_{air}$. $Dm \geq 5$ – the difference of PDD_{air} on that day compared to the monthly mean of PDD_{air} was above 5 °C; $+Prec \geq 10$ – the daily amount of liquid precipitation was 10 mm or more; $+Prec \geq 50$ – the daily amount of liquid precipitation was 50 mm or more. Panel (c): occurrence of velocity peaks for R2b (upper bars) and R7c (lower bars). The colors of the peak periods refer to the strength of the peaks (given as the 25, 50, and 75 % quantile of $diff_p$).

Intra-Annual Variability: Velocity Peaks

In addition to the seasonal variations in velocities, short peaks in velocity of a few days' duration can be observed at some GPS stations. Applying the methods described in Sect. 3.4, velocity peaks were detected for the lower station on rock glacier Breithorn (R2b), and the two fast-moving stations on rock glaciers Dirru R7a and R7c (Figs. 4, 6, and 7). However, at R7a only one peak was detected during the zero curtain period in 2012/13 (11–23 June 2013, $diff_p$ = 8 %). We therefore limited the analysis of peaks to stations R2b and R7c.

Characteristics of Velocity Peaks

In total, 17 peaks were detected for R2b and 20 for R7c (Tables 5 and 7). The velocity periods detected as peaks at the two stations occurred roughly around the same time of the year; for example, in every year and at both stations one or more peaks were detected during the zero curtain period. Eight of these peaks, observed at R2b and R7c were synchronous, or at least overlapping. In addition, one peak at R7c started directly after a peak at R2b. Nevertheless, most of the individual peaks (R2b: N = 10; R7c: N = 13) did not occur simultaneously at both stations, and sometimes strong peaks were detected at one station but not at the other station (e.g. 9–11 May 2012 at R2b; Fig. 7). Peaks were detected between March and December and most frequently between May and October (N = 32). For R7c, the number of peaks per month was highest for June (N = 5) and July (N = 4). For R2b, most peaks occurred in September (N = 4) and May (N = 3). The duration of the peaks for R2b ranged from 3 to 15 days (with a median (md) of 4 days) and from 1 day to 20 days for R7c (md = 5). The number of peaks per individual study year varied between 5 and 7 for R2b and 5 and 8 for R7c (Table 5). At both stations, most peaks occurred in 2013/14 (R2b: N = 7; R7c: N = 8). At R7c, during each year, and for R2b in 2012/13 and 2013/14, the number of peaks was higher during the snow-free than the snow cover or zero curtain period (Table 5). Even if the number of peaks is normalized by the duration of the periods, most peaks occurred during the snowfree period (except in 2011/12 at R7c). At R2b, in 2011/12 the number of peaks was highest during the snow cover period.

The strength of the peaks ($diff_p$) was stronger at R7c (with a mean μd_{diffp} = 237 % and a median mdd_{diffp} = 47 %) than at R2b (μ_{diffp} = 20 %, md_{diffp} = 11 %; Table 7). At R7c, $diff_p$ ranged from 7 to 2424 % and at R2b from 6 to 98 %. At R2b, the mean of $diff_p$ was highest in 2011/12 (μd_{diffp} = 38 %) and lowest in 2013/14 (μ_{diffp} = 11 %). At R7c, the mean of $diff_p$ was highest in 2013/14 (mean=484 %) and lowest in 2012/13 (mean = 68 %). On average over all peaks, the median of $diff_p$ was highest for peaks during the zero curtain period at both stations (Table 5). The mean strength of the peaks, however, was on average stronger over all peaks for R2b stronger for the snow cover (μd_{diffp} = 38 %) than for the zero curtain (μd_{diffp} = 21 %) or snow-free period (μd_{diffp} = 10 %). At R7c, generally the mean of $diff_p$ was stronger during the snowfree (μd_{diffp} = 268 %) than the snow cover period (μd_{diffp} = 113 %), but no peaks occurred during the snow cover period outside the zero curtain period. For both stations, the weakest peaks per period occurred during the snow-free period (Table 5). At R7c, in 2012/13 and in 2013/14 the strongest peaks also occurred during the snow-free period, whereas in 2011/12 the strongest peak at R7c occurred during the zero curtain period. At R2b, the strongest peak in 2011/12 occurred during the snow cover, in 2012/13 during the zero curtain, and in 2013/14 during the snow-free period. The average strength of the peaks (median of $diff_p$) was highest during the zero curtain period at R2b in 2012/13 and 2013/14 and at R7c in 2011/12 and 2013/14 (Table 5). In 2011/12 at R2b the average strength (median) was highest during the snow cover period, and at R7c in 2013/14 it was highest during the snow-free period.

The cumulative horizontal displacement during peaks was equal to 10 % (2.1 m, 99 days) of the total displacement for R2b and 25 % (3.6 m, 124 days) for R7c. For R2b, around 10 % (6.8°) of the total inclination occurred during peaks. For R7c, inclination was generally smaller, but inclinometer measurements at R7c were not available for the whole period. At R2b, the displacement during the strongest peak ($diff_p$ = 98 %), which lasted for 3 days, was 9 cm. During this peak, the mast at R2b tilted by 2.5° and the elevation changed by 11 cm. However, the change in elevation corrected for the mast tilt was only 1 cm. At R7c, the displacement during the strongest peak ($diff_p$ = 2424 %), lasting for 1 day, was 49 cm and the change in elevation was 26 cm.

The deviation from the main movement direction ($diff_{azi}$) was in general small but stronger during peak periods (R2b: $\mu diff_{azi}$ = 17°; R7c: $\mu diff_{azi}$ = 10°) than during no-peak periods (R2b: $\mu diff_{azi}$ = 6°; R7c: $\mu diff_{azi}$ = 4°), especially for R2b (Fig. 4). During the strongest peaks, $diff_{azi}$ was 27° for R7c and 123° for R2b (towards its orographic right margin). The strong $diff_{azi}$ during the strongest peak points to a rotation of the station (around the Z axis that could not be detected with the measurement set-up). However, excluding the strongest peak at R2b, the mean of $diff_{azi}$ of all peak periods ($\mu diff_{azi}$ = 3°) becomes even smaller than for all no-peak periods ($\mu diff_{azi}$ = 6 °). For R7c, excluding the strongest peak has nearly no influence of the mean of $diff_{azi}$ of peak periods ($\mu diff_{azi}$ = 10°).

Relation to Meteorological Factors

During the snow-free periods, 10 peaks are detected for R2b and 16 for R7c (N = 26; Table 5). For most of these peaks, strong liquid precipitation (≥ 10 mm) was measured during the peak (N = 19) and/or a few days before the peak (N = 23, Fig. 7, Table 7). Two peaks where liquid precipitation during or shortly before the peak was below 10 mm occurred during a period (in June) when part of the study area was still snow covered. Furthermore, during most of the peaks where precipitation during the peak was below 10 mm new snow that melted again within a few days (N = 4) was observed, in combination with a rain on snow event (ROS). Only for one peak (at R7c, 27 June–1 July 2014) was precipitation below 10 mm, and no new snow was observed. During this peak a few remaining snow patches from winter snow cover were visible on webcam images and the cumulative amount of precipitation was 9 mm, indicating ROS. During several days (N = 6) on which the daily amount of precipitation was high (≥ 10 mm), no peak was observed (Fig. 7).

The positive relationship between the occurrence of a peak during the snow-free period and the maximum of the daily amount of precipitation during the peak is also visible in spineplots (e.g. Fig. 8 and Figs. S2 and S3 in the Supplement). However, when precipitation was exceptionally high (maximum daily amount ≥ 40 mm), a peak was not always detected (Fig. 7). In addition, for both stations the number of peaks was higher when new snow was observed during or 7 days before the peak (see, e.g., Fig. 8), but few no-peak periods occurred when new snow was detected.

During the snow cover period (N = 11), six peaks occurred during the zero curtain period, in combination with an ROS (Fig. 7 and Table 7). The five peaks outside the zero curtain period all occurred at R2b and can either be related to new snow and strong melt (N = 3), ROS (N = 1), or to a combination of these (N = 1; Fig. 7 and Table 7) 3 . However, except for two peaks (in December 2011 and December 2012), all peaks during the snow cover period occurred during a period when a zero curtain could be detected within the study site at least for one station. For both stations, during the snow cover period a positive relationship between the number of days during zero curtain or increasing PDDair during the peak and the number of peaks exists (see, e.g., Figs. S2 and S3). In addition, for R7b, the number of peaks increases with mean and maximum amount of precipitation during the peak.

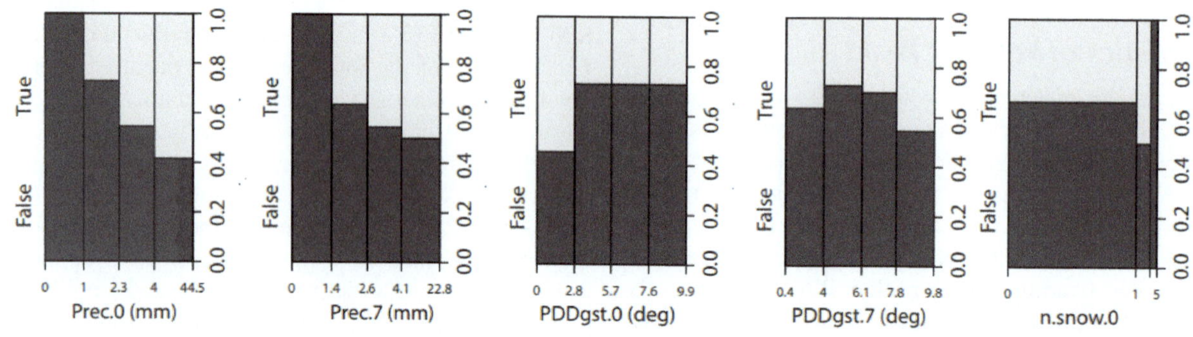

Figure 8. Spineplots for periods during the snow-free period of R7c for selected variables (see Table 3) showing frequencies of the occurrence of a peak (light grey) as opposed to the absence (dark grey) of a peak conditional in relation to potential explanatory variables. Bar widths refer to the 25 % quantiles of the explanatory variables, except for new snow (n.snow). Note: spineplots for all potential explanatory variables of both positions (R2b and R7c) during the snow cover and snow-free period can be found in the Supplement to this paper (Figs. S2 and S3).

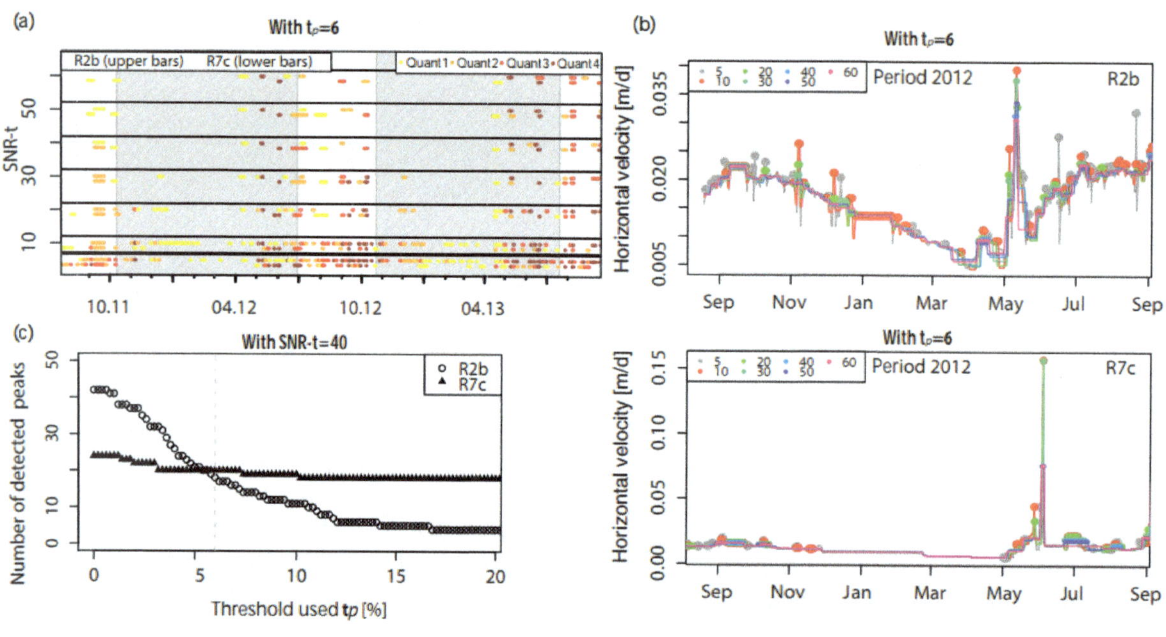

Figure 9. Detected peaks depend on the chosen SNR threshold (SNR-t) for velocity estimations and threshold (t_p) for peak detection. Panel (a): sensitivity of detected peaks for different SNR-t (t_p = 6 %) for R2b and R7c. The colours refer to the strength of the peak ($diff_p$; the 25, 50, and 75 % quantile of $diff_p$). The period with an insulating winter snow cover is shadowed in light grey. Panel (b): estimated horizontal velocities and detected peaks (points) applying different SNR-t for R2b (upper) and R7c (lower). Panel (c): sensitivity of the detected peaks for thresholds t_p for R2b (points) and R7c (triangles). The x axis refers to different values applied for tp. In this study, we apply t_p = 6 and SNR-t = 40.

Based on those observations and the P values of the Wilcoxon rank-sum test (Table S1 in the Supplement), the potential variables for the full logistic regression models were as follows: mean and maximum amount of precipitation, PDD_{gst}, and PDD_{air} (during as well as 7 days before the peak) and the indices for snow cover (snow.ind) and zero curtain (zc.ind).

The AUROC values of the final models (N = 16; see, e.g., Table 6) ranged from 0.69 to 0.85. Hence, according to Hosmer et al. (2013), these models were fair to good. Although different variables for precipitation and temperature were included in the models, the estimates for precipitation and temperature were similar. The highest AUROC was obtained if liquid precipitation and PDD_{gst} during the peak were included in the full model (Table 6). An overview of all final logistic regression models, the variables included, and the AUROC values is given in Table S2.

All the final models included the variables' precipitation and position. While precipitation was significant in most of the final models (N = 14), position was only significant in one of the final models. Additionally, final models mostly contained the interaction terms of precipitation with snow cover (N = 13), with position (N = 11), with zero curtain (N = 11), and/or with temperature (N = 9). Temperature was included in 11 of the final models but was only significant in 1 final model.

Figure 10. Visible cracks (left, from September 2011) and depressions (middle left, from June 2013) at the front of the rock glacier Breithorn, visible water outflow (middle right) at the front of rock glacier Dirru (from June 2013), and a webcam image of the front of rock glacier Bre-ithorn (right, from 14 June 2013).

Table 6. Model coefficients (and standard errors in parentheses) of the final model with highest AUROC values (final model 1) and the final model with highest AUROC without temperature (final model 2). In addition, the corresponding AUROC and AIC values are given. An overview of all variables and their meaning is given in Table 3. In addition, an index with the GPS position (R2b or R7c) is applied for potential differences between the two rock glaciers

	final model 1	final model 2
(Intercept)	−0.06 (0.70)	−1.27 (0.36)^{c}
Prec.0	1.80 (0.44)^{c}	0.74 (0.24)^{b}
PDD_{gst}.0	−0.21 (0.10)^{a}	
zc.ind	2.58 (1.29)^{a}	
snow.ind	−3.02 (1.34)^{a}	
posR2b	−0.69 (0.47)	−0.83 (0.47)
Prec.0:PDD_{gst}.0	−0.20 (0.05)^{c}	
Prec.0:zc.ind	1.18 (0.38)^{b}	
Prec.0:snow.ind	−1.85 (0.44)^{c}	
Prec.0:posR2b		−0.73 (0.24)^{b}
AUROC	0.85	0.75
AIC	148.5	168.53

Significance of Wald test: $^{a} < 0.05$,; $^{b} < 0.01$; $^{c} < 0.001$.

The coefficients of the final models indicate the following (values refers to the final model with the highest AUROC; Table 6): chances of the occurrence of a peak are generally higher during the snow-free period than during winter conditions and higher for R7c than for R2b (a reduction of the odds for R2b of approximately 0.2). Zero curtain has the strongest influence on odds; during zero curtain the odds for the occurrence of a peak increase by more than 14 times, but un-certainties are large (Table 6). The estimates for precipitation were mostly higher for the mean than the maximum amount of precipitation and generally higher if only the days during the peak were included. In all final models the influence of precipitation on the odds is higher during the snow-free than the snow cover period and stronger for R7c than for R2b. For example, during the snow-free period, an increase in log(Prec.0) by one unit leads to an increase in the odds by a factor of 6 for R7c and a factor of 3 for R2b. A change in temperature (by 1 C) generally has a smaller influence on the odds than a change in precipitation by one unit.

DISCUSSION

The following discussion has three distinct elements. Firstly, we set out the methodological strengths and weaknesses of our approach (Sect. 5.1). The second and third parts of our discussion focus on the research questions set out at the beginning of the paper: thus, we explore how the observed main patterns of short-term variability in this study compare with observations from previous studies (Sect. 5.2) and dis-cuss potential controlling factors and the related processes (Sect. 5.3).

Advantages and Limitations of Our Approach

The observed intra-annual variability strongly depends on the temporal resolution of the velocity estimations. In comparison to most previous studies, our data have a much higher temporal resolution (Table 1; see, e.g., Haeberli, 1985; Per-ruchoud and Delaloye, 2007). Furthermore, in comparison to the small number of earlier studies with similarly high temporal resolution (daily), our measurements have (a) a much longer temporal coverage and (b) better spatial coverage due to the inclusion of three neighbouring rock glaciers (earlier studies with high temporal resolution only included 1– 3 point measurements on a single rock glacier; Roer, 2005; Krainer and He, 2006; Buchli et al., 2013; Table 1). This paper extends the data described in Wirz et al. (2015) with an additional year's worth of observations.

A key advantage of the SNRT method is that the smoothing windows directly depend on the SNR of the data (Wirz et al., 2014). This, however, means that the temporal resolution of velocity estimations may differ between individual stations. Where velocities are lower, the velocity periods (smoothing windows) need to be larger to get an SNR above SNR-t. For the relatively slow stations R6a and R6b, with an average duration of the velocity periods of 1 month for R6a and 2 months for R6b, the temporal resolutions of velocities are similar to most previous studies on intra-annual variability in rock glaciers (e.g. Perruchoud and Delaloye, 2007; Liu et al., 2013). However, for the other stations, velocity estimations presented in this study have a much higher temporal resolution (ranging from a few days to 2 weeks). The temporal resolution of the position measurements is similar to those in Buchli et al. (2013) and Krainer and He (2006), but the temporal coverage in these studies was less than 1 year and thus periods with similar seasonal conditions occurred only once.

Table 7. Detected peaks. Main characteristics, including duration of peak (Dur.), the velocity (Vel.), diffp, during the period of winter snow cover (snow) and during zero curtain (zc) are given. In addition, supposed main triggers are given: f.snow – fresh snow that melts again within a few days; rain.p – intense (> 10 mm) liquid precipitation during the peak-period; rain.bp – intense liquid precipitation before the peak (within ≤ 7 days before the start of the peak-period); high.T – high temperature in summer; melt – strong snowmelt in winter or spring; ROS – rain on snow event

Pos.	From	To	Dur.	Vel.	$Diff_p$	Snow	zc	f.snow	rain.p	rain.bp	high.T	Melt	ROS
R7c	10 September 2011	23 September 2011	14	0.02	10	F	F	x	x				x
R2b	10 September 2011	23 September 2011	14	0.02	11	F	F	x	x				x
R7c	5 October 2011	11 October 2011	7	0.01	7	F	F	x	x				
R2b	12 December 2011	18 December 2011	7	0.02	9	T	F	x				x	
R2b	13 April 2012	19 April 2012	7	0.01	65	T	F	x				x	x
R2b	9 May 2012	11 May 2012	3	0.03	98	T	F	x				x	(x)
R7c	3 June 2012	5 June 2012	3	0.08	314	T	T					x	(x)
R2b	13 June 2012	18 June 2012	6	0.02	7	F	F					x	(x)
R7c	23 June 2012	12 July 2012	20	0.02	37	F	F		x		x		
R7c	3 August 2012	9 August 2012	7	0.01	21	F	F		x				
R7c	30 August 2012	4 September 2012	6	0.02	37	F	F	x	x				
R2b	1 September 2012	4 September 2012	4	0.02	11	F	F	x	x				
R7c	6 October 2012	10 October 2012	5	0.02	46	F	F		x	x			
R2b	26 December 2012	30 December 2012	5	0.02	7	T	F					x	x
R7c	16 April 2013	22 April 2013	7	0.01	52	T	T	x				x	
R2b	9 May 2013	12 May 2013	4	0.03	30	T	T					x	x
R7c	10 May 2013	14 May 2013	5	0.02	50	T	T					x	x
R2b	15 June 2013	18 June 2013	4	0.03	6	F	F					x	(x)
R7c	19 June 2013	21 June 2013	3	0.04	155	F	F					x	x
R7c	26 July 2013	30 July 2013	5	0.02	49	F	F		x				
R7c	8 August 2013	11 August 2013	4	0.02	90	F	F		x	x			
R2b	8 August 2013	11 August 2013	4	0.03	14	F	F		x				
R2b	7 September 2013	11 September 2013	5	0.03	17	F	F		x				
R7c	14 September 2013	18 September 2013	5	0.02	46	F	F	x					x
R2b	15 September 2013	18 September 2013	4	0.03	10	F	F	x					x
R7c	22 October 2013	25 October 2013	4	0.03	125	F	F	(x)					x
R2b	22 October 2013	25 October 2013	4	0.03	11	F	F	(x)					x
R2b	15 March 2014	29 March 2014	15	0.01	12	T	F	x				x	
R2b	20 May 2014	24 May 2014	5	0.02	12	T	T					x	x
R7c	24 May 2014	28 May 2014	5	0.03	33	T	T					x	x
R7c	17 June 2014	18 June 2014	2	0.06	102	F	F		x				x
R7c	27 June 2014	1 July 2014	5	0.03	22	F	F		(x)				x
R7c	8 July 2014	8 July 2014	1	0.49	2424	F	F	x					
R7c	21 July 2014	21 July 2014	1	0.27	1075	F	F	x					
R2b	27 July 2014	30 July 2014	4	0.03	7	F	F		x	(x)			
R7c	28 July 2014	11 August 2014	15	0.03	41	F	F		x	x			
R2b	11 August 2014	14 August 2014	4	0.03	8	F	F		x				

A main limitation of our data is that we have few point measurements on individual landforms at the surface, which are (a) not necessarily representative of the entire landform, or parts of it (e.g. rock glacier tongue) and (b) a result of the cumulative displacement potentially occurring at differ-ent depths (Fig. 1). However, surface velocity measurements with a high spatial resolution have shown that the velocity field is comparatively homogenous (at least along the central flow line;

see, e.g., Kääb et al., 2003). Furthermore, remote-sensing data show that movement extends across the whole active lobe of Dirru (R7a and R7c, Strozzi et al., 2009a) and Breithorn (R2a and R2b Strozzi et al., 2009b). In addition, surface velocity measurements form a good approximation of the movement at depth (Lambiel and Delaloye, 2004), as the main movement within a rock glacier is concentrated within (few) thin shear horizons (see, e.g., Arenson et al., 2002).

Nevertheless, movement observed at the surface may occur at different depths (Fig. 1, cf. Roer et al., 2008), and thus, observed displacement at the surface may result from (1) a displacement within one (Arenson et al., 2002) or more shear horizons (Delaloye et al., 2013), (2) internal deformation above the shear horizon(s) (this is mostly small however; Arenson et al., 2002), (3) displacement within or at the base of the active layer (Buchli et al., 2013), (4) sliding of the front similar to a rotational slide (Roer et al., 2008); and/or (5) displacement or rotation of a single boulder (Wirz et al., 2014).

Short-Term Variability in Velocities in the Context of Previous Research

The observed patterns of the short-term variability in the rock glaciers within this study area generally correspond well with observations in earlier studies.

The observed increase in mean annual velocities over the 3 years and at all stations is in good agreement with the observed average increase in velocities of +32 % from summer 2012 to summer 2014 observed at eight rock glaciers in the Valais and Bernese Alps by the Swiss permafrost monitoring network (PERMOS, 2015). In these earlier studies, the observed long-term acceleration seems, with a phase lag of 1 year, to correlate with an increase in MAGST (see, e.g., Hoelzle et al., 1998; Delaloye et al., 2010; PERMOS, 2013). However, our data set is too small to allow conclusions to be drawn on variation on multi-annual timescales but provides further indications that winter temperatures and short-term melt periods may be influential with regard to mean annual velocities.

Observed intra-annual patterns on the rock glaciers within this study area are similar to observations made in previous studies on intra-annual variability (Table 1): the seasonal cycle and the timing of minima and maxima in velocity observed in this study agree well with previous observations on rock glaciers (Table 1; Haeberli, 1985; Roer, 2005; Kääb et al., 2005; Perruchoud and Delaloye, 2007; Buchli et al., 2013), although differences between individual rock glaciers and from year to year have been reported (Delaloye et al., 2010). Similarly, as described by Haeberli (1985), the maxi-mum in velocity occurred earlier at the lower station than at the upper station on the same rock glacier (Fig. 3).

Compared to most earlier studies, the higher temporal resolution of our measurements (for most of the stations) al-lows us to detect more detail in the intra-annual variability in rock glacier velocities, such as the timing of seasonal acceleration, the asymmetric seasonal cycle, and the velocity peaks. In previous studies, a deceleration in winter was also observed to be smoother than acceleration in spring (Table 1, see, e.g., Perruchoud and Delaloye, 2007). However, as for the stations on the rock glacier Steintälli, such an asymmetric annual cycle was not always detected in previous studies (e.g. on Muragl rock glacier: Arenson et al., 2002; Kääb, 2005; Krainer and He, 2006). Both the rock glacier Muragl and Steintälli have low velocities of less than 0.5 m a ^{1}, and hence, the lack of an asymmetric annual cycle may result from a coarser temporal resolution of velocity estimations (monthly; Sect. 4.1).

Previously, short abrupt velocity peaks in rock glacier velocities were only detected on one rapidly moving rock glacier in a nearby valley during snowmelt (Furggwanghorn rock glacier with MAV> 3 m a $^{-1}$: Buchli et al., 2013). However, in most former studies, temporal resolution was too low to be able to detect such short-lived velocity peaks. The coarser temporal resolution of the velocity estimations (median duration of 15 d, Sect. 5.1) might also explain why for rock glacier Steintälli (R6a, R6b) and at the upper stations of the rock glacier Breithorn (R2a) and Dirru (R7a) almost no such peaks, lasting only a few days, were detected.

Effects of Meteorological Factors on Short-Term Velocity Variations and Hypotheses on Underlying Processes

In the following we discuss factors identified from our analysis and link these to potential processes driving the observed temporal patterns of rock glacier movements. In doing so we focus on intra-annual and short-term variations. Since both velocities and meteorological factors are based on surface measurements, while movement and related processes also occur at depth, it is necessary to make an inference from our surface point measurements to the behavior of the complete landform. In linking velocity's response to meteorological forcing, we use lag times as one potential indicator for understanding and determining processes occurring at depth.

Based on the observations of this study we hypothesize that peaks in velocities were mainly caused by strong water infiltration from heavy precipitation or strong snowmelt during or a few days be-fore the peak.

Velocity peaks were only detected at the two fast-moving rock glacier tongues (R2b and R7c). Most probably, the lack of detection of velocity peaks at the other stations is the result of the coarser temporal resolution of the velocity estimates (≥ 2 weeks) compared to the resolution at R2b and R7c (5 days, applying SNRT with SNR-t = 40).

All of the velocity peaks detected at R2b and R7c occurred outside cold winter conditions and during or shortly after high water input, driven by heavy precipitation, snowmelt (indicated by ROS or positive temperatures during the snow cover period; cf. Marks et al., 2001), or late summer snowfall melting rapidly within a few days (Fig. 7). Both qualitative and statistical analyses show that the influence of temperature is weaker and indicate that likely no peak was solely caused by high temperatures without associated water input.

Water infiltration from heavy precipitation or snowmelt may lead to an increase in pore water pressure and hence a reduction in shear resistance (Terzaghi, 1943), thus triggering increased velocities. In addition, water infiltration can lead to a rapid warming of the ground at depth (through thermal advection and release of latent heat; Rist and Phillips, 2005), in turn leading to an increase in the strain rates of the ice in the frozen matrix (see, e.g., Yamamoto, 2013) and a higher unfrozen water content (especially if temperatures were close to 0 C; Anderson et al., 1973; Arenson et al., 2002); both factors facilitate slope movement.

At both stations, maximum velocities at R2b and R7b within 1 year were mostly observed during peaks that occurred during the zero curtain period in combination with rain on snow. Similarly, at the Furggwanghorn rock glacier, two velocity peaks were observed during the snowmelt period which were suggested to be caused by snow meltwater that filtered in through the active layer, ran off along the permafrost table, and, thus, reduced the shear resistance in this zone (Buchli et al., 2013). The short duration of the peaks further suggests that the water that has filtered in runs off within a short time. Indeed, the few existing studies on the hydrology of rock glaciers have all shown that water is trans-mitted through the rock glacier within a few hours (Krainer and Mostler, 2002; Ikeda et al., 2008; Buchli et al., 2013). They further observed that water flows in direct contact with the ice (Krainer et al., 2007), which might indicate ground ice acting as a flow barrier. Furthermore, the observed short time lag between the water input (e.g. precipitation) and the start of the velocity peak indicates that the water that had filtered in reached the shear horizon(s) within a short time and/or that the corresponding shear horizon lies at rather shallow depths (e.g. at the active layer base; Buchli et al., 2013). The velocity peaks at the Furggwanghorn rock glacier were caused by a displacement at the active layer base (at 4 m depth; Buchli et al., 2013). However, as in this study only measurements at the surface are available, it is unknown where the measured movement actually took place and if the water that had faltered in reached the corresponding shear horizon (Fig. 1).

Only approximately half of the peaks occurred simultaneously at the two stations, more frequently during the snow-free period. In addition, velocity peaks were not always detected after or during strong precipitation events or rain on snow events during the zero curtain period, and no clear reasons were found why no velocity peak occurred in these cases. One potential explanation for these observations, especially during the snowmelt period, might be that the amount of water infiltration caused by snowmelt can vary significantly even within a catchment (Grünewald et al., 2010). The timing and intensity of heavy rain in summer can also vary significantly even within a small catchment (see, e.g., Goodrich et al., 1995). Indeed, the measured precipitation during single rain fall events in summer 2014 sometimes varied significantly between the two weather stations within the study site. These observations, however, also indicate that the exact circumstances causing a velocity peak are still largely unknown.

Furthermore, differences in the development of the drainage system within the rock glacier may explain the variable response to similar forcing events. Speed-up events of short duration are a well-known phenomenon for glaciers and are generally assumed to result from high water input into a poorly connected drainage system and to terminate as soon as the drainage system becomes more channelized and more efficient (Kamb, 1987; Nienow et al., 1998). Two observations indicate in our case that at least a partly channelized drainage system existed before the velocities started to increase: (a) the location of the water outflow visible at the front of rock glacier Dirru during periods of strong melt (see, e.g., Fig. 10), and hence the location of related drainage channels, remained more or less constant in its position over the 3 years; (b) the spring outflow at the front of Dirru always became visible before (or at the same time as) velocities started to accelerate.

Regarding water infiltration, high velocities in combination with steep topography may lead to a positive feedback through the opening of fissures and depressions, as has been observed in this study (Fig. 10) and elsewhere (e.g. Roer et al., 2008; Delaloye et al., 2013; Buchli et al., 2013), and to more open pore spaces (dilatancy effect). This could enhance permeability of the ground and favor fast water infiltration and hence lead to further acceleration in velocities.

The findings of this study suggest that rock glaciers have a seasonal cycle in velocities, with seasonal acceleration apparently strongly influenced by meltwater and the amount of seasonal deceleration influenced by winter temperatures.

Seasonal decrease in velocity was typically smooth and occurred with a phase lag of a few weeks to months behind decreasing air temperature and GST (Fig. 3). The smooth deceleration in winter may, beside a phase-lagged cooling through heat conduction, be the result of a decrease in water content within a rock glacier (and shear horizon in particular) as a result of strongly reduced water input (percolation) during winter conditions (nearly no liquid precipitation or snowmelt that might lead to drainage; Coe et al., 2003); it may also be influenced by slow refreezing of pore water as a result of heat loss (see, e.g., Ikeda et al., 2008). The observation that both observed minimum temperatures and minimum velocities in winter increased over the 3 study years provides further support for the suggestion that winter temperatures influence deceleration of rock glacier movements in winter, as previously proposed by various authors (Ikeda et al., 2008; Kääb et al., 2007; Delaloye et al., 2010). Commonly, higher GSTs in winter are associated with a thicker insulating snow cover (see, e.g., Ikeda et al., 2008). This scenario, however, is unlikely for this study as at the MeteoSwiss station in Grächen, the thickest winter snow cover (mean in January and February) occurred in 2012.

In contrast to the smooth deceleration in winter, an abrupt and strong acceleration in spring was observed at most stations; this commenced during snowmelt periods (zero curtain) and was typically related to short-term speed-up events. These observations indicate that in contrast to winter deceleration, acceleration in spring is strongly influenced by water infiltration (from snowmelt) as previously indicated (see, e.g., Krummenacher et al., 2008; Ikeda et al., 2008; Buchli et al., 2013). As a rock glacier body is likely to be permeable (see, e.g., Arenson et al., 2002), it seems plausible that during the snowmelt period water infiltration also occurs in the frozen body beneath the active layer and thus reaches greater depths and potentially the shear horizon (see, e.g., Fig. 1). Dilatancy effects during dry winter conditions might additionally lead to the formation of open pore spaces, which in spring might act as preferential flow paths (Ikeda et al., 2008). Immediate water infiltration to depth might further explain why, in spring, the differences between the individual stations in the start of the acceleration were generally smaller compared to the differences in the start of deceleration, which is likely governed by slow heat conduction.

CONCLUSIONS AND FURTHER WORK

Within this study, continuous GPS measurements on three rock glaciers with distinct morphology from summer 2011 to summer 2014 have been analyzed in relation to meteorological factors, such as GST, air temperature, precipitation, and snow(melt). The presented data set is unique regarding the high temporal resolution of the measurements (daily) in combination with the comparably long temporal (3 years) and spatial coverage (six stations on three independent rock glaciers). The high temporal resolution (daily) of the data allows us to investigate the influence of meteorological factors, and their changes, on the short-term variability in rock glacier movement. We performed detailed qualitative and statistical analyses on the intra-annual variability in rock glacier movement, with a clear focus on the short-term peaks, and its relation to meteorological factors. The main findings of this study are as follows:

- At two steep and fast-moving rock glacier tongues in addition to the seasonal cycle, short velocity peaks occurred immediately after strong water input from snowmelt (in combination with ROS) and heavy precipitation. This rapid and high-magnitude response in rock glacier surface velocities suggests high water input events with rapid infiltration into the rock glacier body as a forcing mechanism.
- The seasonal variability in the rock glacier movement is strongly influenced by snow meltwater infiltration and winter temperatures, as indicated by the observation that both velocities (minimal velocities in winter and MAV) and winter temperatures increased over the observation period and that spring acceleration is mostly rapid and abrupt (peaks) and always started during the snowmelt period.

Our measurements, and the statistical analyses, revealed that there are large similarities but also clear differences in the reaction to changes in environmental factors and the resulting short-term variability, both between different rock glaciers but also different parts (e.g. tongue vs. middle) of the same landform.

The high resolution and spatial proximity of GPS and meteorological measurements allowed us to study and demonstrate the influence of meteorological factors on the short-term variability in the movements for three different rock glaciers. However, it is likely that the surface movement on rock glaciers derives from movements occurring at different depths, which would require subsurface measurements.

AUTHOR CONTRIBUTIONS

Measurement design, installations, and maintenance: J. Beutel, S. Gruber, and V. Wirz; study concept: V. Wirz, S. Gruber, R. Purves, and J. Beutel; analyses: V. Wirz in collaboration with S. Gubler, S. Gruber, and R. Purves; inter-pretation of results: V. Wirz, A. Vieli, I. Gärtner-Roer, S. Gruber, and R. Purves; manuscript: V. Wirz wrote the paper, J. Beutel, I. Gärtner-Roer, S. Gruber, R. Purves, and A. Vieli edited and commented on the manuscript.

ACKNOWLEDGEMENTS

This project (as part of the project X-Sense 1 and X-sense 2) was scientifically evaluated by the SNSF, financed by the Swiss Confederation, and funded by Nano-Tera.ch and by the faculty of science of the University of Zurich. This work was also aided by the Grid Computing Competence Center (GC3, www.gc3.uzh.ch) with computational infrastructure and assistance, including customized libraries (gc_gps and GC3Pie) and user support. We thank R. Delaloye (Physical Geography at the University of Fribourg) for fruitful discussions and for providing webcam images. This study would not have been possible without the collaboration with colleagues from the project X − Sense, notably B. Buchli (Computer Engineering and Networks Lab of ETH Zurich), who contributed significantly to the design of the GPS stations, P. Limpach (Geodesy and Geodynamics Lab of ETH Zurich), who processed the daily GPS solutions; S. Weber and T. Gsell (Computer Engineering and Networks Lab of ETH Zurich), who contributed significantly to the installation and maintenance of the field site; and H. Raetzo (Federal Office for the Environment), who provided valuable insights and discussion concerning placement of GPS stations. The software iAssist (Keller et al., 2010) was used to program and read out the iButtons.

REFERENCES

1. Anderson, D., Tice, A., and McKim, H.: The unfrozen water and the apparent specific heat capacity of frozen soils, in: Proceedings of the Second International Conference on Permafrost, 13–28 July 1973, Washington, DC, USA, 289–295, 1973.
2. Arenson, L., Hoelzle, M., and Springman, S.: Borehole deformation measurements and internal structure of some rock glaciers in Switzerland, Permafrost Periglac., 13, 117–135, doi:10.1002/ppp.414, 2002.
3. Barsch, D.: Nature and importance of mass-wasting by rock glaciers in alpine permafrost environments, Earth Surf. Proc. Land., 2, 231–245, doi:10.1002/esp.3290020213, 1977.
4. Barsch, D.: Permafrost creep and rock glaciers, Permafrost Periglac., 3, 175–188, doi:10.1002/ppp.561, 1992.
5. Barsch, D. and Hell, G.: Photogrammetrische Bewegungsmessun-gen am Blockgletscher Murtèl I, Oberengadin, Schweizer Alpen, Zeitschrift für Gletscherkunde und Glazialgeologie, 11, 111– 142, 1975.
6. Baur, F.: Constructing confidence sets using rank statistics, J. Am. Stat. Assoc., 67, 687–690, 1972.
7. Bernknopf, R. L., Campbell, R. H., Brookshire, D. S., and Shapiro, C. D.: A probabilistic approach to landslide hazard map-ping in Cincinnati, Ohio, with applications for economic evaluation, Bulletin of the Association of Engineering Geologists, 25, 39–56, 1988.
8. Beutel, J., Buchli, B., Ferrari, F., Keller, M., Thiele, L., and Zimmerling, M.: X-Sense: sensing in extreme environments, in: Proceedings of Design, Automation and Test in Europe, 2011, 1–6, IEEE, Grenoble, France, 1460–1465, 2011.
9. Boeckli, L., Brenning, A., Gruber, S., and Noetzli, J.: Permafrost distribution in the European Alps: calculation and evaluation of an index map and summary statistics, The Cryosphere, 6, 807– 820, doi:10.5194/tc-6-807-

2012, 2012.

10. Brenning, A. and Trombotto, D.: Logistic regression modeling of rock glacier and glacier distribution: topographic and climatic controls in the semi-arid Andes, Geomorphology, 81, 141–154, doi:10.1016/j.geomorph.2006.04.003, 2006.
11. Buchli, B., Sutton, F., and Beutel, J.: GPS-equipped wireless sensor network node for high-accuracy positioning applications, in: Wireless Sensor Networks, Springer, Trento, Italy, 179–195, 2012.
12. Buchli, T., Merz, K., Zhou, X., Kinzelbach, W., and Springman, S.: Characterization and monitoring of the Furggwanghorn rock glacier, Turtmann valley, Switzerland: results from 2010 to 2012, Vadose Zone J., 12, 1–15, doi:10.2136/vzj2012.0067, 2013.
13. Can, T., Nefeslioglu, H., Gokceoglu, C., Sonmez, H., and Duman, T.: Susceptibility assessments of shallow earth-flows triggered by heavy rainfall at three catchments by logistic regression analyses, Geomorphology, 72, 250–271, doi:10.1016/j.geomorph.2005.05.011, 2005.
14. Coe, J., Ellis, W., Godt, J., Savage, W., Savage, J., Michael, J., Ki-bler, J., Powers, P., Lidke, D., and Debray, S.: Seasonal movement of the Slumgullion landslide determined from Global Positioning System surveys and field instrumentation, Eng. Geol., 68, 67–101, doi:10.1016/S0013-7952(02)00199-0, 2003.
15. Dach, R., Hugentobler, U., Fridez, P., and Meindl, M.: Bernese GPS Software, Version 5.0, Astronomical Institute, University of Bern, Bern, Switzerland, 2007.
16. Delaloye, R., Kaufmann, M., Bodin, X., Hausmann, H., Ikeda, A., Kääb, A., Kellerer-Pirklbauer, A., Krainer, K., Lambiel, C., Mi-hajlovic, D., Roer, I., and Thibert, E.: Recent interannual vari-ations of rock glacier creep in the European Alps, in: Proceed-ings of the 9th International Conference on Permafrost, 28 June– 3 July 2008, Fairbanks, Alaska, USA, 343–348, 2008a.
17. Delaloye, R., Strozzi, T., Lambiel, C., Perruchoud, E., and Raetzo, H.: Landslide-like development of rock glaciers detected with ERS-1/2 SAR interferometry, in: Proceedings of the 8th In-ternational Conference on Permafrost, 21–25 July 2003, Zurich, Switzerland, 26–30, 2008b.
18. Delaloye, R., Lambiel, C., and Gärtner-Roer, I.: Overview of rock glacier kinematics research in the Swiss Alps, Geogr. Helv., 65, 135–145, doi:10.5194/gh-65-135-2010, 2010.
19. Delaloye, R., Barboux, C., Morard, S., Abbet, D., and Gruber, V.: Rapidly moving rock glaciers in Mattertal, in: Mattertal – ein Tal in Bewegung, Publikation zur Jahrestagung der Schweizerischen Geomorphologischen Gesellschaft, St. Niklaus, Switzerland, 21– 31, 2013.
20. Francou, B. and Reynaud, L.: 10 year surficial velocities on a rock glacier (Laurichard, French Alps), Permafrost Periglac., 3, 209– 213, doi:10.1002/ppp.3430030306, 1992.
21. Friendly, M.: Mosaic displays for multi-way contingency tables, J. Am. Stat. Assoc., 89, 190–200, 1994.
22. Goodrich, D., Faurres, J., Woolhiser, D., Lane, L., and Sorooshian, S.: Measurement and analysis of small-scale convective storm rainfall variability, J. Hydrol., 173, 283–308, doi:10.1016/0022-1694(95)02703-R, 1995.
23. Graf, C., Deubelbeiss, Y., Bühler, Y., Meier, L., McArdell, B., Christen, M., and Bartelt, P.: Gefahrenkartierung Matter-tal: Grundlagenbeschaffung und numerische Modellierung von Murgängen, in: Mattertal – ein Tal in Bewegung, Publika-tion zur Jahrestagung der Schweizerischen Geomorphologischen Gesellschaft, St. Niklaus, Switzerland, vol. 29, 85–112, 2013.
24. Grünewald, T., Schirmer, M., Mott, R., and Lehning, M.: Spa-tial and temporal variability of snow depth and ablation rates in a small mountain catchment, The Cryosphere, 4, 215–225, doi:10.5194/tc-4-215-2010, 2010.
25. Gubler, S., Fiddes, J., Keller, M., and Gruber, S.: Scale-dependent measurement and analysis of ground surface temperature variability in alpine terrain, The Cryosphere, 5, 431–443, doi:10.5194/tc-5-431-2011, 2011.
26. Haeberli, W.: Creep of mountain permafrost: internal structure and flow of alpine rock glaciers, Mit. Ver. Was., 77, 5–142, 1985.
27. Haeberli, W., Hoelzle, M., Kääb, A., Keller, F., Vonder Mühll, D., and Wagner, S.: Ten years after drilling through the permafrost of the active rock glacier Murtèl, Eastern Swiss Alps: answered questions and new perspectives, in: Proceedings of the 7th International Conference on Permafrost, 23–27 June, Nordicana, Yellowknife, Canada, 403–410, 1998.
28. Haeberli, W., Hallet, B., Arenson, L., Elconin, R., Hum-lum, O., Kääb, A., Kaufmann, V., Ladanyi, B., Matsuoka, N., Springman, S., and Vonder Mühll, D.: Permafrost creep and rock glacier dynamics, Permafrost Periglac., 17, 189–214, doi:10.1002/ppp.561, 2006.
29. Hausmann, H., Krainer, K., Brückl, E., and Mostler, W.: Creep of two Alpine rock glaciers – observation and modelling (Ötztal-and Stubai Alps, Austria), Grazer Schriften der Geographie und Raumforschung, 43, 145–150, 2007.
30. Hoelzle, M., Wagner, S., Kääb, A., and Vonder Mühll, D.: Surface movement and internal deformation of ice-rock mixtures within rock glaciers at Pontresina-Schafberg, Upper Engadin, Switzer-land, in: Proceedings of the Seventh International Conference on Permafrost, 23–27 June 1998, Yellowknife, Northwest Territories, Canada, 465–471, 1998.
31. Hosmer, J., David, W., Lemeshow, S., and Sturdivant, R.: Applied Logistic Regression, 3rd edn., John Wiley and Sons, New York, USA, 2013.
32. Humlum, O., Christiansen, H., and Juliussen, H.: Avalanche-derived rock glaciers in Svalbard, Permafrost Periglac., 18, 75– 88, doi:10.1002/ppp.580, 2007.
33. Ikeda, A., Matsuoka, N., and Kääb, A.: Fast deformation of perennially frozen debris in a warm rock-glacier in the Swiss Alps: an effect of liquid water, J. Geophys. Res., 113, 1–12, doi:10.1029/2007JF000859, 2008.
34. Johnson, J. and Nickling, W.: Englacial temperature and deformation of a rock glacier in the Kluane Range, Yukon Territory, Canada, Can. J. Earth Sci., 16, 2275–

2283, 1979.

35. Jomelli, V., Delval, C., Grancher, D., Escande, S., Brunstein, D., Hetu, B., Filion, L., and Pech, P.: Probabilistic analysis of recent snow avalanche activity and weatherin the French Alps, Cold Reg. Sci. Technol., 47, 180–192, doi:10.1016/j.coldregions.2006.08.003, 2007.

36. Kääb, A.: Remote Sensing of Mountain Glaciers and Permafrost Creep, vol. 48, Schriftenreihe Physische Geographie, Department of Geography, University of Zurich, Zurich, Switzerland, 2005.

37. Kääb, A., Kaufmann, V., Ladstädter, R., and Eiken, T.: Rock glacier dynamics: implications from high-resolution measurements of surface velocity fields, in: Proceedings of the 8th International Conference on Permafrost, 21–25 July 2003, Zurich, Switzer-land, 501–506, 2003.

38. Kääb, A., and Weber, M.: Development of transverse ridges on rock glaciers: field measurements and laboratory experiments, Per-mafrost Periglac., 15, 379–391, doi:10.1002/ppp.506, 2004.

39. Kääb, A., Huggel, C., Fischer, L., Guex, S., Paul, F., Roer, I., Salz-mann, N., Schlaefli, S., Schmutz, K., Schneider, D., Strozzi, T., and Weidmann, Y.: Remote sensing of glacier- and permafrost-related hazards in high mountains: an overview, Nat. Hazards Earth Syst. Sci., 5, 527–554, doi:10.5194/nhess-5-527-2005, 2005.

40. Kääb, A., Frauenfelder, R., and Roer, I.: On the response of rock glacier creep to surface temperature increase, Global Planet. Change, 56, 172–187, doi:10.1016/j.gloplacha.2006.07.005, 2007.

41. Kamb, B.: Glacier surge mechanism based on linked cavity configu-ration of the basal water conduit system, J. Geophys. Res.-Solid, 92, 9083–9100, doi:10.1029/JB092iB09p09083, 1987.

42. Kaufmann, V., and Ladstädter, R.: Mapping of the 3D surface motion field of Doesen rock glacier (Ankogel Group, Austria) and its spatio-temporal change (1954–1998) by means of dig-ital photogrammetry, in: Proceedings ot the 9th International Symposium on High Mountain Remoste Sensing Cartography (HMRSC-IX), Graz, 14–15 September 2006, Grazer Schriften der Geographie und Raumforschung, 43, 134–144, 2007.

43. Keller, M., Hungerbuehler, G., Knecht, O., Sheikh, S., Beutel, J., Gubler, S., Fiddes, J., and Gruber, S.: iAssist: rapid deployment and maintenance of tiny sensing systems, in: Proceedings of the 8th ACM Conference on Embedded Networked Sensor Systems, SenSys '10, 3–5 November 2010, 401–402, ACM, New York, USA, doi:10.1145/1869983.1870043, 2010.

44. Konrad, S., Humphrey, N., Steig, E., Clark, D. J. P., and Pf-effer, W.: Rock glacier dynamics and paleoclimatic implications, Geology, 27, 1131–1134, doi:10.1130/0091-7613(1999)027<1131:RGDAPI>2.3.CO;2, 1999.

45. Krainer, K. and He, X.: Flow velocities of active rock glaciers in the Austrian Alps, Geogr. Ann. A, 88, 267–280, doi:10.1111/j.0435-3676.2006.00300.x, 2006.

46. Krainer, K. and Mostler, W.: Hydrology of active rock glaciers: ex-amples from the Austrian Alps, Arct. Antarct. Alp. Res., 142– 149, 2002.

47. Krainer, K., Mostler, W., and Spötl, C.: Discharge from active rock glaciers, Austrian Alps: a stable isotope approach, Aust. J. Earth Sci., 100, 102–112, 2007.

48. Krummenacher, B., Mihajlovic, D., Nussbaum, A., and Staub, B.: 20 Jahre Furggentälti, Permafrostuntersuchungen auf der Gemmi, Geographica Bernensia G80, Geographisches Institut der Universität Bern, Bern, Switzerland, 2008.

49. Labhart, T.: Geologie der Schweiz, 3rd edn., Ott Verlag, Thun, Switzerland, 1995.

50. Lambiel, C. and Delaloye, R.: Contribution of real-time kinematic GPS in the study of creeping mountain permafrost: examples from the Western Swiss Alps, Permafrost Periglac., 15, 229–241, doi:10.1002/ppp.496, 2004.

51. Lambiel, C., Delaloye, R., and Perruchoud, E.: Yearly and season-ally variations of surface velocities on creeping permafrost bod-ies. Cases studies in the Valais Alps, in: 3rd Swiss Geoscience Meeting, Zürich, 18–19 November 2005, 1–2, 2005.

52. Laube, P. and Purves, R. S.: How fast is a cow? Cross-scale analysis of movement data, Transactions in GIS, 15, 401–418, doi:10.1111/j.1467-9671.2011.01256.x, 2011.

53. Limpach, P. and Grimm, D.: Rock glacier monitoring with low-cost GPS receivers, in: Abstract Volume 7th Swiss Geoscience Meet-ing, 20–21 November 2009, Neuchatel, Switzerland, 247–248, 2009.

54. Liu, L., Millar, C. I., Westfall, R. D., and Zebker, H. A.: Surface motion of active rock glaciers in the Sierra Nevada, California, USA: inventory and a case study using InSAR, The Cryosphere, 7, 1109–1119, doi:10.5194/tc-7-1109-2013, 2013.

55. Lugon, R., Delaloye, R., and Perruchoud, E.: Aléas naturels en zone de pergélisol et détection des Alpes valaisannes, Suisse, Rapport final, Programme INTERREG IIIa France-Suisse Haut-Rhône, Institut universitaire Kurt Bösch, Sion, Suisse; Géosciences, In-stitut de Géographie, Université de Fribourg, Suisse, 2008.

56. Marks, D., Link, T., Winstral, A., and Garen, D.: Simulating snowmelt processes during rain-on-snow over a semi-arid mountain basin, Ann. Glaciol., 32, 195–202, doi:10.3189/172756401781819751, 2001.

57. Mihajlovic, D., Kölbing, D., Kunz, I., Schwab, S., Kienholz, H., Budmiger, K., Imhof, M., and Krummenacher, B.: Developing new methods for monitoring periglacial phenomena, in: Proceedings of the 8th International Conference on Permafrost, 21– 25 July 2003, Zurich, Switzerland, 765–770, 2003.

58. Nienow, P., Sharp, M., and Willis, I.: Seasonal changes in the morphology of the subglacial drainage system, Haut Glacier d'Arolla, Switzerland, Earth Surf. Proc. Land., 23, 825– 843, doi:10.1002/(SICI)1096-9837(199809)23:9<825::AID-ESP893>3.0.CO;2-2, 1998.

59. Ohlmacher, G. C. and Davis, J. C.: Using multiple

logistic regres-sion and GIS technology to predict landslide hazard in north-east Kansas, USA, Eng. Geol., 69, 331–343, doi:10.1016/S0013- 7952(03)00069-3, 2003.

60. Outcalt, S., Nelson, F., and Hinkel, K.: The zero-curtain effect: heat and mass transfer across an isothermal region in freezing soil, Water Resour. Res., 26, 1509–1516, doi:10.1029/WR026i007p01509, 1990.
61. PERMOS: Permafrost in Switzerland 2008/2009 and 2009/2010, edited by: Noetzli, J., Glaciological Report (Permafrost) No. 10/11 of the Cryospheric Commission of the Swiss Academy of Sciences, Zurich, Switzerland, 2013.
62. PERMOS: Key messages on permafrost in Switzerland 2013/2014, unpublished report by the PERMOS Scientific Committee, Zurich, Switzerland, 7 pp., 2015.
63. Perruchoud, E. and Delaloye, R.: Short-term changes in surface velocities on the Becs-de-Bosson rock glacier (western Swiss Alps), Grazer Schriften der Geographie und Raumforschung, 43, 131–136, 2007.
64. Potter, N.: Ice-cored rock glacier, Galena Creek, northern Absaroka Mountains, Wyoming, Geol. Soc. Am. Bull., 83, 3025–3058, doi:10.1130/0016-7606(1972)83[3025:IRGGCN]2.0.CO;2, 1972.
65. Rist, A. and Phillips, M.: First results of investigations on hy-drothermal processes within the active layer above alpine permafrost in steep terrain, Norsk Geogr. Tidsskr., 59, 177–183, doi:10.1080/00291950510020574, 2005.
66. Robin, X., Turck, N., Hainard, A., Tiberti, N., Lisacek, F., Sanchez, J., and Müller, M.: pROC: an open-source package for R and S+ to analyse and compare ROC curves, BMC Bioinfor-matics, 12, 1–8, doi:10.1186/1471-2105-12-77, 2011.
67. Roer, I.: rock glacier Kinematics in a High Mountain Geosystem, Ph.D. thesis, Mathematisch-Naturwissenschaftlichen Fakultät der Rheinischen Friedrich-Wilhelms-Universität Bonn, Bonn, Germany, 2005.
68. Roer, I., Kääb, A., and Dikau, R.: rock glacier acceleration in the Turtmann valley (Swiss Alps): probable controls, Norsk Geogr. Tidsskr., 59, 157–163, doi:10.1080/00291950510020655, 2005.
69. Roer, I., Haeberli, W., Avian, M., Kaufmann, V., Delaloye, R., Lambiel, C., and Kääb, A.: Observations and considerations on desta-bilizing active rock glaciers in the European Alps, in: Proceedings of the 9th International Conference on Permafrost, 28 June– 3 July 2008, Fairbanks, Alaska, USA, vol. 2, 1505–1510, 2008.
70. Schmid, M.-O., Gubler, S., Fiddes, J., and Gruber, S.: Inferring snowpack ripening and melt-out from distributed measurements of near-surface ground temperatures, The Cryosphere, 6, 1127– 1139, doi:10.5194/tc-6-1127-2012, 2012.
71. Schoeneich, P., Bodin, X., Echelard, T., Kaufmann, V., Kellerer-Pirklbauer, A., Krysiecki, J., Lieb, G.: Velocity changes of rock glaciers and induced hazards, Engineering Geology for Society and Territory, 1, 223–227, doi:, 2015.
72. Sloan, V. and Dyke, L.: Decadal and millennial velocities of rock glaciers, Selwyn Mountains, Canada, Geogr. Ann. A, 80, 237– 249, 1998.
73. Strozzi, T., Delaloye, R., Raetzo, H., and Wegmüller, U.: RADAR interferometric observations of destabilized rock glaciers, in: Proceedings of the "Fringe 2009 Workshop", 30 November– 4 December 2009, Frascati, Italy, (ESA SP-677, March 2010), 1–5, 2009a.
74. Strozzi, T., Wegmuller, U., Werner, C., Wiesmann, A., Delaloye, R., and Raetzo, H.: Survey of landslide activity and rock glaciers movement in the Swiss Alps with TerraSAR-X, in: Geo-science and Remote Sensing Symposium, 2009 IEEE Interna-tional, IGARSS, 12–17 July 2009, vol. 3, pp. III-53–III-56, doi:10.1109/IGARSS.2009.5418175, 2009b.
75. Terzaghi, K.: Theoretical Soil Mechanics, Wiley, New York, USA, 510 pp., 1943.
76. Venables, W. and Ripley, B.: Modern Applied Statistics with S, 4th edn., ISBN 0-387-95457-0, Springer, New York, USA, 2002.
77. Vitek, J. and Giardino, J.: Rock glaciers, in: Rock Glaciers: A Re-view of the Knowledge Base, edited by: Vitek, J. and Giardino, J., Allen and Unwin Boston, London, UK, 1–26, 1987.
78. Vonder Mühll, D.: Evidence of intrapermafrost groundwater flow beneath an active rock glacier in the Swiss Alps, Permafrost Periglac., 3, 169–173, doi:10.1002/ppp.3430030216, 1992.
79. Wagner, S.: Creep of alpine permafrost, investigated on the Murtèl rock glacier (short communication), Permafrost Periglac., 3, 157–162, doi:10.1002/ppp.3430030214, 1992.
80. Wagner, S.: Dreidimensionale Modellierung zweier Gletscher und Deformationsanalyse von eisreichem Permafrost, Mit. Ver. Was., 1–146, doi:10.3929/ethz-a-000954695, 1994.
81. Wahrhaftig, C. and Cox, A.: Rock glaciers in the Alaska Range, Geol. Soc. Am. Bull., 70, 383–436, doi:10.1130/0016-7606(1959)70[383:RGITAR]2.0.CO;2, 1959.
82. Wirz, V., Limpach, P., Buchli, B., Beutel, J., and Gruber, S.: Temporal characteristics of different cryosphere related slope movements in high mountains, in: Landslide Science and Practice, Springer, Berlin, Heidelberg, Germany, 383–390, 2013.
83. Wirz, V., Beutel, J., Gruber, S., Gubler, S., and Purves, R. S.: Estimating velocity from noisy GPS data for investigating the temporal variability of slope movements, Nat. Hazards Earth Syst. Sci., 14, 2503–2520, doi:10.5194/nhess-14-2503-2014, 2014.
84. Wirz, V., Marten Geertsema, M., Gruber, S., and Purves, R.: Temporal variability of diverse mountain permafrost slope movements derived from multi-year daily GPS data, Mattertal, Switzerland, Landslides, 1–17, doi:10.1007/s10346-014-0544-3, 2015.
85. Xu, W., Yu, W., and Zhang, G.: Prediction method of debris flow by logistic model with two types of rainfall:

a case study in the Sichuan, China, Nat. Hazards, 62, 733–744, doi:10.1007/s11069-011-9988-0, 2012.

86. Yamamoto, Y.: Instabilities in Alpine permafrost: strength and stiff-ness in a warming regime, Ph. D. thesis, Institut for Geotechnical Engeneering, ETH Zurich, Zurich, Switzerland, 2013.

CHAPTER 10

ROCK PHYSICAL EXPERIMENTAL RESEARCH IN TANAN VOLCANIC ROCK RESERVOIR

Lailin Li, Haibo Zhao, Xuehui Han

Keywords: Tuffaceous Rock-Fragment Sandstones, Acoustic Velocity, Porosity, Shale Content, Density, Water Saturation

INTRODUCTION

Haita Basin is a part of Central Asia- Ancient Aulacogen, which belongs to Ergu'Na-Greater Khingan Range tectonic zone. Tongbomiao Formation and Nantun Formation in the lower Cretaceous of Zhalainuoer group develop a large number of oil-bearing tight tuffaceous conglomerate and medium to low porosity and permeability rock-fragment sandstone reservoirs, which forms the main area for improving the production and has great potential for prospecting [1]. At present, due to the lack of adequate understanding of the factors that affect the velocity of the volcano clastic rock in Haita Basin, the seismic and logging data can't be applied to predict the lithology of rock, calculate the porosity and detect the oil and gas effectively. Therefore, a detailed research on the influencing factors of velocity in the volcano clastic rock in Haita Basin is needed.

Until now, many researches have been done on the velocity properties and influencing factors of velocity (lithology, physical properties, fluid properties, density and environmental factors) in the sedimentary reservoirs. For instance, Wyllie (1956) proposed time average equation [2], Simmons (1964) gave the general linear relationship between the acoustic velocity and the component of rock and its density [3], Ma and Jie combined compressional wave velocity and shear wave velocity with density by least square method and proposed general Gardner Equation in 2005 [4], critical porosity was proposed by Nur *et al.* in 1992 [5], Han (1986) established empirical relationship for clasolite to describe the influence of lithology and physical properties on the acoustic velocity [6], Biot (1956) and Gassmann (1951) obtained the equation to describe the influence of fluid on the rock acoustic velocity [7] [8], many other equations are also established based on them to describe fluid effect (such as Mackojizba Formula [9]), Garden (1974) proposed the power function relationship between the acoustic velocity and density [10], Gan (2002) [11] ,Ma and Wu (2006) [12], Mavko (1998) [13], Han and Nur (1986) also carried out the experiments to investigate the influence of pressure and temperature on the velocity and its physical mechanism. However, few researches are done on volcano clastic rock and there are not many related papers. Schön (1996) [14] did a lot of statistical work on the acoustic velocity data of igneous rock and found that, the compressional wave velocity decreases with the increase of SiO_2 in the rock, Khatchikian(1982) [15] combined density, neutron and acoustic logging curves to determine the lithology of igneous rock and calculated the porosity of reservoirs.

The compressional and shear wave velocities of conglomerate and tuffaceous rock-fragment sandstone in Tanan were measured in this paper. The influences of lithology, mineral content, density, porosity, clay content and fluid saturation on the compressional and shear wave velocity are then analyzed, which provide the experimental foundation for employing seismic and logging data to predict the lithology, calculate porosity and recognize the fluid type.

GEOLOGICAL CHARACTERISTICS OF VOLCANO CLASTIC SANDSTONE RESERVOIRS

The core samples are obtained from Nantun Formation and Tongbomiao Formation of Cretaceous in Tanan. Through the thin section analysis, it shows that the core samples belong to tuffaceous conglomerate, tuffaceous rock-fragment sandstone and tuff. The sandstone is mainly composed of quartz, feldspar

Citation: Li, L. , Zhao, H. and Han, X. (2014) Rock Physical Experimental Research in Tanan Volcanic Rock Reservoir. Journal of Applied Mathematics and Physics, 2, 284-295. doi: 10.4236/jamp.2014.26034.

and magmatite debris, the void space in the sandstone is filled by tuff. The samples have low degree of sorting and the degree of weathering is moderate to high, the roundness of the grains is mainly subangular and subangular-subround. The samples are supported by grains or clay, the contact types between the grains are no contact to point contact and point contact to line contact, the porosity types belong to original granular porosity and granular increment porosity, the samples are pore cemented. Among the samples, the tuffaceous rock-fragment sandstone has better physical properties than the other samples (the average porosity is 9.96%, the average permeability is 28.01 mD, the shale content is low), it is the main reservoir of this area. The tuffaceous conglomerate with better physical properties can also be the good reservoirs.

EXPERIMENTAL METHODS

The Measurement of Porosity, Permeability, Clay Content and Density

The porosity and permeability of samples are measured by CMS300 which is made by CORELAB Company of US. Under the simulated formation pressure (effective pressure is 25 MPa), the average porosity of the samples ranges from 0.02% to 21.19%, the average value is 6.21%, the permeability is distributed between 0.01 mD and 793 mD, the average value is 12.61 mD. The shale content is determined by MASTERSIZER-2000 Laser Particle Size Analyzer, the shale content ranges from 14.2% to 38.6%, the average value is 24.18%. The density of core samples is calculated using the mass of the core samples, the matrix volume and void space volume measured by CMS300.

$$\rho = \frac{m}{V_g + V_{por}} \quad (1)$$

where ρ is the density, g/cm³; m is the mass of the samples; V_g, V_{por} are matrix volume and void space volume respectively, cm³.

Measurement of Acoustic Velocity

The acoustic velocities of core samples are measured by Multi Parameter Measurement System Under HPHT made by China University of Petroleum (East China). The measurement system is showed in Figure 1(a). From the Figure 1(a), we can see that the system is composed of four parts: hydraulic pump, temperature control unit, sample length measurement unit (micro displacement sensor) and data acquisition (Figure 1(b)), the velocity of the core samples can be measured under the simulated formation temperature (≤150°C), and pressure (confining pressure, axle pressure and effective pressure ≤70 MPa). Ultrasonic Pulse Transmission Method is employed in the measurement of compressional and shear wave velocity. The acoustic velocity of the medium is calculated through the ratio of the length of transmission L to the arrival time t subtracted the system zero time t_0, *i.e.*

$$V = L/(t - t_0) \quad (2)$$

where V is the acoustic velocity, *m/s*; L is the length of the core samples, m; t is the arrival time of the sound wave; to is the zero time of the system, s. The frequencies of the wafers for compressional and shear wave are 1.2 MHZ and 0.6 MHZ, respectively. The frequencies of detectors for compressional and shear wave are 0.25 MHZ and 0.12 MHZ, respectively. The relative uncertainty for the measurement of compressional wave velocity is 0.3% - 0.9%, that for shear wave velocity is 0.8% - 1.4%. In this experiment, the compressional and shear wave velocities of the dry, brine bearing (the salinity is 5000 mg/L) and oil bearing(the density of the simulated oil is 0.798g/cm³, the viscosity is 3.52 m*Pa*s*) samples are measured under the formation condition (tempera-ture is 90°C, confining pressure is 50 MPa, the porosity pressure is 25 MPa).

THE ANALYSIS OF THE FACTORS THAT AFFECT THE ACOUSTIC VELOCITY

The Influence of Mineral Content and Lithology on the Acoustic Velocity

The influences of quartz, feldspar, magmatite debris and tuff on the compressional and shear wave velocities of dry core samples are presented in **Figure 2**. The lithology and mineral content are determined by thin section. The results show that the compressional and shear wave velocities decrease with the increase of the content of quartz and feldspar as a whole, but they increase as the content of magmatite debris increase. The content of tuff is found to have little relationship with the acoustic velocities. Among all the core samples, the velocities of the tight tuffaceous conglomerate are found to be obviously higher than those of the tuffaceous conglomerate and tuffaceous rock-fragment sandstones with good physical properties, through careful analysis, it is believed that it is mainly because of the low content of quartz (≤10%) and high content of magmatite debris (≥60%) in the tight tuffaceous conglomerate. Some of the data points in the **Figure 2** do not fit the allover trend, it is maybe due to the different geometry, distribution, cementation of the minerals and the porosity between the mineral grains.

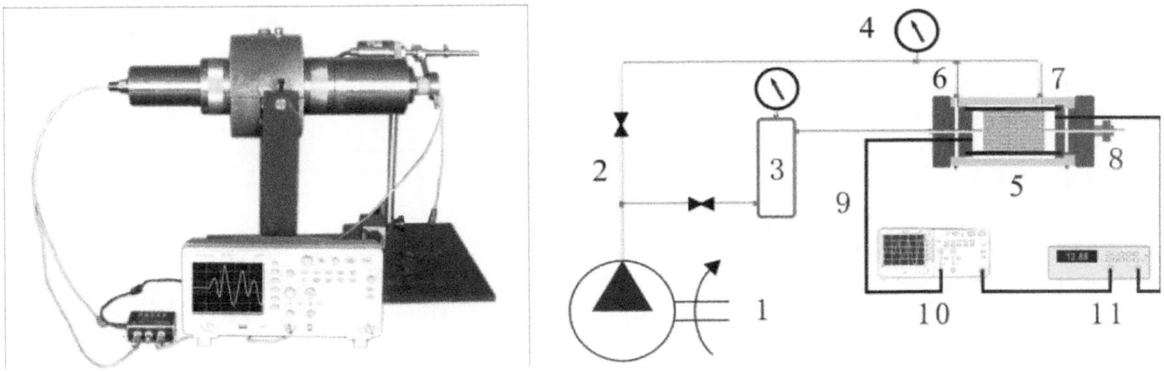

Figure 1. Sketch map of measurement system for acoustic parameters of core samples. 1—hydraulic pump; 2—fluid pipeline; 3—fluid container; 4—Pressure gauge; 5—hassler holder; 6—entry of Axial pressure; 7—entry of Central pressure; 8—Micro -displacement Sensor; 9—Acoustic data line; 10—Agilent Oscil - loscope;11—Pulse generator.

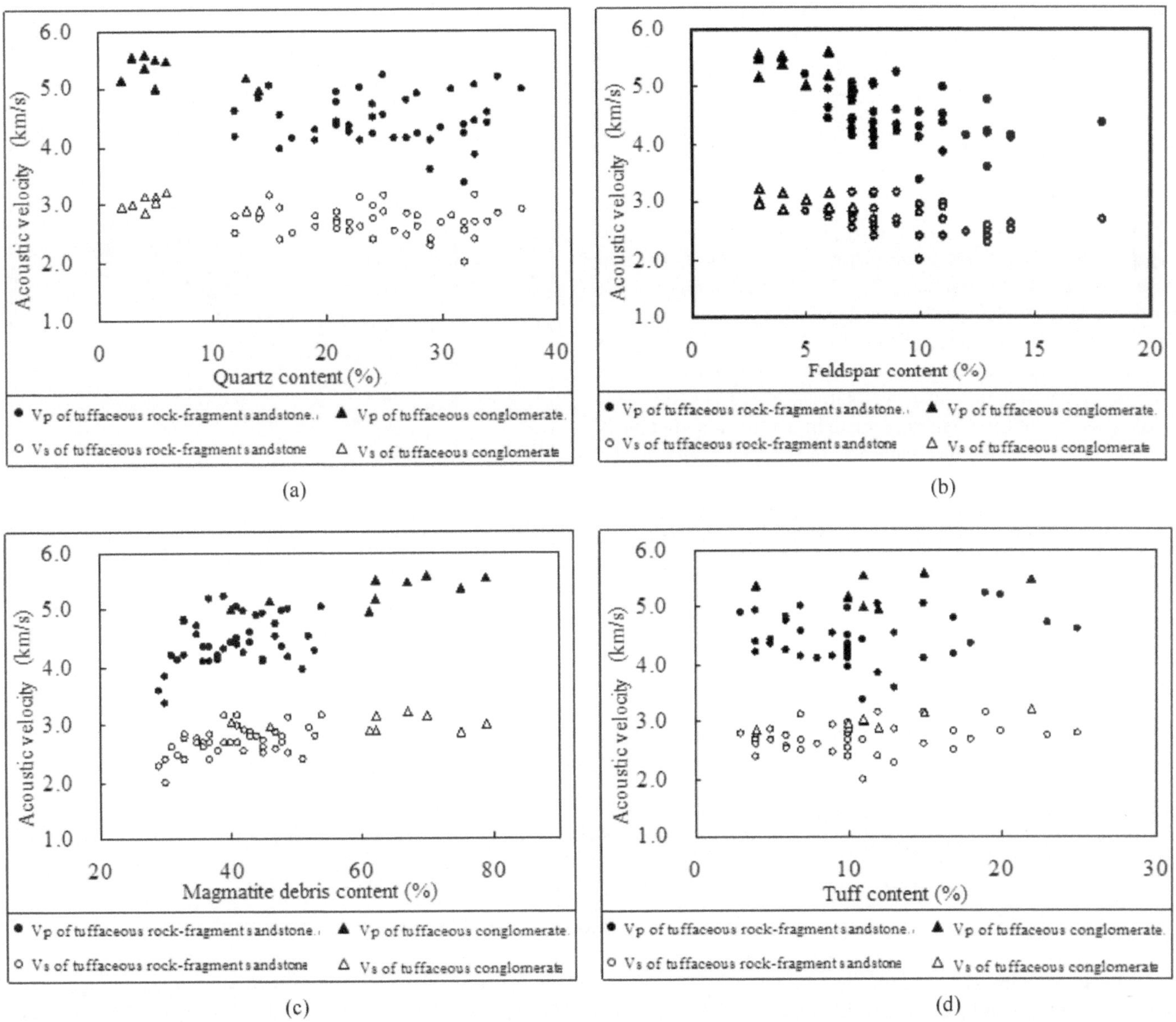

Figure 2. Compressional and shear wave velocities versus content of Quartz (a); Feldspar (b); Magmatic rock debris (c) and Tuffaceous miscellaneous base (d).

The Influence of Density of Core Samples on the Velocities

Figure 3 shows the cross-plot of the compressional and shear wave velocities versus the density of core samples ρ . The results show that the compressional and shear wave velocities of the core samples increase with their density. The data points with relative high compressional and shear wave velocities in **Figure 3** are tuffaceous conglomerate (it is showed in the ellipse of the upper right side of **Figure 3**). The tuffaceous rock-fragment sandstones and tuffaceous conglomerate can be separated by the compressional velocities as a whole and the value is about 5300 m/s. **Table 1** is the relationship between the density of tuffaceous rock-fragment sandstones ρ and the acoustic velocities. It is found that the difference between the correlation coefficient of the fitting formula obtained from general Gardner Formula and that obtained from Gardenr Formula is not large, the former just improves a little. With the lack of shear wave velocities, we can only use compressional wave velocities to inverse the density ρ , it can also get good results. Ma and Xie (2005) also got similar results.

The Influence of Porosity and Shale Content on the Acoustic Velocity

Through the literature review [16]-[19], we can see that the compressional and shear wave velocities of samples decrease with the increase of porosity, the influence of shale content on the acoustic velocities is much more complicated, for the sandstone with good consolidation, the increase of clay content will lead to the decrease of acoustic velocities, but it will cause a slight increase in the acoustic velocities for the sandstone with weak con-solidation. Generally speaking, the shale content is the leading reason for the scattering of the data points in the velocity-porosity cross-plot. According to the linear empirical relationship gave by Han (1986) [20], the relationship between the compressional and shear wave velocities and the porosity-shale content is fitted (**Table 2**).

$$V = a + b\phi + cV_{sh} \qquad (3)$$

where V is the acoustic km/s; f is the porosity of the samples, %; V_{sh} is the shale content of samples, %; *a, b, c* are the constants obtained by the experiments.

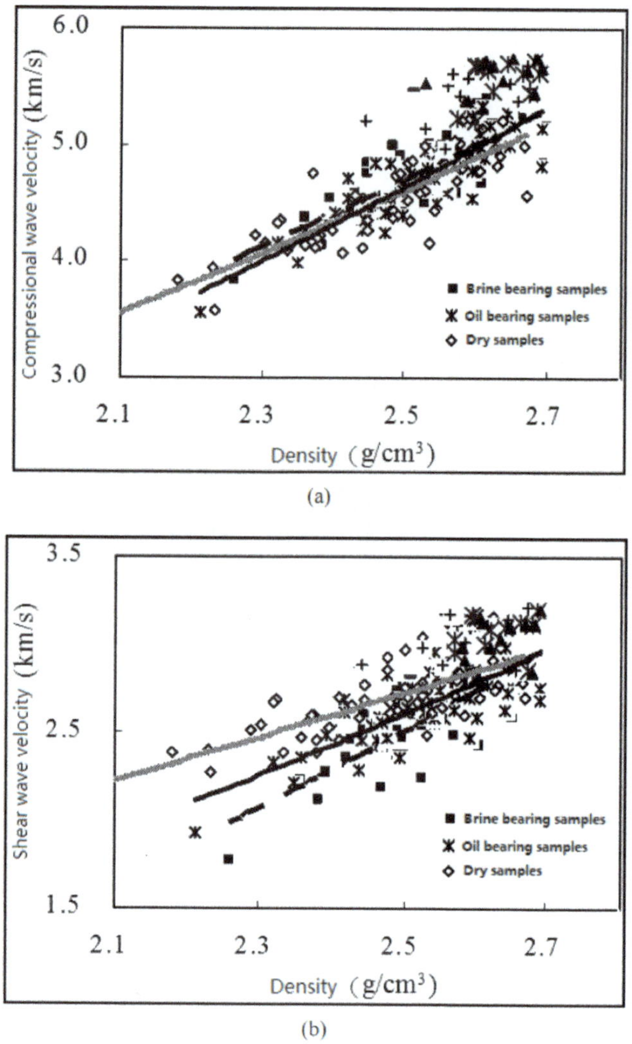

Figure 3. Compressional wave (a) and shear wave (b) velocities versus density.

Table 1. The fitting equation between density and compressional or shear wave velocities

Status of Samples	Fitting Formula (V_P)	Coefficient	Fitting Formula (V_s)	Coefficient	General Gardner	Coefficient
Dry	$\rho = 1.095\,V_p^{0.537}$	R = 0.86	$\rho = 1.506V_s^{0.482}$	R = 0.85	$\rho = 1.218\,V_p^{0.318}V_S^{0.222}$	R = 0.88
Brine bearing	$\rho = 1.332\,V_p^{0.408}$	R = 0.81	$\rho = 1.995V_s^{0.246}$	R = 0.82	$\rho = 1.697\,V_p^{0.157}V_S^{0.158}$	R = 0.87
Oil bearing	$\rho = 1.314\,V_p^{0.418}$	R = 0.87	$\rho = 1.839\,V_s^{0.314}$	R = 0.82	$\rho = 1.306\,V_p^{0.427}V_S^{0.006}$	R = 0.87

Note: The unit for acoustic velocity is *m/s*, that for the density of the samples is g/cm³.

The study shows that there is an obvious negative linear correlation between tuffaceous rock-fragment sand-stone and the porosity, shale content ($R \geq 0.85$), the influence of shale content on the compressional and shear wave velocities is only about 1/5 - 1/10 of that of porosity. Therefore, the influence of shale content on acoustic velocities is very weak and can be ignored. This result is different from Han's which is based on the experimental data from 75 shale sandstone samples in Gulf of Mexico, his result shows that the influence of shale content is about 1/3 of that of porosity. Through careful analysis, the main reason for the difference may be that the shale in the volcanic clastic rock is mainly composed of tuff, the main component of tuff is volcanic debris, whose acoustic velocity is much higher than that of clay and is near to that of matrix, this will weaken the influence of shale content to a certain degree. Until now, there are only experimental results for the tuffaceous rock-fragment sandstone in the cretaceous of Haita Basin, it needs further research on this aspect.

Table 2. Fitting formula of density and compressional and shear wave velocities on porisity and clay content

Status	Expression	Correlation coefficient
Dry sample	$V_p = 5.12 - 6.98\varphi - 0.4V_{sh}$	R = 0.86
	$V_s = 3.33 - 4.91\varphi - 0.82V_{sh}$	R = 0.87
Brine bearing	$V_p = 5.45 - 6.12\varphi - 0.5V_{sh}$	R = 0.87
	$V_s = 3.28 - 5.7\varphi - 0.75V_{sh}$	R = 0.89
Oil bearing	$V_p = 5.46 - 7.18\varphi - 0.65V_{sh}$	R = 0.90
	$V_s = 3.31 - 5.1\varphi - 0.83V_{sh}$	R = 0.88

The Influence of Pore Fluid on the Acoustic Velocity

The influence of pore fluid on the acoustic velocity is investigated through fluid substitution. The theories usually used in the analysis are Biot Theory and Gassman Equation. The difference in compressional and shear wave velocities between brine-bearing and oil-bearing versus porosity is showed in Figure 4.

From Figure 4, we can see that the absolute value of the difference in compressional and shear wave velocities between brine-bearing and oil-bearing samples increases with the porosity. The compressional wave velocity of brine-bearing samples is higher than that of oil-bearing, the absolute change rate ranges from 0 – 300 m/s, the relative change rate (relative to the compressional velocities of oil-bearing samples) is near 6% at most, the samples are sensitive to the fluid substitution. The shear wave velocities of the brine-bearing samples are relative small than that of oil-bearing samples, the absolute change rate ranges from 0 - 120 m/s, the relative change rate is small, the response of fluid substitution is not obvious.

According to Biot Theory and Gassmann Equation, the change of compressional velocity after fluid substitution is mainly because of the difference in the bulk modulus of the fluid, the bulk modulus of the water is higher than that of the oil, so the brine-bearing core samples have higher velocities than the oil-bearing core samples. For the change of the shear wave velocities, Wang and Nur (1986) attributed it to the change of the density and viscosity in the samples. As the fluid substitution won't cause obvious change in the shear modulus, the decrease of the density or the increase of viscosity will both increase the shear wave velocity.

CONCLUSION AND DISCUSSION

The research is done to investigate the influence of lithology, mineral content, density, porosity, shale content and fluid saturation on the compressional and shear wave velocities, the following conclusions can be obtained.

- The compressional and shear wave velocities increase with increase of the content of magmatite debris, but they decrease with the increase of quartz and feldspar content. It is due to the difference in elastic modulus and density between magmatite and quartz, feldspar.
- Tight tuffaceous conglomerate has low content of quartz (≤10%) and high content of magmatite debris (≥60%), its velocity is obviously higher than the tuffaceous rock-fragment sandstone with good physical pro-perties. We can use 5300 m/s as the lower velocity bound to separate the compressional velocity of tight tuffaceous conglomerate and tuffacous rock-fragment sandstone.
- The compressional and shear wave velocities increase with the increase of density. There is a good power relationship between the acoustic velocity and the density. In the seismic and logging evaluation work, the power statistic relationship established by experiments can be applied to predict the shear wave velocity and reestablish the density logging curve.

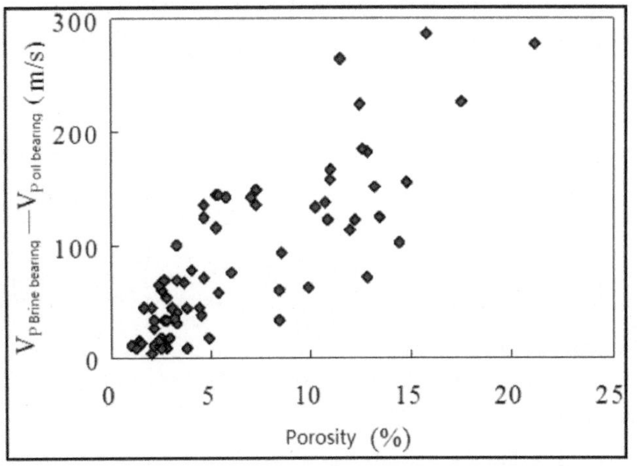

(a)

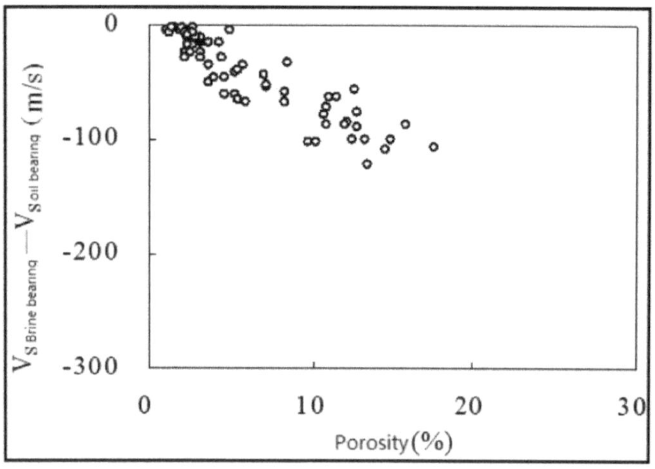

(b)

Figure 4. Difference in compressional wave (a) and Shear wave (b) Velocities between brine-bearing and oil-bearing one Versus Porosity.

- The compressional and shear wave decrease when the porosity or the shale content increase, the acoustic velocity of tuffaceous rock-fragment sandstone has good negative linear relationship with the porosity and shale content. The influence of shale content on the acoustic velocity is only 1/5 - 1/10 of that of porosity, which may be related to the tuff shale in the samples. Therefore, when calculating the porosity of the reservoir in cretaceous of Haita, the influence of shale can be ignored.
- The increase of water saturation will lead to the increase of compressional velocity and the slight decrease of shear wave velocity. The change of acoustic velocity is positive related to the porosity, but the largest change can't exceed 6%. Therefore, if the compressional and shear wave velocities obtained from the seis-mic and logging data has low uncertainty, it is possible to use the seismic and logging data to recognize the oil and gas reservoirs with good physical properties.

REFERENCES

1. Wu, Y., Chen, J.L. and Zhang, Y. (2009) Relationships between Different Types of Structural Zones and Hydrocarbons in Hailaer-Tamtsag Basin. Journal of Daqing Petroleum Institute, 33, 31-35. http://en.cnki.com.cn/Article_en/CJFDTotal-DQSY200903010.htm
2. Wyllie, M.R.J., Gregory, A.R. and Gardner, L.W. (1956) Elastic Wave Velocities in Heterogeneous and Porous Media. GEOPHYSICS, 41-70. http://dx.doi.org/10.1190/1.1438217
3. Simmons, G. (1964) Velocity of Compressional Waves in Various Minerals at Pressure to 10 Kbars. Journal of Geo- physical Research, 69, 1117-1121. http://dx.doi.org//10.1029/JZ069i006p01117/pdf
4. Ma, Z.G. and Xie, J.G. (2005) Relationship among Compressional Wave, Shear Wave Velocities and Density of Rocks. Progress in Geophysics (in Chinese), 20, 905-910. http://dx.doi.org/10.1190/1.1441933
5. Nur, A. (1992) Critical Porosity and the Seismic Velocity in Rocks. EOS Trans Am Geophys Union, 73, 43-66. http://dx.doi.org/10.1190/1.1437977
6. Han, D., Nur, A. and Morgan, D. (1986) Effects of Porosity and Clay Content on Acoustic Properties of Sandstones and Unconsolidated Sediments. GEOPHYSICS, 51, 2093-2107. http://dx.doi.org/10.1190/1.1442062
7. Biot, M.A. (1956) Theory of Propagation of Elastic Waves in A Fluid Saturated Porous Solid, I: Low Frequency Range, and II: Higher-frequency Range. Journal of the Acoustical Society of America, 168-196. http://dx.doi.org/10.1121/1.1908241
8. Gassmann, F. (1951) Elastic Waves through a Packing of Spheres. GEOPHYSICS, 16, 673-685. http://dx.doi.org/10.1190/1.1437718
9. Mavko, G. and Jizba, D. (1991) Estimating Grain-Scale Fluid Effects on Velocity Dispersion in Rock. GEOPHYSICS, 56, 1940-1949. http://dx.doi.org/10.1190/1.1443005
10. Gardner, G.H.F., Gardner, L.W. and Gregory, A.R. (1974) Formation Velocity and Density-The Diagnostic Basics for Stratigraphic Traps. GEOPHYSICS, 39, 770-780. http://dx.doi.org/10.1190/1.1440465
11. Gan, L.D. (2002) 4D Seismic and Its Application to the Monitoring of Water Flooding Reservoir. Ph.D. Thesis, China University of Geosciences (in Chinese), Beijing.
12. Ma, Z.G. and Wu, X.Y. (2006) Effects of Effective Pressure on Wave Velocities on Rocks. Progress in Exploration Geophsics (in Chinese), 29, 183-186. http://www.cnki.com.cn/Article/CJFDTotal-KTDQ200603006.htm
13. Mavko, G., Mukerji, T. and Dvorkin, J. (1998) The Rock Physics Handbook: Tools for Seismic Analysis in Porous Media. Cambridge University Press, 208-210
14. Schön, J.H. (1996) Physical Properties of Rock: Fundamentals and Principles of Petrophysics. Ultrasonic, 311-317. http://dx.doi.org/10.1029/97EO00363
15. Khatchikian, A. (1982) Log Evaluation of Oil-Bearing Igneous Rocks. 23th SPWLA Annual Logging Symposium Transactions, Texas, 6-9 July, AA. https://www.onepetro.org/conference-paper/SPWLA-1982-AA
16. Wang, B.Z. (2008) Seismic Rock Physics and Its Applied Research. Ph.D. Thesis, Chengdu University of Science and Technology (in Chinese), Chengdu.
17. Shi, G. and Yang, D.Q. (2001) The Regression Analysis Study on Velocity and Porosity and Clay Content of Rocks. Acta Scicentiarum Naturalum Universitis Pekinesis (in Chinese), 37, 379-384. http://dx.doi.org/10.3321/j.issn:0479-8023.2001.03.015
18. Klimentos, T. (1991) The Effects of Porosity-Permeability-Clay Content on the Velocity of Compressional Waves. GEOPHYSICS, 56, 1930-1939. http://dx.doi.org/10.1190/1.1443004
19. Dvorkin, J., Mavko, G. and Nur, A. (1995) Squirt Flow in Fully Saturated Rocks. GEOPHYSICS, 60, 97-107. http://dx.doi.org/10.1190/1.1443004
20. Han, D., Nur, A. and Morgan, D. (1986) Effects of Porosity and Clay Content on Wave Velocities in Sandstones. GEOPHYSICS, 51, 2093-2107. http://dx.doi.org/10.1190/1.1442062.

CHAPTER 11

TOPOGRAPHIC ROUGHNESS AS A SIGNATURE OF THE EMERGENCE OF BEDROCK IN ERODING LANDSCAPES

D. T. Milodowski, S. M. Mudd, and E. T. A. Mitchard

INTRODUCTION

The geomorphic transition from hillslopes with a continuous soil mantle to rugged bedrock is a key phase in the evolution of eroding landscapes. Many slowly eroding landscapes feature sediment transport processes that act to diffuse and dampen short wavelength features of the topography, generating smooth, soil-mantled hillslopes (Gilbert, 1909; Carson and Kirkby, 1972). Bedrock becomes exposed at the surface when the rate of erosion exceeds the maximum rate of soil production (Carson and Kirkby, 1972; Heimsath et al., 1997, 2012). This transition is gradual, and spatially variable, reflecting the fact that both soil production and sediment transport are spatially heterogeneous, and typically operate via discrete events (Wilkinson et al., 2005; Strudley et al., 2006a, b; Gabet and Mudd, 2010; Furbish and Roering, 2013). The emergence of bedrock signifies a fundamental change in the dynamics of sediment transport, which become increasingly stochastic as mobile colluvium is stripped away and the hillslope sediment flux becomes detachment limited (e.g. Binnie et al., 2007). Furthermore, the establishment of terrestrial ecosystems is dependent on a hospitable substrate: the mosaic of bedrock and soil that constitutes the hillslope surface imposes a physical template on the development of terrestrial ecosystems (Phillips and Marion, 2004; Pelletier and Rasmussen, 2009; Gabet and Mudd, 2010; Sheffer et al., 2013). The rate of erosion that is sufficient to completely strip soil may therefore represent a limiting threshold for ecosystem development (Graham et al., 2010). In addition, the presence or absence of bedrock outcrop may reveal important information about the availability of nutrients such as phosphorous in soil parent material (Hahm et al., 2014). Equally, the transition between deep and shallower soils, signaled by the appearance of bedrock outcrops, is an ecological gradient allowing for niche specialization, driving biodiversity and diversity within species, influencing ecosystem function, species creation and adaptability (Smith et al., 1997). Quantifying the spatial distribution of rock exposure and its relationship to the ecological and geomorphological characteristics of a landscape thus comprises an important challenge in understanding critical zone dynamics.

The advent of airborne Light Detection And Ranging (LiDAR) as a remote-sensing technology over the last decade or so has driven a revolution in the fields of both geomorphology and ecology by providing high-resolution (< 1 m) observations of both canopy structure and sub-canopy topography, therefore enabling observations to be made at length-scales sufficiently small to analyze the geomorphic characteristics of hillslopes (Roering et al., 2010; Hurst et al., 2012; DiBiase et al., 2012). Higher resolution still (< 1 cm) is possible using terrestrial LiDAR systems, permitting the analysis of multi-scale dimensionality from length scales of centimeters to several metres, enabling the objective classification of point clouds into specific features, such as vegetation and bedrock, with a high degree of accuracy (Brodu and Lague, 2012; Lague et al., 2013). Despite the obvious benefits of high-resolution terrestrial LiDAR scanning, the greater spatial coverage permitted by airborne surveys maintains its utility for landscape scale applications, requiring the development of remote sensing methods with which it is possible to extract information about the geomorphic characteristics of hillslopes, such as the extent of rock exposure, from such comparatively low-resolution data.

DiBiase et al. (2012) used airborne LiDAR data to investigate the impact of increasing erosion rates on hillslope

Citation: Milodowski, D. T., Mudd, S. M., and Mitchard, E. T. A.: Topographic roughness as a signature of the emergence of bedrock in eroding landscapes, Earth Surf. Dynam., 3, 483-499, doi:10.5194/esurf-3-483-2015, 2015.

morphology in the San Gabriel Mountains, CA, demonstrating that slope distributions became increasingly skewed towards higher gradients, as steep, bedrock slopes became increasingly abundant. They successfully developed the Rock Exposure Index (REI) as a topographic metric to map rock exposure in this landscape, defined as areas in which the local gradient exceeds a threshold steepness beyond which soil is no longer retained on the hillslope. DiBiase and Lamb (2013) exploited this metric to quantify sediment storage by vegetation on steep slopes, and thus assess the likely impact of wild fires on hillslope sediment fluxes. Marshall and Roering (2014) used a similar slope-based metric to map erosion resistant sandstone beds in the Oregon Coast Range.

However, slope-based metrics are not universally applicable. For example, when long-term rates of erosion exceed the local maximum rate of soil production, bedrock will be exposed at the surface, irrespective of slope (Carson and Kirkby, 1972; Heimsath et al., 1997, 2012). Within a given setting, rates of soil production may be limited by factors such as climate, vegetation, lithology and soil thickness (e.g. Pelletier and Rasmussen, 2009; Chorover et al., 2011; Goodfellow et al., 2014a). It is evident that in many landscapes rock exposure emerges in places even at low topographic gradients, and is particularly common in regions with thin regolith cover, where tor formation is common (Anderson, 2002; Strudley et al., 2006b), on ridgelines (Gabet et al., 2015), or where bedrock heterogeneities drive small-scale variation in weathering rates (Goodfellow et al., 2014b).

Another method by which rock exposure might be mapped from high-resolution topographic models of hillslopes is through changes in their textural characteristics. On hillslopes mantled by a veneer of soil, sediment transport is driven by the time-integrated effect of a suite of local-scale diffusive processes, including bioturbation, tree throw, dry ravel and rain splash (e.g. Gabet, 2003; Gabet et al., 2003; Yoo et al., 2005; Furbish et al., 2007). The net efficiency of these processes in transporting material increases with topographic gradient – they are diffusion-like (Furbish et al., 2009) – such that they act to dampen the amplitude of local topography, particularly when viewed at length-scales greater than those at which the dominant sediment transport rates operate. The resultant hillslopes therefore typically exhibit smooth, convex surfaces that are ubiquitous to many soil-mantled landscapes (Gilbert, 1909; Culling, 1963, 1965; Carson and Kirkby, 1972; McKean et al., 1993). The emergence of bedrock at the surface potentially drives a significant increase in roughness, because there is a fundamental change in the dynamics of sediment transport at this location within the landscape: sediment transport is detachment limited (Dietrich et al., 2003) and the local relief structure is governed by the characteristics of the bedrock (fracture density and orientation, bedding and foliation, weathering behavior).

In this paper we exploit this idea and develop a new technique to identify areas of rock exposure from high-resolution LiDAR data, based on short-wavelength topographic roughness. This method is validated in two granitoid landscapes by comparing the results to rock exposure mapped independently from high-resolution orthophotographs, highlighting its utility and limitations. Finally, as a case study, we apply the algorithm in two strongly transient landscapes – the first in the Feather River region of the northern Sierra Nevada, California; the second in the Salmon River region SW of the Bitterroot Mountains, Idaho – in order to illustrate the transition from diffusive, soil-mantled hillslopes to rough, bedrock hillslopes as erosion rates increase in both settings.

METHODS – QUANTIFYING SURFACE ROUGHNESS

Sediment fluxes on soil-mantled hillslopes have been shown to be well approximated by a linear relationship with the topographic slope (Carson and Kirkby, 1972), becoming nonlinear as erosion rates increase and steepen hillslopes towards a limiting slope beyond which mobile colluvium is unstable (Roering et al., 1999). The resultant topography is diffusive: hillslope processes act to dampen the amplitude of local micro-topography generating characteristically smooth hillslope topography. Our method starts from the hypothesis that the emergence of bedrock through the soil mantle should be detectable as an increase in the local roughness of the topographic surface, due to a geomorphic process transition away from diffusion-like hillslope processes.

Specifically we analyze surface roughness using the variability of the orientation of local slope normal vectors, using the eigenvalues of an orientation tensor, derived from the vectors normal to the topographic surface. A similar approach has been used in a range of geological applications, notably in earthquake seismology (Fara and Scheidegger, 1963), analyzing trends in geological structural data (Woodcock, 1977) and more recently as a method to objectively locate landslides from high-resolution topographic data (McKean and Roering, 2004). We note here that other metrics describing surface roughness, such as the standard deviation of slope, have been used in other geomorphic contexts, such as LiDAR-based mapping of volcanic deposits (Whelley et al., 2014) and channel bed morphology (Cavalli et al., 2008).

Initially a second order polynomial surface is fitted to a moving data window of 3 × 3 pixels (Evans, 1980). This method of surface approximation to calculate topographic metrics has been widely utilized in the calculation of surface derivatives, predominately slope and curvature, for the extraction of geomorphic features such as hilltops (Hurst et al., 2012), channel networks (Pirotti and Tarolli, 2010; Sofia et al., 2011), landslides (Tarolli et al., 2010; Lin et al., 2013), and anthropogenic features on floodplains (Sofia et al., 2014). Using a larger length-scale would dampen the roughness signal, but may be necessary if the topographic data are noisy (Sofia et al., 2011). The surface can be described by:

$$z = ax^2 + by^2 + cxy + \mathrm{d}x + ey + f, \tag{1}$$

where z is the surface elevation, x and y are horizontal coordinates, and a, b, c, d, e, and f are empirical fitting coefficients. A similar approach was employed by Hurst et al. (2012) to calculate hilltop curvature, who found no significant difference between the results obtained using six or nine term polynomials in their surface fitting algorithm. Consequently we use a six term polynomial as it maximizes computational efficiency. The normal to a surface is given by:

$$\boldsymbol{n} = \nabla(f(x, y) - z). \tag{2}$$

For Eq. (1), using spherical coordinates (r, θ, j) at the origin, the unit normal vector becomes:

$$n = \left(1, \tan^{-1}\left(\sqrt{d^2 - e^2}\right), \tan^{-1}\left(\frac{e}{d}\right)\right). \tag{3}$$

For N surface normal vectors, the orientation matrix, T, can be constructed using the directional cosines l_i, m_i and n_i, as shown below:

$$\mathbf{T} = \begin{pmatrix} \sum_i^N l_i^2 & \sum_i^N l_i m_i & \sum_i^N l_i n_i \\ \sum_i^N m_i l_i & \sum_i^N m_i^2 & \sum_i^N m_i n_i \\ \sum_i^N n_i l_i & \sum_i^N n_i m_i & \sum_i^N n_i^2 \end{pmatrix}. \tag{4}$$

The orientation matrix can be solved to find the three eigenvectors v_1, v_2, v_3 and their corresponding eigenvalues, λ_1, λ_2, λ_3, which describe the degree of clustering of the normal vectors about the principal axes of the distribution (Watson, 1966). Following Woodcock (1977), we normalize the eigenvalues by the number of observations (N):

$$S_1 = \frac{\lambda_1}{N}, \quad S_2 = \frac{\lambda_2}{N}, \quad S_3 = \frac{\lambda_3}{N}. \tag{5}$$

S_1 $(\frac{1}{3} \leq S_1 \leq 1)$ describes the clustering around the major axis, S_2 $(0 \leq S_2 \leq \frac{1}{2})$ the intermediate axis, and S_3 $(0 \leq S_3 \leq \frac{1}{3})$ the minor axis. These normalized eigenvalues can be used to describe the morphology of a given surface (Woodcock, 1977): for a smooth surface, the local surface normal vectors will have similar orientations, thus they will cluster tightly around the major axis, v_1, and S_1 will be large, whereas the degree of clustering around the minor axis, v_3, will thus be very small (low S_3). Conversely, for a rough surface, the normal vectors will be more randomly orientated; there will be a weaker degree of clustering around v_1 (low S_1), whilst the clustering around v_3 will be relatively high (therefore high S_3).

A moving data kernel is passed over the data set to analyze the variability of the surface normal vectors within the local (circular) neighborhood. The radius of this kernel determines the length-scale over which the roughness of the surface is quantified. Identifying the correct length-scale in this case is critical – too large, and long wavelength variations in the topography (i.e. ridge-valley topography) will dominate, obscuring any signal from rock exposure; too small, and then the measured roughness will pick out locally smooth surfaces within an exposure of bedrock. We discuss determining the optimal length-scale in the Validation section (for results, see Sect. 3.4).

VALIDATION OF THE SURFACE ROUGHNESS ALGORITHM

Validation Sites

In order to test the surface roughness metric described above as a measure of rock exposure, we selected two validation sites in western USA (Fig. 1) based on the availability of co-located LiDAR and high-resolution (< 30 cm) orthophotographs. A further requirement for validation sites was that the degree of vegetation cover was minimal, to permit the objective classification of rock outcrop in the imagery (Sect. 3.2). All LiDAR data sets and orthophotographs used in the study are freely available from either the National Science Foundation's OpenTopography service (www.opentopography.org) or from the United States Geological Survey (USGS; earthexplorer.usgs.gov/). Technical details for the data sets have been collated in Table 1.

Table 1. Summary of data sets used during in this study

Airborne LiDAR				
Region	Acquisition date	Areal extent used (km^2)	Point Density (pts m^{-2})	Data set acknowledgement
Rayleigh Peak, CO	May 2010	23	10.1	1
Poway Creek, CA	Jan 2005	1.4	1.4	2
Bald Rock Basin, CA	Sep 2008	4.0	9.8	1
Harrington Creek, ID	Aug 2011	49.0	4.6	1
Orthophotographs				
Region	Acquisition date	Resolution/m	Sensor type	Data set acknowledgement
Rayleigh Peak, CO	Mar 2010	0.30	Colour Near-Infrared	3
Poway Creek, CA	May 2012	0.15	Colour Near-Infrared	3

1: National Center for Airborne Laser Mapping (NCALM – http://www.ncalm.org); 2: USGS Center for LiDAR Information Coordination and Knowledge (CLICK – http://lidar.cr.usgs.gov/; via OpenTopography); 3: USGS (via EarthExplorer http://earthexplorer.usgs.gov/).

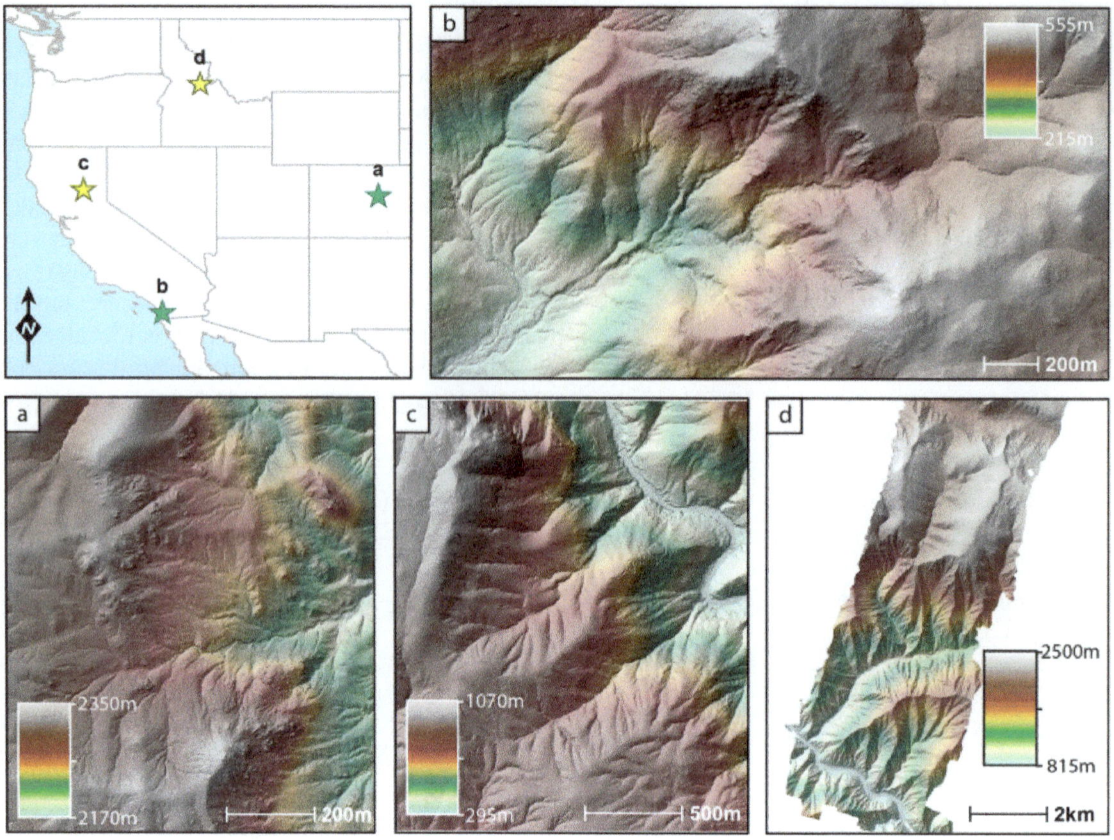

Figure 1. Field sites used in this study; (a) headwaters of the Spring Creek catchment, ~ 2.7 km SW of Rayleigh Peak, in the Colorado Front Ranges; (b) Poway Creek, California; (c) Bald Rock Basin, draining into the Middle Fork Feather River, Californian Sierra Nevada; (d) Harrington Creek, which drains into the Salmon River, Idaho. Sites (a) and (b) were used to validate our algorithm; sites (c) and (d) were subsequently analyzed to investigate the transition from soil-mantled to bedrock hillslopes in transient landscapes.

Rayleigh Peak, Colorado

The first validation site is located in the headwaters of the Spring Creek catchment, in the central Colorado Frontal Range, which drains into the South Platte River 40 km SSW of Denver (Fig. 1a). The climate is semi-arid with frequent intense summer storms. Mean Annual Precipitation (MAP) is 440 mm, and average monthly temperatures varies from a maximum (minimum) of 27.7 (10.8) C in summer to 6.0 (9.0) C in winter (http://www.prismclimate. org). Vegetation comprises grassland and sparse coniferous forest, of which Ponderosa Pine and Douglas Fir are the principal components, the distribution of which is dominated by the impact of the 1996 Buffalo Creek wildfire, in which 79 % of the Spring Creek catchment suffered severe burn damage (Moody and Martin, 2001), so that forest canopy now covers only a small proportion of the landscape. The bedrock geology comprises Pikes Peak Granite (Ruleman et al., 2011), which forms large, blocky outcrops. The degree of rock out-crop at the site varying from almost full exposure on hill-slopes around Rayleigh Peak, which dominates the topography, to fully soil-mantled hillslopes that are now predominately covered by grassland.

Poway Creek, California

The second study site is located in the Poway Creek catchment, located just east of the city of Poway, north of San Diego (Fig. 1b). MAP is 825 mm and temperatures typically range from 29.1 (14.1) C in summer to 11.5 (0.2) C in winter (http://www.prismclimate.org). The bedrock geology is principally composed of granodiorite with daciticandesitic extrusive rocks underlying the eastern margin (Todd et al., 2004). There is a gradient in rock exposure from predominately soil-mantled, grassy hillslopes that are frequently gullied, to abundant rock outcrop in the steep, rugged headwaters. Due to classification errors in the original data set, the LiDAR point cloud was reclassified using the multi-scale curvature algorithm incorporated within the MCC-LiDAR tool (Evans and Hudak, 2007).

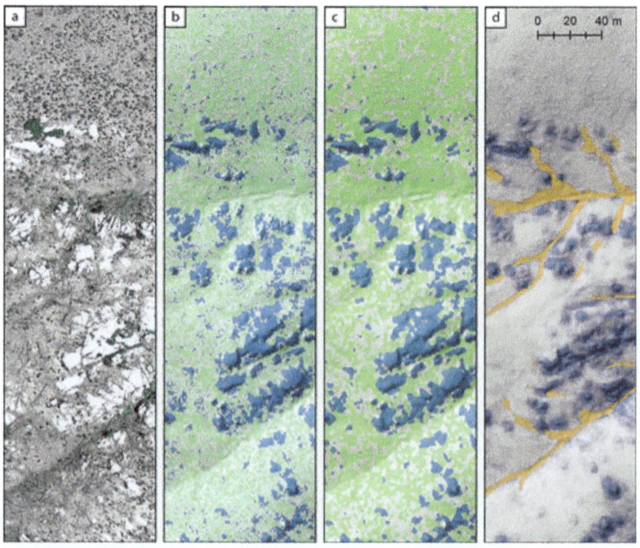

Figure 2. Validation procedure illustrated for the Rayleigh Peak site: **(a)** high-resolution color near-infrared orthophotograph;

(b) results from the SVM classification procedure – rock D blue, soil mantled/vegetation D green; **(c)** classified image following the subsequent majority filter; **(d)** map of S_3, which we use as a mea-sure of surface roughness, measured using a neighborhood window radius of 3 m. Orange pixels mark areas identified as being channelized.

Objective Identification of Rock Exposure from High-Resolution Orthophotographs

The high-resolution orthophotographs were classified using the supervised classification toolbox available within the ENVI 4.8 processing environment. Specifically we utilized the Support Vector Machine classification method (Wu et al., 2004), trained using a series of manually selected sample Regions Of Interest (ROIs) for each class. The classes used to analyze each orthophotograph comprised the following: "Rock", "Vegetation", "Bare Earth" and "Shadow". With the exception of the "Shadow" class, which was not as spatially extensive, each ROI had a minimum of 10 000 pixels. The SVM classification was implemented to analyze the imagery at two pyramid levels, with a Pyramid Reclassification Threshold (i.e. the probability threshold required to reclassify a pixel, if given a different class at a finer resolution) of 0.90. As the avoidance of false positives within our validation data set was of paramount importance, pixels were left unclassified if the confidence level for the final class fell below 95 %. Subsequently a 7 7 pixel majority filter was employed to reduce the noise in the classified image Fig. 2. As our focus is on comparing soil-mantled and rocky hill-slopes, we combine the vegetation and bare earth classes, and treat areas that are in shadow as unclassified.

The quality of the classification scheme for each image was judged based both with a visual inspection of the classification results to ensure that there were no systematic errors located away from the training ROIs, and using the error matrices for each classification, providing a quantitative assessment of the scheme's ability to correctly reproduce the classification of the initial ROIs. At the 95 % confidence interval, the SVM scheme discarded 9.4 % of the ROI pixels as unclassified in the Rayleigh Peak data set and 12.0 % in the Poway Creek data set. At the Rayleigh Peak site, the classification scheme was able to replicate the rock ROIs with a commission error (ratio of non-rock pixels classified as rock to the total number of pixels in the rock ROI) of 0.23 % and an omission error (ratio of rock pixels incorrectly classified to the total number of pixels in the rock ROI) of 0.01 %. At the Poway Creek site, the ROIs were replicated with a commission error of 0.01 % and an omission error of 0.13 %. Across the region as a whole, both of our validation sites, the classification scheme struggled in areas where there are large changes in the saturation of the imagery (Figs. 3 and 4), due to aspect-driven differences in illumination: as a result some areas have an increased proportion of unclassified pixels. This problem is endemic to image classification in high relief terrain, and is very hard to correct even with good topographic data and bi-directional reflectance function (BDRF-driven) models, as there is often no information captured in the brightest and darkest parts of the image (e.g. Teillet et al., 1982; Colby, 1991; Hale and Rock, 2003). Again, this highlights the potential advantages of landscape classification techniques based on the morphological characteristics of the topographic surface. In addition, it is evident that there are still some areas where the image classification provides an incorrect classification. Nevertheless, the classification is sufficiently successful to provide two large test data sets with which to validate our roughness metric. Errors in the validation data sets will, if anything, lead to an underestimate of the accuracy of our topographically derived metric; it is hard to imagine how errors in the classification could inflate the accuracy of the topographic roughness metric, as the data sets are entirely independent and any errors unlikely to be co-located.

Validation Procedure

We used the rock exposure maps from the classifications de-scribed above to perform the validation of the roughness algorithm in each of the four test landscapes. Since channels are often topographically rough, we first restricted our analysis to the hillslope domain. Several methods have been pro-posed to identify channel pixels in high-resolution topography (e.g. Lashermes et al., 2007; Passalacqua et al., 2010; Pelletier, 2013); in each landscape we have used the method of Lashermes et al. (2007), in which the topography is faltered using a Gaussian filter, and then a curvature threshold to define the extent of the channel network is obtained statistically by looking for the departure from the expectations of a Gaussian distribution. This approach produces visibly satisfactory results across the range of landscapes used here.

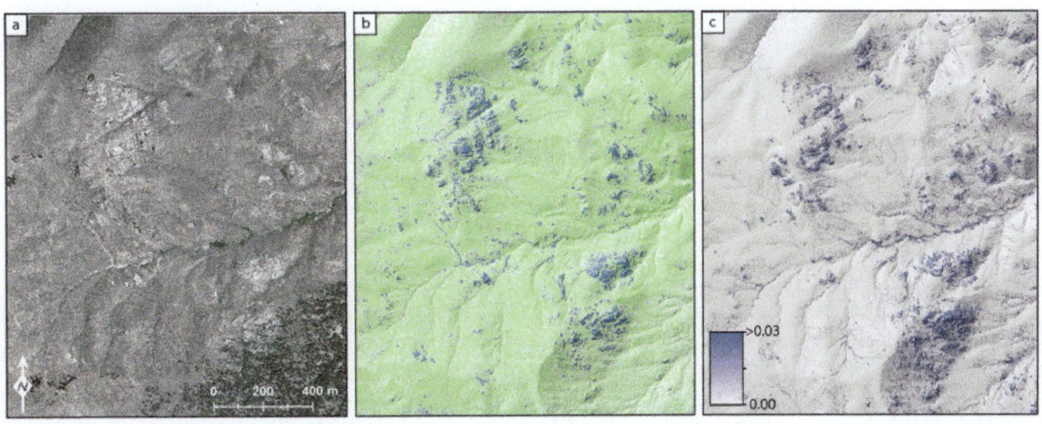

Figure 3. Validation maps for the Rayleigh Peak site: **(a)** high resolution, color-near infrared orthophotograph; **(b)** results from combined classification procedure: rock D blue, soil mantled/vegetation D green; **(c)** map of S_3, which we use as a measure of surface roughness, measured using a neighborhood window radius of 3 m. To maximize the clarity of the maps, channelized portions of the landscape have not been masked.

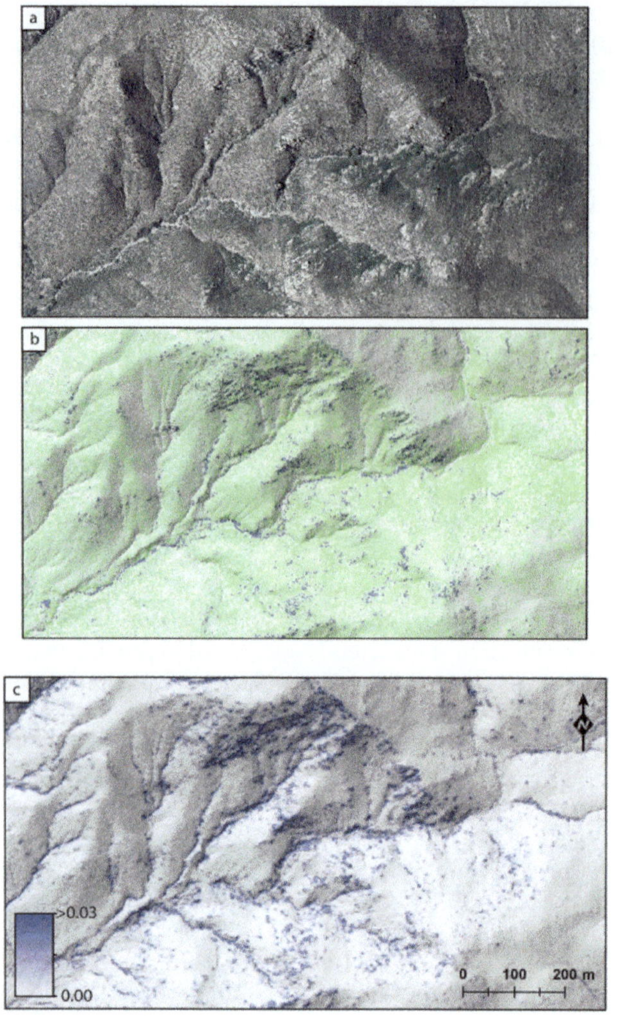

Figure 4. Validation maps for the Poway Creek site: **(a)** high resolution, color-near infrared orthophotograph; **(b)** results from combined classification procedure: rock D blue, soil mantled/vegetation D green; **(c)** map of S_3, which we use as a measure of surface roughness, measured using a neighborhood window radius of 3 m. To maximize the clarity of the maps, channelized portions of the landscape have not been masked.

After isolating the hillslopes, we searched through the parameter space for the S_3 eigenvalue, performing a pixel-pixel comparison with the orthophotograph classifications to ascertain whether the algorithm produced a true positive (TP), false positive (FP), true negative (TN) or false negative (FN) for a given roughness threshold. In order to objectively assess the performance of the algorithm and determine an optimum threshold value to delineate areas with rock exposure, we calculated five test statistics: (i) true positive rate (D TP/(TP C FN)); (ii) false positive rate (D FP/(TNCFP)); (iii) commission error (D FP/(TP C FN)); (iv) omission errors (D FN/(TP C FN)); and (v) the overall accuracy (D (TPCTN)/Total); to objectively assess the performance of the algorithm and determine an optimum thresh-old value to delineate areas with rock exposure. In order to avoid bias in the aforementioned statistics towards either class, the larger of the two classes was randomly subsampled to the same number of test pixels as the smaller of the two before proceeding with the calculations. We repeated this procedure for three neighborhood radii (3, 5 and 7 m) in each of the two field sites to assess the influence of neighborhood size on the measured surface roughness. An important consideration when interpreting the validation results is that the surface roughness represents a spatially aggregated metric, representing a blend of the topographic characteristics within the circumference of the neighborhood window. Consequently, it is unlikely that this metric will discriminate between small areas of patchy soil interspersed between rugged rock outcrops at length scales smaller than the neighborhood window. This effect becomes increasingly significant as the window size increases and is an inevitable out-come from neighborhood statistical approaches. As a result, we eliminate from our validation data set areas that are not classed as rock exposure that lie within 7 m (the largest neighborhood radius used) of mapped rock exposure. For comparison, we also report the same statistics for the full data set.

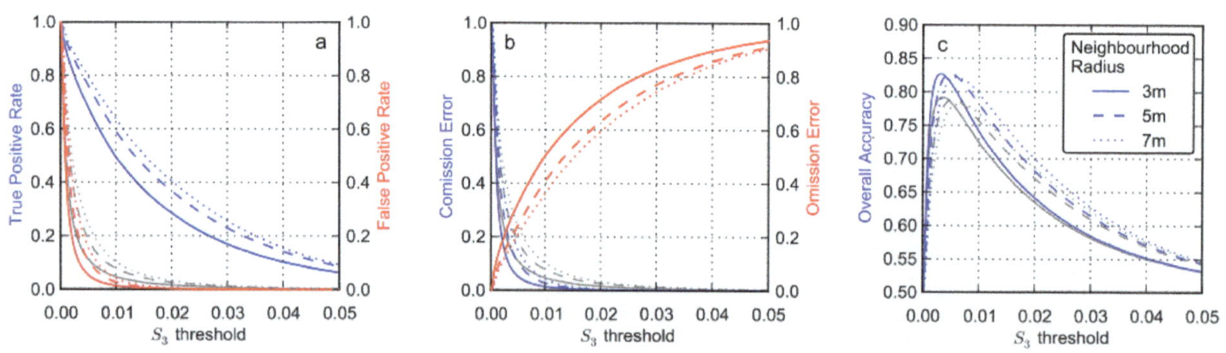

Figure 5. Validation statistics for Rayleigh Peak site as a function of the roughness threshold used to delimit rock exposure for three different neighborhood window radii: **(a)** true positive and false positive rates; **(b)** commission and omission errors; **(c)** overall accuracy. These tests were conducted twice – the red and blue lines illustrate the results from tests in which the pixels classified as soil-mantled pixels were filtered to avoid localities proximal to rock exposure (see text), therefore is more representative of the roughness signature of a pure soil-mantled hillslope; the grey lines illustrate the same tests, but without this prior filtering step.

Validation Results

In both landscapes, the close correspondence between the topographically derived roughness maps against the rock exposure mapped from the high resolution orthophotographs attests to a qualitatively good agreement between the two (Figs. 3 and 4). Hillslopes that are covered by a continuous mantle of soil map consistently as areas that are topographically smooth, having locally consistent normal vector orientations; in contrast the emergence of bedrock drives a significant increase in the roughness of the affected hillslopes that is clearly picked up by our algorithm.

In the Rayleigh Peak example, both areas with widespread rock outcrop and more isolated exposures are picked out (Fig. 3). The primary area of discordance lies in the SW corner of the image. Here the roughness algorithm predicts a much greater extent of rock exposure than the classified image. Inspection of the orthophotograph in this area reveals significant vegetation cover, obscuring areas where there is clearly bedrock, thus severely hampering the optical classification in this location. Areas of enhanced roughness running laterally along the trunk channel, which flows from west to east here, provide another potential false positive in the roughness map; this highly localized roughness signature marks the banks of the incised channel. The validation statistics similarly show a distinct difference between soil-mantled hillslopes and areas with rock exposure (Fig. 5; Table 2). The FPR rapidly decreases as the value of $\boldsymbol{S}_3$ used to discriminate between the two characteristics increases, with a maximum accuracy (taking into account both false positives and false negatives) of > 80 % for ~ 0.003 ≤ S_3, threshold ≤ ~ 0.005.. The TPR also decreases across this interval, which is likely to be driven by areas of rock exposure where the rock surfaces have a low fracture density, therefore appear smooth, and the fact that our test data set is not perfect (see discussion in Sect. 3.2). We stress here that the imperfections in the validation data set derived from the orthophotographs will lead to a conservative estimate of the true accuracy of the roughness algorithm. Critically from the perspective of mapping out areas of rock exposure, the rate at which the TPR decreases with increasing values of $\boldsymbol{S}_{3;\ \mathrm{threshold}}$ is much lower than that of the FPR. Increasing the size of the neighborhood window over which the surface roughness is characterized acts to in-crease the number of true positives for a given threshold, but there is a trade-off, as this improvement is accompanied by an increase in the number of false positives (Figs. 5 and 7; Table 2). This is probably due to the "leakage" of the roughness signal from areas where there is rock exposure into the expanded neighborhoods of proximal soil pixels (Fig. 7), and also due to the fact that the longer wavelength topographic structure imposed by the ridge-valley architecture starts to influence the variability in the distribution of surface normal vectors; the latter case is particularly prevalent in areas that are located close to gullies and channels.

The pattern that emerges from the Poway Creek site is very similar; again, the maps of rock exposure do a qualitatively good job at locating hillslopes with rock outcrops, although the visual comparison is hindered by the spatially variable success of the classification scheme (Fig. 4). Again, the net-work of channels and gullies provides additional sources of roughness in the landscape. The performance in the quantitative tests exhibits very similar patterns to those obtained for the Rayleigh Peak site (Fig. 6; Table 2).

Implications for Use of Topographic Roughness in Other Settings

The fact that the roughness signatures of both validation landscapes display strikingly similar characteristics (Figs. 5 and 6), suggests that surface roughness is a promising tool for mapping the extent of bedrock outcrop on hill-slopes. As with existing methods (e.g. REI; DiBiase et al., 2012), an important caveat is that full calibration is dependent on the a posteriori knowledge of threshold values, obtained, for example, through comparison against rock expo-sure mapped from high-resolution photographs (DiBiase et al., 2012; this study). This is non-trivial in areas with significant vegetation cover due to the difficulty in resolving the ground surface; indeed, in areas with significant tree cover a significant portion of exposed rock is always hidden. Greater uncertainty will arise in areas where prior calibration against orthophotographs is not possible. A further element of caution is required, as our validation sites are limited to low-moderate relief, granitoid settings, but nevertheless, we expect that the methodology can be used judiciously in other landscapes. We provide an illustration of the method in a landscape underlain by layered sedimentary rocks in the Supplement. A number of important considerations are necessary in doing so, given that in many scenarios it will not be possible to use aerial imagery to independently judge the performance of the algorithm.

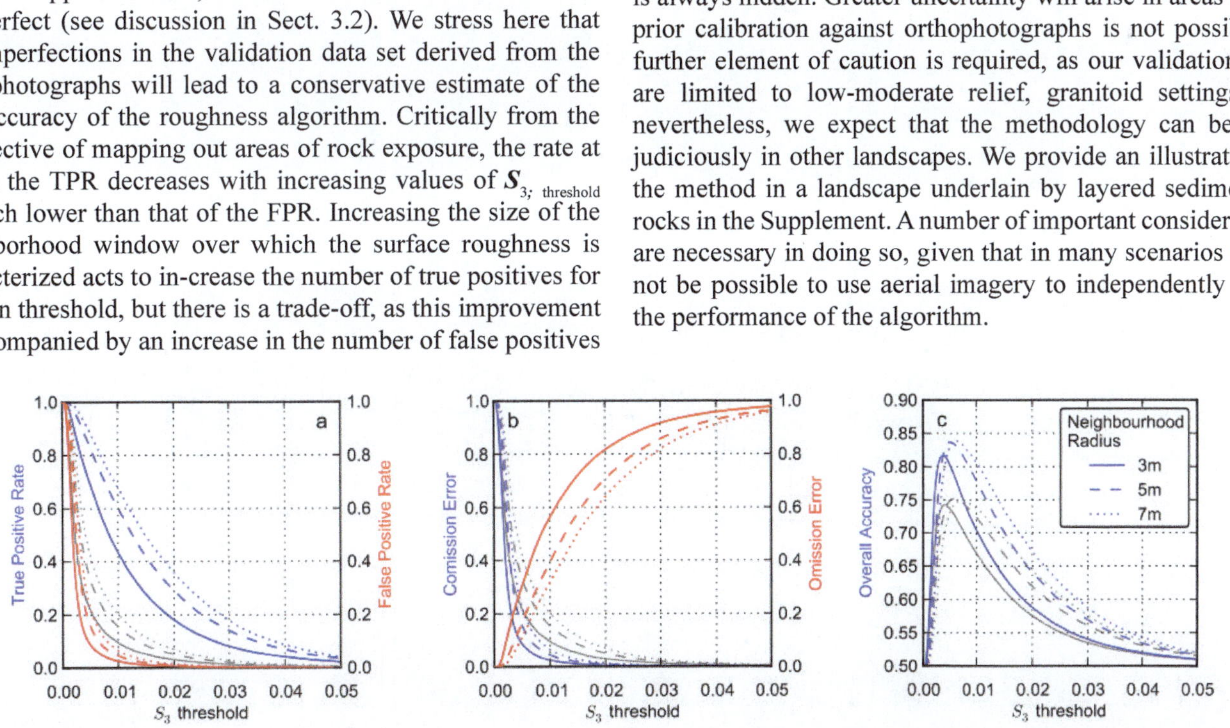

Figure 6. Validation statistics for Poway Creek site as a function of the roughness threshold used to delimit rock exposure for three different neighborhood window radii: **(a)** true positive and false positive rates; **(b)** commission and omission errors; **(c)** overall accuracy. These tests were conducted twice – the red and blue lines illustrate the results from tests in which the soil-mantled samples were filtered to avoid localities proximal to rock exposure, therefore is more representative of the roughness signature of a pure soil-mantled hillslope; the grey lines illustrate the same tests, but without this prior filtering step.

Table 2. Summary of validation results for three different threshold values of the eigenvalue S_3. These represent a subsample from the data displayed in Figs. 5 and 6. TPR D True Positive Rate; FPR D False Positive Rate; CE D Commission Error; OE D Omission Error; OA D Overall Accuracy (for definitions see text). As the surface roughness metric is spatially aggregated, this pixel-wise comparison was conducted avoiding soil-mantled pixels that were located proximal to areas of rock exposure (see text). Including these results in an increase in the false positive rate and commission errors, and corresponding drop in overall accuracy (see also Figs. 4 and 6); however these errors are collocated with areas of rock exposure, and arise as a consequence of this proximity

	TPR			FPR			CE			OE			OA		
Neighbourhood Window Radius	3 m	5 m	7 m	3 m	5 m	7 m	3 m	5 m	7 m	3 m	5 m	7 m	3 m	5 m	7 m
$S_{3,\text{threshold}}$	Rayleigh Peak														
0.005	0.68	0.76	0.80	0.05	0.11	0.16	0.05	0.11	0.16	0.32	0.24	0.20	0.81	0.83	0.82
0.010	0.50	0.59	0.64	0.01	0.03	0.04	0.01	0.03	0.04	0.50	0.41	0.36	0.74	0.78	0.80
0.015	0.37	0.46	0.50	< 0.01	0.01	0.01	< 0.01	0.01	0.01	0.63	0.54	0.49	0.68	0.73	0.75
$S_{3,\text{threshold}}$	Poway Creek														
0.005	0.69	0.83	0.88	0.09	0.15	0.23	0.09	0.15	0.23	0.31	0.17	0.12	0.80	0.84	0.83
0.010	0.43	0.60	0.68	0.03	0.05	0.07	0.03	0.05	0.07	0.57	0.40	0.32	0.70	0.78	0.81
0.015	0.28	0.42	0.50	0.01	0.02	0.03	0.01	0.02	0.03	0.72	0.58	0.50	0.63	0.70	0.73

Firstly, it is evident from Figs. 3–6 that a minor portion of landscapes mapped as rock exposure is topographically smooth. Variations in bedrock morphology present a challenge for the textural classification of topography. Errors may be introduced in areas where a significant proportion of the bedrock has been polished, or where the bedrock is massive and exhibits sparse jointing. The latter case is illustrated by smooth, massive granitoid domes, where the distribution of fractures is dominated by surface parallel exfoliation joints (Migon, 2006). In such cases the textural characteristics of bedrock hillslopes may be indistinguishable from those with a continuous soil mantle. In the case of layered rocks, slopes parallel to the structural fabric may be smooth, whereas slopes that cross-cut the layering will appear rougher. This may drive variable accuracy in the results of textural classification metrics. However, large areas of smooth bedrock should be readily visible in satellite/aerial imagery because such conditions are unlikely to support significant vegetation cover (Graham et al., 2010; Hahm et al., 2014). Furthermore, where smooth surfaces form steep structures, a slope-based metric such as the REI (DiBiase et al., 2012) can easily be employed alongside surface roughness to catch these false negatives. Combinations of topographic metrics in this way may potentially permit more robust feature extraction from high-resolution data.

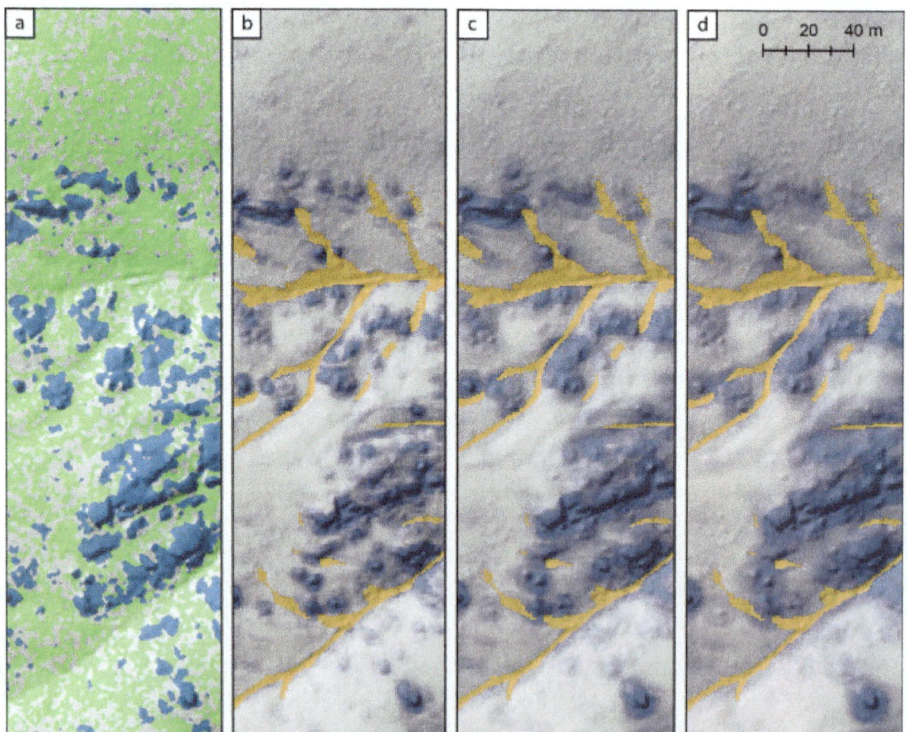

Figure 7. Illustration of the impact of the effect of changing neighborhood window radius on the roughness signal that is measured: **(a)** results from combined classification procedure – rock D blue, soil mantled/vegetation D green; **(b–d)** maps of S_3 using a neighborhood window radius of **(a)** 3 m; **(b)** 5 m; and **(c)** 7 m. Orange pixels mark areas identified as being channelized. Note the increase in the leakage of the roughness signal into proximal areas as the neighborhood radius is increased.

Secondly, bedrock exposure is not unique in adding rough-ness elements to landscapes, as surface roughness may potentially be generated by other processes. At length scales of 11–50 m, topographic roughness may be dominated by the signature of deep seated landslides, if present (Booth et al., 2009), while other features associated with landslides may generate roughness at shorter wavelengths (McKean and Roering, 2004; Tarolli et al., 2010). Roughness at small length scales (typically < 7.5 m) can also be generated via tree throw where this process is prevalent (Roering et al., 2010; Marshall and Roering, 2014). Moreover a degree of familiarity with target landscapes is likely essential in order to critically evaluate the results, although this criteria is not unique to this method. Furthermore, in more complex landscapes with multiple roughness generation mechanisms, the spatial distribution of roughness generated by different processes may still allow useful quantitative information to be extracted (for example, instances of tree throw are likely to be quasi-random, or at least spatially discrete events, whereas exposure of bedrock in hillslopes is likely to generate connected "clusters" of roughness), although we do not extend our analysis in this manner here.

The size of the polynomial surface-fitting window should ideally be comparable to the feature being extracted. In landscapes where other roughening elements are present, or when the LiDAR data are noisy, a larger window can be employed, or the topography can be smoothed, with the limitation that as the degree of smoothing increases, the textural information that distinguished bedrock hillslopes from soil-mantled hillslopes is progressively lost (Albani et al., 2004; Sofia et al., 2013). Finally, the neighborhood size used to quantify surface roughness will dictate the resolution at which you can discriminate between soil and rock outcrop (Fig. 7).

For many applications, whether making an assessment of shallow landslide hazard, or testing hypotheses concerning the transition from soil-mantled bedrock topography, avoiding false negatives is of paramount importance. For neighborhood radii of 3–5 m, a threshold value of S_3 D 0.01 limits the occurrence of false positives to < 5 % (Fig. 5), decreasing to < 2 % for S_3 D 0.015. Omission errors decrease substantially by increasing the radius of the neighborhood window, but there is a trade-off against an increasing frequency of commission errors (Figs. 5 and 6).

In Fig. 8, we illustrate an alternative approach to mapping rock exposure using the surface roughness metric introduced above. Specifically we assess the fraction of pixels within a local neighborhood that have a value of S_3 greater than a specified threshold value. Employing a sufficiently high threshold, we can thus express the expected rock exposure within that neighborhood. This provides a conservative estimate of the degree of rock outcrop for a given portion of hillslope. In all cases, there is a positive correlation between the rock exposure mapped from the orthophotographs and the roughness of the topographic surface (Fig. 8). However, when the S_3 threshold is set too low, the frequency of false positives leads to an overestimation of the rock exposure in a given portion of the landscape, as expected from our previous analysis (Figs. 4–7). In the Rayleigh Peak site, there is a good agreement between the degree of rock exposure mapped by the two methods using an S_3 threshold of 0.010, if roughness is quantified with a neighborhood radius of 3 m, and 0.015 if quantified with a neighborhood radius of 5 m. Again this conforms to the expectations arising from the validation tests (Fig. 5). In Poway Creek, there appears to be a systematic over-estimation of the rock exposure. The Poway Creek catchment presents a more challenging landscape to classify for three reasons: (i) gullies are common, and many of the channels show evidence of recent incision; the channel banks in these incised localities generate false positives due to the sharp break in slope. There may be bedrock ex-posed in the terrace walls, but if present may be obscured by overhanging vegetation. (ii) Changing insolation conditions across the image made classification using the optical data more difficult (Fig. 5). (iii) The original LiDAR point cloud was relatively sparse (Table 1), as a consequence of which discrimination of ground returns from those hitting low lying shrubs is more difficult.

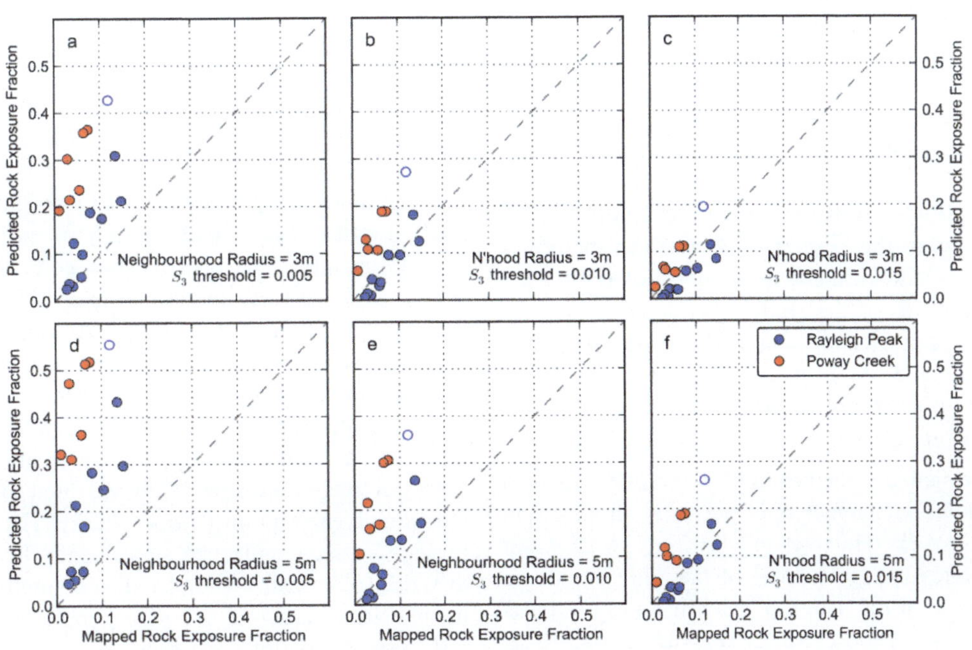

Figure 8. A comparison of the rock exposure classified from the orthophotographs against the expected fraction of rock exposure predicted using different thresholds of the surface roughness metric, S_3, for a series of the validation sites near Rayleigh Peak, Colorado, and Po-

way Creek, California. Each data point represents the rock exposure mapped within a 401 m 401 m square region within a regularly spaced grid. **(a–c)** S_3 mapped using a neighborhood radius of 3 m; **(d–f)** S_3 mapped using a neighborhood radius of 5 m. The hollow symbol outlined in blue is from the SE corner of the Rayleigh Peak site, where the rock exposure mapped from the orthophotographs significantly under-predicts the true degree of rock exposure due to a combination vegetation cover and variable insolation conditions.

As a general point we emphasize that although the high-resolution orthophotographs provide the best means of objectively testing our algorithm, the resulting validation data sets are not perfect, and classification errors will result in under-estimation of the success of the roughness metric.

APPLICATION OF THE ROUGHNESS ALGORITHM TO TRANSIENT LANDSCAPES – INVESTIGATING THE SOIL-BEDROCK TRANSITION IN BALD ROCK BASIN, CALIFORNIA, AND HARRINGTON CREEK, IDAHO

Study Sites

We investigate the variations in hillslope characteristics exhibited in two landscapes – Bald Rock Basin, in the Californian Sierra Nevada, and the Harrington Creek catchment, a tributary of the Salmon River, Idaho – which both exhibit strongly transient states of landscape evolution, under different climate regimes.

Bald Rock Basin, California

The Bald Rock Basin catchment drains into Middle Fork Feather River, in the north-western Sierra Nevada Mountains, California (Fig. 1c). The regional climate in this locality is strongly seasonal, with maximum (minimum) temperatures range from 30 (12) C in the summer to 9 (1) C in the winter, and mean annual precipitation typically 1750 mm, a substantial majority of which falls between October and April, whereas the summer months are dry (http://www. prismclimate.org). Geologically, the catchment is underlain by the Bald Rock Pluton, a trondhjemite-tonalite intrusion of mid-late Mesozoic age (Saucedo and Wagner, 1992). The landscape is close to fully vegetated by mixed conifer forest that is typical of the mid-elevation Sierra Nevada (Barbour and Billings, 2000). The notable exception to this is Bald Rock Dome, which rises precipitously from the Feather River Canyon to form a broad, smooth, bare bedrock dome to the north of Bald Rock Basin. Although outside of the study catchment, it hints at the possibility of significant compositional or structural heterogeneity within the pluton that is imposing a localized bottom-up restriction on forest growth in some parts of the landscape (Hahm et al., 2014).

Landscape transience in the Feather River region is driven by a wave of fluvial incision that is presently propagating up the channel network (Hurst et al., 2012). The resultant range of erosion rates spans an order of magnitude, placing fundamental controls on the nature of the hillslopes (Hurst et al., 2012, 2013a), soils (Yoo et al., 2011; Attal et al., 2014; Gabet et al., 2015) and biosphere (Milodowski et al., 2015). Rates of erosion in the inner canyon, driven by fluvial incision along the main-stem Feather River, reach ~250 mm kyr $^{-1}$ (Riebe et al., 2000; Hurst et al., 2012). Bald Rock Basin has not fully adjusted to this elevated rate of fluvial incision, with a prominent topographic knickpoint marking the transition to lower relief topography that is eroding much more slowly at 30–40 mm ky r $^{-1}$ (Riebe et al., 2000; Hurst et al., 2012).

Moving across this gradient in erosion rates, hillslope form changes from being low-gradient and convex to steep and planar in the rejuvenated parts of the landscape below the knickpoint (Hurst et al., 2012), consistent with the expectations of models of non-linear, diffusion-like sediment trans-port (Roering et al., 1999). Within Bald Rock Basin itself, Yoo et al. (2011) investigated changes in substrate characteristics between a series of transects across this transition, indicating that the increase in erosion rate drives a reduction in the residence time of material within the weathering zone, highlighted by a decrease in the extent of weathering of both the soil and saprolite. Consistent with these observations, a more detailed inventory of soil grain size distributions from soil pits throughout Bald Rock Basin indicate a marked increase in the coarser grain fraction in more rapidly eroding parts of the basin (Attal et al., 2014). We use the surface roughness algorithm introduced above to expand on this earlier work and further characterize changes in the bedrock exposure across the geomorphic transition.

Harrington Creek, Idaho

The Harrington Creek catchment drains into Main Salmon River, around 40 km SSW of the Bitterroot Mountains, Idaho (Fig. 1d). The regional climate is continental, with maxi-mum (minimum) temperatures ranging from 26.2 (6.2) C in the summer to 0.0 (10.8) C in the winter, whereas precipitation is more evenly distributed throughout the year, with mean annual precipitation typically 630 mm (http: //www.prismclimate. org). Vegetation in the catchment comprises coniferous forest with variable canopy cover (Barbour and Billings, 2000). The catchment is underlain by plutonic rocks related to the Idaho Batholith, with small inclusions of Eocene dykes of rhyolitic-dacitic composition (Lewis and Stanford, 2002). Analysis of fission tracks in apatite and zircon grains from the Idaho Batholith suggest that exhumation rates have varied from 0.03–0.1 mm yr ^{1} between 50– 10 Ma to 0.32 0.10 mm yr^{1} from 10 Ma-present, associated with canyon-forming fluvial incision along the Salmon River (Sweetkind and Blackwell, 1989; Ferrier et al., 2012). Point measurements of regolith production rates, based on cosmogenic ^{10}Be concentrations, suggest erosion rates integrated over 10^3–10^4 years of up to 0.12 mm yr ^{1} (Ferrier et al., 2012). Associated with this fluvial incision are a series of knickpoints that are propagating up the tributaries of the Salmon River, including Harrington Creek, which mark the transition from a slowly eroding, relict landscape to steep, rapidly eroding, rejuvenated topography that is actively adjusting to the elevated incision rates below the fluvial knick-point (Wood, 2013). The Harrington Creek region has been subject to significantly less research relative to Bald Rock Basin; we use the same methods for this site to investigate changes in the geomorphic characteristics of the hillslopes across this transition.

Topographic Analysis

Changing bedrock exposure across the knickzones was mapped utilizing the surface roughness method as described in Sect. 2, using a circular neighborhood with a radius of 3 m, which was shown to perform well, with limited false positives, in our previous validation (Sect. 3). Topographic gradient was also measured using the slope of the best fit-ting six term polynomial surface, defined by a least squares regression to a circular neighborhood with 7 m radius (e.g. Hurst et al., 2012). In order to map changes in hillslope characteristics along the length of the trunk channel, we use longitudinal swath profiles, following a similar approach to the implementation of the generalized swath profile algorithm described by Hergarten et al. (2014), to map each point on the hillslope to the nearest location in the channel network. This method allows frequently used swath profile analysis to be undertaken using curvilinear features, such as river channels, as the baseline rather than requiring linear features. The trunk channels themselves were defined using the DrEICH algorithm (Clubb et al., 2014), which searches for the up-stream limit of the topographic signature of fluvial incision to define the fluvial network within the channelized domain. To first order, the longitudinal swath profiles should link hill-slopes to the section of channel that sets their lower boundary condition, enabling us to link geomorphic changes in fluvial incision.

Results

In both Bald Rock Basin (Figs. 9 and 10) and Harrington Creek (Figs. 11 and 12), there are clearly distinct, contrasting topographic domains separated by major knickpoints. Moving across this transition, hillslope morphology changes from low gradient, convex hillslopes (modal gradients above principal knickpoints are 0.5 and 0.4 within the headwaters of Bald Rock Basin and Harrington Creek respectively), to steep, planar hillslopes downstream of the knickpoints: respective modal gradients are 0.9 and 0.8. However, in addition to the changes in the hillslope profile across this transition there are concomitant textural changes to the hill-slopes pertaining to the widespread emergence of bedrock. In both landscapes, the low gradient headwaters are also characterized by smooth topography indicative of a continuous soil mantle: within Bald Rock Basin, 1.5 % of hillslope pixels have $S_3 > 0.010$; < 1 % have $S_3 > 0.015$; within Harrington Creek, 3 % have $S_3 > 0.010$; 1.5 % have $S_3 > 0.015$. In contrast, in the rejuvenated parts of the landscape, the increased dominance of bedrock is indicated by elevated topographic roughness: in the lower reaches of Bald Rock Basin 15 % of hillslope pixels have $S_3 > 0.010$; 7 % have $S_3 > 0.015$, while in the equivalent parts of the Harrington Creek drainage, 29 % have $S_3 > 0.010$; 19 % have $S_3 > 0.015$. Critically, the emergence of bedrock is not uniform across the steeper parts of the landscape. Rather, the steep hillslopes present a rugged patchwork of bedrock outcrops and discontinuous soil cover.

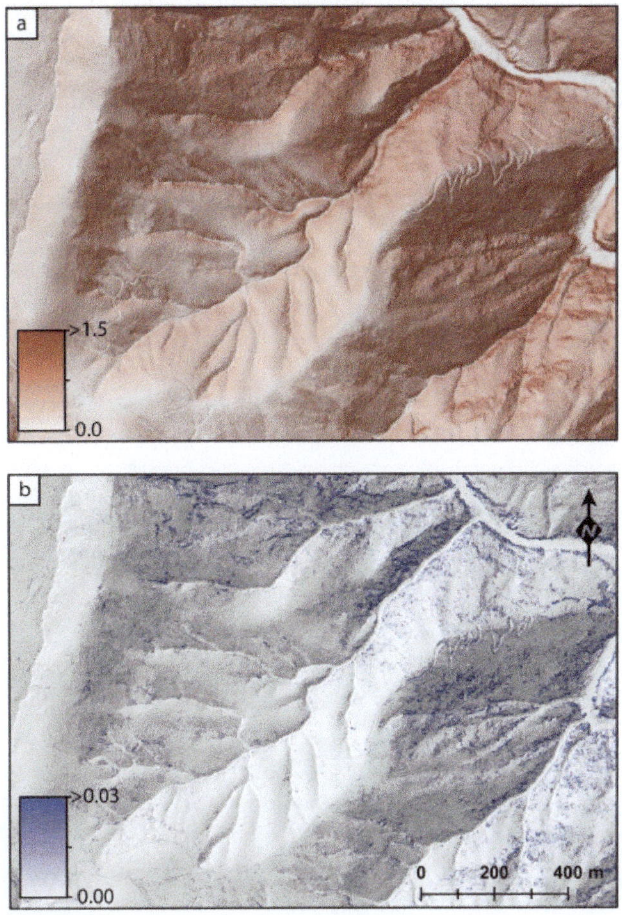

Figure 9. Maps displaying **(a)** topographic slope, and **(b)** S_3 for Bald Rock Basin, Californian Sierra Nevada. The Middle Fork Feather River is located in the NE corner of each map, flowing from NW to SE.

Likewise, across the upper part of Bald Rock Basin, there are a number of isolated patches of elevated roughness that can be picked out from the prevailing smooth terrain (Fig. 9). Field Inspection of these selected "rough spots" indicated that they corresponded to isolated rock outcrops, whereas instances of tree throw mounds, which could also generate roughness at short wavelengths, were comparatively rare.

Discussion

In both Bald Rock Basin and Harrington Creek, topographic knickpoints mark the domain transitions between a slowly eroding "relict" landscape, and rejuvenated topography responding to elevated rates of fluvial incision (Hurst et al., 2012, 2013a; Wood, 2013). Both landscapes exhibit similar hillslope responses to this geomorphic forcing. In this contribution we have deployed our new roughness algorithm to quantify the dynamics of the soil to bedrock transition. Specifically, in both landscapes the transition from soil-mantled to bedrock hillslopes is gradual and patchy. Furthermore, the steep hillslopes do not appear to be completely stripped of soil;

the persistence of topographically smooth areas that manage to sustain a forest canopy (Milodowski et al., 2015) indicates that patchy soil cover persists at high erosion rates. In the Feather River Region, aboveground biomass hosted by the hillslopes has been shown to decrease with in-creasing erosion rates (Milodowski et al., 2015), but biogenic soil production is still able to keep pace with elevated rates of erosion to maintain a partial soil mantle. This is in agreement with observations from soil depth transects within the basin that show little difference in soil depths measured above the knickpoint, ranging from 40–80 cm, to those measured be-low the knickpoint, which ranged from 30–60 cm (Yoo et al., 2011).

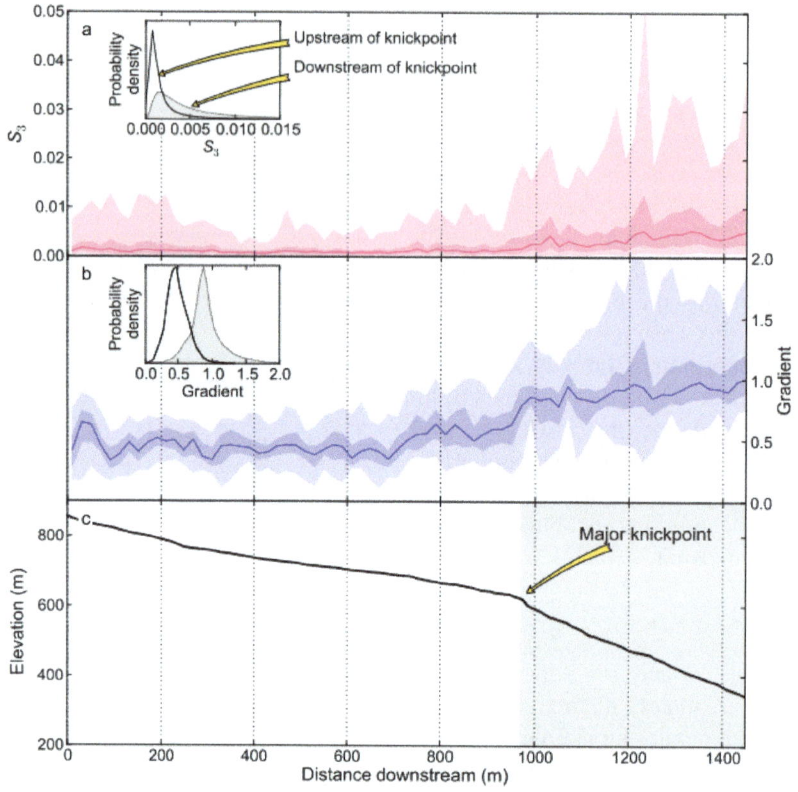

Figure 10. Changes in topographic characteristics along a longitudinal swath centered on the trunk channel draining Bald Rock Basin: **(a)** surface roughness, S_3; **(b)** topographic gradient; **(c)** the longitudinal channel profile. The principal knickpoint has been high-lighted, with the inset histograms summarizing the distributions of the topographic metrics above and below. Upstream of the major knickpoint, smaller deviations from the typical graded profile indicate a series of smaller knickpoints. The swath has a half width of 250 m, and has been binned into 50 m intervals. In plates **(a)** and **(b)**, the median has been plotted with the shaded intervals bounded by the 25–75th quantiles and 2.5–97.5th quantiles. S_3 was calculated using a 3 m radius neighborhood window.

The nature if the soil-bedrock transition observed at these two sites aligns closely with the observations from the San Gabriel Mountains in California (DiBiase et al., 2012). A gradual, patchy transition is significant because it is at odds with the expectations from widely used models of soil production, in which the rate of production decays exponentially with depth from a maximum production rate for a bare bedrock surface (e.g. Heimsath et al., 1997), which in this framework represents a threshold erosion rate defining a sharp transition from soil-mantled to bedrock topography. The patchy transition observed may be driven in part by structural or compositional controls on the rate at which bedrock breaks down to form mobile regolith, but can also be rationalized by models of soil production that consider the processes driving soil production and sediment transport as occurring in discrete events (Strudley et al., 2006a, b; Ga-bet and Mudd, 2010). Understanding whether these patches are stationary in time or dynamic is important in understanding the longer term evolution of steep landscapes and how this evolution is shaped by the coupling of geomorphic and ecological processes. Finally, while clearly important from a hillslope perspective, there are broader implications for landscape evolution: the dynamics of sediment transport in bedrock landscapes are very different to those in soil-mantled landscapes (e.g. Binnie et al., 2007; Dietrich et al., 2003), impacting on the calibre (Attal et al., 2015; Whittaker et al., 2010) and temporal variability (Hovius et al., 2000) of sediment supplied to the channel network; therefore the nature of the soil-bedrock transition impacts on the nature of hillslope-channel coupling, modulating the fluvial response to changes in base level.

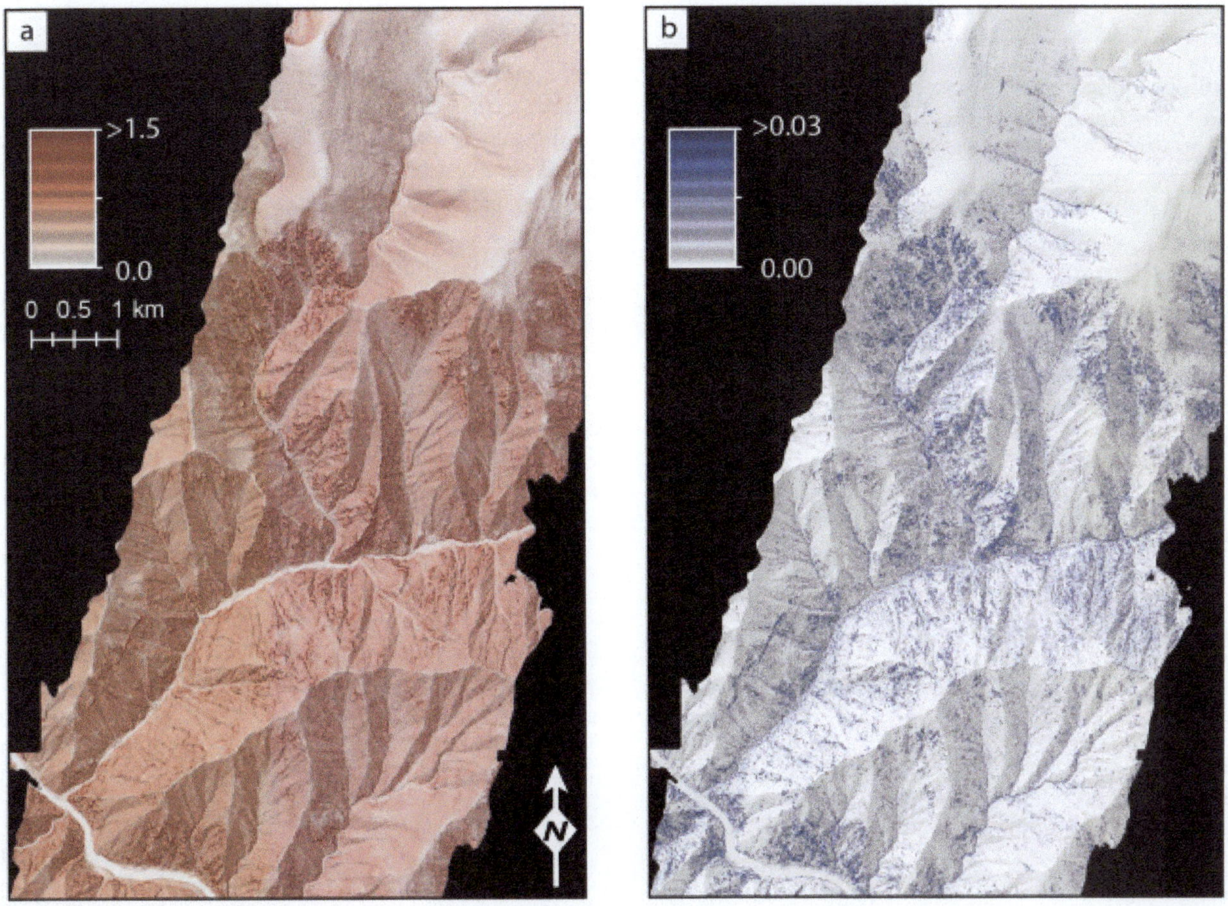

Figure 11. Maps displaying **(a)** topographic slope, and **(b)** S_3, for a sub-catchment of Harrington Creek, Idaho. S_3 was calculated using a 3 m radius neighborhood window.

OVERALL DISCUSSION AND CONCLUSIONS

The structure of topographic relief is controlled by different processes operating at different spatial scales (Perron et al., 2008): at wavelengths greater than 100 m, topography is dominated by the spacing of ridges and valleys (Perron et al., 2008, 2009); at the sub-hillslope length-scale, other processes generate detectable topographic signatures (e.g. McKean and Roering, 2004; Roering et al., 2010). Booth et al. (2009) exploited spectral analysis to show that areas affected by deep-seated landslides exhibit significantly greater power at intermediate wavelengths (11–50 m), enabling the objective classification of regions in which deep-seated landslides were prevalent. At shorter length-scales, Roering et al. (2010) suggested that roughness generated at small length-scales (< 7.5 m) in the Oregon Coast Ranges could be attributed to the presence of tree throw mounds; similar analysis of topographic profiles extracted from contrasting catchments in the same setting found a lack of spectral power at these short wavelengths for resistant bedrock hillslopes in comparison to soil-mantled hillslopes, attributed to a diminished biotic contribution to weathering (Marshall and Roering, 2014).

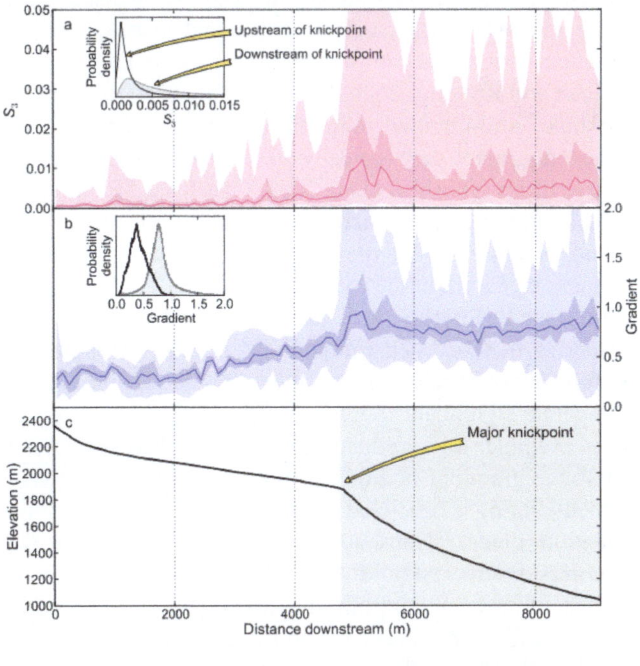

Figure 12. Changes in topographic characteristics along a longitudinal swath centered on the trunk channel draining the principal tributary to Harrington Creek: **(a)** surface roughness, S_3; **(b)** topographic gradient; **(c)** the longitudinal channel profile. The principal knickpoint has been highlighted, with the inset histograms summarizing the distributions of the topographic metrics above and be-low. Upstream of the major knickpoint, smaller deviations from the typical graded profile indicate a series of smaller knickpoints. The swath has a half width of 350 m, and has been binned into 50 m intervals. In plates **(a)**

and **(b)**, the median has been plotted along with the shaded intervals bounded by the 25–75th quantiles and 2.5– 97.5th quantiles.

We propose that short wavelength surface roughness, quantified using the same roughness algorithms introduced by McKean and Roering (2004) can be used to make inferences about hillslope characteristics specifically pertaining to the exposure of bedrock. Comparison against rock exposure measured independently and objectively from high-resolution orthophotographs from multiple landscapes suggests that the emergence of bedrock in hillslopes produces a detectable topographic signature that distinguishes it from hillslopes that have a continuous soil mantle. We applied this technique to forested landscapes in California and Idaho, highlighting the ability of LiDAR surveys to resolve high-resolution features of the topography through canopy. For obvious reasons validation is simpler in un-vegetated terrain but prior to the introduction of below-canopy UAVs for data col-lection, we suggest our method is adequate for use in vegetated terrain. Users should validate using field observations to avoid false positives from, for example, tree throw mounds. Thus we propose surface roughness as a new method for mapping rock exposure from LiDAR data that complements previously published metrics (DiBiase et al., 2012), and is likely to be of particular benefit in landscapes in which rock outcrops are present at topographic gradients lower than the angle of repose.

We caveat this finding with the statement that rock exposure is not the only mechanism of generating topographic roughness at short length-scales; for example, gullying and slumping provide two mechanisms by which the smooth parabolic morphology associated with ideal, diffusive soil-mantled hillslopes may be modified (Tarolli and Dalla Fontana, 2009); likewise small-scale features associated with deep-seated landslides, such as folds and scarps, generate a roughness signal at similar length-scales to rock outcrop (McKean and Roering, 2004; Tarolli et al., 2010). In addition, while many soil-mantled sediment transport processes act to diffuse topography, they typically do so through discrete events (e.g. tree throw) (Furbish et al., 2009; Gabet and Mudd, 2010), and while the fingerprint that these individual events leave on the landscape is transient, they pro-vide a potentially important roughness signature at the relevant length-scale for that mode of disturbance (Roering et al., 2010). An additional factor to consider is that bedrock morphology is itself variable, and therefore certain mechanisms of generating rock exposure may not generate significant roughness; this would be exemplified by, for example, low gradient, glacially polished surfaces, or by massive granitoid bedrock with very low fracture density in which jointing is restricted to approximately surface parallel exfoliation planes. Consequently, interpretation of surface rough-ness metrics should critically take into account the presence of other geomorphic processes that are potentially operating within the landscape and the characteristics of the bedrock itself. Indeed, this principal applies to the interpretation of any topographic metric obtained from remotely sensed data; in complex geomorphic settings, isolation of specific hills-lope characteristics from a single textural attributes may be impossible at the data resolution presently available from air-borne surveys; ultimately a combination of metrics, covering a broader range of morphological characteristics may well be necessary.

The characterization of hillslopes is of importance across a diverse range of surface processes research, providing a better understanding of controls on hydrological flow routing, sediment production and transport processes and ecosystem development. The utility of topographic data to aid this endeavour is strongly dependent on the resolution of these data sets. In the case of hillslope characteristics, such as rock exposure, roughness is expressed at the meter scale; using 1 m resolution digital elevation models, it is possible to examine variations in hillslope form at sufficient levels of detail that it is possible to distinguish between soil and bedrock hill-slopes; this information is rapidly lost as the data resolution is coarsened (DiBiase et al., 2012). However these shorter length scales are particularly susceptible to noise in the data set (Albani et al., 2004; Sofia et al., 2013). This highlights the requirement for high quality, high resolution, which permit accurate classification of vegetation and ground returns prior to surface creation. LiDAR surveys with higher shot spacing are therefore likely to provide a disproportionately greater level of detail on hillslope characteristics (Brodu and Lague, 2012), and this should be taken into account when planning airborne surveys. In particular, the continued development of unmanned aerial vehicles (UAVs) as a platform for airborne LiDAR collection will increasingly make higher resolution surveys accessible to the research community (e.g. Lin et al., 2011).

Finally, from our analysis of the geomorphic changes associated with changing rates of erosion in two different landscapes reveals a number of significant conclusions regarding the nature of the soil-bedrock transition. In both cases, the transition from soil-mantled hillslopes to bedrock dominated hillslopes is clearly gradual, with areas of patchy soil coverage persistent on steep, rapidly eroding hillslopes. A "patchy" transition from soil-mantled to bedrock hillslopes challenges prevailing modelling approaches towards soil production, but is in agreement with conclusions from previous studies of soil production in rapidly eroding landscapes – the European Alps (Norton et al., 2008), San Gabriel Mountains, California (Heimsath et al., 2012) and Southern Alps, New Zealand (Larsen et al., 2014) – each of which observe the coexistence of soil and bedrock on rapidly eroding hillslopes. This has been attributed in part to efficient biogenic soil production (Larsen et al., 2014), which facilitates the rapid generation and stabilization of soil between landslide events, and lithological susceptibility to weathering processes (Norton et al., 2008). The hypothesis of a biogenically mediated soil-bedrock transition is supported by the observation in these landscapes that patchy vegetation cover persists on the steeper hillslopes where trees have maintained a foothold, and is in agreement with expectations from numerical modelling of soil production by discrete events (Gabet and Mudd, 2010). Capturing the salient aspects of these models within larger-scale landscape evolution models rep-resents a key challenge in simulating the evolution of mixed-bedrock landscapes that are typical of many upland settings.

AUTHOR CONTRIBUTIONS

D. T. Milodowski and S. M. Mudd the al-gorithms and wrote the code. D. T. Milodowski, S. M. Mudd and E. T. A. Mitchard performed the analysis and wrote the paper.

ACKNOWLEDGEMENTS

This research was funded by a NERC studentship (NERC DTG NE/152830X/1 and NE/J500021/1; DTM), in addition to the Harkness Award from the University of Cambridge (DTM). ETAM is funded by a NERC Fellowship (NE/I021217/1). SMM is supported by U.S. Army Research Office contract number W911NF-13-1-0478. The authors would like to thank Emmanuel Gabet, Dimitri Lague, Stuart Grieve, and Fiona Clubb for valuable discussions that facilitated the development of this research.

REFERENCES

1. Albani, M., Klinkenberg, B., Andison, D. W., and Kimmins, J. P.: The choice of window size in approximating topographic sur-faces from Digital Elevation Models, Int. J. Geogr. Inf. Sci., 18, 577–593, doi:10.1080/13658810410001701987, 2004.
2. Anderson, R. S.: Modeling the tor-dotted crests, bedrock edges, and parabolic profiles of high alpine surfaces of the Wind River Range, Wyoming, Geomorphology, 46, 35–58, doi:10.1016/S0169-555X(02)00053-3, 2002.
3. Attal, M., Mudd, S. M., Hurst, M. D., Weinman, B., Yoo, K., and Naylor, M.: Impact of change in erosion rate and landscape steep-ness on hillslope and fluvial sediments grain size in the Feather River basin (Sierra Nevada, California), Earth Surf. Dynam., 3, 201–222, doi:10.5194/esurf-3-201-2015, 2015.
4. Barbour, M. G. and Billings, W. D.: North American Terrestrial Vegetation, edited by: Barbour, M. G. and Billings, W. D., Cam-bridge University Press, Cambridge, UK, New York, NY, USA, 2000.
5. Binnie, S. A., Phillips, W. M., Summerfield, M. A., and Fi-field, L. K.: Tectonic uplift, threshold hillslopes, and denuda-tion rates in a developing mountain range, Geology, 35, 743–746, doi:10.1130/G23641A.1, 2007.
6. Booth, A. M., Roering, J. J., and Perron, J. T.: Auto-mated landslide mapping using spectral analysis and high-resolution topographic data: Puget Sound lowlands, Washing-ton, and Portland Hills, Oregon, Geomorphology, 109, 132–147, doi:10.1016/j.geomorph.2009.02.027, 2009.
7. Brodu, N. and Lague, D.: 3D terrestrial lidar data classification of complex natural scenes using a multi-scale dimensionality cri-terion: Applications in geomorphology, ISPRS J. Photogramm., 68, 121–134, doi:10.1016/j.isprsjprs.2012.01.006, 2012.
8. Carson, M. A. and Kirkby, M. J.: Hillslope form and process, Cam-bridge University Press, Cambridge, UK, 1972.
9. Cavalli, M., Tarolli, P., Marchi, L., and Dalla Fontana, G.: The effectiveness of airborne LiDAR data in the recog-nition of channel-bed morphology, CATENA, 73, 249–260, doi:10.1016/j.catena.2007.11.001, 2008.
10. Chorover, J., Troch, P. A., Rasmussen, C., Brooks, P. D., Pelletier, J. D., Breshars, D. D., Huxman, T. E., Kurc, S. A., Lohse, K. A., McIntosh, J. C., Meixner, T., Schaap, M. G., Litvak, M. E., Perdrial, J., Harpold, A., and Durcik, M.: How Water, Carbon, and Energy Drive Critical Zone Evolution: The Jemez–Santa Catalina Critical Zone Observatory, Vadose Zone J., 10, 884–899, doi:10.2136/vzj2010.0132, 2011.
11. Clubb, F. J., Mudd, S. M., Milodowski, D. T., Hurst, M. D., and Slater, L. J.: Objective extraction of channel heads from high-resolution topographic data, Water Resour. Res., 50, 4283–4304, doi:10.1002/2013WR015167, 2014.
12. Colby, J.: Topographic Normalization in Rugged Terrain, Pho-togramm. Eng. Rem., 57, 531–537, 1991.
13. Culling, W.: Soil Creep and the Development of Hillside Slopes, J. Geol., 71, 127–161, 1963.
14. Culling, W. E. H.: Theory of Erosion on Soil-Covered Slopes, J. Geol., 73, 230–254, 1965.
15. DiBiase, R. A., Heimsath, A. M., and Whipple, K. X.: Hillslope response to tectonic forcing in threshold landscapes, Earth Surf. Proc. Land., 37, 855–865, doi:10.1002/esp.3205, 2012.
16. DiBiase, R. A. and Lamb, M. P.: Vegetation and wildfire controls on sediment yield in bedrock landscapes, Geophys. Res. Lett., 40, 1093–1097, doi:10.1002/grl.50277, 2013.
17. Dietrich, W. E., Bellugi, D. G., Sklar, L. S., Stock, J. D., Heimsath, A. M., and Roering, J. J.: Geomorphic transport laws for pre-dicting landscape form and dynamics, Geophys. Monogr.-Am. Geophys. UNION, 135, 103–132, 2003.
18. Evans, I. S.: An integrated system of terrain analysis and slope map-ping, Z. Geomorphol, 36, 274–295, 1980.
19. Evans, J. S. and Hudak, A. T.: A Multiscale Curvature Al-gorithm for Classifying Discrete Return LiDAR in Forested Environments, IEEE T. Geosci. Remote, 45, 1029–1038, doi:10.1109/TGRS.2006.890412, 2007.
20. Fara, H. D. and Scheidegger, A. E.: An eigenvalue method for the statistical evaluation of fault plane solutions of earthquakes, B. Seismol. Soc. Am., 53, 811–816, 1963.
21. Ferrier, K. L., Kirchner, J. W., and Finkel, R. C.: Weak influ-ences of climate and mineral supply rates on chemical ero-sion rates: measurements along two altitudinal transects in the Idaho Batholith, J. Geophys. Res.-Earth, 117, F02026, doi:10.1029/2011JF002231, 2012.
22. Furbish, D. J., Hamner, K. K., Schmeeckle, M., Borosund, M. N., and Mudd, S. M.: Rain splash of dry sand revealed by high-speed imaging and sticky paper splash targets, J. Geophys. Res.-Earth, 112, F01001, doi:10.1029/2006JF000498, 2007.
23. Furbish, D. J., Haff, P. K., Dietrich, W. E., and Heimsath, A. M.: Statistical description of slope-dependent soil transport and the diffusion-like coefficient, J. Geophys. Res., 114, F00A05, doi:10.1029/2009JF001267, 2009.
24. Furbish, D. J. and Roering, J. J.: Sediment disentrainment and the concept of local vs. nonlocal transport on hillslopes, J. Geophys. Res.-Earth, 118, 937–952, doi:10.1002/jgrf.20071, 2013.
25. Gabet, E. J.: Sediment transport by dry ravel, J. Geophys. Res.-Sol. Ea., 108, 2049, doi:10.1029/2001JB001686, 2003.

26. Gabet, E. J., Reichman, O. J., and Seabloom, E. W.: The effects of bioturbation on soil processes and sediment transport, Annu. Rev. Earth Pl. Sc., 31, 249–273, 2003.
27. Gabet, E. J. and Mudd, S. M.: Bedrock erosion by root frac-ture and tree throw: A coupled biogeomorphic model to explore the humped soil production function and the per-sistence of hillslope soils, J. Geophys. Res., 115, F04005, doi:10.1029/2009JF001526, 2010.
28. Gabet, E. J., Mudd, S. M., Milodowski, D. T., Yoo, K., Hurst, M. D., and Dosseto, A.: Local topography and erosion rate control regolith thickness along a ridgeline in the Sierra Nevada, California, Earth Surf. Proc. Land., 40, 1779–1790, doi:10.1002/esp.3754, 2015.
29. Gilbert, G.: The convexity of hilltops, J. Geol., 17, 344–350, 1909. Goodfellow, B. W., Chadwick, O. A., and Hilley, G. E.: Depth and character of rock weathering across a basaltic-hosted climose-quence on Hawai'i, Earth Surf. Proc. Land., 39, 381–398, doi:10.1002/esp.3505, 2014a.
30. Goodfellow, B. W., Skelton, A., Martel, S. J., Stroeven, A. P., Jansson, K. N., and Hättestrand, C.: Controls of tor formation, Cairngorm Mountains, Scotland, J. Geophys. Res.-Earth., 119, 2013JF002862, doi:10.1002/2013JF002862, 2014b.
31. Graham, R., Rossi, A., and Hubbert, R.: Rock to regolith conver-sion: producing hospitable substrates for terrestrial ecosystems, GSA Today, 20, 4–9, doi:10.1130/GSAT57A.1, 2010.
32. Hahm, W. J., Riebe, C. S., Lukens, C. E. and Araki, S.: Bedrock composition regulates mountain ecosystems and land-scape evolution, P. Natl. Acad. Sci. USA, 111, 3338–3343, doi:10.1073/pnas.1315667111, 2014.
33. Hale, S. R. and Rock, B. N.: Impact of topographic normalization on land-cover classification accuracy, Photogramm. Eng. Rem. S., 69, 785–791, doi:10.14358/PERS.69.7.785, 2003.
34. Heimsath, A., Dietrich, W., Nishiizumi, K., and Finkel, R.: The soil production function and landscape equilibrium, Nature, 388, 358–361, doi:10.1038/41056, 1997.
35. Heimsath, A. M., DiBiase, R. A., and Whipple, K. X.: Soil produc-tion limits and the transition to bedrock-dominated landscapes, Nat. Geosci., 5, 210–214, doi:10.1038/ngeo1380, 2012.
36. Hergarten, S., Robl, J., and Stüwe, K.: Extracting topographic swath profiles across curved geomorphic features, Earth Surf. Dynam., 2, 97–104, doi:10.5194/esurf-2-97-2014, 2014.
37. Hovius, N., Stark, C. P., Hao-Tsu, C., and Jiun-Chuan, L.: Supply and removal of sediment in a landslide-dominated mountain belt: Central Range, Taiwan, J. Geol., 108, 73–89, 2000.
38. Hurst, M. D., Mudd, S. M., Walcott, R., Attal, M., and Yoo, K.: Using hilltop curvature to derive the spatial distribution of erosion rates, J. Geophys. Res.-Earth, 117, F02017, doi:10.1029/2011JF002057, 2012.
39. Hurst, M. D., Mudd, S. M., Yoo, K., Attal, M., and Walcott, R.: In-fluence of lithology on hillslope morphology and response to tec-tonic forcing in the northern Sierra Nevada of California, J. Geo-phys. Res.-Earth, 118, 832–851, doi:10.1002/jgrf.20049, 2013a.
40. Hurst, M. D., Mudd, S. M., Attal, M., and Hilley, G.: Hillslopes Record the Growth and Decay of Landscapes, Science, 341, 868– 871, doi:10.1126/science.1241791, 2013b.
41. Lague, D., Brodu, N., and Leroux, J.: Accurate 3D comparison of complex topography with terrestrial laser scanner: Application to the Rangitikei canyon (N-Z), ISPRS J. Photogramm., 82, 10–26, doi:10.1016/j.isprsjprs.2013.04.009, 2013.
42. Larsen, I. J., Almond, P. C., Eger, A., Stone, J. O., Montgomery, D. R., and Malcolm, B.: Rapid Soil Production and Weather-ing in the Western Alps, New Zealand, Science, 343, 637–640, doi:10.1126/science.1244908, 2014.
43. Lashermes, B., Foufoula-Georgiou, E., and Dietrich, W. E.: Channel network extraction from high resolution topog-raphy using wavelets, Geophys. Res. Lett., 34, L23S04, doi:10.1029/2007GL031140, 2007.
44. Lewis, R. S. and Stanford, L. R.: Geologic map compilation of the western half of the Nez Perce Pass 30 60 min quadrangle, Idaho Geological Survey, Idaho, USA, 2002.
45. Lin, Y., Hyyppa, J., and Jaakkola, A.: Mini-UAV-Borne LIDAR for Fine-Scale Mapping, IEEE Geosci. Remote S., 8, 426–430, doi:10.1109/LGRS.2010.2079913, 2011.
46. Lin, C.-W., Tseng, C.-M., Tseng, Y.-H., Fei, L.-Y., Hsieh, Y.-C., and Tarolli, P.: Recognition of large scale deep-seated landslides in forest areas of Taiwan using high resolution topography, J. Asian Earth Sci., 62, 389–400, doi:10.1016/j.jseaes.2012.10.022, 2013.
47. Marshall, J. A. and Roering, J. J.: Diagenetic variation in the Ore-gon Coast Range: Implications for rock strength, soil production, hillslope form, and landscape evolution, J. Geophys. Res.-Earth, 119, 1395–1417, doi:10.1002/2013JF003004, 2014.
48. McKean, J. A., Dietrich, W. E., Finkel, R. C., Southon, J. R., and Caffee, M. W.: Quantification of soil production and downs-lope creep rates from cosmogenic 10Be accumulations on a hillslope profile, Geology, 21, 343–346, doi:10.1130/0091-7613(1993)021<0343:QOSPAD>2.3.CO;2, 1993.
49. McKean, J. and Roering, J.: Objective landslide detection and sur-face morphology mapping using high-resolution airborne laser altimetry, Geomorphology, 57, 331–351, doi:10.1016/S0169-555X(03)00164-8, 2004.
50. Migon, P.: Granite Landscapes of the World, Oxford University Press, Oxford, UK, 2006.
51. Milodowski, D. T., Mudd, S. M., and Mitchard, E. T. A.: Erosion rates as a potential bottom-up control of forest structural char-acteristics in the Sierra Nevada Mountains, Ecology, 96, 31–38, doi:10.1890/14-0649.1, 2015.
52. Moody, J. A. and Martin, D. A.: Initial hydrologic and geomorphic response following a wildfire in the Colorado Front Range, Earth Surf. Proc. Land., 26,

1049–1070, doi:10.1002/esp.253, 2001.

53. Norton, K. P., von Blanckenburg, F., Schlunegger, F., Schwab, M., and Kubik, P. W.: Cosmogenic nuclide-based investigation of spatial erosion and hillslope channel coupling in the tran-sient foreland of the Swiss Alps, Geomorphology, 95, 474–486, doi:10.1016/j.geomorph.2007.07.013, 2008.

54. Passalacqua, P., Trung, T. D., Foufoula-Georgiou, E., Sapiro, G., and Dietrich, W. E.: A geometric framework for channel network extraction from lidar: Nonlinear diffusion and geodesic paths, J. Geophys. Res., 115, F01002, doi:201010.1029/2009JF001254, 2010.

55. Pelletier, J. D.: A robust, two-parameter method for the extraction of drainage networks from high-resolution digital elevation models (DEMs): Evaluation using synthetic and real-world DEMs, Wa-ter Resour. Res., 49, 75–89, doi:10.1029/2012WR012452, 2013.

56. Pelletier, J. D. and Rasmussen, C.: Quantifying the climatic and tectonic controls on hillslope steepness and erosion rate, Litho-sphere, 1, 73–80, doi:10.1130/L3.1, 2009.

57. Perron, J. T., Kirchner, J. W., and Dietrich, W. E.: Spectral sig-natures of characteristic spatial scales and nonfractal struc-ture in landscapes, J. Geophys. Res.-Earth, 113, F04003, doi:10.1029/2007JF000866, 2008.

58. Perron, J. T., Kirchner, J. W., and Dietrich, W. E.: Formation of evenly spaced ridges and valleys, Nature, 460, 502–505, doi:10.1038/nature08174, 2009.

59. Phillips, J. D. and Marion, D. A.: Pedological memory in for-est soil development, Forest Ecol. Manag., 188, 363–380, doi:10.1016/j.foreco.2003.08.007, 2004.

60. Pirotti, F. and Tarolli, P.: Suitability of LiDAR point density and derived landform curvature maps for channel network extraction, Hydrol. Process., 24, 1187–1197, doi:10.1002/hyp.7582, 2010.

61. Riebe, C. S., Kirchner, J. W., Granger, D. E., and Finkel, R. C.: Erosional equilibrium and disequilibrium in the Sierra Nevada, inferred from cosmogenic 26Al and 10Be in alluvial sediment, Geology, 28, 803–806, 2000.

62. Roering, J. J., Kirchner, J. W., and Dietrich, W. E.: Evidence for nonlinear, diffusive sediment transport on hillslopes and implications for landscape morphology, Water Resour. Res., 35, 853– 870, doi:10.1029/1998WR900090, 1999.

63. Roering, J. J., Marshall, J., Booth, A. M., Mort, M., and Jin, Q.: Evidence for biotic controls on topography and soil production, Earth Planet. Sc. Lett., 298, 183–190, doi:10.1016/j.epsl.2010.07.040, 2010.

64. Ruleman, C. A., Bohannon, R. G., Bryant, B., Shroba, R. R., and Premo, W. R.: Geologic Map of the Bailey 30^0 60^0 Quadrangle, North-Central Colorado, US Geological Survey, Denver, Col-orado, USA, 2011.

65. Saucedo, G. J. and Wagner, D. L.: Geologic Map of the Chico Quadrangle, California Department of Conservation, Division of Mines and Geology, Sacramento, California, USA, 1992.

66. Sheffer, E., von Hardenberg, J., Yizhaq, H., Shachak, M., and Meron, E.: Emerged or imposed: a theory on the role of physical templates and self-organisation for vegetation patchiness, Ecol. Lett., 16, 127–139, doi:10.1111/ele.12027, 2013.

67. Smith, T. B., Wayne, R. K., Girman, D. J., and Bruford, M. W.: A Role for Ecotones in Generating Rainforest Biodiversity, Science, 276, 1855–1857, doi:10.1126/science.276.5320.1855, 1997.

68. Sofia, G., Tarolli, P., Cazorzi, F., and Dalla Fontana, G.: An objective approach for feature extraction: distribution analy-sis and statistical descriptors for scale choice and channel net-work identification, Hydrol. Earth Syst. Sci., 15, 1387–1402, doi:10.5194/hess-15-1387-2011, 2011.

69. Sofia, G., Pirotti, F., and Tarolli, P.: Variations in multiscale cur-vature distribution and signatures of LiDAR DTM errors, Earth Surf. Proc. Land., 38, 1116–1134, doi:10.1002/esp.3363, 2013.

70. Sofia, G., Fontana, G. D., and Tarolli, P.: High-resolution topogra-phy and anthropogenic feature extraction: testing geomorphome-tric parameters in floodplains, Hydrol. Process., 28, 2046–2061, doi:10.1002/hyp.9727, 2014.

71. Strudley, M. W., Murray, A. B., and Haff, P. K.: Emergence of pediments, tors, and piedmont junctions from a bedrock weathering–regolith thickness feedback, Geology, 34, 805–808, doi:10.1130/G22482.1, 2006a.

72. Strudley, M. W., Murray, A. B., and Haff, P. K.: Regolith thick-ness instability and the formation of tors in arid environments, J. Geophys. Res.-Earth, 111, F03010, doi:10.1029/2005JF000405, 2006b.

73. Sweetkind, D. S. and Blackwell, D. D.: Fission-track evidence of the Cenozoic thermal history of the Idaho batholith, Tectono-physics, 157, 241–250, doi:10.1016/0040-1951(89)90142-X, 1989.

74. Tarolli, P. and Dalla Fontana, G.: Hillslope-to-valley transition mor-phology: New opportunities from high resolution DTMs, Geo-morphology, 113, 47–56, doi:10.1016/j.geomorph.2009.02.006, 2009.

75. Tarolli, P., Sofia, G., and Fontana, G. D.: Geomorphic features ex-traction from high-resolution topography: landslide crowns and bank erosion, Nat. Hazards, 61, 65–83, doi:10.1007/s11069-010-9695-2, 2010.

76. Teillet, P. M., Guindon, B., and Goodenough, D. G.: On the slope-aspect correction of multispectral scanner data, Can. J. Remote Sens., 8, 84–106, 1982.

77. Todd, V. R., Alvarez, R. M., and Techni Graphic Systems, Inc.: Preliminary geologic map of the El Cajon 30^0 60^0 quadrangle, southern California, US Geological Survey, Menlo Park, Califor-nia, USA, 2004.

78. Watson, G. S.: Statistics of orientation data, J. Geol., 74, 786–797, 1966.

79. Whelley, P. L., Glaze, L. S., Calder, E. S., and Harding, D. J.: LiDAR-Derived Surface Roughness Texture Mapping: Application to Mount St. Helens Pumice Plain Deposit Analysis, IEEE T. Geosci. Remote, 52, 426–438, doi:10.1109/TGRS.2013.2241443, 2014.

80. Whittaker, A. C., Attal, M., and Allenn, P. A.: Characterising the origin, nature and fate of sediment exported from catch-ments perturbed by active

tectonics, Basin Res., 22, 809–828, doi:10.1111/j.1365-2117.2009.00447.x, 2010.

81. Wilkinson, M. T., Chappell, J., Humphreys, G. S., Fifield, K., Smith, B., and Hesse, P.: Soil production in heath and forest, Blue Mountains, Australia: influence of lithology and palaeoclimate, Earth Surf. Proc. Land., 30, 923–934, doi:10.1002/esp.1254, 2005.

82. Woodcock, N. H.: Specification of Fabric Shapes Using an Eigenvalue Method, Geol. Soc. Am. Bull., 88, 1231–1236, doi:10.1130/0016-7606(1977)88<1231:SOFSUA>2.0.CO;2, 1977.

83. Wood, R.: Transient hillslope response to an incision wave sweep-ing up a watershed: a case study from the Salmon River, Mas-ters Theses, available at: http://scholarworks.sjsu.edu/etd_theses/ 4322 (last access: 14 October 2015), 2013.

84. Wu, T. F., Lin, C. J., and Weng, R. C.: Probability estimates for multi-class classification by pairwise coupling, J. Mach. Learn. Res., 5, 975–1005, 2004.

85. Yoo, K., Amundson, R., Heimsath, A. M., and Dietrich, W. E.: Process-based model linking pocket gopher (Thomomys bottae) activity to sediment transport and soil thickness, Geology, 33, 917–920, doi:10.1130/G21831.1, 2005.

86. Yoo, K., Weinman, B., Mudd, S. M., Hurst, M., Attal, M., and Ma-her, K.: Evolution of hillslope soils: The geomorphic theater and the geochemical play, Appl. Geochem., 26, S149–S153, 2011.

CHAPTER 12

EFFICACY OF BEDROCK EROSION BY SUBGLACIAL WATER FLOW

F. Beaud, G. E. Flowers, and J. G. Venditti

INTRODUCTION

Textbook descriptions of glacial erosion detail mechanisms of abrasion and quarrying, but mention erosion by subglacial meltwater as a potential, unquantified, additional incision mechanism (e.g. Bennett and Glasser, 2009; Anderson and Anderson, 2010). This imbalance reflects the deficiency in our understanding of the latter. In fact, subglacial meltwater loaded with sediment has been inferred to carve meter-scale channels in bedrock (e.g. Glasser and Bennett, 2004), often called Nye channels (N-channels; Weertman, 1972) and kilometer-scale tunnel valleys (e.g. Ó Cofaigh, 1996; Glasser and Bennett, 2004; Jørgensen and Sandersen, 2006; Dürst Stucki et al., 2010; Kehew et al., 2012), and more recently its role has been invoked as a necessary mechanism in the carving and deepening of inner gorges (Dürst Stucki et al., 2012; Jansen et al., 2014).

Although the ability of subglacial water flow to flush subglacial sediment is well established (e.g. Humphrey and Raymond, 1994; Gurnell et al., 1996; Hallet et al., 1996; Willis et al., 1996; Alley et al., 1997; Koppes and Hallet, 2002, 2006; Riihimaki et al., 2005; Koppes and Montgomery, 2009), there has been little work quantifying subglacial sediment transport (Creyts et al., 2013), and no work on bedrock erosion by subglacial meltwater. Most studies of glacial erosion that use measurements of proglacial sediment yield rely on the hypothesis that subglacial meltwater flow is the most important process removing sediment from the glacier bed (e.g. Gurnell et al., 1996; Koppes and Hallet, 2002, 2006; Orwin and Smart, 2004; Riihimaki et al., 2005).

Citation: Beaud, F., Flowers, G. E., and Venditti, J. G.: Efficacy of bedrock erosion by subglacial water flow, Earth Surf. Dynam., 4, 125-145, doi:10.5194/esurf-4-125-2016, 2016.

In numerical models of glacial erosion, sediment transport by subglacial water flow is usually neglected (e.g. MacGregor et al., 2000, 2009; Brocklehurst and Whipple, 2002; Tomkin and Braun, 2002; Anderson et al., 2006; Herman and Braun, 2008; Egholm et al., 2009, 2011a, b, 2012; Herman et al., 2011; Beaud et al., 2014), often under the assumption that sediment is removed instantaneously. Bedrock erosion resulting from subglacial water is likewise neglected. A better understanding of the processes that lead to the formation of tunnel valleys or inner gorges is also important for the evaluation of deep geological repositories for nuclear waste in regions facing a potential future glaciation (e.g. Iverson and Person, 2012).

Subglacial water flows through two main types of drainage systems: distributed and channelized. In numerical models (see Flowers, 2015, for a review), a distributed drainage system is typically represented by a network of connected cavities (e.g. Lliboutry, 1968; Iken, 1981; Kamb, 1987; Schoof, 2010; Bartholomaus et al., 2011), a macroporous sheet of sediment (e.g. Clarke, 1996; Creyts and Schoof, 2009), or a water film (e.g. Weertman, 1972; Le Brocq et al., 2009). These representations reflect field observations of an increase in water pressure with discharge (e.g. Iken and Bindschadler, 1986; Nienow et al., 1998). A channelized drainage system is most often described by a single Röthlisberger channel or a network thereof (e.g. Röthlisberger, 1972; Nye, 1976; Flowers et al., 2004; Kessler and Anderson, 2004; Schoof, 2010; Hewitt et al., 2012). Water velocities are relatively high in the channelized system and, under steady-state conditions, water pressure decreases with increasing discharge (Röthlisberger, 1972). Conduits carved both in sediment and ice, so-called canals (e.g. Walder and Fowler, 1994; Ng, 2000; KyrkeSmith and Fowler, 2014), have properties closer to those of a distributed system as pressure often increases with discharge (Walder and Fowler, 1994). In winter, the distributed drainage system evacuates most basal water, and water pressures tend to be relatively high. As surface melt becomes significant, the water input becomes too large for the distributed drainage system alone, water pressure increases, and R-channels start to form. Once an efficient drainage system is established, meltwater is routed relatively quickly downstream and baseline water pressures are generally lower than in winter, with large daily fluctuations. As surface melt decreases, channels are reduced in size by ice creep and may eventually close. We hypothesize that this cycle may have a significant effect on bedrock erosion by subglacial meltwater flow (e.g. Willis et al., 1996; Swift et al., 2005).

Three main processes produce bedrock erosion in rivers: abrasion, macro-abrasion, and quarrying (e.g. Whipple et al., 2000, 2013). Abrasion is the result of particles entrained by the flow (saltating or in suspension) colliding with the bedrock and is governed by the tools and cover effect, whereby particles (i.e. tools) entrained by the flow impact exposed bedrock but can also shield it if they are immobile (i.e. cover; e.g. Whipple et al., 2000, 2013; Sklar and Dietrich, 2001, 2004; Turowski et al., 2007; Lamb et al., 2008; Cook et al., 2013; Scheingross et al., 2014). Macro-abrasion and quarrying both result from dislodgement of blocks and require a relatively high joint density in the bedrock. Macroabrasion occurs when blocks are dislodged as a result of the impact of moving particles, while quarrying is the result of dislodgement by pressure gradients caused by water flow (Whipple et al., 2000, 2013). Over highly jointed bedrock, quarrying and macro-abrasion can produce large canyons under extreme flow conditions (e.g. Bretz, 1969; Lamb and Fonstad, 2010; Baynes et al., 2015). In this study, we limit our analysis to abrasion (Whipple et al., 2000, 2013; Cook et al., 2013) and use well-established models to estimate the erosion due to total sediment load (Lamb et al., 2008) and saltating load only (Sklar and Dietrich, 2004).

Alley et al. (1997) suggest that the sediment transport capacity of an R-channel is most affected by changes in water discharge and hydraulic potential gradient and further posit that, in most cases, the hydraulic potential gradient increases downstream (due to steep ice-surface slopes close to the terminus), so that the transport capacity should also increase. Dürst Stucki et al. (2012) calculate an erosional potential of subglacial water based on the hydraulic potential gradient under a valley glacier and find that the erosional potential increases toward the terminus and could explain the deepening of inner gorges during a glaciation (e.g. Jansen et al., 2014). Both studies are, however, quite speculative regarding the processes behind subglacial meltwater erosion. Creyts et al. (2013) were the first to couple subglacial water flow in a distributed drainage system, basal refreezing, and sediment transport in a numerical model to explore the evolution of bed slopes adverse to ice flow close to the terminus. They demonstrate strong feedbacks between sediment deposition/entrainment and hydraulic conditions as well as the importance of daily fluctuations in water input on the sediment flux.

In proglacial studies of seasonal sediment yield, hysteresis between sediment and water discharge is often observed (e.g. Willis et al., 1996; Swift et al., 2005; Mao et al., 2014). It is usually attributed to changes in sediment availability, tapping of new areas of the bed by the developing drainage system, or increased mobilization caused by sudden changes in the subglacial hydraulic system (e.g. Willis et al., 1996). An event during which sediment transport peaks before discharge is usually defined as clockwise hysteresis and is interpreted as the manifestation of an unlimited sediment source (e.g. Mao et al., 2014). The opposite is true for anticlockwise hysteresis. In a study of bedload transport by a proglacial stream in the Italian Alps, Mao et al. (2014) identify a transition throughout the melt season from hysteresis dominated by clockwise events to hysteresis dominated by anticlockwise events. The authors infer that this transition is due to the activation of different sediment sources across the drainage basin.

Evidence of the erosional action of subglacial meltwater flow is widespread in formerly glaciated regions and appears in the form of N-channels, tunnel valleys, and inner gorges. Tunnel valleys are large (a few hundred metres to kilometres wide and up to tens of kilometres long) channel-like features found within the limits of former continental ice sheets (e.g. Ó Cofaigh, 1996; Glasser and Bennett, 2004; Jørgensen and Sandersen, 2006; Dürst Stucki et al., 2010; Kehew et al., 2012) or in Antarctica (e.g. Denton and Sugden, 2005; Rose et al., 2014) in substrata varying from loosely consolidated sediment to bedrock. Their formation is attributed to the action of pressurized subglacial meltwater, and three particular mechanisms have been proposed (e.g. Ó Cofaigh, 1996; Glasser and Bennett, 2004; Kehew et al., 2012): (1) sediment creep toward preferential groundwater flow paths, (2) carving by sediment-loaded subglacial water flow, and (3) erosion caused by large subglacial water floods.

Inner gorges are narrow canyons incised at the bottom of an otherwise U-shaped valley (e.g. Montgomery and Korup, 2011), and are found extensively in formerly glaciated mountain ranges like the Alps (e.g. Montgomery and Korup, 2011; Dürst Stucki et al., 2012). The origin of inner gorges was originally entirely attributed to postglacial fluvial erosion, although Montgomery and Korup (2011) conclude that such features persist through repeated glaciations instead of being reset by glacial erosion. For example, Valla et al. (2010) find that fluvial incision rates on the order of a centimeter per year occurred during at least the past 4000 years in a gorge in the French Western Alps. Recently Dürst Stucki et al. (2012) showed that pressurized water flow is necessary to explain the longitudinal profile of an inner gorge in the foothills of the Alps, and Jansen et al. (2014) infer from cosmogenic nuclide exposure that the timing of carving of seven inner gorges in the Baltic Shield matches the timing of glacial cover. Inner gorges are therefore most likely the combined product of fluvial erosion during interglacial periods and subglacial meltwater erosion during glacial periods, although the importance of fluvial vs. glacial conditions is probably dependent on surrounding topography. Nye channels, tunnel valleys, and inner gorges share some characteristics suggesting a common genetic origin, although the specific combination of processes responsible for their evolution may differ slightly.

We implement a one-dimensional (1-D) model of subglacial water flow in which a network of cavities is dynamically coupled to one or a few channels. We then compute the shear stress exerted on the bed and use it to compute transport stages and bedrock erosion rates by abrasion caused by particle impacts. We compare the results of models that treat erosion by saltating particles only (Sklar and Dietrich, 2004) and erosion by both the saltating and suspended particles (Lamb et al., 2008). We use the word "erosion" only to describe the carving of bedrock, while "transport", "mobilization", or "entrainment" refer to the movement of unconsolidated sediments.

We first perform steady-state simulations with the channelized drainage system alone to demonstrate basic model behavior. We investigate the role of ice geometry, surface melt, and sediment supply. We then introduce the coupled (distributed and channelized) hydraulic system and water forcing with sub-seasonal fluctuations to test the importance of transients in the subglacial drainage system. Finally, we discuss the implications of our results for the formation of N-channels and consequently of tunnel valleys over bedrock and the persistence of inner gorges through repeated glaciations. Our specific research questions are as follows. (1) What are the major controls on subglacial meltwater erosion? (2) How important is subglacial meltwater erosion compared to overall glacial erosion? (3) Can ordinary seasonal melt processes lead to subglacial bedrock channel incision (and potentially the formation of an incipient tunnel valley or persistence of an inner gorge)? (4) What are the implications of the water flow regime in channels for hysteresis and sediment transport?

MODELLING APPROACH

Subglacial Water Flow

We use a 1-D model of coupled cavity and channelized drainage styled after previous models (e.g. Werder et al., 2013), with the numerical distensibility parameter of Clarke (2003) and the water exchange term of Hewitt and Fowler (2008). The system of equations describing R-channels admits both a wave-like solution and the solution for water flow which introduces numerical stiffness. To circumvent this problem, Clarke (2003) proposed to treat water as a slightly compressible fluid and he introduced a numerical distensibility parameter, γ . The water exchange term is required as the channelized and distributed drainage systems are otherwise defined as two independent systems (Flowers et al., 2004; Hewitt and Fowler, 2008).

Channelized Drainage

As in most subglacial drainage models, we assume the channelized system to be a network of semi-circular Röthlisberger channels (Röthlisberger, 1972; Nye, 1976). The channels are assumed to be linear (along the x coordinate) and parallel to one another if more than one is considered. The conservation of water mass is given by (Clarke, 2003)

$$-\gamma S\frac{\partial \phi_{\mathrm{ch}}}{\partial t}=\frac{\partial Q_{\mathrm{ch}}}{\partial x}+\frac{\Xi-\Pi}{L}\left(\frac{1}{\rho_{\mathrm{i}}}-\frac{1}{\rho_{\mathrm{w}}}\right)$$
$$-v_{\mathrm{cc}}-(\dot{b}_{\mathrm{ch}}+\varphi)W, \qquad (1)$$

where S is the cross-sectional area of the R-channel, f_{ch} is the hydraulic potential, Q_{ch} is the water discharge, ρ_w and ρ_i are the densities of water and ice respectively, v_{cc} is the rate of creep closure of the channel walls, $\dot{b}_{\mathrm{ch}}$ is a water source term to the channel, j is the exchange rate between the distributed and channelized drainage systems, W is the width of the catchment drained by the R-channel, γ is a numerical compressibility parameter (Clarke, 2003), x is the coordinate along the flow line, and t is time. Note that we assume changes in bed topography with time are small enough to be neglected. Therefore $\partial\phi_{\mathrm{b}}/\partial t=0$ and $\partial p_{\mathrm{ch}}/\partial t=\partial(\phi_{\mathrm{ch}}-\phi_{\mathrm{b}})/\partial t=\partial\phi_{\mathrm{ch}}/\partial t$, where p_{ch} is the water pressure in the channel and f_b the hydraulic potential at the bed. The evolution of channel cross-sectional area S with time is

$$\frac{\partial S}{\partial t}=\frac{\Xi-\Pi}{\rho_{\mathrm{i}}L}-v_{\mathrm{cc}}, \qquad (2)$$

where Ξ is the dissipation of potential energy, 5 is the energy to maintain the water at the pressure melting point, L is the latent heat of fusion, and thus $(\Xi-\Pi)/(\rho_{\mathrm{i}}L)$ is the rate of channel opening by viscous heat dissipation. We refer the reader to the Supplement for the description of Ξ, Π, v_{cc}, and Q_{ch}.

Distributed Drainage

The distributed drainage system is treated as a network of connected cavities (Kamb, 1987) undergoing turbulent flow. Assuming that water can be stored englacially or subglacially, the conservation of water mass can be written (Werder et al., 2013)

$$\frac{e_{\mathrm{v}}}{\rho_{\mathrm{w}}g}\frac{\partial\phi_{\mathrm{ca}}}{\partial t}+\nabla\cdot\boldsymbol{q}_{\mathrm{ca}}+v_{\mathrm{o}}-v_{\mathrm{c}}-(\dot{b}_{\mathrm{ca}}-\varphi)=0, \qquad (3)$$

where f_{ca} is the hydraulic potential in the cavity network, e_v is the englacial void ratio, q_{ca} is the water flux in the cavity network, g is the gravitational acceleration, v_o and v_c are the opening and closure rate, respectively, and $\dot{b}_{\mathrm{ca}}$ is a water source term to the cavity network. The evolution of the average cavity height h_{ca} with time is

$$\frac{\partial h_{\mathrm{ca}}}{\partial t}=v_{\mathrm{o}}-v_{\mathrm{c}}. \qquad (4)$$

We refer the reader to the Supplement for the description of v_o, v_c, and q_{ca}.

Coupled Channelized and Distributed Drainage

To couple the two drainage systems we use a term to describe water exchange between cavities and channels as a function of their respective pressure differences (e.g. Flowers et al., 2004; Hewitt and Fowler, 2008):

$$\varphi=k_{\mathrm{ex}}(N_{\mathrm{ch}}-N_{\mathrm{ca}}), \qquad (5)$$

where kex is an exchange coefficient (Table 1) and N_{ch} and N_{ca} are the effective pressures in the channelized and cavity systems, respectively (see Supplement). The system formed by Eqs. (1), (2), (3), and (4) is solved with the MATLAB pdepe solver.

Table 1. Summary of hydrological model parameters

Parameter	Description	Value
M	Number of grid points	101
dx	Cell size	500 m
dt	Time step	600 s
W	System width	1000 m
X_L	Glacier length	50 km
ρ_w	Water density	$1000\ \mathrm{kg\,m^{-3}}$
ρ_i	Ice density	$910\ \mathrm{kg\,m^{-3}}$
L	Latent heat of fusion	$3.34 \times 10^5\ \mathrm{J\,kg^{-1}}$
n	Flow-law exponent	3
k_{ex}	Exchange coefficient	$5 \times 10^{-13}\ \mathrm{m^{3/2}\,kg^{-1/2}}$
e_v	Englacial void ratio	1×10^{-3}
γ	Numerical compressibility	$10^{-9}\ \mathrm{Pa^{-1}}$

Shear Stress Partitioning

Following Clarke (2003) we write the shear stress exerted on the bed by subglacial water flow as

$$\tau_b = \frac{1}{8} f_b \rho_w u^2, \tag{6}$$

where f_b is the Darcy–Weisbach friction coefficient for the bed and u is the depth-averaged water flow velocity defined as $u = Q_{ch}/S$ in a channel and $u = q_{ca}/h_{ca}$ in the network of cavities. Details pertaining to the shear stress calculation are described in the Supplement.

Subglacial Water Flow Model Simplifications

We implement the model in a 1-D continuum such that the drainage system is assumed to be fully connected. Distributed and channelized drainage systems are assumed to be saturated. The bed is impermeable and undeformable. We also neglect the routing of supra- and englacial water, the effect of particles or refreezing on water flow constriction, and the feedbacks between sliding speed and water pressure.

Boundary and Initial Conditions

Atmospheric pressure defines the downstream boundary condition for both the network of cavities and the R-channels: $\phi_{ch}(x = X_L, t) = \phi_b + 1000\ \mathrm{Pa}$. For most simulations, a no- flux condition is applied at the upstream boundary such that $Q_{ch}(x = 0, t) = 0$, $q_{ca}(x = 0, t) = 0$. Otherwise, for the simulation in which discharge is constant throughout the domain, we apply a Neumann boundary condition at the upstream node of the channel: $Q_{ch}(x = 0, t) = 18\ \mathrm{m^3\,s^{-1}}$. We use a 1-year model spin-up for all the transient simulations as we find no significant difference in results using 1- and 2-year spin-ups.

Erosion by Saltating and Total Load

In river reaches where bedrock fracture density is low (i.e. blocks are larger than 1–2 m in size), abrasion is the primary erosional process (Whipple et al., 2000, 2013). Particles that are mobilized by the flow and move by saltation or in suspension can impact the bed, and the energy released upon impact can cause erosion. To erode its bed, a river therefore requires tools (particles available for transport) and an exposed bed.

Saltation Erosion Model (SEM) (Sklar and Dietrich, 2004)

The rate of erosion caused by a saltating load can be expressed as

$$\dot{e}_{salt} = I_r V_i F_e, \tag{7}$$

where I_r is the rate of particle impact with the bed, V_i is the volume of bed material removed upon impact, and F_e the fraction of the bed exposed. The impact rate (I_r) is a function of the number of particles and their saltation trajectories. The more particles and the shorter the saltation length, the higher the impact rate. The volume removed upon impact (V_i) is a function of the energy released on impact and therefore of the particle mass, its speed normal to the bed, and the physical properties of the bed and the particle. The fraction exposed (F_e) determines the areal extent of bedrock vulnerable to erosion and represents the cover effect. All of the quantities described here are given per unit width. Values of V_i, I_r, and F_e are computed as

$$V_i = \frac{\pi \rho_s D^3 w_{si}^2 Y}{6 k_v \sigma_T^2}, \tag{8}$$

where ρ_s is the sediment density, D is the particle diameter, w_{si} is the particle velocity upon impact, Y is the Young's modulus of the bedrock, k_v is an empirical rock erodibility coefficient, and σ_T is the rock tensile strength;

$$I_r = \frac{6 q_s}{\pi D^3 L_s}, \tag{9}$$

where q_s is the sediment supply rate and L_s is the hop length of particles;

$$F_e = 1 - \frac{q_s}{q_{tc}}, \text{ for } q_s \leq q_{tc}, \tag{10}$$

where q_{tc} is the sediment transport capacity. Substituting Eqs. (8), (9), and (10) into (7) and simplifying yields

$$\dot{e}_{salt} = \frac{\rho_s q_s w_{si}^2 Y}{L_s k_v \sigma_T^2}\left(1 - \frac{q_s}{q_{tc}}\right). \tag{11}$$

Note that when $q_s/qt_c > 1$ the erosion rate becomes $\dot{e}_{salt} = 0$ because the bed is completely covered with sediment. Constants and parameters are listed in Table 2, and the formulations of L_s and w_{si} are described in the Supplement. We follow Sklar and Dietrich (2004) in using the Fernandez-Luque and van Beek (1976) formulation of transport capacity:

$$q_{tc} = 5.7(rgD^3)^{1/2}(\tau^* - \tau_c^*)^{3/2}, \tag{12}$$

where $r = \rho_s/\rho_w - 1$ is the buoyant density of sediment and τ^*_c the critical value of the Shields stress. The Shields stress τ^* is computed as

$$\tau^* = \frac{\tau_b}{(\rho_s - \rho_w) g D}. \tag{13}$$

Total Load Erosion Model (TLEM) (Lamb et al., 2008)

The total load erosion model (TLEM) is based on the original model of Sklar and Dietrich (2004) and extended by Lamb et al. (2008) to account for the impact of particles in suspension close to the bed. The total load erosion rate is calculated as

$$\dot{e}_{\mathrm{tot}} = \frac{A_1 \rho_s Y}{k_v \sigma_T^2} c_b w_{i,\mathrm{eff}}^3 \left(1 - \frac{q_b}{q_{tc}}\right), \quad (14)$$

where $A_1 < 1$ is a coefficient to account for the upward lifting of particles close to the bed by turbulent eddies (here $A_1 = 0.36$), c_b is the near-bed sediment concentration, $w_{i,\mathrm{eff}}$ is the effective impact velocity, and q_b is the volumetric flux of sediment transported as bedload. The calculation of these quantities is explained in the Supplement. By analogy with the saltation-erosion model, c_b is similar to the impact rate I_r, $w_{i,\mathrm{eff}}$ to the volume removed upon impact V_i, and $(1 - q_b/q_{tc})$ to the fraction of the bed exposed (F_e). Similarly to the SEM, when $q_b/q_{tc} > 1$ the erosion rate becomes $\dot{e}_{\mathrm{tot}} = 0$ because the bed is completely covered with sediment. Constants and parameters used for the TLEM are the same as in the SEM and are listed in Table 2.

The major differences between the TLEM and the SEM are that (1) the effective impact velocity $w_{i,\mathrm{eff}}$(TLEM) increases monotonically with transport stage as it accounts for the effect of turbulence causing particle collisions with the bed, whereas w_{si} tends toward 0 at large transport stages (and therefore $\dot{e}_{\mathrm{salt}}$ approaches 0) as particles trajectories become parallel to the bed, and (2) the fraction of the bed exposed in the TLEM $(1 - q_b/q_{tc})$ uses the bedload flux (q_b) as contributing to the cover. This means that as the transport stage increases, the fraction of bedrock exposed in the TLEM is higher than in the SEM, provided a significant portion of the sediment is transported in suspension.

Erosion Model Simplifications

We calculate erosion rates with a constant particle diameter (D = 60 mm) and assume that this particle size is representative of the median grain size in an R-channel. This choice is consistent with Riihimaki et al. (2005), who report a median grain size corresponding to very coarse gravel in the proglacial area of Bench Glacier, Alaska. We assume that the bedrock has a uniform resistance to erosion, erosion occurs uniformly across the width of a channel, abrasion is the main erosion mechanism, and the cover effect is linear. A linear cover effect means that as long as the bed is partially exposed, newly deposited particles are assumed to cover bedrock rather than previously deposited sediment, hence the exponent of 1 on the "cover" fraction q_s/q_{tc} (see Turowski et al., 2007). We also neglect sediment transport and the effect of downstream fining.

Table 2. Summary of parameter values for the saltation (Sect. 2.2.1) and total load (Sect. 2.2.2) erosion models

Parameter	Description	Value
A_1	Coefficient for upward lifting of particles*	0.36
D	Diameter of particles	60 mm
σ_T	Rock tensile strength*	7.0 MPa
Y	Young's modulus of bedrock*	5.0×10^4 MPa
k_v	Rock erodibility coefficient*	1.0×10^6
τ_c^*	Critical Shields stress*	0.03
ρ_s	Sediment density	$2650\ \mathrm{kg\,m^{-3}}$
$q_{s,\mathrm{ref}}$	Ref. sediment supply per unit width (steady state – transient)	$3.6 - 9.1 \times 10^{-3} \mathrm{m^2\,s^{-1}}$

* Parameter values from Lamb et al. (2008).

MODELLING STRATEGY AND RATIONALE

The model outlined above involves numerous variables and parameters leading to feedbacks. Applications of the model of erosion by abrasion in rivers (e.g. Sklar and Dietrich, 2004, 2006, 2008; Turowski et al., 2007; Lamb et al., 2008; Chatanantavet and Parker, 2009; Nelson and Seminara, 2011; Egholm et al., 2013) show that the primary dependencies are the transport stage, the relative sediment supply, and the hydraulic potential gradient. Numerical modelling studies of subglacial water flow emphasize the importance of the frequency and amplitude of the water input forcing and of the ice and bed geometry (e.g. Flowers, 2008; Creyts and Schoof, 2009; Schoof, 2010; Hewitt, 2013; Werder et al., 2013; Beaud et al., 2014), although few studies have discussed the shear stress on channel walls (e.g. Clarke, 2003). In an effort to identify the key variables, parameters, and feedbacks, we start with simple experiments and build up the complexity. The simulations are separated in two subsections: (1) steady-state decoupled simulations of a channelized drainage system only (Table 3) and (2) transient simulations of a coupled channelized and distributed drainage system (Table 4).

In the steady-state simulations (Table 3), we first assess the erosion pattern resulting from a constant discharge along the glacier bed. Then we introduce a water forcing that increases with decreasing ice-surface elevation and test the effect of ice geometry and sediment supply. Additional experiments assessing the effect of water flow through a network of cavities, water input, sediment size, and channel wall and bed roughness are shown in the Supplement. In the transient simulations (Table 4) we first analyse the role of a synthetic melt season, then use water forcing following realistic melt seasons, and test the role of ice geometry, channel density, and sediment supply.

As per the theory on which the SEM (Sklar and Dietrich, 2004) and TLEM (Lamb et al., 2008) are based, sediment supply, through the tools and cover effects, exerts a major control on erosion rates and patterns by water flow beneath glaciers. Although it is possible to estimate sediment supply to rivers (e.g. Gurnell et al., 1996; Willis et al., 1996; Kirchner et al., 2001; Orwin and Smart, 2004; Sklar and Dietrich, 2004; Riihimaki et al., 2005; Turowski et al., 2009; Mao et al., 2014), no method exists to quantify subglacial sediment supply or transport. In glacierized catchments, measurements are usually made in the proglacial stream, relatively close to the terminus (e.g. Warburton, 1990; Gurnell et al., 1996; Willis et al., 1996; Orwin and Smart, 2004; Riihimaki et al., 2005; Swift et al., 2005; Mao et al., 2014), where water flow is subaerial and potentially influenced by channel dynamics between the glacier terminus and the measurement station (Warburton, 1990; Orwin and Smart, 2004; Mao et al., 2014). In the steady-state simulations we impose a sediment supply that leads to an interesting and diverse range of simulations illustrating most processes and their feedbacks, whereas in the transient simulations we choose a sediment supply that leads to a sediment yield at the last node equivalent to a few millimetres of glacial erosion per year.

We use a flat bed and a parabolic surface for all but two geometries: STP and WDG (Fig. 1). The STP geometry aims at reproducing the steep front of an advancing ice sheet, while the WDG geometry has a wedge shape that resembles the profile of a thinning and retreating ice sheet margin. We assume that water input is distributed uniformly (no moulins) along the bed except for the simulation S_MOULIN. For almost all simulations we fix the sediment supply per unit width; the total sediment supply therefore increases with channel size. This is similar to the assumption of a uniform till distribution across the width of the glacier in which the channelized water flow sources its sediment.

RESULTS

All quantities related to the erosion model are given per unit width of flow. The cross-sectional area of a subglacial channel (S), and thus its width $(W_{\mathrm{ch}} = 2\sqrt{2S/\pi})$, changes more rapidly with distance along the flow path than a typical subaerial river. We need to account for these changes when displaying the results, and therefore introduce three quantities: the total transport capacity $Q_{\mathrm{tc}} = q_{\mathrm{tc}} W_{\mathrm{ch}}\,(\mathrm{m^3\,s^{-1}})$, the total erosion computed with the TLEM $E_{\mathrm{tot}} = \dot{e}_{\mathrm{tot}} W_{\mathrm{ch}}\,(\mathrm{m^2\,s^{-1}})$, and the total erosion computed with the SEM $E_{\mathrm{salt}} = \dot{e}_{\mathrm{salt}} W_{\mathrm{ch}}\,(\mathrm{m^2\,s^{-1}})$.

Table 3. Summary of steady-state simulations. For simulations in which meltwater input is a function of ice-surface elevation z_s, we compute $f(z_s(x)) = \dot{b}_{\mathrm{ss\,max}} \times (1 - (z_s(x) - z_{s,\mathrm{min}})/z_{s,\mathrm{max}})$, where $\dot{b}_{\mathrm{ss\,max}} = 8.5 \times 10^{-7}\,\mathrm{m\,s^{-1}}$ is the maximum meltwater input rate to the channelized drainage system, and $z_{s,\mathrm{min}}$ and $z_{s,\mathrm{max}}$ are respectively the minimum and maximum ice-surface elevations. Note that $\dot{b}_{\mathrm{ss\,max}} = 8.5 \times 10^{-7}\,\mathrm{m\,s^{-1}}$ corresponds to 7.6 cm of ice melt per day assuming $\rho_i = 910\ \mathrm{kg\ m^{-3}}$. The reference sediment supply used for the steady-state simulations is $q_{s,\mathrm{ref}} = 3.6 \times 10^{-3}\ \mathrm{m^2\ s^{-1}}$ (Table 2)

Simulation	Purpose	Forcing	Difference from reference run	Section
S_MOULIN	R-channel only	$Q_{\mathrm{ch}}(x=0,t) = 18\,\mathrm{m^3\,s^{-1}}$	Localized input upstream boundary	4.1.1
S_REF	Reference	$\dot{b}_{\mathrm{ch,ref}}(x,t) = f(z_s(x))$		4.1.2
S_WDG	Ice geometry	$\dot{b}_{\mathrm{ch}}(x,t) = \dot{b}_{\mathrm{ch,ref}}$	Constant ice-surface slope $d\phi_0/dx = \mathrm{cst}$	4.1.2
S_STP	Ice geometry	$\dot{b}_{\mathrm{ch}}(x,t) = \dot{b}_{\mathrm{ch,ref}}$	Steeper and thicker terminus	4.1.2
S_1300	Ice geometry	$\dot{b}_{\mathrm{ch}}(x,t) = \dot{b}_{\mathrm{ch,ref}}$	$z_{s,\mathrm{max}} = 1300\,\mathrm{m}$	4.1.2
S_700	Ice geometry	$\dot{b}_{\mathrm{ch}}(x,t) = \dot{b}_{\mathrm{ch,ref}}$	$z_{s,\mathrm{max}} = 700\,\mathrm{m}$	4.1.2
S_SSP	Sediment supply	$\dot{b}_{\mathrm{ch}}(x,t) = \dot{b}_{\mathrm{ch,ref}}$	$q_s = q_{s,\mathrm{ref}}/20 - q_{s,\mathrm{ref}} \times 25$	4.1.3

Table 4. Summary of transient simulations. In this series of simulations the basal sliding speed is $u_b = 5\ \mathrm{m\ a^{-1}}$ and the reference sediment supply is $q_{s,\mathrm{ref}} = 9.1 \times 10^{-3}\mathrm{m^2\ s^{-1}}$ (Table 2). Water is fed to the network of cavities rather than the channel, hence $\dot{b}_{\mathrm{ch}} = 0\,\mathrm{m\,s^{-1}}$, and the function of surface elevation is that of the steady-state simulations (Table 3)

Simulation	Purpose	Description	Section
T_REF	Reference transient	Reference geometry and synthetic water input	4.2.1
T_2007	Realistic forcing	2007 surface-melt time series	4.2.2
T_2008	Realistic forcing	2008 surface-melt time series	4.2.2
T_1300	Ice geometry	$z_{s,\mathrm{max}} = 1300\,\mathrm{m}$	4.2.3
T_700	Ice geometry	$z_{s,\mathrm{max}} = 700\,\mathrm{m}$	4.2.3
T_W500	Drainage catchment width	$W = 500\,\mathrm{m}$, i.e. 2 R-channels	4.2.4
T_W333	Drainage catchment width	$W = 1000/3\,\mathrm{m}$, i.e. 3 R-channels	4.2.4
T_W250	Drainage catchment width	$W = 250\,\mathrm{m}$, i.e. 4 R-channels	4.2.4
T_SSP/4	Sediment supply	$q_s = q_{s,\mathrm{ref}}/4$	4.2.5
T_SSP/2	Sediment supply	$q_s = q_{s,\mathrm{ref}}/2$	4.2.5
T_SSPOPT	Sediment supply	Optimized erosion; $q_s = 0.6 \times q_{\mathrm{tc}}$	4.2.5

Steady-State Decoupled Simulations

We examine basic steady-state behavior of key model variables such as R-channel cross-sectional area (***S***), transport capacity (q_{tc}), impact velocity (w_{si} and $w_{i;eff}$), transport stage (τ^*/τ_c^*), and relative sediment supply ($q_s = q_{tc}$). In the steady-state simulations, the drainage system is composed of a single R-channel. Simulations are terminated once dependent variables (***S*** and f_{ch}) reach steady state. In this series of simulations, unless stated otherwise, we impose the sediment supply $q_s = 3.6 \times 10^{-3} m^2 s^{-1}$ to produce an interesting and diverse range of model behavior.

R-Channel with Constant Discharge

In the S_MOULIN simulation we use the reference glacier geometry (Fig. 1, REF) and the water input is imposed at the uppermost node only, i.e. as if a moulin were feeding the system. This permits us to drive the system with a constant water discharge and to analyze the resulting relation between discharge, channel cross-sectional area, velocity, instantaneous erosion rates, and transport capacity patterns.

Although discharge is constant (Fig. 2a), the cross-sectional area of the R-channel changes along the profile (Fig. 2a) due to the ice geometry (Fig. 1). Over the first 46 km, the cross-sectional area decreases in response to the steepening hydraulic potential gradient (Fig. 2b), the latter being a function of ice-surface slope (Fig. 1). Close to the terminus (last 4 km of the profile), the ice thins significantly, the hydraulic potential gradient shallows (by a factor of 3), and the cross-sectional area increases. The average water velocity assumes the opposite pattern (Fig. 2b). Because the grain size is kept constant, $q_{tc} \propto (\tau^* - \tau_c^*)^{3/2}$, the rate of sediment transport (Fig. 2c) is amplified relative to the velocity (q_{tc} drops by a factor of 6, while ***u*** is reduced by about 50 %); both have maxima at 46 km and decrease sharply near the terminus. In this simulation the sediment transport capacity is always larger than the supply rate (Fig. 2c), exposing most of the bed ($F_e > 0.5$).

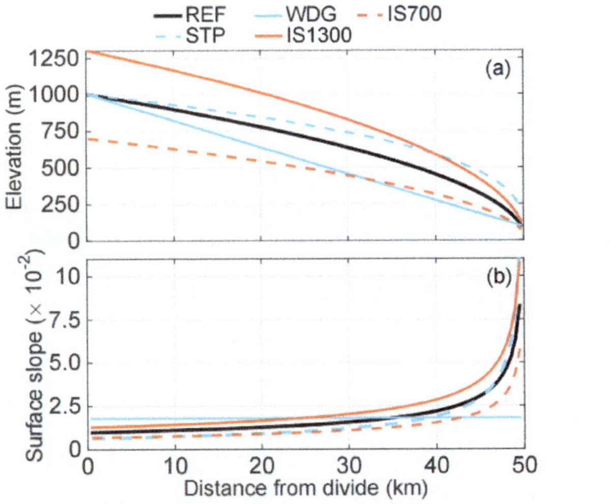

Figure 1. (a) Different ice geometries considered (see Tables 3 and 4) and (b) corresponding ice-surface slopes. The geometries REF, IS1300, and IS700 are parabolas such that the surface elevation is given by $z_s(x) = z_{s,max}\sqrt{x}$, where $z_{s,max}$ is the maximum ice-surface elevation and x ∈ [1,0]. In order to obtain the steeper front in STP we use a cubic root instead of the square root: $z_s(x) = z_{s,max}x^{1/3}$. The wedge-like geometry WDG is defined by a straight line with the same elevation change as REF.

In both the TLEM and SEM, near-bed sediment concentration and impact rate are described as a function of the sediment supply and the hop trajectory of a particle, so they are similar (Fig. 2d). As the velocity increases (km 0–46) the sediment is transported faster and further from the bed and the near-bed sediment concentration and impact rate decrease.

Impact velocities (Fig. 2e) vary depending on the model. In the SEM the impact velocity tends toward zero as the hop length increases and the particles approach transport in suspension, which leads to a local minimum around km 46. The TLEM accounts for the effect of turbulent eddies on the trajectory of particles close to the bed, and thus the effective impact velocity is commensurate with the velocity and peaks around km 46.

Erosion rates (Fig. 2f) in both the SEM and TLEM show a minimum at 46 km and a maximum close to the terminus. The minimum and maximum correspond respectively to the minimum and sharp rise in near-bed sediment concentration or impact rate. Even under a constant discharge, ice-surface slope and channel size produce a peak in velocity just upstream of the terminus. At this peak, the flow has the power to lift particles far enough from the bed to reduce erosion in both the SEM and TLEM. Erosion rates are higher with the TLEM than the SEM due to the difference in impact velocities when the sediment transport regime approaches suspension. For the transport stages we obtain, impact velocities with the TLEM are consistently higher than with the SEM. Moreover, for relatively large transport stages, more of the bed is exposed in the TLEM: when some fraction of the total load travels in suspension, $q_s > q_b$.

Ice Geometry

The surface slope of a glacier is a first-order control on subglacial water flow (Eqs. 1–2). We experiment with thicker ice (IS1300, Fig. 1), thinner ice (IS700, Fig. 1), and steeper ice-surface slopes close to the terminus (STP, Fig. 1), all compared to the reference ice geometry (REF, Fig. 1). Shallower surface slopes close to the terminus are tested with a wedge-like geometry of constant surface slope (WDG, Fig. 1). All these simulations (Fig. 3; see Table 3) employ the surface-melt profile of S_REF and therefore yield nearly identical discharge profiles (Fig. 3a). The water input is a function of the ice-surface elevation (Table 3); hence, the discharge is largest close to the terminus.

The thinner the ice close to the terminus the larger the channel (Fig. 3b) and the lower the transport stage (Fig. 3c). In the case of S_STP, the combination of a particularly steep hydraulic potential gradient and relatively thick ice near the terminus inhibits channel enlargement over the last 5–10 km as observed in other simulations (Fig. 3b). Since the dis-charge profile is the same for the simulations presented, the smaller the channel, the faster the flow and hence the larger the transport stage. The maximum transport stage for the simulation with the steepest terminus (S_STP, $\tau^*/\tau_c^* \approx 18.5$, Fig. 3c) is almost 4 times larger than that for the simulation with the shallowest terminus (S_WDG, $\tau^*/\tau_c^* \approx 5.5$, Fig. 3c; see Fig. 1 and Table

3). Over the first 25 km of the profile the simulation with a constant ice-surface slope (S_WDG) shows the steepest hydraulic potential gradient, creating comparatively higher transport stages than other models.

Erosion begins around km 4 for S_WDG (Fig. 3d), and around km 19–20 for S_STP and S_700 (Fig. 3d), because for the prescribed sediment supply, the bed becomes exposed for transport stages $\tau^*/\tau_c^* \gtrsim 2.5$. All simulations but S_WDG have a local maximum in total erosion ($\boldsymbol{E}_{tot}$, Fig. 3d) between km 30 and 37 and a local minimum between km 45 and 49. The sediment supply per unit width is constant; therefore, the relative sediment supply decreases as the transport stage in-creases and thus the erosion rate per unit width decreases when $\boldsymbol{q}_s = \boldsymbol{q}_{tc} < 0{:}5$ because the number of tools decreases. For simulation S_WDG the increase in channel size compensates for the small decrease in erosion rate per unit width (not shown) because the transport stage remains relatively low ($\tau^*/\tau_c^* <$ 6). Over the last few kilometers of the profile, total erosion increases again for the simulations in which trans-port stage drops to moderate values (S_REF and S_1300). Total erosion drops sharply at the terminus for S_700 as the transport stage drops below 2.5. When the transport stage re-mains relatively high (τ^*/τ_c^* > 15) the number of tools remains low, as does total erosion (S_STP, Fig. 3c and 3d). Ice geometry exerts a primary influence on transport stage and erosion patterns via its control on hydraulic potential gradients and channel size. Simulation S_WDG yields the most erosion, despite relatively low transport stages, illustrating the importance of the tools effect (e.g. Sklar and Dietrich, 2006).

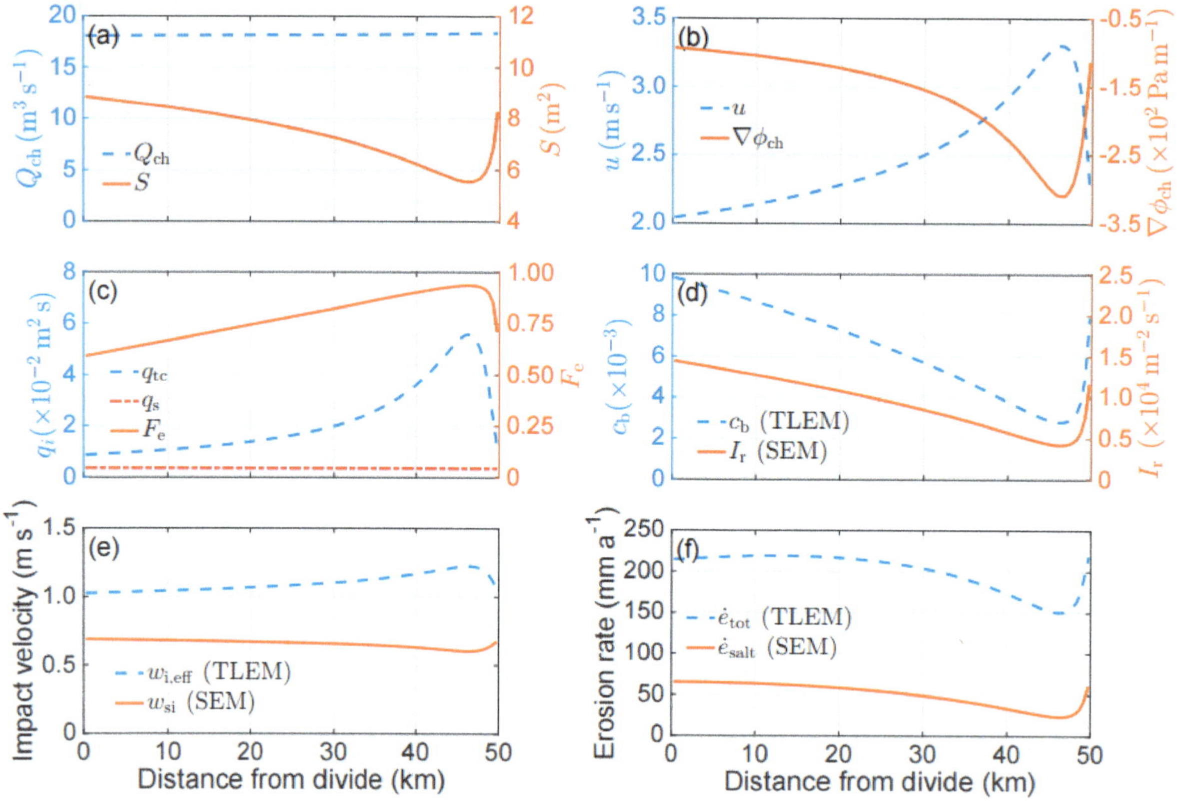

Figure 2. Relationship between the hydraulic and erosion variables in a steady-state R-channel with constant discharge (Table 3, S_MOULIN; Sect. 4.1.1). Components of TLEM and SEM are also compared. (a) R-channel discharge, Q_{ch}, and cross-sectional area S; (b) velocity, u, and gradient in hydraulic potential, ∇f_{ch}; (c) sediment transport capacity, q_{tc}, prescribed sediment supply rate, q_s, and fraction of bed exposed, Fe; (d) near-bed sediment concentration, c_b (TLEM), and impact rate, I_r (SEM); (e) impact velocity in the TLEM, $w_{i,eff}$, and the SEM, w_{si}; and (f) erosion rate calculated with the TLEM, $\dot{e}_{tot}$, and the SEM, $\dot{e}_{salt}$.

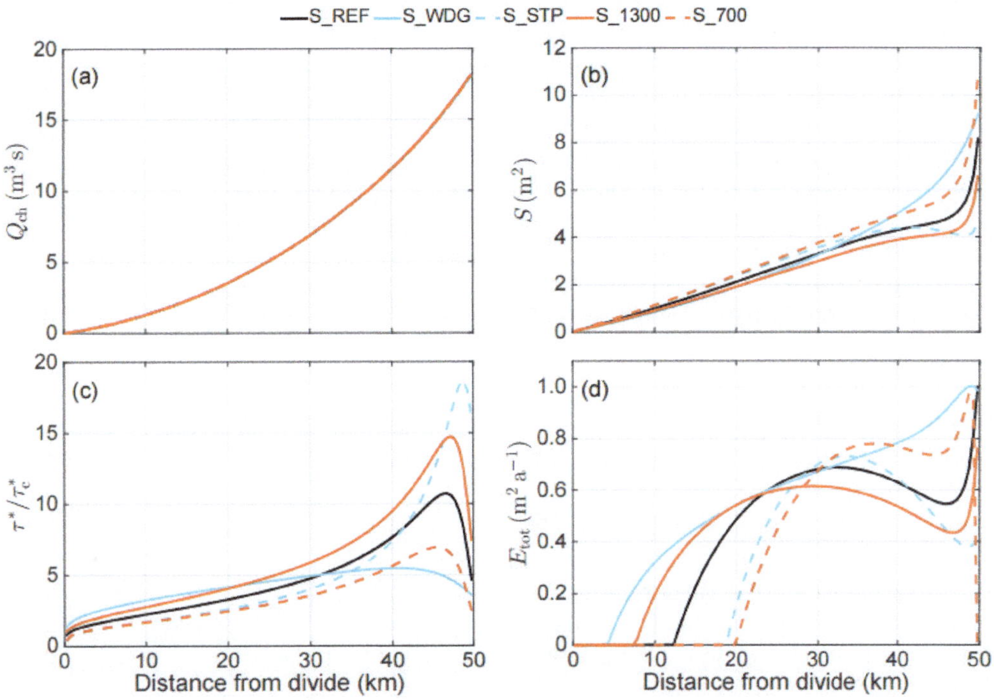

Figure 3. Comparison of steady-state simulations with varying ice geometries, reference water input and drainage through a single Rchannel (Table 3). (a) Discharge in the channel, Q_{ch}, (all curves overlap); (b) channel cross-sectional area, S; (c) transport stage, τ^*/τ_c^*; and (d) total erosion computed with the TLEM (E_{tot}).

Sediment Supply

Tools and cover compete so that both a lack and an overabundance of tools hinder erosion. Sklar and Dietrich (2004) and Lamb et al. (2008) have shown that erosion peaks for a given flow regime at an optimum relative sediment supply. We thus investigate how varying sediment supply ($q_s = 1.8 \times 10^{-4}$ to $8.9 \times 10^{-2}\ \mathrm{m^2\,s^{-1}}$) affects the rates and patterns of subglacial meltwater erosion, while the subglacial hydraulic regime re-mains that of S_REF (Fig. 3; Table 3).

For sediment supply rates $q_s \leq q_{s,ref} \times 2$ (Fig. 4a) the patterns of total erosion with the TLEM and SEM (Fig. 4b and 4c) show the same features as in Fig. 3: no erosion in the uppermost part of the profile followed by a local maximum around mid-profile, a local minimum around km 46–47 and a sharp rise in the last 3 km. For larger sediment supply $q_s \geq q_{s,ref} \times 5$, the patterns change and have a single maximum only. If the relative sediment supply satisfies $q_s/q_{tc} > 0.3$ (Fig. 4a) the number of tools remains high enough for total erosion to increase with transport stage ($\tau * /\tau * c$) as the increase in channel size compensates for the small drop in erosion per unit width $\dot{e}_{tot}$ when $q_s/q_{tc} < 0.5$.

Erosion occurs over 48 km of the bed (S_SSP, $\boldsymbol{q}_{s;ref}/20$) in both the SEM and TLEM for the lowest sediment supply, whereas the bed is almost completely shielded for the second largest sediment supply (S_SSP, $\boldsymbol{q}_{s;ref} \times 20$) and erosion occurs only between km 46 and 48 in the TLEM (Fig. 4). The SEM treats the whole sediment supply as participating to the cover effect, and the bed is shielded as soon as $qs/qtc \leq 1$. The TLEM discriminates between transport as bedload and in suspension; therefore, if $qb < qs$ the bed can remain partially exposed even for a relative sediment supply $qs/qtc \geq 1$. The erosion window is therefore larger for the TLEM (Fig. 6 in Lamb et al., 2008). The fact that erosion rates computed with the TLEM are higher than those computed with the SEM is inherent to the model formulation (Lamb et al., 2008).

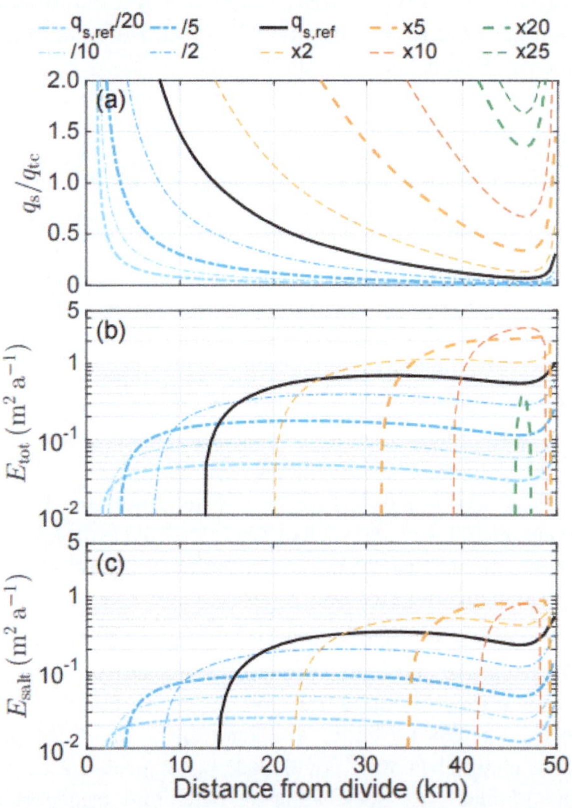

Figure 4. Influence of sediment supply rate q_s (S_SSP, Table 3) on total erosion with the TLEM and SEM for the same hydrology as the reference simulation (S_REF, Table 3). The legend applies to all panels. (a) Relative sediment supply (q_s/q_{tc}), (b) total erosion rate computed with the TLEM (E_{tot}), and (c) total erosion rate computed with the SEM (E_{salt}). Note the logarithmic scale on the y axis of (b) and (c).

An interesting conclusion arising from Fig. 4 is that total erosion is significant over most of the bed (Fig. 4b and 4c) when the sediment supply rate is relatively low ($q_s \leq q_{s,ref} \times 2$; Fig. 4a). Total erosion becomes more localized at higher relative sediment supply rates. Changing the particle diameter (D), instead of sediment supply (q_s), leads to similar changes in relative sediment supply and therefore in erosion patterns (see Supplement). The transport stages calculated are large enough to produce significant differences between the SEM and TLEM for large relative sediment supplies. Hereafter, we focus on the TLEM only in the results as it is more appropriate for the flow conditions encountered beneath glaciers.

Transient Coupled Simulations

Arguably, most erosion and sediment transport in fluvial systems occur during flood events (e.g. Whipple et al., 2000, 2013; Kirchner et al., 2001; Sklar and Dietrich, 2004; Lamb et al., 2008; Turowski et al., 2009; Lamb and Fonstad, 2010; Cook et al., 2013). In glacial environments large daily variations in meltwater input to the glacier bed can be likened to periodic flooding (e.g. Willis et al., 1996). We perform a series of transient simulations with a cavity network coupled to an R-channel (Eqs. 1, 2, 3 and 4) to explore how the transience in subglacial water flow is affected by changes in ice thickness, surface melt, and sediment supply, and how this transience impacts instantaneous and annually integrated erosion rates. In this series of simulations, we choose the sediment supply ($q_{s,ref} = 9.1 \times 10^{-3} m^2 s^{-1}$) such that the modelled sediment yield ($\int \min(q_s(X_L)W_{ch}(X_L), Q_{tc}(X_L))dt$) corresponds to an inferred basin-wide erosion rate of a few millimetres of erosion per year (e.g. Gurnell et al., 1996; Hallet et al., 1996; Koppes and Montgomery, 2009).

Reference Model

The reference model uses a synthetic forcing in the form of water supply to the distributed system ($\dot{b}_{ca}$, Eq. (3); sinusoid with a period of 120 days on which we superimpose daily fluctuations) and the reference glacier geometry (Fig. 1, REF; Table 4, T_REF). The sediment supply rate per unit width is assumed to be uniform along the bed. With this simple test we explore how the transience in water input and response of the subglacial drainage system affect transport stage and erosion rate. Modelling subglacial water flow through coupled distributed and channelized drainage systems has been described extensively in recent literature (see Table 3 in Flowers, 2015), so we omit a discussion of the drainage system itself and focus instead on how subglacial hydrology affects transport stage and erosion.

Since the sediment supply is fixed, transport stage (Fig. 5a) and relative sediment supply (Fig. 5b) are anti-correlated. The time transgression in transport stage and relative sediment supply is a result of the up-glacier incision of Rchannels. Once a channel is well developed, water pressure decreases in the channel, as does water velocity and transport stage (Fig. 5a). Daily fluctuations are only detectable close to the terminus, where a channel is relatively well developed. Similar to what we find in the steady-state simulations, the largest transport stages are found a few kilometres up-glacier from the terminus. The bed remains shielded over the first 25 km from the ice divide. The time window during which erosion occurs ($q_s/q_{tc} \lesssim 1 - 1.5$) is longest at km 45 and decreases up- and down-glacier. The size of the R-channel at the terminus (km 50) after day 80 becomes large enough that the transport stage (Fig. 5a) is insufficient to maintain an exposed bed. Further upstream from km 45, the hydraulic potential gradient shallows and the discharge decreases so that the bed is exposed for a shorter time.

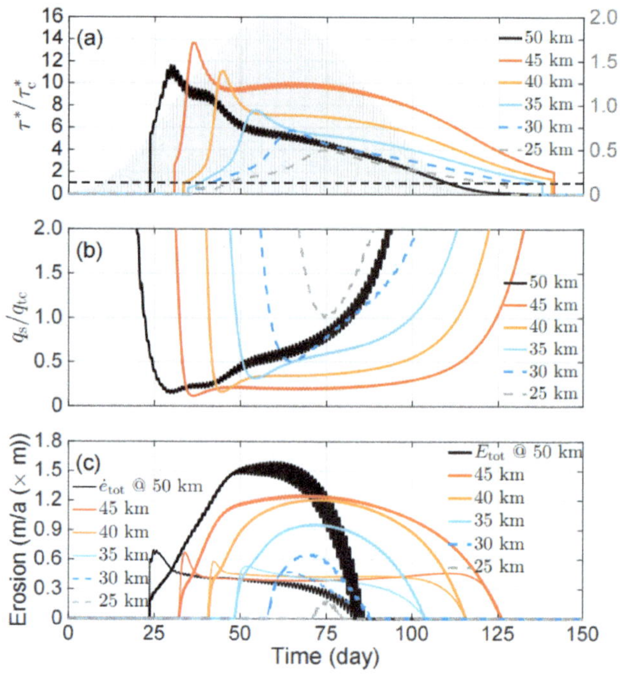

Figure 5. Time series at six different distances from the divide (km 50, 45, 40, 35, 30 and 25) for simulation T_REF (Table 4). (a) Transport stage ($\tau*/\tau*_c$) and normalized meltwater input ($\dot{b}_{ca}/\dot{b}_{ss\,max}$) at the terminus (light grey). The dashed line represents the threshold for sediment motion: $\tau*/\tau*_c = 1$. (b) Relative sediment supply q_s/q_{tc}. (c) Erosion per unit width $\dot{e}_{tot}$ ($m\ a^{-1}$, thin lines) and total erosion E_{tot} ($m^2\ a^{-1}$, thick lines) computed with the TLEM.

Erosion per unit width ($\dot{e}_{tot}$, Fig. 5c) peaks when the relative sediment supply satisfies $0.25 \leq q_s/q_{tc} \leq 0.4$, i.e. at the onset of R-channel formation, even before the peak in transport stage for the lowermost 15 km (km 35–50 in Fig. 5). The peak in erosion is followed by relatively constant values and eventually an abrupt drop. At the terminus, erosion ceases due to low shear stress in the relatively large channel, while upstream erosion ceases due to declining water supply. Note that erosion can occur after the melt season ends (day 120) at km 45 because water remains stored englacially and subglacially (Eq. 3). Further up-glacier (km 25–30) the maximum erosion per unit width coincides with the minimum in relative sediment supply, as the latter remains larger than 0.5 and enough tools are available close to the bed.

Patterns of total erosion (E_{tot}, Fig. 5c, thick lines) all peak between day 65 and 75, except at the terminus (km 50). The initial peak in erosion per unit width occurs while the channel is small, and thus total erosion is largely controlled by channel size over most of the record. Simulations with transient meltwater input highlight the role of channel size in controlling transport stages and erosion close to the terminus.

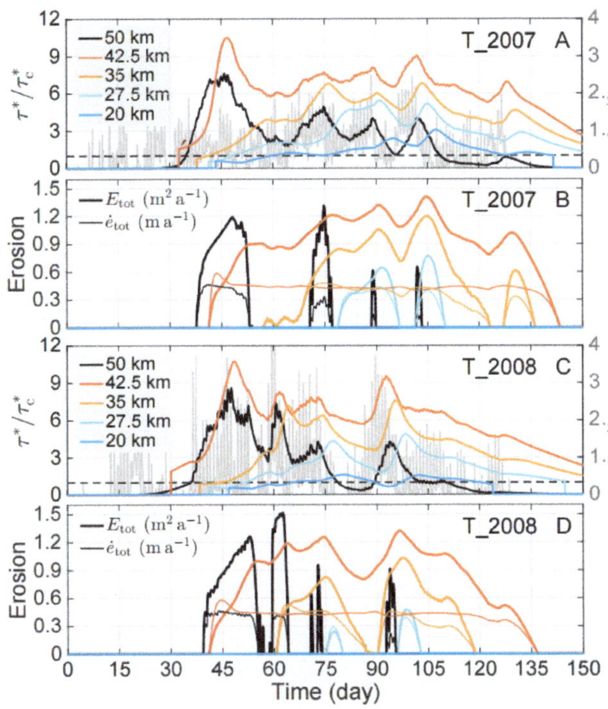

Figure 6. Time series of transport stage and erosion rates for two realistic water input time series (Table 4, T_2007 and T_2008) at five different distances from the divide (km 20, 27.5, 35, 42.5, and 50). (a) Transport stage ($\tau*/\tau*_c$) and normalized meltwater input $(\dot{b}_{ca}/\dot{b}_{ss\,max})$ at the terminus (light grey) for simulation T_2007. (b) Erosion rate per unit width $\dot{e}_{tot}$ (thin lines) and total erosion E_{tot} (thick lines) for simulation T_2007. (c) Transport stage ($\tau*/\tau*_c$) and normalized meltwater input $(\dot{b}_{ca}/\dot{b}_{ss\,max})$ at the terminus (light grey) for simulation T_2008. (d) Erosion rate per unit width $\dot{e}_{tot}$ (thin lines) and total erosion E_{tot} (thick lines) for simulation T_2008. The dashed lines in (a) and (c) represent the threshold for sediment motion: $\tau*/\tau*_c = 1$.

Surface 004Delt

We vary the amount of water reaching the bed to explore the differences between using synthetic and realistic melt records. The realistic melt records come from the ablation area of an unnamed glacier in the Saint Elias Mountains, Yukon, Canada, in 2007 and 2008 (Wheler et al., 2014). We scale the 2007 melt record (T_2007, Table 4) so that the total volume of water is identical to the synthetic input. The melt time series from 2008 (T_2008, Table 4) is then scaled such that the ratio of 2007 to 2008 melt volumes is preserved. This test is intended to highlight the importance of total melt volume and the temporal structure of meltwater input.

When we apply the realistic forcing from 2007 (T_2007), the transport stage exhibits four to five peaks (Fig. 6a) at km 35, 42.5, and 50. At these three locations, the first peak occurs once enough water is supplied to the bed to form a channel (Fig. 6a; after 45 days at 42.5 and 50 km and after 75 days at 35 km). The subsequent peaks in transport stage (after day 60 at 42.5 and 50 km and after day 80 at 35 km) follow periods of high melt. After day 45, the transport stage remains highest at 42.5 km because creep closure prevents R-channels from becoming too large and water discharge is high enough to maintain high velocities. At km 20 and 27.5 it takes about 60 days for the first peak in transport stage to occur. The subsequent peaks at these two locations (around day 90, 105, and 129) lag high melt periods even further.

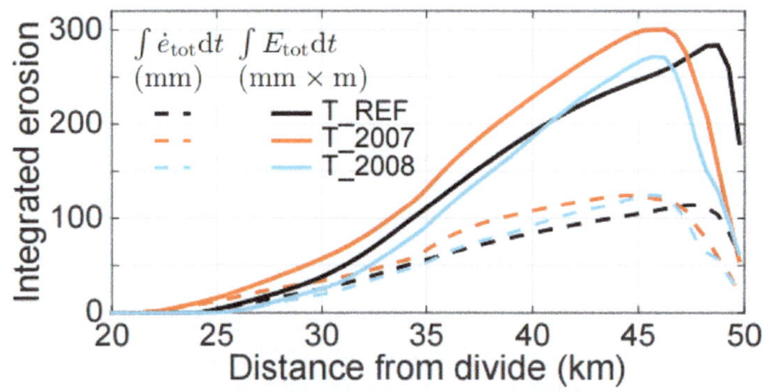

Figure 7. Comparison of time-integrated erosion per unit width $(\int \dot{e}_{tot} dt)$ and total erosion $(\int E_{tot} dt)$ for different water inputs. The first 20 km of the profile are not shown because the bed is alluviated and erosion rates are negligible.

Given the prescribed sediment supply rate, the bed is only exposed at transport stages larger than 3.5. Erosion (Fig. 6b; $\dot{e}_{tot}$ and E_{tot}) at km 27.5 and 50 therefore only occurs during peaks in transport stage; at km 20 the bed is always covered. Erosion rate per unit width ($\dot{e}_{tot}$, Fig. 6b) plateaus at moderate transport stages, thus it remains relatively constant once the bed is partially exposed. On the other hand, total erosion (E_{tot}, Fig. 6b) peaks with transport stage (Fig. 6a). Similar results are obtained for T_2008 (Fig. 6c and 6d), where a different melt time series is employed. While the amplitudes of the fluctuations of meltwater input are a few times larger in T_2008 than T_2007 (Fig. 6a and 6c), the total melt in T_2008 is about 80 % that of T_2007.

These realistic meltwater forcings produce episodic variations in transport stage and erosion rate that suggest multi-day fluctuations in meltwater input are important. These multi-day variations in water input also lead to a succession of channel enlargement events (Fig. 6a and 6c), represented by multiple peaks in transport stage ($\tau*/\tau*c$). On the timescale of several days, creep closure near the terminus is low enough that a channel is sustained between the melt events, leading to an up-glacier migration of relatively large transport stages and integrated erosion. Thus, the pressure in the channel close to the terminus, and hence transport stage, is low. This results in the integrated total erosion $(\int E_{tot} dt)$ being about 3 times lower for realistic inputs (Fig. 7, T_2007 and T_2008) than for the synthetic one (Fig. 7, T_REF) at the terminus. For the same total water input (T_REF and T_2007), the realistic melt season produces more erosion averaged over the glacier bed than the synthetic input (8×10^{-2} mm for T_2007 vs. ~ 7×10^{-2} mm for T_REF).

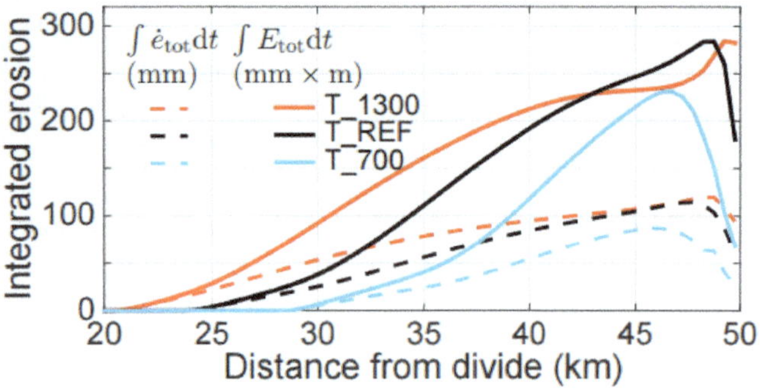

Figure 8. Comparison of time-integrated erosion per unit width ($\int \dot{e}_{tot}dt$) and total erosion ($\int E_{tot}dt$) for different ice geometries. The first 20 km of the profile are not shown because the bed is alluviated and erosion rates are negligible

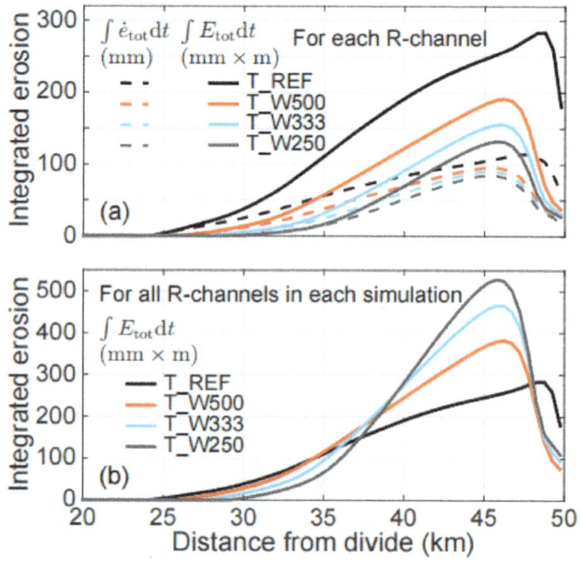

Figure 9. Influence of drainage catchment width (W) on timeintegrated erosion. (a) Time-integrated erosion per unit width ($\int \dot{e}_{tot}dt$) and total erosion ($\int E_{tot}dt$) for an individual R-channel. (b) Time-integrated total erosion ($\int E_{tot}dt$) summed over all Rchannels in each simulation. The first 20 km of the profile are not shown because the bed is alluviated and erosion rates are negligible.

Ice Geometry

Studies of sediment yield from glacierized catchments (e.g. Hallet et al., 1996; Koppes and Hallet, 2002, 2006; Koppes and Montgomery, 2009) conclude that glaciers are more erosive during retreat than during advance due to the amount of meltwater production. Glacier thinning (or thickening) during a phase of retreat (or advance) will also impact the development of the subglacial drainage system and hence its ability to flush sediments and erode the bed. In this series of model tests we hold the sediment supply fixed and vary the glacier geometry by changing the maximum ice thickness (Fig. 1, T_REF, T_1300, T_700, Table 4), while the water input remains the same. As we have already described the principal mechanisms responsible for fluctuations in erosion rates in previous sections, we now focus on annually integrated erosion.

For all ice geometries tested in Fig. 8 (T_1300, T_700 and T_REF, Table 4), significant erosion only occurs downglacier of km 20. The thicker the ice, the further up-glacier significant erosion (both ($\int \dot{e}_{tot}dt$) and ($\int E_{tot}dt$)) occurs (up to km 21 for T_1300 and km 29 for T_700). In these tests, thicker ice also means steeper surface slopes (Fig. 1). Since water input is identical for these simulations, steeper surface slopes lead to faster water flow and the possibility of initiating sediment motion further up glacier. At the terminus almost 4 times as much erosion occurs for T_1300 than T_700 because thick ice prevents the growth of a large channel.

Subglacial Drainage Catchment Width of a Channel

Hydraulic properties of the distributed drainage system determine the density of channels that form (e.g. Werder et al., 2013). The smaller the channel spacing, the lower the discharge in a single channel. A smaller channel, at equilibrium, yields larger water pressures, so we expect that more water would be evacuated through the cavity network. In this test we fix the total glacier width at 1000 m and allow two, three, or four channels to form such that channel catchment widths (W) are, respectively, 500, 333, and 250 m (T_W500, T_W333, T_W250; Table 4).

If we consider the erosion in a single R-channel per simulation (Fig. 9a), the smaller the drainage catchment width, the smaller the discharge, and the smaller the time-integrated erosion (Fig. 9a; ($\int \dot{e}_{tot}dt$) and ($\int E_{tot}dt$)). The feedback causing erosion rate per unit width to decrease at large transport stages (see Fig. 2) is such that even for the simulations where the drainage catchment width is relatively small, erosion rates are comparable (($\int \dot{e}_{tot}dt$), Fig. 9a) despite the lower transport stages. The differences are, however, relatively large for the annually integrated total erosion (maximum ($\int E_{tot}dt$) for T_REF is more than twice that of T_W250, Fig. 9a) because of the effect of channel size.

The hierarchy in total integrated erosion is inverted when all R-channels within a fixed glacier width are accounted for (Fig. 9b, between km ~ 37 and ~ 48). Once the number of R-channels present is taken into account, integrated total erosion is largest for the smallest channel catchment (almost twice as large for T_W250 than T_REF, Fig. 9b). In this case, numerous small channels therefore produce more erosion than few large ones.

At the terminus (km 50), the simulations with a catchment width per channel smaller than that in T_REF show values of annually integrated erosion ($\int E_{tot}dt$) of about half that of the reference simulation (Fig. 9b). The relatively smaller Rchannels in these simulations remain more pressurized and thus drain less water from the cavity network (Eq. 1); the relative discharge in the cavity network near the terminus is therefore larger than in the reference simulation (T_REF), further diminishing transport stage near the terminus (see Figs. 2–3).

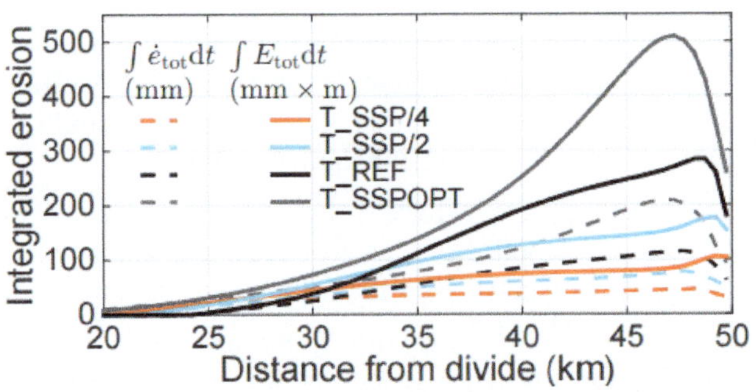

Figure 10. Time-integrated erosion per unit width (R e˙totdt) and time-integrated total erosion ($\int E_{tot}dt$) as a function of sediment supply rate (q_s) (Table 4). The first 20 km of the profile are not shown because the bed is alluviated and erosion rates are negligible.

Sediment Supply

In the present model the values and patterns of sediment supply are amongst the key unknowns. Most till is produced subglacially (e.g. Sanders et al., 2013) and the amount and size distribution of till depends on the history and patterns of production (quarrying) and comminution (abrasion). We test the sensitivity of erosion rates and patterns to different values of input sediment supply. In two simulations (T_SSP/2 and T_SSP/4; Table 4) the sediment supply rate per unit width is constant in space and time and is taken as a fraction of the reference supply rate in T_REF (Table 4). The largest erosion rate in the SEM occurs when the relative sediment supply is $q_s/q_{tc} = 0.5$. For the TLEM and transport stages $\tau* /\tau*_c < 100$, the maximum erosion rate is obtained for a relative sediment supply of $0.5 \le q_s/q_{tc} < 0.8$ (Lamb et al., 2008). We determine a ratio q_s/q_{tc} close to optimum and examine the resulting erosion rates and patterns (T_SSPOPT; Table 4). This provides us an upper bound on subglacial meltwater erosion rates. The hydraulic conditions in this suite of simulations are that of T_REF (Fig. 5).

Decreasing the sediment supply leads to a decrease in the maximum integrated erosion (Fig. 10, T_SSP/2 and T_SSP/4; ($\int \dot{e}_{tot}dt$) and ($\int E_{tot}dt$)) and to the bed being eroded further up-glacier. For a relatively low sediment supply (T_SSP/4), the peak in annually integrated erosion per unit width (($\int \dot{e}_{tot}dt$), Fig. 10) is hardly discernible and the peak in annually integrated total erosion (($\int E_{tot}dt$), Fig. 10) is controlled by channel size. In order to estimate the maximum erosion that can occur under the given hydraulic conditions and sediment size, we optimize the sediment supply rate by expressing it as a function of the transport capacity ($q_s/q_{tc} \approx 0.6$). The resulting patterns of annually integrated erosion (($\int \dot{e}_{tot}dt$) and ($\int E_{tot}dt$), Fig. 10, T_SSPOPT) mimic the transport stage patterns (see $\tau* /\tau*_c$, Fig, 5a), and peak at nearly twice the values of T_REF.

DISCUSSION

Significance of Model Simplifications

We have detailed the simplifications and underlying assumptions of the model while describing the model and the strategy; we therefore focus on the potential implications of the most important simplifications. At the onset of the melt sea-son, sliding is expected to accelerate as a response to increased water supply to a distributed drainage system (e.g. Iken, 1981; Hooke et al., 1989; Mair et al., 2003; Anderson et al., 2004; Sole et al., 2011; Meierbachtol et al., 2013; Hewitt, 2013; Hoffman and Price, 2014) which would promote cavity enlargement and water flow through the distributed rather than the incipient channelized drainage system. This sliding feedback alone could produce a small decrease in water pressure (e.g. Hoffman and Price, 2014) and hence a decrease in transport stage.

In this study, we treat only the case of bedrock erosion by abrasion and we neglect the effect of quarrying. Although the latter can lead to erosion rates up to an order of magnitude larger than abrasion, it requires that the bedrock be highly jointed (Whipple et al., 2000, 2013). Quarrying is a two-step process: (1) loosening of blocks around pre-existing cracks (or possibly opening of new cracks) and (2) mobilization and transport of loose blocks (Whipple et al., 2000, 2013; Chatanantavet and Parker, 2009; Dubinski and Wohl, 2013; Lamb et al., 2015). The depth of loose cracks could be related to sediment availability (Chatanantavet and Parker, 2009) and mobilization and transport of quarried blocks scale with the transport stage (Dubinski and Wohl, 2013; Lamb et al., 2015). Therefore we expect that the patterns of quarrying would be similar to the transport stage, yet limited by the thickness of the loosened layer.

We compute erosion with only a single particle size that is assumed to be the median of size of the sediment mixture (e.g. Sklar and Dietrich, 2004, 2006, 2008; Turowski et al., 2007; Lamb et al., 2008; Chatanantavet and Parker, 2009; Nelson and Seminara, 2011). The SEM (Sklar and Dietrich, 2004) was generalized for a grain size distribution by Egholm et al. (2013), a study in which they, however, omit a discussion of the implications of the generalization of the SEM. As for the TLEM, Lamb et al. (2008) suggest that a generalization to grain size distribution would require re-evaluation of some of the equations to account for the interactions be-tween particles of different sizes within the bedload layer. A decrease in median sediment size would probably result in an erosion profile more spread out along the bed and an in-crease in median sediment size would result in a localization of erosion (see Supplement). The changes in erosion would be quantitatively similar to a decrease in sediment supply (q_s) and thus a decrease in relative sediment supply (q_s=q_{tc}), which strongly controls erosion patterns.

We make the assumption of a supply-limited glacier bed and hence neglect the effect of sediment transport and the interactions between sediment thickness and water flow. The mobilization and particularly deposition of sediment affect the flow regime by enlarging or reducing the cross section of flow (Creyts et al., 2013). On a timescale of days, when sediment is mobilized, the cross section of flow is enlarged and could result in a drop in channel water pressure and a corresponding loss of flow strength. The opposite effect, leading to flow strengthening, could occur when sediment is deposited. We do not treat the case of transport-limited conditions, where the channelized drainage system would more closely resemble canals (Walder and Fowler, 1994; Ng, 2000; Kyrke-Smith and Fowler, 2014). To implement sediment transport adequately, it is necessary to improve existing models for subglacial water flow through canals (Walder and Fowler, 1994; Ng, 2000; Kyrke-Smith and Fowler, 2014) with time-evolving effective pressure. Alley et al. (1997), however, argue that in the case of subglacial water flow through canals, the water pressure remains relatively high and very little water would be drained from the distributed system, limiting the capacity of canals to transport sediment. In contrast, the steady-state water pressure in an R-channel decreases with increasing discharge, favoring water flow in the channelized system and enhancing transport and erosion.

Accounting for the production of sediment and the evolution of particle diameter at the glacier bed would also largely influence sediment supply patterns. For example, we can speculate that if the sediment sources are localized in areas of more easily eroded bedrock (e.g. Dühnforth et al., 2010), tools would only be present downstream from these areas. If, instead of fixing the sediment supply per unit width, we fix the total sediment supply (simulation not shown), tools are less available at peak flows, reducing erosion, whereas the cover effect is enhanced for a relatively small channel. We also tested a simple power-law downstream fining function (e.g. Sklar and Dietrich, 2006; simulation not shown). The results were very similar to those obtained with a decrease in relative sediment supply, because particles were smaller than the reference size of 60 mm in the region of the bed where channels form. Another means of obtaining insight into sediment supply rates and patterns would be through the use of a comprehensive model of glacial erosion, i.e. a model encompassing transient subglacial hydrology (between distributed and channelized systems), ice dynamics, glacial abrasion, and quarrying. Such a model is, however, yet to be developed as patterns of glacial erosion remain poorly understood (see Beaud et al., 2014). Finally, subglacial water flow evacuates a significant volume of sediment despite the small area over which R-channels operate and the tendency of these channels to remain stably positioned in association with moulins (e.g. Gulley et al., 2012). The mechanism by which large volumes of sediment are delivered to the channels remains elusive. More work is therefore required to quantify subglacial sediment production patterns.

What are the major controls on subglacial meltwater erosion?

We rank the transient simulations by glacier-area-averaged erosion rate in Fig. 11a. Because we prescribe water input rates sufficient to form a channelized drainage system, it stands out from the model formulation that sediment supply is the most important parameter. A lack or overabundance of tools inhibits erosion. In our results this is shown by the fact that T_SSPOPT (sediment supply optimized for erosion; Table 4) produces the most erosion and T_SSP/4 (smallest sediment supply; Table 4) the least. Changing the ice geometry also leads to a relatively large range of averaged erosion rates as T_1300 (thick ice; Table 4) yields twice as much erosion as T_700 (thin ice; Table 4; Fig. 11a). Larger hydraulic potential gradients in T_1300 cause the shear stress to be large enough over larger portions of the bed to create erosion (Fig. 9). Subglacial drainage catchment width, within the range tested, plays a lesser role than sediment supply or ice geometry although the averaged erosion rate in T_W250 (four channels; Table 4) is 30 % more than that of T_REF. The fact that T_2007 (realistic melt season from 2007 record; Table 4) produces more averaged erosion than T_REF suggests that an increase in the multi-day variability of the water input enhances erosion.

The relations are different for apparent erosion rate (Fig. 11b), here defined as the equivalent thickness of bed material evacuated by the integrated sediment flux ($\int \min(q_s W_{ch}, Q_{tc}) dt$) at the terminus (km 50). The apparent erosion rate corresponds to the quantity estimated by studies of sediment yield in proglacial channels, lake, or fjords. Relatively large hydraulic potential gradients and relatively thick ice close to the terminus, both of which inhibit R-channel growth, compete against the loss of trans-port capacity. Therefore the largest apparent erosion occurs for the thickest ice (T_1300, Fig. 3). Interestingly, the lowest drainage density (T_REF) yields more apparent erosion than the highest (T_W250). Discharge through the cavity network close to the terminus increases with R-channel density; the smaller the channel, the larger the water pressure and the lower the pressure gradient between the two systems. This feedback, in addition to the discharge in the R-channel being smaller due to the R-channel drainage catchment size, reduces the transport stage close to the terminus.

The results in Fig. 11b suggest that, despite the increase in apparent erosion that accompanies an increase in meltwater input (e.g. the total melt in T_2007 is about 1.25 times that of T_2008), the thinning associated with the retreat of an ice mass would have a competing effect by decreasing the hydraulic potential gradient (see Fig. 3). The flushing power of subglacial water flow is conducive to the removal of sub-glacial sediment enabling glacial abrasion and quarrying to be efficient. Our results suggest that the subglacial drainage conditions most favorable for glacial erosion occur where significant surface melt and relatively steep surface slopes occur simultaneously, i.e. during an ice sheet maximum advance or during early phases of retreat. This corroborates the hypothesis of Jørgensen and Sandersen (2006) that some Danish tunnel valleys were excavated during the stagnation of the Scandinavian ice sheet. However, these findings challenge the hypothesis that glaciers deliver more sediment to proglacial areas during retreat than during advance (e.g. Hal-let et al., 1996; Koppes and Hallet, 2002, 2006; Koppes and Montgomery, 2009), yet more work is required to explore this hypothesis. The lack of flow strength in the upper reaches of the glacier (upstream from km 20 for most simulations) suggests that subglacial sediment in the accumulation area is transported almost solely by entrainment due to sliding at the ice–bed interface.

In the steady-state simulations we find that significant erosion can occur in a network of cavities (see Supplement). In the transient simulations, however, the coupling with R-channels prevents large shear stresses from developing in the distributed drainage system, and the threshold for sediment motion is not even reached for particles of 1 mm diameter. We thus argue that bedrock erosion in the distributed drainage system is limited unless specific conditions are satisfied, for example a subglacial flood or a surge.

How important is subglacial meltwater erosion compared to overall glacial erosion?

In most literature on modelling landscape evolution by glacial erosion it is assumed that subglacial meltwater efficiently removes sediment from the glacier bed, while its effect on bedrock erosion is neglected (e.g. MacGregor et al., 2000, 2009; Brocklehurst and Whipple, 2002; Tomkin and Braun, 2002; Anderson et al., 2006; Herman and Braun, 2008; Egholm et al., 2009, 2011a, b; Herman et al., 2011). On the other hand, in formerly glaciated landscapes, erosional features like tunnel valleys (e.g. Glasser and Bennett, 2004; Denton and Sugden, 2005; Dürst Stucki et al., 2010, 2012; Kehew et al., 2012; Jansen et al., 2014) indicate that sub-glacial water can produce significant bedrock erosion. The results we obtain with our simple ice geometries and water input forcing indicate that the areally averaged bedrock erosion produced by subglacial water flow is on the order of 10 $^{-1}$ 10 $^{-2}$ mm a $^{-1}$ (Fig. 11), while glacial erosion rates are most often on the order of 1 10 mm a $^{-1}$ (e.g. Gurnell et al., 1996; Hallet et al., 1996; Koppes and Montgomery, 2009; Riihimaki et al., 2005). Bedrock erosion by abrasion from sediment-bearing subglacial water appears negligible compared to reported erosion rates in proglacial areas. Our results corroborate the assumption that subglacial meltwater efficiently removes sediment from the bed and we postulate that this flushing action is necessary for glacial abrasion and quarrying to access an exposed bed and remain efficient.

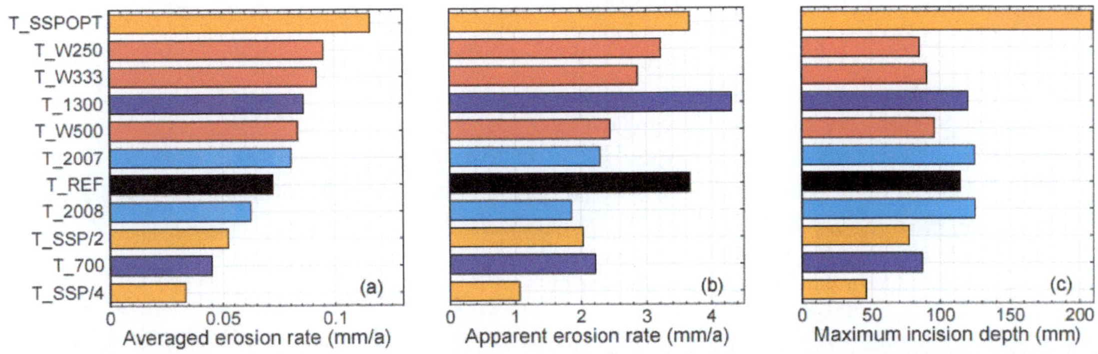

Figure 11. Synthesis of transient simulations (Table 4) through comparison of the following quantities calculated for one model year. (a) Erosion rate averaged over the whole glacier bed $(\int\int E_{tot} dt dx)$. (b) Apparent erosion rate calculated as the volume of sediment that is transported across the last grid node, i.e. terminus, $(\int \min(q_s W_{ch}, Q_{tc}) dt)$ averaged over the glacier area. This quantity corresponds to what one would measure as the sediment flux in a proglacial stream. (c) Maximum incision depth (max $(\int \dot{e}_{tot} dt)$). Simulations are ranked by averaged erosion rate and the colors represent different simulation suites: black for "reference", blue for "water input", purple for "ice geometry", red for "drainage width", and orange for "sediment supply" (Table 4).

Can ordinary seasonal melt processes lead to subglacial bedrock channel incision?

We find maximum modelled vertical bedrock incision ranging from ~50 to ~200 mm a $^{-1}$ (Fig. 11c). Assuming that over a period of 20 years climate is relatively steady and the bedrock does not change significantly, the location of moulins would remain relatively fixed laterally and so would the channel paths (Gulley et al., 2012). Using the lowest incision rate (T_SSP/4), an N-channel almost a meter deep and a few meters wide could be carved near an ice sheet margin in 20 years. A similar N-channel would be carved in only five years assuming the largest incision rate (T_SSPOPT).

Landforms created by former continental ice sheets indicate that subglacial waterways can occupy persistent paths throughout a deglaciation. Eskers deposited by the retreating Laurentide ice sheet can be traced for up to several hundred kilometers and show a dendritic pattern almost as far upstream as the former divide (e.g. Storrar et al., 2014). Some tunnel valleys show several cut-and-fill structures suggesting different excavation events; moreover, tunnel valleys carved during different glaciations tend to follow the same paths (e.g. Jørgensen and Sandersen, 2006). Eskers also commonly lie inside tunnel valleys (e.g. Jørgensen and Sandersen, 2006; Kehew et al., 2012). Assuming an incision rate of 100 mm a $^{-1}$ (e.g. Fig. 7), a simple volume calculation suggests that it would take about 15 000 years to carve a 30 m deep and 100 m wide V-shaped tunnel valley, similar to the dimensions of tunnel valleys observed in Ireland (Knight, 2003).

In the context of an alpine glacier, valley geometry tends to focus subglacial water flow paths toward the thalweg. Assuming that a glacier occupies topography strongly imprinted by fluvial processes, erosion by subglacial water flow may tend to preserve if not enhance the pre-existing fluvial features along the valley centerline. For an alpine glacier eroding its bed at a pace of 2 mm a $^{-1}$ (e.g. Hallet et al., 1996; Riihimaki et al., 2005; Koppes and Montgomery, 2009), in the case of simulation T_2008 (Table 3, apparent erosion of ~ 2 mm a^{-1}; ice geometry comparable to that of a large valley glacier), the maximum incision depth in one year is ~ 125 mm (Fig. 11c). The relief of a canyon with the maximum width of the N-channel (~ 4.5 m, for T_2008) would increase by more than ~ 120 mm a^{-1} (rate of vertical bedrock incision minus rate of surrounding glacial erosion). If the canyon were 5 times as wide (~ 22.5 m), the maximum rate of relief increase would still be ~ 24 mm a^{-1}, about twice the measured incision rates in a meters-wide gorge in the French Western Alps (e.g. Valla et al., 2010), highlighting the erosional power of localized subglacial meltwater action.

What are the implications of the water flow regime in channels for hysteresis and sediment transport?

We calculate the direction of daily hysteresis between modelled transport stage (Fig. 12a) and water discharge (Fig. 12b) at four locations within the last 10 km of the glacier pro-file (Fig. 12c) for simulation T_2008 (Table 4; Fig 6). Over-all, hysteresis is dominated by clockwise events, with anti-clockwise events only occurring during the second half of the melt season. Clockwise hysteresis correlates well with the rising limb of multi-day water discharge and transport stage peaks, while anticlockwise hysteresis correlates with the falling limb, particularly at km 46 (Fig. 12). During the rising limb of a multi-day melt event, changes in channel size are dominated by enlargement; the pressure in the channel therefore peaks before the discharge, as do the averaged water flow velocity and transport stage. During the falling limb of the melt event, if the channelized drainage is relatively well established, changes in channel size are dominated by closure, and the pressure peak can occur after the peak in dis-charge. In the case of a proglacial stream carrying a sediment load smaller than its transport capacity, peaks in transport stage would act as mobilizing events propagating sediment pulses downstream. We therefore surmise that the direction of hysteresis in sediment transport and discharge is not necessarily linked to changes in sediment supply conditions or the tapping of new sediment sources, but may be the result of changes in subglacial sediment mobilization in the vicinity of the glacier terminus.

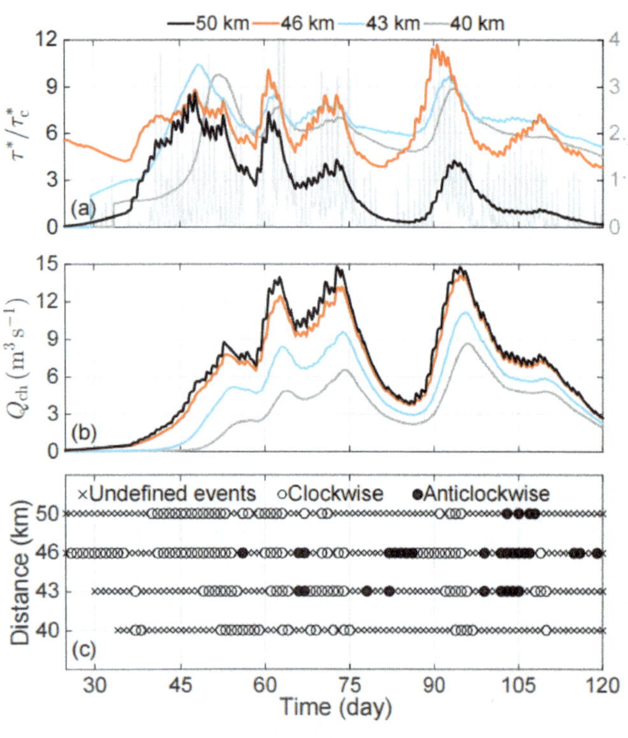

Figure 12. Hysteresis between water discharge and transport stage for simulation T_2008 (see Table 4 and Fig. 6). Time series at four locations (km 40, 43, 46, and 50) distributed over the last 10 km of the glacier profile of (a) transport stage (τ^*/τ_c^*) and normalized water input $(\dot{b}_{ca}/\dot{b}_{ss\,max})$ at the terminus (light grey); (b) discharge in the channel, Q_{ch}; and (c) calculated direction of the daily hysteresis when transport stage is plotted against water discharge. The hysteresis is clockwise when transport stage peaks before discharge over a daily cycle. Undefined events represent days where the fluctuations in transport stage or discharge are either simultaneous or not strong enough to produce hysteresis.

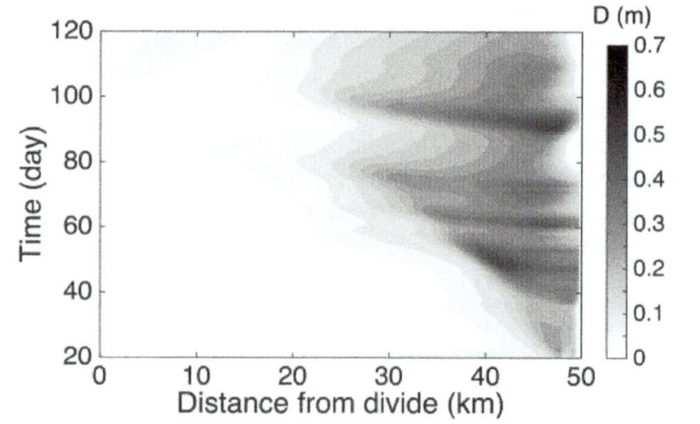

Figure 13. Maximum particle diameter for which movement would be initiated in simulation T_2008 assuming $\tau_c^* = 0.03$ (see Table 4 and Fig. 6).

In coarse-bedded streams, grain hiding has a significant effect on sediment transport (e.g. Yager et al., 2012; Scheingross et al., 2013), as mobile grains can be trapped behind larger immobile particles. We calculate the maximum particle diameter for which movement would be initiated in simulation T_2008 (Fig. 13) and find that boulders of up to 70 cm in diameter can be transported within the last 10 km of the profile and would correspond to flood-like conditions in rivers. The analogy to river systems might have influenced the interpretation of glacial deposits such as eskers, where the presence of boulders or lack of fines is often used to in-fer emplacement during flood events (e.g. Brennand, 1994; Burke et al., 2012).

For simulation T_2008, we find that transport stage exhibits a sharp decrease close to the terminus (Figs. 5a, 6a and 6c) which leads to a correspondingly sharp decrease in the size of particles transported (Fig. 13) and could lead to a bottleneck in sediment transport. This bottleneck effect could lead to the deposition of sediment, filling the channel to-ward the end of the melt season. A similar process, but operational over a longer timescale, would be consistent with time-transgressive deposition of eskers near the mouths of R-channels beneath retreating ice margins (e.g. Brennand, 1994; Burke et al., 2012).

CONCLUSIONS

This study is the first attempt to quantify bedrock erosion rates by transient subglacial water flow with a numerical model. We implement a 1-D model of subglacial drainage in which a network of cavities and R-channels interact. We compute the shear stress exerted on the bed and the resulting bedrock erosion by abrasion (saltation erosion, after Sklar and Dietrich, 2004, and total load erosion after Lamb et al., 2008). Because of the large calculated transport stage we argue that, in the case of subglacial meltwater erosion, it is probably more appropriate to use the TLEM than the SEM. Assuming that a significant amount of meltwater is produced and reaches the bed, the main drivers of subglacial water erosion that we isolate are the rate of sediment supply, particularly the relative sediment supply, and ice geometry.

From this exercise, we conclude the following:

- Bedrock erosion and transport stage in the subglacial drainage system do not scale directly with water dis-charge. Instead, transport stage and erosion are related to the hydraulic potential gradient and hence a combination of water discharge, ice-surface slope, and channel (or cavity) cross-sectional area. In our simulations, this combination of discharge, slope, and channel cross-sectional area leads to a drop in transport stage close to the terminus as water pressure approaches atmospheric.
- Erosion rates due to the action of subglacial water flow averaged over the whole glacier bed are negligible com-pared to the rates of glacial erosion necessary to produce the sediment supply rates we impose.
- In our transient simulations, a bedrock channel a few to several decimeters in depth could be carved over a single melt season as erosion is concentrated at the base of R-channels.
- The vertical incision rates we calculate are a few to several times larger than published rates of fluvial incision in gorges. Therefore, this mechanism may explain the gradual excavation of tunnel valleys in bedrock and the preservation or even initiation of inner gorges.

Though we have demonstrated the potential for subglacial water flow to incise bedrock on seasonal timescales, site-specific and quantitative assessments of its importance will require more realistic 2-D hydrology models (e.g. Hewitt, 2013; Werder et al., 2013) and simulations over timescales of glacial advance and retreat.

ACKNOWLEDGEMENTS

We are grateful to the Swiss National Cooperative for the Disposal of Radioactive Waste (Nagra) for providing funding within the framework of investigating ice-age effects related to the long-term safety of radioactive waste disposal. We are particularly grateful to Urs Fischer for overseeing this project and we would like to thank Michelle Koppes for insightful discussions. We would also like to thank John D. Jansen; the anonymous reviewer; and the handling associate editor, Arjen Stroeven, for their comments. Funding was also provided by the Natural Sciences and Engineering Research Council of Canada (NSERC), the Canada Research Chairs Program, and Simon Fraser University (SFU).

REFERENCES

1. Alley, R. B., Cuffey, K. M., Evenson, E. B., Strasser, J. C., Lawson, D. E., and Larson, G. J.: How glaciers entrain and transport basal sediment: Physical constraints, Quaternary Sci. Rev., 16, 1017– 1038, 1997.
2. Anderson, R. S. and Anderson, S. P.: Geomorphology: the mechanics and chemistry of landscapes, Cambridge University Press, Cambridge, UK, 212–269, 2010.
3. Anderson, R. S., Anderson, S. P., MacGregor, K. R., Waddington, E. D., O'Neel, S., Riihimaki, C. A., and Loso, M. G.: Strong feedbacks between hydrology and sliding of a small alpine glacier, J. Geophys. Res., 109, F03005, doi:10.1029/2004JF000120, 2004.
4. Anderson, R. S., Molnar, P., and Kessler, M. A.: Features of glacial valley profiles simply explained, J. Geophys. Res., 111, F01004, doi:10.1029/2005JF000344, 2006.
5. Bartholomaus, T. C., Anderson, R. S., and Anderson, S. P.: Growth and collapse of the distributed subglacial hydrologic system of Kennicott Glacier, Alaska, USA, and its effects on basal motion, J. Glaciol., 57, 985–1002, 2011.
6. Baynes, E., Attal, M., Niedermann, S., Kirstein, L. A., Dug-more, A. J., and Naylor, M.: Erosion during extreme flood events dominates Holocene canyon evolution in north-east Iceland, P. Natl. Acad. Sci. USA, 112, 2255–2360, doi:10.1073/pnas.1415443112, 2015.
7. Beaud, F., Flowers, G. E., and Pimentel, S.: Seasonal-scale abrasion and quarrying patterns from a two-dimensional ice-flow model coupled to distributed and channelized subglacial drainage, Geomorphology, 219, 176–191, 2014.
8. Bennett, M. M. and Glasser, N. F.: Glacial geology: ice sheets and landforms, John Wiley & Sons Ltd, The Atrium, Chichester, West Sussex, UK, second edition, 109–133, 2009.
9. Brennand, T. A.: Macroforms, large bedforms and rhythmic sedimentary sequences in subglacial eskers, south-central Ontario: implications for esker genesis and meltwater regime, Sediment. Geol., 91, 9–55, 1994.
10. Bretz, J. H.: The Lake Missoula floods and the channeled scabland, J. Geol., 77, 505–543, 1969.
11. Brocklehurst, S. H. and Whipple, K. X.: Glacial erosion and relief production in the Eastern Sierra Nevada, California, Geo-morphology, 42, 1–24, 2002.
12. Burke, M., Brennand, T. A., and Perkins, A. J.: Transient sub-glacial hydrology of a thin ice sheet: insights from the Chasm esker, British Columbia, Canada, Quaternary Sci. Rev., 58, 30– 55, 2012.
13. Chatanantavet, P. and Parker, G.: Physically based modeling of bedrock incision by abrasion, plucking, and macroabrasion, J. Geophys. Res., 114, F04018, doi:10.1029/2008JF001044, 2009.
14. Clarke, G. K. C.: Lumped-element analysis of subglacial hydraulic circuits, J. Geophys. Res., 101, 17547–17559, 1996.
15. Clarke, G. K. C.: Hydraulics of subglacial outburst floods: new in-sights from the Spring-Hutter formulation, J. Glaciol., 49, 299– 313, 2003.
16. Cook, K. L., Turowski, J. M., and Hovius, N.: A demonstration of the importance of bedload transport for fluvial bedrock erosion and knickpoint propagation, Earth Surf. Proc. Land., 38, 683– 695, 2013.
17. Creyts, T. T. and Schoof, C. G.: Drainage through sub-glacial water sheets, J. Geophys. Res., 114, F04008, doi:10.1029/2008JF001215, 2009.
18. Creyts, T. T., Clarke, G. K. C., and Church, M.: Evolution of sub-glacial overdeepenings in response to sediment

redistribution and glaciohydraulic supercooling, J. Geophys. Res.-Earth, 118, 1–24, 2013.

19. Denton, G. H. and Sugden, D. E.: Meltwater Features That Suggest Miocene Ice-Sheet Overriding of the Transantarctic Mountains in Victoria Land, Antarctica, Geogr. Ann. A, 87, 67–85, 2005.
20. Dubinski, I. M. and Wohl, E.: Relationships between block quarrying, bed shear stress, and stream power: a physical model of block quarrying of a jointed bedrock channel, Geomorphology, 180, 66–81, doi:10.1016/j.geomorph.2012.09.007, 2013.
21. Dühnforth, M., Anderson, R. S., Ward, D., and Stock, G. M.: Bedrock fracture control of glacial erosion processes and rates, Geology, 38, 423–426, 2010.
22. Dürst Stucki, M., Reber, R., and Schlunegger, F.: Subglacial tunnel valleys in the Alpine foreland: an example from Bern, Switzer-land, Swiss J. Geosci., 103, 363–374, 2010.
23. Dürst Stucki, M., Schlunegger, F., Christener, F., Otto, J.-C., and Götz, J.: Deepening of inner gorges through subglacial meltwater—An example from the UNESCO Entlebuch area, Switzer-land, Geomorphology, 139, 506–517, 2012.
24. Egholm, D. L., Nielsen, S. B., Pedersen, V. K., and Lesemann, J.-E.: Glacial effects limiting mountain height, Nature, 460, 884–888, 2009.
25. Egholm, D. L., Knudsen, M. F., Clark, C. D., and Lesemann, J. E.: Modeling the flow of glaciers in steep terrains: the integrated second-order shallow ice approximation (iSOSIA), J. Geophys. Res., 116, F02012, doi:10.1029/2010JF001900, 2011a.
26. Egholm, D. L., Pedersen, V. K., Knudsen, M. F., and Larsen, N. K.: On the importance of higher order ice dynamics for glacial land-scape evolution, Geomorphology, 141–142, 67–80, 2011b.
27. Egholm, D. L., Pedersen, V. K., Knudsen, M. F., and Larsen, N. K.: Coupling the flow of ice, water, and sediment in a glacial land-scape evolution model, Geomorphology, 141–142, 47–66, 2012.
28. Egholm, D. L., Knudsen, M. F., and Sandiford, M.: Lifespan of mountain ranges scaled by feedbacks between landsliding and erosion by rivers, Nature, 498, 475–478, 2013.
29. Fernandez-Luque, R. and van Beek, R.: Erosion and transport of bed-load sediment, J. Hydraul. Res., 14, 127–144, 1976.
30. Flowers, G. E.: Subglacial modulation of the hydrograph from glacierized basins, Hydrol. Process., 22, 3903–3918, 2008.
31. Flowers, G. E.: Modelling water flow under glaciers and ice sheets, Proc. R. Soc. Lon. Ser.-A, 471, 1–41, doi:10.1098/rspa.2014.0907, 2015.
32. Flowers, G. E., Björnsson, H., Pálsson, F., and Clarke, G. K. C.: A coupled sheet-conduit mechanism for jökulhlaup propagation, Geophys. Res. Lett., 31, L05401, doi:10.1029/2003GL019088, 2004.
33. Glasser, N. F. and Bennett, M. R.: Glacial erosional landforms: origins and significance for palaeoglaciology, Prog. Phys. Geog., 28, 43–75, 2004.
34. Gulley, J. D., Grabiec, M., Martin, J. B., Jania, J., Catania, G., and Glowacki, P.: The effect of discrete recharge by moulins and het-erogeneity in flow-path efficiency at glacier beds on subglacial hydrology, J. Glaciol., 58, 926–940, 2012.
35. Gurnell, A., Hannah, D., and Lawler, D.: Suspended sediment yield from glacier basins, IAHS Publications-Series of Proceedings and Reports-Intern Assoc Hydrological Sciences, 236, 97–104, 1996.
36. Hallet, B., Hunter, L., and Bogen, J.: Rates of erosion and sediment evacuation by glaciers: A review of field data and their implications, Global Planet. Change, 12, 213–235, 1996.
37. Herman, F. and Braun, J.: Evolution of the glacial landscape of the Southern Alps of New Zealand: Insights from a glacial erosion model., J. Geophys. Res.-Earth, 113, F02009, doi:10.1029/2007JF000807, 2008.
38. Herman, F., Beaud, F., Champagnac, J. D., Lemieux, J. M., and Sternai, P.: Glacial hydrology and erosion patterns: A mechanism for carving glacial valleys, Earth Planet Sc. Lett., 310, 498–508, 2011.
39. Hewitt, I. and Fowler, A.: Seasonal waves on glaciers, Hydrol. Process., 22, 3919–3930, 2008.
40. Hewitt, I. J.: Seasonal changes in ice sheet motion due to melt water lubrication, Earth Planet Sc. Lett., 371, 16–25, doi:10.1016/j.epsl.2013.04.022, 2013.
41. Hewitt, I. J., Schoof, C., and Werder, M. A.: Flotation and free sur-face flow in a model for subglacial drainage. Part 2. Channel flow, J. Fluid Mech., 702, 157–187, 2012.
42. Hoffman, M. and Price, S.: Feedbacks between coupled subglacial hydrology and glacier dynamics, J. Geophys. Res.-Earth, 119, 414–436, 2014.
43. Hooke, R. L., Calla, P., Holmlund, P., Nilsson, M., and Stroeven, A.: A 3 year record of seasonal variations in surface velocity, Storglaciären, Sweden, J. Glaciol., 35, 235–247, 1989.
44. Humphrey, N. F. and Raymond, C. F.: Hydrology, erosion and sediment production in a surging glacier: Variegated Glacier, Alaska, 1982-83, J. Glaciol., 40, 539–552, 1994.
45. Iken, A.: The effect of the subglacial water pressure on the sliding velocity of a glacier in an idealized numerical model, J. Glaciol., 27, 407–421, 1981.
46. Iken, A. and Bindschadler, R. A.: Combined measurements of sub-glacial water pressure and surface velocity of Findelengletscher, Switzerland: conclusions about drainage system and sliding mechanism, J. Glaciol., 32, 101–119, 1986.
47. Iverson, N. and Person, M.: Glacier-bed geomorphic processes and hydrologic conditions relevant to nuclear waste disposal, Geoflu-ids, 12, 38–57, 2012.
48. Jansen, J. D., Codilean, A. T., Stroeven, A. P., Fabel, D., Hättes-trand, C., Kleman, J., Harbor, J. M., Heyman, J., Kubik, P. W., and Xu, S.: Inner gorges cut by subglacial meltwater during Fennoscandian ice sheet decay, Nature communications, 5, 3815, doi:10.1038/ncomms4815, 2014.

49. Jørgensen, F. and Sandersen, P. B.: Buried and open tunnel valleys in Denmark—erosion beneath multiple ice sheets, Quaternary Sci. Rev., 25, 1339–1363, 2006.

50. Kamb, B.: Glacier Surge Mechanism Based on Linked Cavity Con-figuration of the Basal Water Conduit System, J. Geophys. Res., 92, 9083–9100, 1987.

51. Kehew, A. E., Piotrowski, J. A., and Jørgensen, F.: Tunnel valleys: Concepts and controversies – A review, Earth-Sci. Rev., 113, 33– 58, 2012.

52. Kessler, M. A. and Anderson, R. S.: Testing a numerical glacial hydrological model using spring speed-up events and outburst floods, Geophys. Res. Lett., 18, L18503, doi:10.1029/2004GL020622, 2004.

53. Kirchner, J. W., Finkel, R. C., Riebe, C. S., Granger, D. E., Clayton, J. L., King, J. G., and Megahan, W. F.: Mountain erosion over 10 yr, 10 k.y., and 10 m.y. time scales, Geology, 29, 591–594, 2001.

54. Knight, J.: Bedform patterns, subglacial meltwater events, and Late Devensian ice sheet dynamics in north-central Ireland, Global Planet. Change, 35, 237–253, 2003.

55. Koppes, M. and Hallet, B.: Erosion rates during rapid deglacia-tion in Icy Bay, Alaska, J. Geophys. Res., 111, F02023, doi:10.1029/2005JF000349, 2006.

56. Koppes, M. N. and Hallet, B.: Influence of rapid glacial retreat on the rate of erosion by tidewater glaciers, Geology, 30, 47–50, 2002.

57. Koppes, M. N. and Montgomery, D. R.: The relative efficacy of fluvial and glacial erosion over modern to orogenic timescales, Nat. Geosci., 2, 644–647, 2009.

58. Kyrke-Smith, T. M. and Fowler, A. C.: Subglacial swamps, P. R. Soc. A, 470, 20140340, doi:10.1098/rspa.2014.0340, 2014.

59. Lamb, M. P. and Fonstad, M. A.: Rapid formation of a modern bedrock canyon by a single flood event, Nat. Geosci., 3, 477– 481, 2010.

60. Lamb, M. P., Dietrich, W. E., and Sklar, L. S.: A model for fluvial bedrock incision by impacting suspended and bed load sediment, J. Geophys. Res., 113, F03025, doi:10.1029/2007JF000915, 2008.

61. Lamb, M. P., Finnegan, N. J., Scheingross, J. S., and Sklar, L. S.: New insights into the mechanics of fluvial bedrock erosion through flume experiments and theory, Geomorphology, 244, 33–55, doi:10.1016/j.geomorph.2015.03.003, 2015.

62. Le Brocq, A. M., Payne, A. J., Siegert, M. J., and Alley, R. B.: A subglacial water-flow model for West Antarctica, J. Glaciol., 55, 879–888, 2009.

63. Lliboutry, L.: General theory of subglacial cavitation and sliding of temperate glaciers, J. Glaciol., 7, 21–58, 1968.

64. MacGregor, K. R., Anderson, R. S., Anderson, S. P., and Wadding-ton, E. D.: Numerical simulations of glacial longitudinal profile evolution, Geology, 28, 1031–1034, 2000.

65. MacGregor, K. R., Anderson, R. S., and Waddington, E. D.: Numer-ical modeling of glacial erosion and headwall processes in alpine valleys, Geomorphology, 103, 189–204, 2009.

66. Mair, D., Willis, I., Fischer, U. H., Hubbard, B., Nienow, P., and Hubbard, A.: Hydrological controls on patterns of surface, inter-nal and basal motion during three "spring events": Haut Glacier d'Arolla, Switzerland, J. Glaciol., 49, 555–567, 2003.

67. Mao, L., Dell'Agnese, A., Huincache, C., Penna, D., Engel, M., Niedrist, G., and Comiti, F.: Bedload hysteresis in a glacier-fed mountain river, Earth Surf. Proc. Land., 39, 964–976, 2014.

68. Meierbachtol, T., Harper, J., and Humphrey, N.: Basal Drainage System Response to Increasing Surface Melt on the Greenland Ice Sheet, Science, 341, 777–779, 2013.

69. Montgomery, D. R. and Korup, O.: Preservation of inner gorges through repeated Alpine glaciations, Nat. Geosci., 4, 62–67, 2011.

70. Nelson, P. A. and Seminara, G.: Modeling the evolution of bedrock channel shape with erosion from saltating bed load, Geophys. Res. Lett., 38, L17406, doi:10.1029/2011GL048628, 2011.

71. Ng, F. S. L.: Canals under sediment-based ice sheets, Ann. Glaciol., 30, 146–152, 2000.

72. Nienow, P., Sharp, M., and Willis, I.: Seasonal changes in the morphology of the subglacial drainage system, Haut Glacier d'Arolla, Switzerland, Earth Surf. Proc. Land., 23, 825–843, 1998.

73. Nye, J. F.: Water flow in glaciers: jökulhlaups, tunnels and veins, J. Glaciol., 17, 181–207, 1976.

74. C. Cofaigh, C.: Tunnel valley genesis, Prog. Phys. Geog., 20, 1–19, 1996.

75. Orwin, J. F. and Smart, C.: Short-term spatial and temporal patterns of suspended sediment transfer in proglacial channels, Small River Glacier, Canada, Hydrol. Process., 18, 1521–1542, 2004.

76. Riihimaki, C. A., MacGregor, K. R., Anderson, R. S., Anderson, S. P., and Loso, M. G.: Sediment evacuation and glacial erosion rates at a small alpine glacier, J. Geophys. Res., 110, F03003, doi:10.1029/2004JF000189, 2005.

77. Rose, K. C., Ross, N., Bingham, R. G., Corr, H. F., Ferraccioli, F., Jordan, T. A., Le Brocq, A. M., Rippin, D. M., and Siegert, M. J.: A temperate former West Antarctic ice sheet suggested by an extensive zone of subglacial meltwater channels, Geology, 42, 971–974, 2014.

78. Röthlisberger, H.: Water pressure in intra– and subglacial channels, J. Glaciol., 11, 177–203, 1972.

79. Sanders, J. W., Cuffey, K. M., MacGregor, K. R., and Collins, B. D.: The sediment budget of an alpine cirque, Geol. Soc. Am. Bull., 125, 229–248, 2013.

80. Scheingross, J. S., Winchell, E. W., Lamb, M. P., and Dietrich, W. E.: Influence of bed patchiness, slope, grain hiding, and form drag on gravel mobilization in very steep streams, J. Geophys. Res.-Earth, 118, 1–20, 2013.

81. Scheingross, J. S., Brun, F., Lo, D. Y., Omerdin, K., and Lamb, M. P.: Experimental evidence for fluvial bedrock incision by suspended and bedload sediment, Geology, 42, 523–526, 2014.

82. Schoof, C.: Ice-sheet acceleration driven by melt supply variability, Nature, 468, 803–806, 2010.

83. Sklar, L. S. and Dietrich, W. E.: Sediment and rock strength controls on river incision into bedrock, Geology, 29, 1087–1090, 2001.

84. Sklar, L. S. and Dietrich, W. E.: A mechanistic model for river incision into bedrock by saltating bed load, Water Resour. Res., 40, W06301, doi:10.1029/2003WR002496, 2004.

85. Sklar, L. S. and Dietrich, W. E.: The role of sediment in controlling steady-state bedrock channel slope: Implications of the saltation– abrasion incision model, Geomorphology, 82, 58–83, 2006.

86. Sklar, L. S. and Dietrich, W. E.: Implications of the saltation– abrasion bedrock incision model for steady-state river longitudinal profile relief and concavity, Earth Surf. Proc. Land., 33, 1129–1151, 2008.

87. Sole, A. J., Mair, D. W. F., Nienow, P. W., Bartholomew, I., King, M. A., Burke, M. J., and Joughin, I.: Seasonal speedup of a Greenland marine-terminating outlet glacier forced by surface melt–induced changes in subglacial hydrology, J. Geophys. Res.-Earth, 116, F03014, doi:10.1029/2010JF001948, 2011.

88. Storrar, R. D., Stokes, C. R., and Evans, D. J. A.: Increased channelization of subglacial drainage during deglaciation of the Laurentide Ice Sheet, Geology, 42, 239–242, 2014.

89. Swift, D. A., Nienow, P. W., and Hoey, T. B.: Basal sediment evacuation by subglacial meltwater: suspended sediment trans-port from Haut Glacier d'Arolla, Switzerland, Earth Surf. Proc. Land., 30, 867–883, 2005.

90. Tomkin, J. H. and Braun, J.: The influence of alpine glaciation on the relief of tectonically active mountain belts, Am. J. Sci., 302, 169–190, 2002.

91. Turowski, J. M., Lague, D., and Hovius, N.: Cover effect in bedrock abrasion: A new derivation and its implications for the model-ing of bedrock channel morphology, J. Geophys. Res.-Earth, 112, F04006, doi:10.1029/2006JF000697, 2007.

92. Turowski, J. M., Yager, E. M., Badoux, A., Rickenmann, D., and Molnar, P.: The impact of exceptional events on erosion, bedload transport and channel stability in a step-pool channel, Earth Surf. Proc. Land., 34, 1661–1673, 2009.

93. Valla, P. G., Van Der Beek, P. A., and Carcaillet, J.: Dating bedrock gorge incision in the French Western Alps (Ecrins-Pelvoux massif) using cosmogenic 10Be, Terra Nova, 22, 18–25, 2010.

94. Walder, J. S. and Fowler, A.: Channelized subglacial drainage over a deformable bed, J. Glaciol., 40, 3–15, 1994.

95. Warburton, J.: An alpine proglacial fluvial sediment budget, Geogr. Ann. A, 72, 261–272, 1990.

96. Weertman, J.: General theory of water flow at the base of a glacier or ice sheet, Rev. Geophys., 10, 287–333, 1972.

97. Werder, M. A., Hewitt, I. J., Schoof, C., and Flowers, G. E.: Mod-eling channelized and distributed drainage in two dimensions, J. Geophys. Res.-Earth, 118, 2140–2158, 2013.

98. Wheler, B. A., MacDougall, A. H., Flowers, G. E., Petersen, E. I., Whitfield, P. H., and Kohfeld, K. E.: Effects of temperature forcing provenance and extrapolation on the performance of an empirical glacier-melt model, Arct. Antarct. Alp. Res., 46, 379–393, 2014.

99. Whipple, K. X., Hancock, G. S., and Anderson, R. S.: River incision into bedrock: Mechanics and relative efficacy of plucking, abrasion, and cavitation, Geol. Soc. Am. Bull., 112, 490–503, 2000.

100. Whipple, K. X., DiBiase, R. A., and Crosby, B. T.: Bedrock rivers, in: Treatise on Geomorphology, edited by: Shroder, J. and Wohl, E., vol. 9, chap. Fluvial Geomorphology, 550–573, Academic Press, San Diego, CA, 2013.

101. Willis, I. C., Richards, K. S., and Sharp, M. J.: Links between proglacial stream suspended sediment dynamics, glacier hydrology and glacier motion at Midtdalsbreen, Norway, Hydrol. Process., 10, 629–648, 1996.

102. Yager, E. M., Dietrich, W. E., Kirchner, J. W., and McArdell, B. W.: Prediction of sediment transport in step-pool channels, Water Re-sour. Res., 48, W01541, doi:10.1029/2011WR010829, 2012.

CHAPTER 13

ROCK SLIDE-DEBRIS: CHARACTERISTICS, DYNAMICS, AND IMPLICATIONS FOR HAZARD AND RISK ASSESSMENT

R. H. Guthrie, P. Friele, K. Allstadt, N. Roberts, S. G. Evans, K. B. Delaney, D. Roche, J. J. Clague, and M. Jakob

INTRODUCTION

At 03:27 local time (PDT), on 6 August 2010, a 48.5×10^6 m^3 landslide was initiated on Mount Meager, at 50.623◦ N/123.501° W in the Coast Mountains of British Columbia (Fig. 1). The landslide originated with the collapse of the southern peak of Mount Meager (2554 m a.s.l.) onto the saturated flank of the mountain where it incorporated a significant volume of material, forming a highly mobile, very rapid debris flow (Fig. 2). The landslide travelled downslope at high velocity over Capricorn Glacier, and then down the entire 7 km length of Capricorn Creek. At the mouth of Capricorn Creek the landslide debris encountered the opposing wall of the Meager Creek – Capricorn Creek confluence; causing a dramatic split in the moving debris that travelled both upstream, filling and damming Meager Creek, and downstream into the wider Lillooet River valley, where it briefly dammed the Lillooet River.

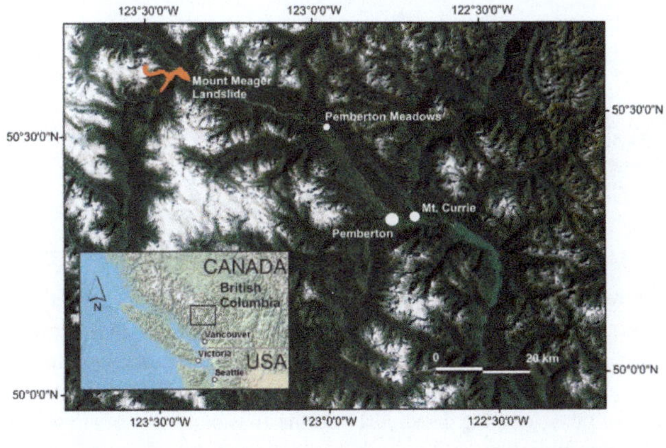

Figure 1. Location map of the Mount Meager landslide. Three communities lie down-valley from the Mount Meager complex (at the head of the landslide): Pemberton Meadows at about 32 km, Pemberton at about 65 km, and Mt. Currie just beyond that.

Citation: Guthrie, R. H., Friele, P., Allstadt, K., Roberts, N., Evans, S. G., Delaney, K. B., Roche, D., Clague, J. J., and Jakob, M.: Rock slide-debris: Characteristics, dynamics, and implications for hazard and risk assessment, Nat. Hazards Earth Syst. Sci., 12, 1277-1294, doi:10.5194/nhess-12-1277-2012, 2012.

This landslide is significant in that it is one of the largest to have occurred worldwide since 1945 and is one of the three largest to have occurred in the Canadian Cordillera in historical time (Evans, 2006; Guthrie and Evans, 2007; Lipovsky et al., 2008). Though there were no deaths or serious injuries associated with the event, approximately 1500 people were evacuated from their homes as part of the emergency response in anticipation of the outburst of a landslide dam that formed in Meager Creek as a result of the landslide. Landslides from the Mount Meager Volcanic Complex (MMVC) have the potential to impact nearby communities, including the village of Pemberton, a town of slightly more than 2100 people, approximately 65 km southeast of the MMVC, and rural communities along the Lillooet River.

In this paper we summarize the main characteristics (initial failure, subsequent behavior, and landslide damming) of the 6 August 2010 landslide. We examine the dynamics of the landslide interpreted from seismic records, and its behavior as simulated in a dynamic analysis model (DAN-W) for velocity and extent of motion. This report is based on data acquired during field work both immediately after the event and over subsequent months, as well as data from postevent high resolution satellite imagery.

THE MOUNT MEAGER VOLCANIC COMPLEX

Geology and Geomorphology

Mount Meager is located in the Province of British Columbia, 150 km north of Vancouver and approximately 65 km northwest of the town of Pemberton (Fig. 1). It is part of the MMVC, which consists of about 20 km3 of volcanic rocks, ranging in age from Pliocene to Holocene (Read, 1978, 1990). The MMVC lies at the northern limit of the Cascade magmatic arc, which includes other prominent volcanoes in the Pacific Northwest such as Mounts Baker, Rainier, and St. Helens in Washington State, Mount Hood in Oregon, and Mount Shasta in California. Volcanic activity at these centers is a product of subduction of the oceanic Juan de Fuca plate beneath continental North America (Hickson, 1994).

The MMVC consists of several coalesced stratovolcanoes, of which Mount Meager is the youngest. The oldest known eruption dates to approximately 2.2 Ma ago; the youngest eruption occurred about 2350 years ago (Clague et al., 1995) near the northeast flank of Plinth Peak, when a rhyodacitic welded ash tuff and lava flow blocked Lillooet River (Read, 1990; Stasiuk et al., 1996; Stewart 2002), and produced a pumiceous air fall deposit across part of southern British Columbia and into westernmost Alberta (Nasmith et al., 1967; Westgate and Dreimanis, 1967).

The MMVC consists of highly fractured, weak, and hydrothermally altered volcanic rocks (Read, 1990), the lowermost of which were assigned by Read (1978) to the Capricorn Formation, which comprises 200 m of porphyritic rhyodacite breccia and tuff, and 200 m of porphyritic rhyodacite flows. Above the Capricorn Formation and forming the steep summit flank is a porphyritic dacite plug of the Plinth Formation. The extrusive units support inclined flow layering dipping 35–65° to the east-southeast.

Groundwater daylights in the MMVC as hot springs and bedrock seeps, and the lowermost extrusive rocks of the Capricorn Formation and portions of the intrusive plug have been hydrothermally altered to an ochre colour.

All except perhaps the top of Mount Meager was covered by the Cordilleran ice sheet during the last Pleistocene glaciation (Clague, 1981). Glacial erosion, followed by deglaciation between about 14 000 and 11 000 yr ago, subsequent advances of alpine glaciers, twentieth-century thinning and retreat of glaciers and recent thaw of alpine permafrost have all contributed to widespread instability within the massif (Holm et al., 2004). A small glacier is located at the head of Capricorn Creek adjacent to the source slope of the 2010 landslide. It was much more extensive during the Little Ice Age than today, extending across the lower part of the failed slope and about 2.5 km down Capricorn Creek (Holm et al., 2004). The glacier scoured and steepened valley sides in the upper part of the valley, and an apron of sediments accumulated against the ice beneath the summit flank.

Figure 2. The 48.5 × 10^6 m^3 rock slide – debris flow from Mount Meager, August 06, 2010. The outline of the initial failure is shown in (A) as a white dashed line. The landslide is approximately south-facing. The lateral shear surfaces are approximately 70 vertical meters high; a helicopter provides scale (inside the white circle). Note the considerable groundwater emerging through weathered bedrock on the western margin and from several other places including the vertical bedrock backscarp. A remnant of the south peak of Mount Meager remains at the apex of the backscarp, where tension cracks persist (some visible in B). Ongoing rock fall from the remaining peak and the sidewalls continued for several months after the main event. The immediate run-out, viewed from the peak looking down, around the first bend of Capricorn Creek is shown in (B). The full extent of the landslide is very difficult to see in any single picture, but (C) gives a sense of the extent to which the run-out inundated several valleys.

Subsequent thinning and retreat of the glacier debutressed the sidewalls in the upper part of the catchment, inducing instability (Holm et al., 2004).

Landslide History

The MMVC is arguably the most landslide-prone region in Canada (Friele et al., 2008; Friele and Clague 2009); at least 25 landslides ≥0.5 × 10^6 m^3 are known to have occurred in the last 10 000 yr (Table 1). Numerous landslides, less than 0.5 × 10^6 m^3 in volume, have been identified from Mount Meager (Jordan, 1994; Jakob, 1996). However, such an inventory is unlikely to be complete even in historical times, and a sampling bias is certain to occur. At the same time, it is likely that several large prehistoric landslides have occurred that have not yet been identified, and thus the prehistoric inventory should not be considered complete.

Table 1. The history of landslides ≥0.5 × 106 m3 from the Mount Meager Complex (adapted and updated from Friele et al., 2008)

Event	Source	Date[1]	Volume[2] ($\times 10^6 m^3$)	Reference
Prehistoric				
Rock avalanche/debris flow	Pylon Pk.	−7900	450*	Friele and Clague (2004)
Rock avalanche/debris flow	Job Ck.	−6250	500*	Friele et al. (2005)
Rock avalanche/debris flow	Capricorn Ck.	−5250	5*	McNeely and McCuaig (1991)
Rock avalanche/debris flow/hyper concentrated flow	Pylon Pk	−4400	200	Friele and Clague (2004); Friele et al. (2005)
Rock avalanche/debris flow	Job Ck.	−2600	500*	Friele et al. (2005); Simpson et al. (2006)
Pyroclastic flow	Syn-eruptive	−2400	440	Stasiuk et al. (1996); Stewart (2002)
Rock avalanche/outburst flood/debris flow/hyper concentrated flow	Syn-eruptive	−2400	200	Stasiuk et al. (1996); Stewart (2002)
Rock avalanche	Syn- to post-eruptive	−2400	44	Stasiuk et al. (1996); Stewart (2002)
Debris flow	Devastation Ck.	−2170	12	McNeely and McCuaig (1991)
Debris flow	Job Ck.	−2240	1	Friele et al. (2008)
Debris flow	Angel Ck.	−1920	0.5*	McNeely and McCuaig (1991)
Debris flow	Job Ck.	−1860	1	McNeely and McCuaig (1991)
Debris flow	Job Ck.	−870	9*	Jordan (1994)
Debris flow	Job Ck.	−630	1	Friele et al. (2008)
Debris flow	No Good Ck.	−370	5*	McNeely and McCuaig (1991)
Historic (age AD)				
Debris flow	Capricorn Ck.	1850	1.3	Jakob (1996); McNeely and McCuaig (1991)
Debris flow	Capricorn Ck.	1903	30	Jakob (1996)
Debris flow	Devastation Ck.	1931	3	Carter (1932); Decker et al. (1977); Jordan (1994)
Rock avalanche	Capricorn Ck.	1933	0.5*	Croft (1983)
Rock avalanche	Devastation Ck.	1947	3	Evans unpublished data
Rock avalanche	Devastation Ck.	1975	12	Mokievsky-Zubok (1977); Evans (2001)
Rock avalanche	Mt. Meager	1986	0.5*	Evans (1987)
Debris flow	Capricorn Ck.	1998	1.3	Bovis and Jakob (2000)
Debris flow	Capricorn Ck.	2009	0.5	Friele (unpublished data)
Rock slide/debris flow	Capricorn Ck.	2010	48.5	This study

[1] Prehistoric dates are radiocarbon ages BP. Cited ages are those that closely constrain an event, however, considerable ranges may be present. For more details see Friele and Clague (2004), or the original sources. [2] Volumes denoted by * are the median volume of a wider range.

6 AUGUST 2010 MOUNT MEAGER LANDSLIDE

Initial Failure

The 2010 Mount Meager landslide originated with the collapse of its southern peak (2554 m a.s.l.) onto its failing saturated flank where it incorporated a significant volume of material, forming a highly mobile, very rapid debris flow. The debris flow travelled the entire 7 km length of Capricorn Creek and inundated both the Meager Creek valley and the Lillooet River valley.

The overall path height (H) is 2183 m and the total path length (*L*) is 12.7 km. This yields ***H /L*** of 0.172 and a fahrboschung" (travel angle) of 9.75° (Fig. 3).

The landslide had no distinct seismic or meteorological trigger but was recorded on seismographs in British Columbia and Washington State. It occurred as a result of several important preconditions. The Mount Meager Volcanic Complex is subject to frequent landslide activity indicating conditions especially favorable to large-scale slope movement. These conditions include poor quality or structurally weak volcanic and volcaniclastic bedrock, a Holocene history of glacial unloading, recent explosive volcanism, and Little Ice Age glacial activity (slope loading, unloading, and oversteepening).

Evidence indicates that groundwater played a key role in the 2010 failure. Prior to failure the flanks of Mount Meager were subject to high pore pressures indicated by extensive surface seepage observed throughout the failure surface and along the lateral shears following the 2010 event (Fig. 2a). The largest visible bedrock spring occurred along the west lateral scarp and was the location of at least two previous landslides, occurring in 2009 and 1998 (Fig. 4). Water supply was exacerbated by summer melt of snow and ice, causing even greater saturation of slopes.

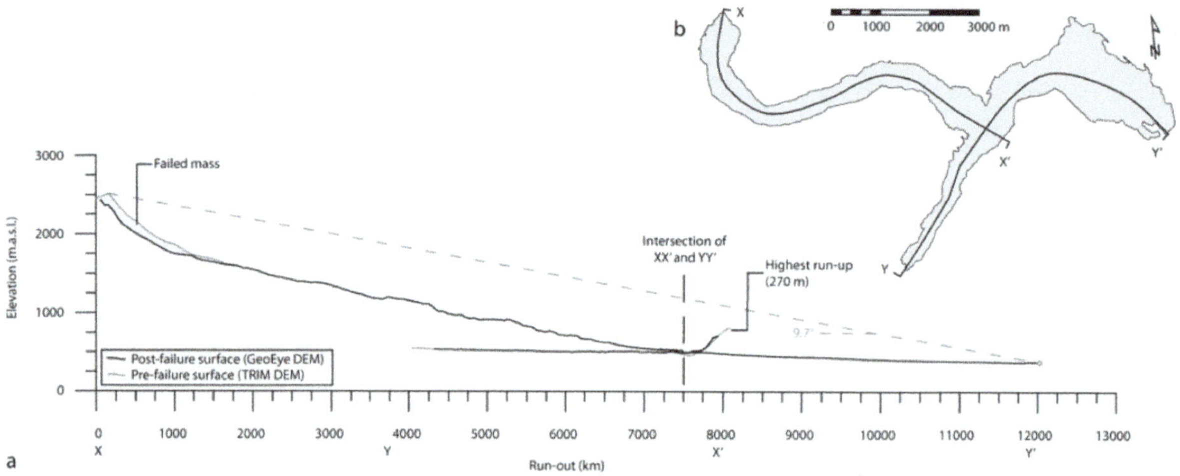

Figure 3. Profile of the Mount Meager landslide.

Figure 4. Mount Meager before the 2010 landslide. The sources and upper paths of the 2009 and 1995 Mount Meager debris flows are outlined and coincide roughly with the present-day location of the lateral shear. The darker color of the 2009 event is from groundwater saturation, and the orange staining of rock on the lower slopes indicates long-term exposure to high groundwater levels and hydrothermally altered volcanic rock (photograph by Dave Southam).

Based on our interpretation of seismic records (see below), initial failure is thought to have begun as a series of major rock falls from the high secondary peak, or gendarme, of Mount Meager, beginning at 03:27:30 h (PDT) on the morning of 6 August 2010. The rock falls fell approximately 500 m onto the weak and heavily saturated south flank of Mount Meager which then began to disintegrate as it was subjected to impact-generated undrained loading. On the seismic record (see below) the impact of the collapsed peak was measured less than 45 seconds after the first movements and resulted in the extremely rapid horizontal displacement of the materials composing the flank. Interpretation of seis-mic records (see below) indicates that the material left the initiation zone at a velocity of approximately 87 m s^{-1}, leaving behind distinct lateral scarps and a high vertical bedrock head scarp (Fig. 2).

Landslide volume Analysis

A robust volume analysis of the initial failure mass was accomplished by comparing pre- and post-event digital elevation models (DEMs) of the landslide (Fig. 5). Topographic data representing the pre-event surface was obtained from a 25 m resolution Terrain Resource Inventory Mapping (TRIM) DEM, the British Columbia provincial standard base map. The post-event surface (DEM) was created using two sets of 5 m resolution GeoEye stereo images obtained 22 Au-gust and 21 September 2010. Each DEM represents a surface model that, while internally consistent, may contain differences due to image off-sets, resolution, and ground control points. GeoEye used only the satellite telemetry for control, and was therefore projected in space slightly above the surface represented by the TRIM DEM. This was corrected by overlaying the GeoEye surface onto the TRIM surface, so that the average difference in ground elevations outside the landslide track was zero.

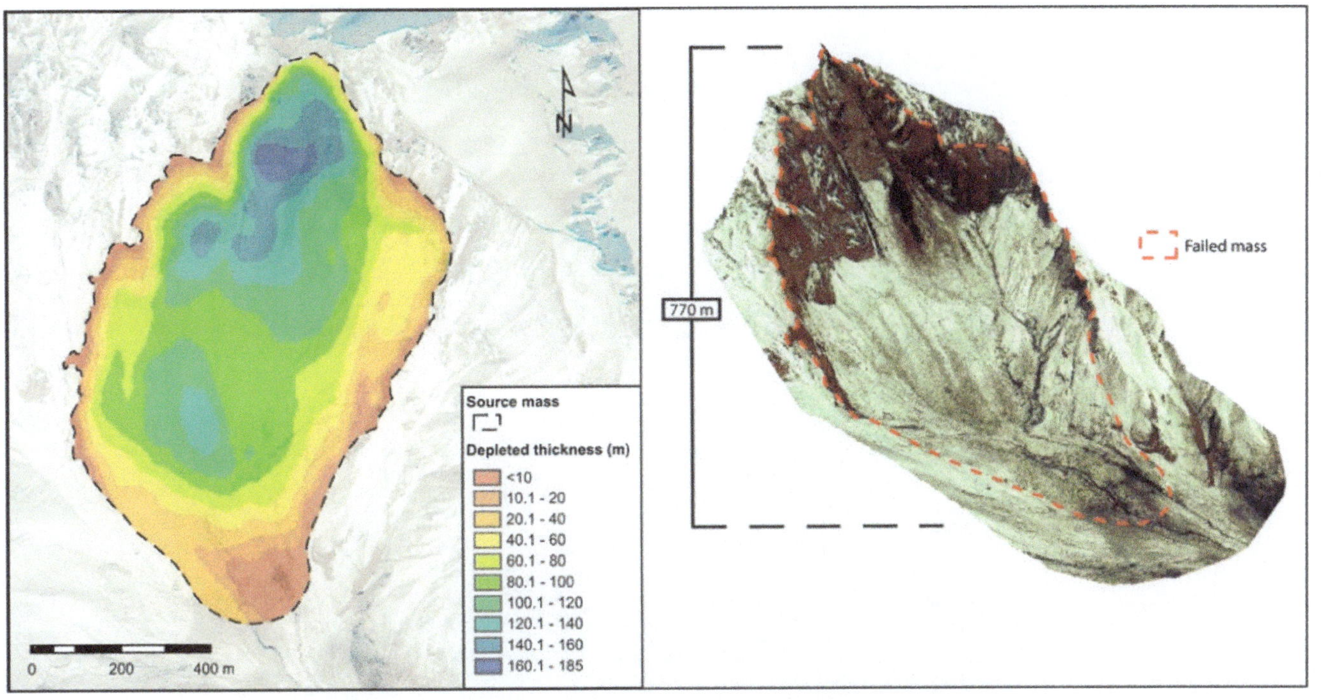

Figure 5. The difference in depth between the TRIM (before) and GeoEye DEMs used to calculate volume. The strong lateral shears, high backscarp and bulge of the pre-failure flank are all readily interpreted from the differencing map. The landslide outline was accurately delineated based on new imagery (right side of figure). The total calculated volume is 48.5×10^6 m^3.

A difference calculation was then performed on the two surfaces inside the landslide track, using raster math in a Geographic Information System (GIS), and summing the total difference for each pixel in the initiation zone. The resulting volume calculated for the initial failure is 48.5×10^6 m^3.

Field estimates of deposit thickness were made at several locations following the landslide, then matched to morphological units, and corroborated by comparing cross sections of the TRIM DEM with the GeoEye DEM. The total volume of the deposit was calculated at 41×10^6 m^3, including bulking. The volume estimate from the field is about 15 % lower than expected, or, if approximately 6 M m^3 of bulking is estimated, then it represents a difference of almost 30 %.

Several explanations may be considered: (1) the initial volume is from a limited area (~0.4 km^2*)* of exceptionally deep scour. It is not particularly sensitive to a change in depth of a few meters; however, it is sensitive to even a small shift in the outline of the initiation zone. Fortunately, this error is controllable in the production of an accurate out-line of the head scarp (Fig. 5). In contrast, the deposit volume was determined across a broad area and it is therefore about 20× more sensitive to a change in depth estimation. (2) The volume estimation of the deposit, involved consider-ably more subjective analysis and expert judgment at each point across a landscape that was ~9 km^2. To minimize this, cross-sections of the two surfaces were also used in careful corroboration of field estimates. (3) The volume contained in Capricorn valley, V-shaped in 2009 and now truncated in profile with new deposition in 2010 was underestimated. (4) The TRIM DEM was built in 1987. Several geomorphological events have occurred since then that could each affect the valley bottom, but not substantially affect the initiation zone, resulting in a real-life surface that was different than the DEM. These events include two previous landslides in Capricorn Creek (1998 and 2009), and ongoing reworking of material from Meager Creek, Capricorn Creek and Lillooet River, since 1987. With these limitations in mind, we con-sider the deposit volume to be in reasonable agreement with the initial volume.

Our estimate of the initial volume of the 2010 Mount Meager landslide is thus approximately 48.5×10^6 m^3. With this volume it is one of the largest landslides to have occurred worldwide since 1945 (Evans, 2006) and one of the three largest landslides to have occurred in the Canadian Cordillera in historical time. It is comparable in volume to the 1965 Hope Slide (Mathews and McTaggart, 1978).

We note that the characteristics of the 6 August Mount Meager event are largely depositional outside of the initiation zone.

Transformation into Debris Flow and Run Out

The disintegrated landslide mass left the initiation zone in a largely horizontal trajectory and traversed Capricorn Glacier, causing almost no damage to the glacier itself (note the drainage features of Capricorn Glacier, perpendicular to the landslide flow, in Fig. 6a), and flowed into the first bend of the Capricorn Creek valley (Fig. 6b). By this time the landslide was typical of a debris flow as the initial failure mass had completely disintegrated and had become fluidized by mixing with the pore water contained in the failed mass.

Figure 6. Looking down the initial landslide path (helicopter in white circle for scale). Note the ice of Capricorn Glacier, intact with longitudinal melt lines perpendicular to the direction of flow **(A)**. Overspray is evident beyond the path edge, particularly on the snow. The lower portion of the valley was in-filled by deposits from the avalanche, which was subsequently evacuated along the left side of the valley (shown by white dotted lines) in one or more liquid slurries. Looking further down the landslide run-out, through the first bend **(B)**, deposition is evident even on the valley sidewalls. Images **(C)** and **(D)** show valley floor deposition as viewed from below and **(E)** looks back up at the initiation zone. White dotted lines in **(C)** correspond to those in A and the white circle in **(D)** shows a person for scale. Ice persists in the valley floor at the base of the deposit **(A)** and remains largely undamaged, though directly over-ridden (indicated by white arrow).

We interpret the rapid transformation to a debris flow to be a consequence of: (1) elevated groundwater conditions leading up to the event; (2) poor rock quality caused by prolonged We interpret the rapid transformation to a debris flow to be a consequence of: (1) elevated groundwater conditions leading up to the event; (2) poor rock quality caused by prolonged exposure to hydrothermally derived groundwater for years prior to the event; (3) the sudden transfer of energy from the collapse of the mountain peak; and (4) a rapid increase in hydrostatic pressure as a result of the impact (undrained loading of the slope).

Analysis and Interpretation of Seismic Signature

The Meager Creek landslide generated long-period seismic waves that were visible at stations from southern California to northern Alaska, up to 2800 km away. A sample of vertical-component broadband seismograms of this event is displayed on Fig. 7. No known earthquake was associated with the failure, but the event itself was assigned an equivalent local magnitude of 2.6 by the Canadian Seismic Net-work. Landslides are distinguished from seismic signals by a characteristically high ratio of long period to short period energy waves (Weichert et al., 1994; McSaveney and Downes, 2002). The long-period surface waves generated by the Mount Meager event were much stronger than would be expected from an earthquake of the same magnitude.

Landslides do not always generate identifiable seismic sig-natures due in part to their slower source process and poor ground coupling as compared to earthquakes. Seismic energy conversion rates for similar events are estimated to be as low as 0.01 % of the kinetic energy and 1 % of the potential energy released by the slide (Berrocal et al., 1978). It requires an extremely energetic source to generate waves high enough in amplitude to be visible for thousands of kilo-meters. The seismogenic nature of this landslide is a result of the large volume of material involved and the extremely rapid velocities.

The seismic signal recorded from the Mount Meager land-slide can be examined in two frequency bands to separate the waves coming from different sources. Figure 8 is a plot of the vertical component seismogram and envelope for two frequency bands at the closest station, WSLR, a three-component broadband station. This station is operated by the Canadian National Seismograph Network, located 70 km southeast of Mount Meager. The signal is filtered to separate the short period and long period waves, and also plotted as an envelope to show the overall shape of a signal by including only the amplitude information from the seismogram. This is helpful in picking arrivals and separating sources that cannot easily be seen in the seismogram. The long-period portion, filtered to include only frequencies below 0.2 Hz (>5 s period), is generated by large and slower sources such as the acceleration of the slide mass, and turns and large bumps in the path of the debris flow (Kanamori and Given 1982; Dahlen 1993; Julian et al., 1998; Deparis et al., 2008). The short-period signals, higher than 1 Hz (<1 s period) are generated by smaller scale source processes such as chaotic flow within the debris flow and at the base of the debris flow as it flows over smaller features (Brodsky et al., 2003; Huang et al., 2007). The short period (>1 Hz) motions had an emergent onset, spindle-shaped signal, and low amplitudes in relation to their duration, typical of rock slides, debris avalanches and debris flows (Norris, 1994). The short period portion of the signal emerges about 25 s after the onset of the long period signal, but before the main long period impact signal, suggesting that disintegration of the mass began before the complete collapse of the secondary peak of Mount Meager onto the footslope.

Phases of landslide activity were inferred over the next 200 s based on analysis of seismic wave type, amplitude, and relative timing and corroborated by field observations. The pulses, their source description, and time from the signal on-set are described in Table 2, with locations shown on Fig. 9. Identified pulses are nearly identical at other broadband stations, with some variation due to the radiation pattern, showing that these are source effects, not site effects or path effects.

The most well-defined of the long-period source events are A, B, and F. Event A is interpreted to be initial failures from the upper slopes onto the flanks below. The dominant period of these waves is about 17 to 20 s, indicating a source process of about an equivalent length (Kanamori and Given 1982). This was confirmed by preliminary waveform modeling. Event B is higher in amplitude than Event A and is composed primarily of Rayleigh waves, suggesting a dominantly vertical force. This is interpreted as the collapse of the secondary peak onto the flanks below, mobilizing and disaggregating the material below into a high-velocity debris flow. The period of these Rayleigh waves was 10 s, indicating a shorter source process than the initial failures (A), consistent with the expected signature of a collapse of the secondary peak of Mount Meager. These initial massive failures (A and B) are roughly equivalent to a single force source mechanism in the opposite direction of slide movement as a block accelerates, and in the same direction as slope movement as a block decelerates (Kanamori and Given, 1982; Dahlen, 1993; Julian et al., 1998). The polarization of the first arrivals of the Rayleigh waves and the variability in the amplitudes of both types of surface waves with azimuth indicate the correct radiation pattern for a single force. Preliminary waveform modeling supports the single force mechanism and suggests forces on the order of 10^9 N for event A and 10^9–10^{10} N for event B.

Events C through E can roughly be correlated with turns in the path of the debris flow, a product of seismic waves generated by rotational accelerations. These events also occur when the short period energy is at its highest, suggesting that they coincide with peak debris flow velocities.

Event F is defined by a pulse nearly as high in amplitude as Event B, indicating another massive impact force. Site evidence showed that the debris flow burst out of Capricorn Creek valley into the facing wall of the Meager Creek valley, impacting a steep slope on the opposite side and running up 270 vertical meters before splitting up and down Meager Creek valley. This impact on the valley wall is the interpreted source for Event F.

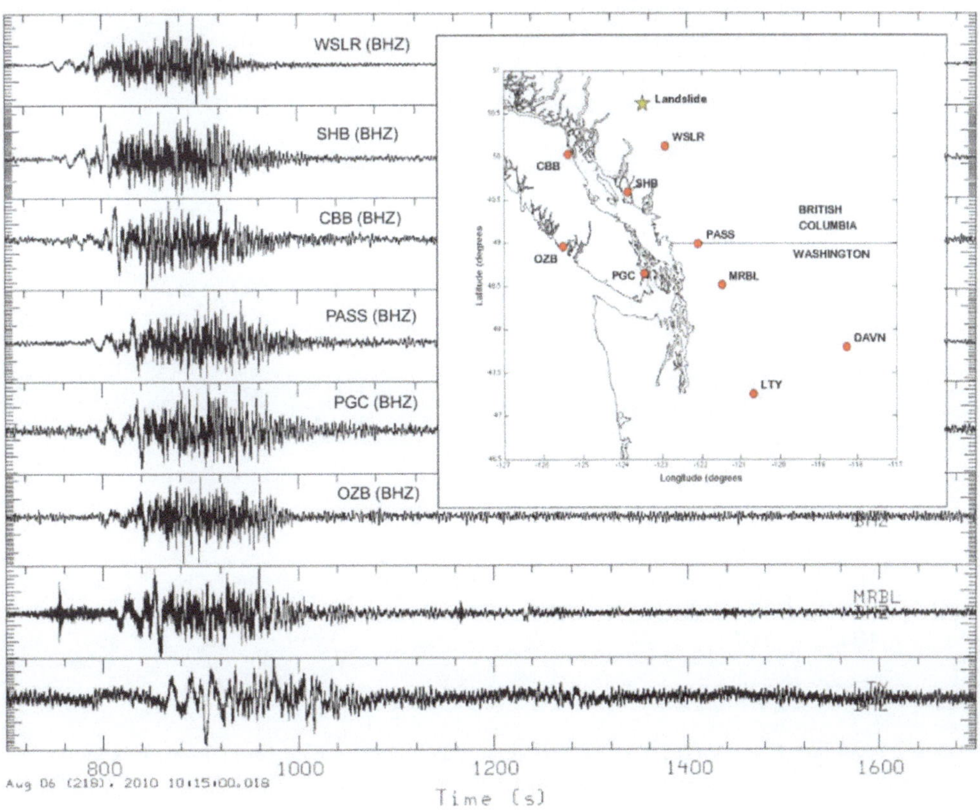

Figure 7. Raw seismic signals from Mount Meager landslide recorded across southwestern Canada and northern Washington State (CNSN and PNSN data). Station and landslide location indicated on inset map.

Using these well-defined time markers and calculating path distance, we can make velocity estimates of the debris flow at various stages along the path. We calculate an average debris flow velocity of 67 m s^{-1}. This is measured from the secondary peak failure (B) to the impact of the debris flow at a right angle to Meager Creek (F), covering a distance of about 7.3 km in 109 s. Rough velocity estimates between each individual point are given on Table 2, including error margins on distance and timing. The error margins are due to uncertainties in defining points, selecting distances between points because these are not point sources, uncertainties in picking arrival times for these long period pulses, and differences in surface waves velocities at different periods. It is more difficult to accurately define the end of the debris flow, as the energy fades gradually on the seismic record.

A second long-period seismic pulse occurs about 7 min after the main collapse began. The waveform of Event H is similar to event A, but smaller in amplitude, and is interpreted as a subsequent smaller volume rock fall or rock slide from the now over-steepened face of Mount Meager, an "aftershock" of sorts. Field work over the next month revealed that smaller rock falls were frequent and ongoing from the source zone.

There were no precursory seismic events recorded prior to the main landslide. However, the seismic stations are 70 km away or farther, so small precursors might have escaped our observation.

Simulation using DAN-W

A first-order dynamic back analysis of the Mount Meager rock avalanche was completed using the two-dimensional numerical simulation model DAN-W (Hungr, 1995; Hungr and McDougall, 2009). DAN-W has been used to simulate the behavior (i.e. run-out distance, velocity along path, and deposition) of a number of catastrophic landslides including rock avalanches, debris avalanches, and debris flows (e.g. Hungr and Evans, 1996; Evans et al., 2001, 2007; Hungr and Evans, 2004; Sosio et al., 2008; among others).

A Voellmy rheology (Hungr, 1995) was selected to model the basal resistance of the moving mass. The Voellmy basal resistance model includes frictional (*f*) and turbulence (ξ) parameters, which are estimated on the basis of field conditions to produce a simulation that best fits observed behaviour

$$\tau_{zx} = \sigma_{zf} + pgv^2/\xi \qquad (1)$$

where τ_{zx} is the basal friction model, σ_z is the bed normal stress at the base of the flow, ***f*** is the frictional coefficient, ρ is the material density, ***g*** is the gravitational constant, ***v*** is the averaged depth flow velocity, and ξ is the turbulence term.

The reader is referred to Hungr (1995, 2008) for details about the input parameters. Output parameters include path velocity and time.

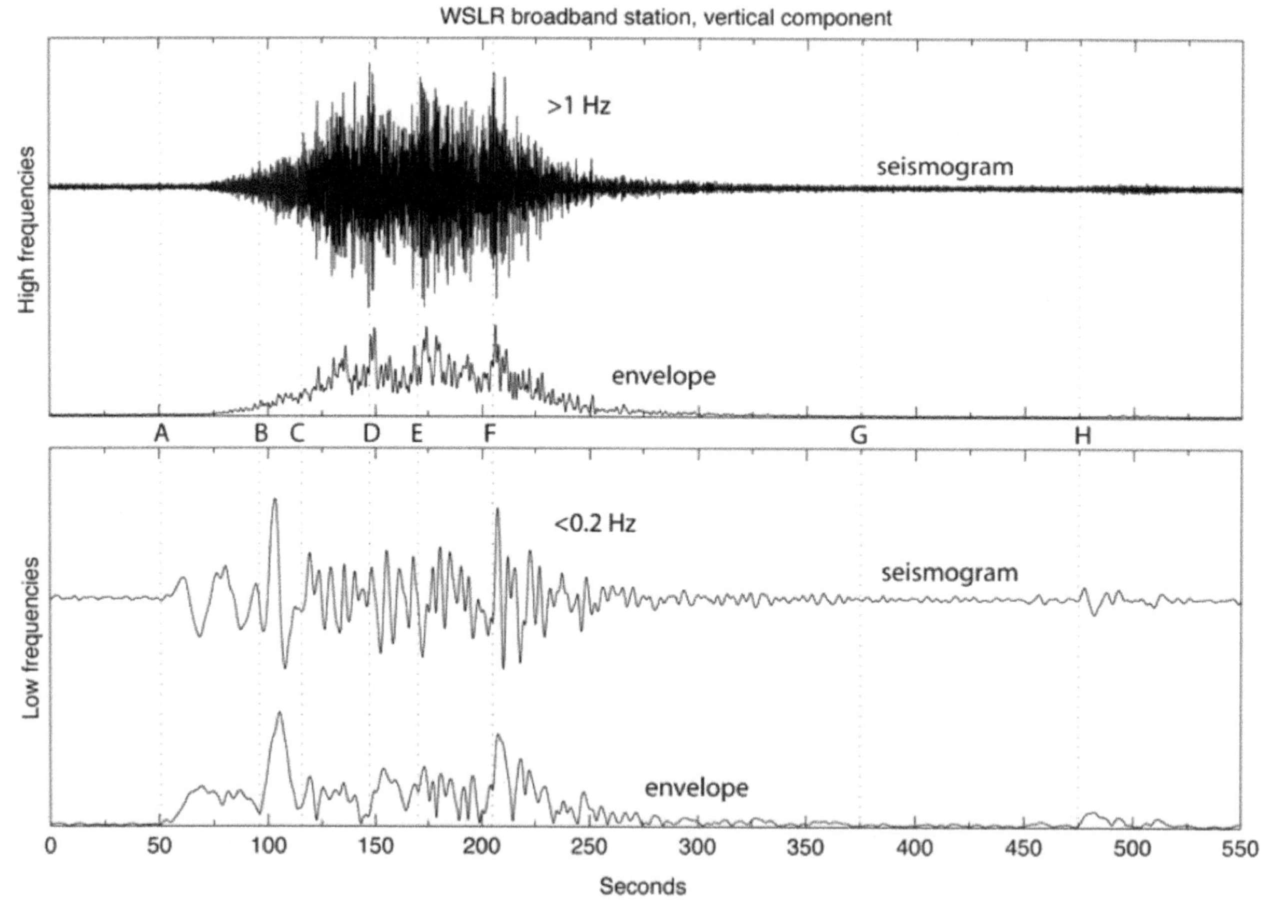

Figure 8. WSLR broadband recording of the landslide single-pass highpass filtered (>1 Hz top) and lowpass filtered (<0.2 Hz, bottom) and plotted with envelopes. Dashed vertical lines are labeled with letters that correspond to Fig. 9 and discussion in text.

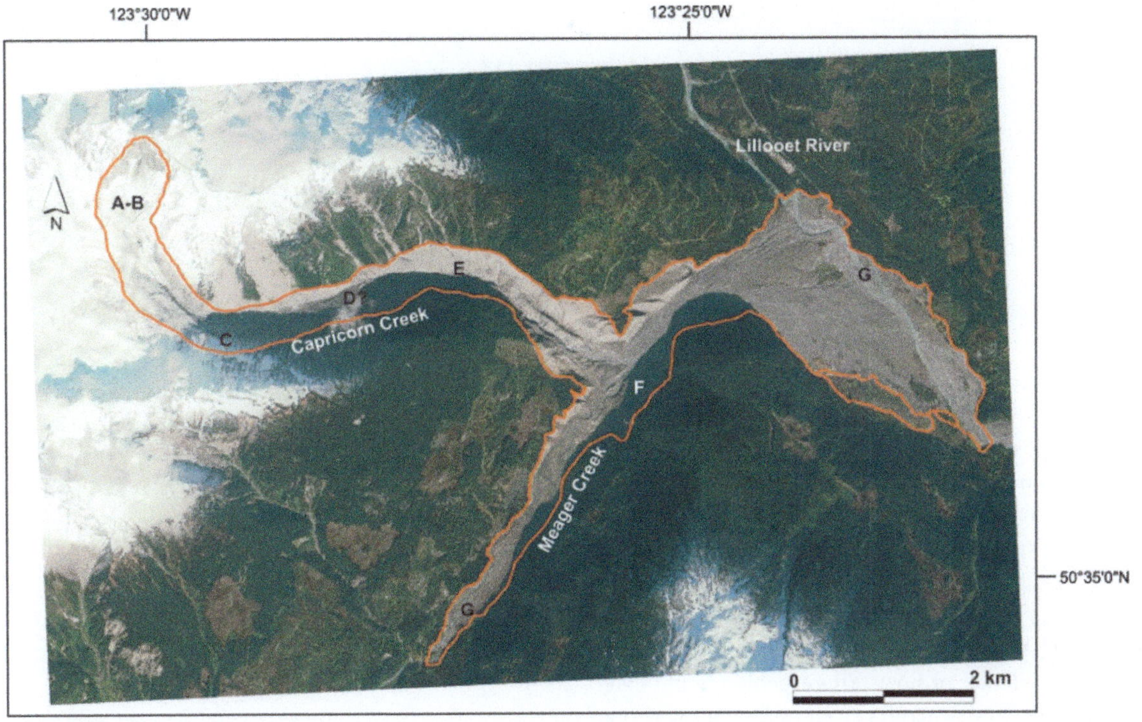

Figure 9. The complete outline of the Mount Meager landslide showing the initiation zone (A–B), the two major bends (C and E), the facing wall of Meager Creek (F), and the bifurcated flow that travelled up Meager Creek, and across the Lillooet River (G). The image was taken following the breach of the Meager Creek dam, and fluvial reworking and considerable incision of the dam itself are evident.

Table 2. Time markers along landslide path according to the seismic signature at stations in the Pacific Northwest (see Figs. 7–9 for path and station locations and seismic trace associated with each marker)

Source event	Description	Time from slide onset (s)	Error (s)	Distance from previous point (km)	Range of distances (error margin) (km)	Average velocity[1] ($m\,s^{-1}$)	Velocity range[1] ($m\,s^{-1}$)
A	Initial failures from high on slopes	0	±1	–	–	–	–
B	Secondary peak collapses and mobilizes flanks below into debris flow	45.3	±2	0	–	–	–
C	Debris flow turns first corner	64.9	±2	1.7	1.5–2.5	87	70–142
D	Debris flow flows over bump and slight turn in path	96.5	±4	1.7	1.7–3	54	54–109
E	Debris flow turns last corner in Capricorn Creek	119.3	±5	1.4	0.8–1.9	61	29–107
F	Front of debris flow strikes wall across Meager Creek	154.1	±0.5	2.2	1.8–2.5	63	51–73
G	Seismic signal drops below noise level, apparent end of debris flow	324.3	±10	3	3–3.5	18	17–22
H	Second smaller rockslide released from Mount Meager	424.1	±0.5	–	–	–	–

[1]From previous event location to current event.

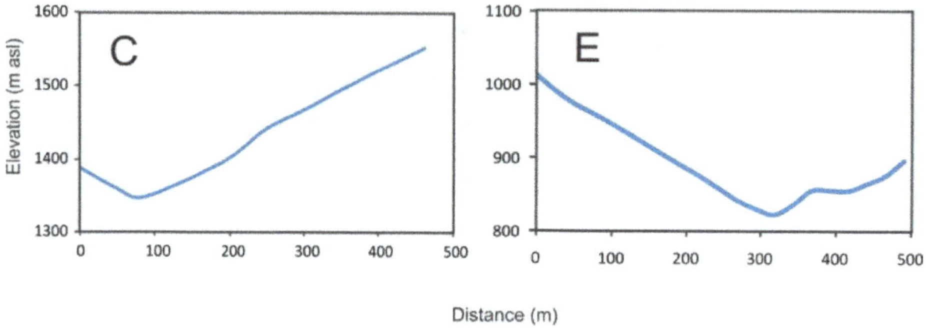

Figure 10. Cross-sectional profiles from the upper bend (C) and the lower bend (E) used for velocity calculations in Fig. 9.

The BC TRIM DEM was used for pre-event surface topography and the GeoEye DEM was used to generate a post-event surface (see Sect. 3.2 for details). These two datasets defined the source region and initial volume (48.5 Mm ^{3}) of the landslide.

Path width from the GeoEye image (Fig. 9) was entered directly into DAN-W along with topography and model parameters. Field evidence of post-failure behavior indicated limited path-erosion and therefore entrainment was not considered in this simulation.

Four surface materials were defined for four segments along the path of the landslide. The values of friction and turbulence parameters for each of these segments that best simulated the landslide behavior are listed in Table 3. We note that the values of the Voellmy parameters for the surface materials in the four path segments fall within the range used in previous successful simulations of similar movements (e.g. Hungr and Evans, 1996; Ayotte and Hungr, 2000; Evans et al., 2009) and generally reflect the transformation of an initial fragmenting rockslide into a fast-moving flow.

Simulations of velocity verus distance and velocity versus time relationships were compared to estimation of velocity from superelevation of debris in the two major bends (C and E on Fig. 10), run-up velocity at the Capricorn Creek – Meager Creek confluence, and velocity measured in the seismic analysis.

Superelevation velocities were measured using

$$V_{min} = ghr/w \qquad (2)$$

where V_{min} = the minimum velocity in m s^{-1}, g = the gravitational constant, h = the run-up height, r = is the radius of curvature in a bend, and w = the width of the path Run-up velocity was also measured at the Capricorn Creek– Meager Creek confluence using

$$V_{min} = (2gh)^{0.5} \qquad (3)$$

where V_{min} = the minimum velocity in m s^{-1}, g = the gravitational constant, and h = the run-up height.

Table 3. Friction and turbulence parameters along path segments used in DAN-W

Segment	Path distance (m)		Friction	Turbulance
	Start	End	(f)	(ξ)
1	0	1407	0.06	750
2	1408	6452	0.1	900
3	6453	7793	0.02	1250
4	7792	12 453	0.02	550

Results of the superelevation analysis indicate minimum flow velocities of 62 m s^{-1} in bend 1 and 73 m s^{-1} in bend 2. These compare to 84.3 m s^{-1} and 81.2 m s^{-1} for the respective bends in the DAN-W simulation, and 87 m s^{-1} and 61 m s^{-1}, respectively, on the seismic trace. The simulated velocities are higher than the velocities calculated from the profile alone, but match the seismic trace quite closely for the first bend. It is likely that a bedrock wall at the beginning of bend 1 limited the superelevation, making the profile velocity sensitive to where in the curve the measurement was taken. We had difficulty matching DAN-W simulated velocity values with the inferred field velocities based on debris behavior, whilst at the same time matching the path geometry requirement of run-out to the Lillooet River. Velocities approximating field estimates in bends in Capricorn Creek did not produce the observed long run out and higher-than-field-estimate velocities were necessary to achieve approximate run-out distance in the simulation. Figures 11 and 12 show the results of the DAN-W analysis.

The geometry of the run-up at the confluence of Capricorn and Meager creeks suggests a minimum velocity of 63 m s^{-1} compared to a DAN-W simulated estimate of 53.3 m s^{-1} (Fig. 11).

Overall, the simulation gave a mean velocity to the confluence of Capricorn and Meager creeks of 68.4 m s^{-1} and a peak velocity of 91.2 m s^{-1}. This compares favorably to the mean velocity of 67 m s^{-1} derived from the seismic traces.

Velocities following impact with the facing wall of Meager Creek were reduced as the debris bifucated and decelerated to the landslide's distal limit at Lillooet River. DAN-W indicates a mean velocity of 9.5 m s^{-1} between the landslide dam and the distal limit. We propose that this comparatively low velocity results from sudden deceleration following (1) the impact with the south side of Meager Creek at the Capricorn-Meager confluence, and (2) the sudden increase in the width of the path as the debris spread out over the Meager Creek fan into the Lillooet River valley. The sudden deceleration is evident in Figs. 11 and 12.

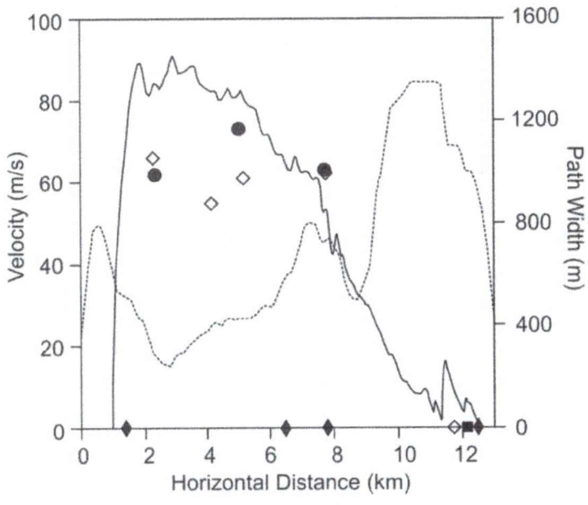

Figure 11. The plot of frontal velocity versus path distance (solid line) and path width versus path distance (dashed line) for the 2010 Mount Meager landslide. Velocities are compared to estimates from superelevation and run-up data (black dots) and seismic analysis (open diamonds). The black square is actual field run-out and filled black diamonds denote the segments corresponding to the four materials in Table 3.

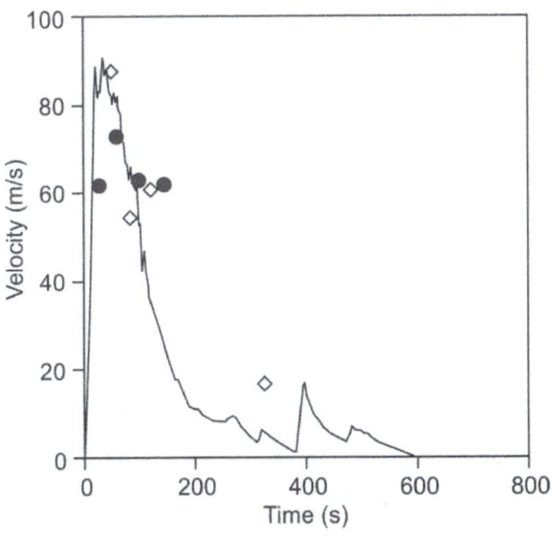

Figure 12. DAN-W simulation plot of velocity versus time. Velocities are compared to estimates from superelevation in bends and run-up geometry (black dots) and seismic analysis (diamonds).

With reference to velocity versus time relationships, there is broad correspondence between the results of DAN-W and the seismic trace for the initial part of the path up to the run-up at the confluence of Capricorn and Meager creeks. We note, however, that in Fig. 12, a large discrepancy is evident in run-out times between the seismic (324 s) and DAN-W analyses (621 s). This may be explained by the low sensitivity threshold of the seismic recordings. We hypothesize that once the moving mass slowed to below 10 m s^{-1}, it was no longer detectable by the seismographs above back-ground seismic noise. The DAN-W analysis however, continues the movement until the velocities reach 0 m s^{-1}, significantly extending the overall run-out time.

Figure 13. The Meager Creek barrier after the dam outburst flood at the confluence of Capricorn and Meager creeks.

Deposit Morphology

Capricorn Creek was predominantly V-shaped prior to the landslide in 2010. Following the 2010 event, the V-shaped valley floor was modified to form a flat-bottomed narrow trapezoid as material was deposited along the path. Deposition began immediately below the initiation zone of the Mount Meager landslide, in-filling the valley bottom and draping the valley sidewalls otherwise stripped bare of vegetation during the event (Fig. 6a–e). Debris incision along the valley floor occurred in later stages of the event, eroded by a liquid slurry as the debris flow reduced in size (inside the white dotted lines of Fig. 6a and c).

At the confluence of Capricorn and Meager creeks, the landslide encountered the opposing rock wall of the Meager Creek valley causing a vertical run up of 270 m and leaving a roughly triangular landslide dam (the Meager Creek barrier) approximately 700 m × 500 m at the base, with a thickness of 30–50 m (Figs. 13 and 14). The landslide dam completely blocked Meager Creek for 19 hours, causing the formation of a landslide dam lake, and ultimately failing into the Lillooet River valley.

The path of the landslide along Meager Creek and Lillooet River was divided into six zones, each with a distinct geomorphic expression (Figs. 14 and 15): (1) spray zone; (2) trim zone; (3) Meager Creek deposit upstream of Capricorn Creek (upstream flow); (4) Meager Creek barrier deposit; (5) the downstream flow in Lillooet River valley including; (6) a thick deposition plug; and (7) late-stage debris flow zone forming the south edge of the deposit (after flow sector).

The spray zone consists of wind-thrown trees and thin de-position of debris ejected laterally as the debris front burst from the confining source valley into Meager Creek valley. The spray zone is about 500–600 m long on both sides of Capricorn Creek mouth on the north wall of Meager Creek valley.

The trim zone lies between the thick debris deposit and the forest edge. Here, in the area eroded by the passing of the frontal lobe, the pre-landslide surface was scoured of vegetation and covered with debris, thin on the side slopes, and thick in the valley bottom where the old V-shaped gulley was in-filled.

The deposit in Meager Creek upstream of the mouth of Capricorn Creek (the upstream flow) consists mainly of mixed sediment and debris filling an incised channel 5–15 m deep, 100–200 m wide, and extending 3250 m upstream from the Meager Creek barrier.

The barrier consists of compressed ridges formed of mixed debris and large grey blocks; the debris directly below the south valley wall, however, is hummocky rather than ridged, having run up the slope and fallen back. The Meager Creek barrier was substantially altered by ongoing debris flows incising the deep valley floor deposits and the edge of the barrier, and then ultimately by the dam outburst flood de-scribed below.

The main deposit downstream along Meager Creek and Lillooet River was largely eroded by the outburst flood; its initial morphology was reconstructed from field observations and oblique air photographs from 6 August 2010. The landslide partially blocked the Lillooet River, causing a 10–15 m rise in water levels on the upstream side that destroyed a bridge at the landslide margin. As it spread across the Lillooet River valley, the flow was divided and slowed by an elevated rise in the valley floor from previous landslide de-posits. The former deposits were overwhelmed by debris from 2010 event, and were buried under a deep plug of debris. The flow crossed Lillooet River valley, narrowly missed a forest recreation campsite on the other side and isolated campers who required rescue.

The debris flow spread across the Lillooet River valley, leaving behind abundant ponded surface water, hummocks, blocks, and compression ridges. Ongoing debris flows subsequent to the main event in-filled the south side of the valley, and reworking of the deposit began almost immediately following the landslide dam failure.

The morphology and architecture of the deposit is typical of large volcanic landslides, with hummocky topography, longitudinal flow banding and areas of transverse compression ridges (Siebert, 1984; Glicken, 1991; Dufresne and Davies, 2009). The low relief areas are underlain by matrix-supported, poorly-sorted debris with grain sizes ranging from clay to boulder-size. Individual hummocks stand up to several meters above the landslide surface, and are cored by blocks of the original rock mass.

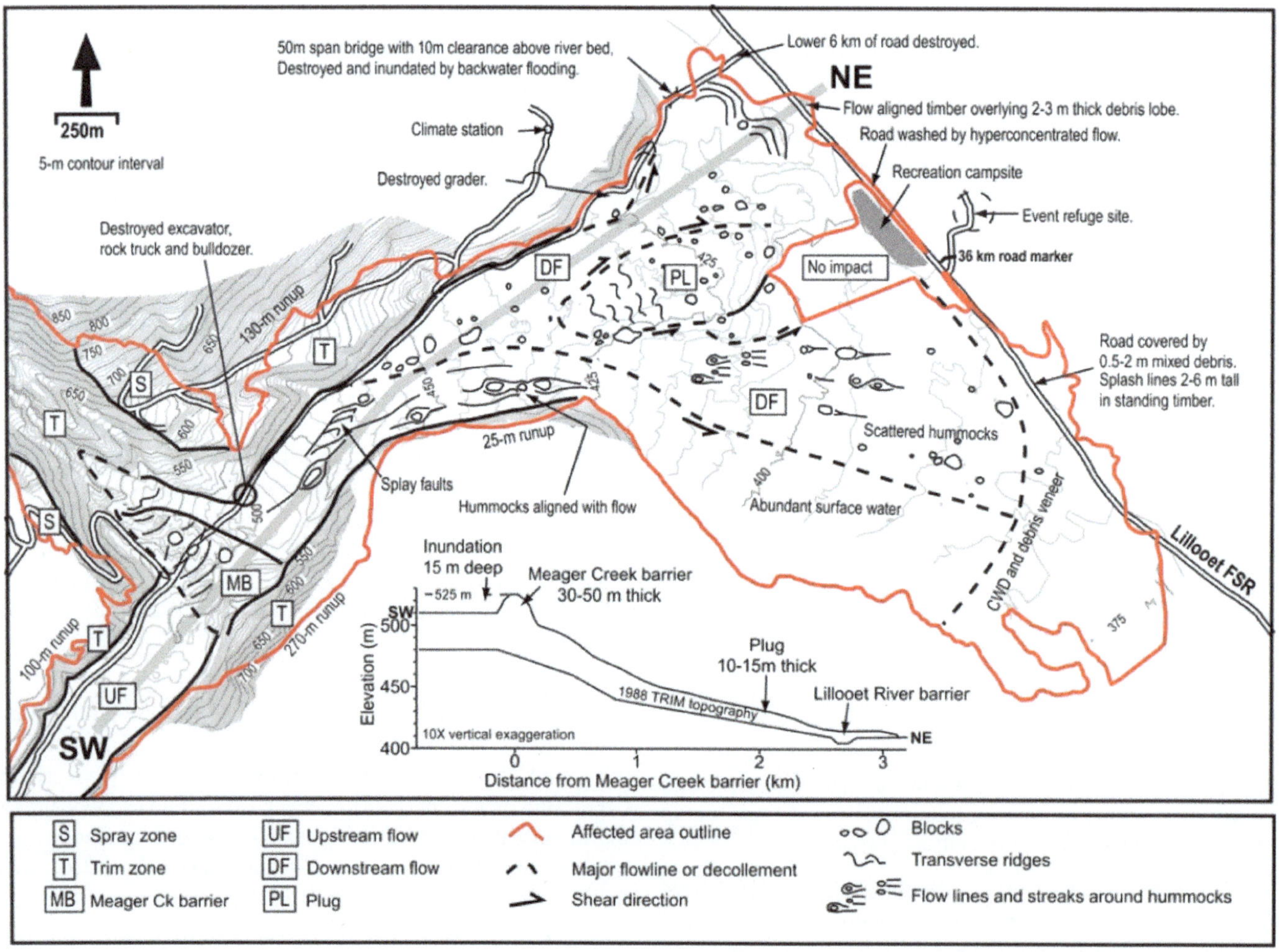

Figure 14. Map of the landslide deposit in the valleys of lowermost Meager Creek and Lillooet River (see text for description).

Failure of the Meager Creek Landslide Dam and Outburst Flood

The landslide completely blocked Meager Creek and partially blocked the Lillooet River at its confluence with Meager Creek. The interruption in the flow of both rivers was recorded at Water Survey of Canada (WSC) hydrometric station 08MG005, located on Lillooet River, upstream of Pemberton, 61.3 km downstream from the confluence of the two rivers. The WSC gauge recorded the hydrometric response of both dams, and provides a unique opportunity to analyze the impacts of the landslide on the river system. Figure 16 shows the Lillooet River hydrograph for 6–7 August 2010.

The flow of Meager Creek was determined by comparing the flow at the Lillooet River WSC station following the event, to the estimated contribution of Meager Creek by watershed area. Meager Creek flow was thereby estimated at 36–38 $m^3 s^{-1}$. The landslide dam impounded between 2.6 and 2.7 × 10^6 m^3 of new water over the next 20 h (Roche et al., 2011).

By 20:30 h on 6 August, it was clear that overtopping of the dam was imminent and approximately 1500 residents in the lower Lillooet Valley were asked to evacuate.

We estimate a total floodwave volume of 2.9 × 10^6 m^3 following overtopping of the Meager Creek barrier. This amount includes introduced water (stored in the dam or in sediments along the path) and sediment content in the volume calculation.

The lake overtopped the landslide dam at around 23:30 h on 6 August and reached a peak discharge of several hundred $m^3 s^{-1}$ at about 01:00 h on 7 August. The floodwave travelled downstream toward Pemberton with an average velocity of approximately 2.5 m s^{-1}. The peak passed the Hurley River Forest Service Road bridge (38 km downstream) at about 04:30 h and the Water Survey of Canada gauge (61 km downstream) at 07:45 h. Flows attenuated in velocity and volume with distance; however, high flows persisted and discharge at Pemberton, almost 7 h after the peak remained about 200 $m^3 s^{-1}$ above the pre-wave condition.

Figure 15. Deposition features in lower Meager Creek. PL = plug; DF = downstream flow; UF = upstream flow; T = trim zone; S = spray zone. Image **(A)** is at the Capricorn–Meager confluence (photo by B. Makarewicz); **(B)** shows the downstream flow against the east wall of the Lillooet River valley; **(C)** and **(D)** are upstream of the Meager Barrier (photo C by D. B. Steers); and **(E)** is the downstream flow and plug.

Socio-Economic and Environmental Impacts

Despite the remote location of the 2010 Mount Meager landslide, the event had considerable socio-economic impact. Approximately 110 000 m^3 of wood was stripped away from the slopes of Capricorn Creek and Meager Creek and the Lillooet River valley bottom and either pulverized into fine organic material, or transported as large woody debris into the river system. The wood was a mixture of western hemlock, amabilis and subalpine fir, western red cedar, and to a lesser extent lodgepole pine and balsam poplar. Wood species and volume were obtained from recent forest cover maps, and multiplied by the average log market value for August 2010 (Province of British Columbia, 2011). The total potential loss based on the markets at the time of the event was $8.7 M CAD. In addition, two forest service bridges and road construction equipment were destroyed, along with several kilometers of roads including almost 6 km of the Meager Creek forest service road. The landslide and the subsequent threat of a dam outburst flood on Meager Creek caused the evacuation of approximately 1500 residents in the lower Lillooet River valley for one night, and rescue efforts for several campers and workers in the vicinity of Mount Meager.

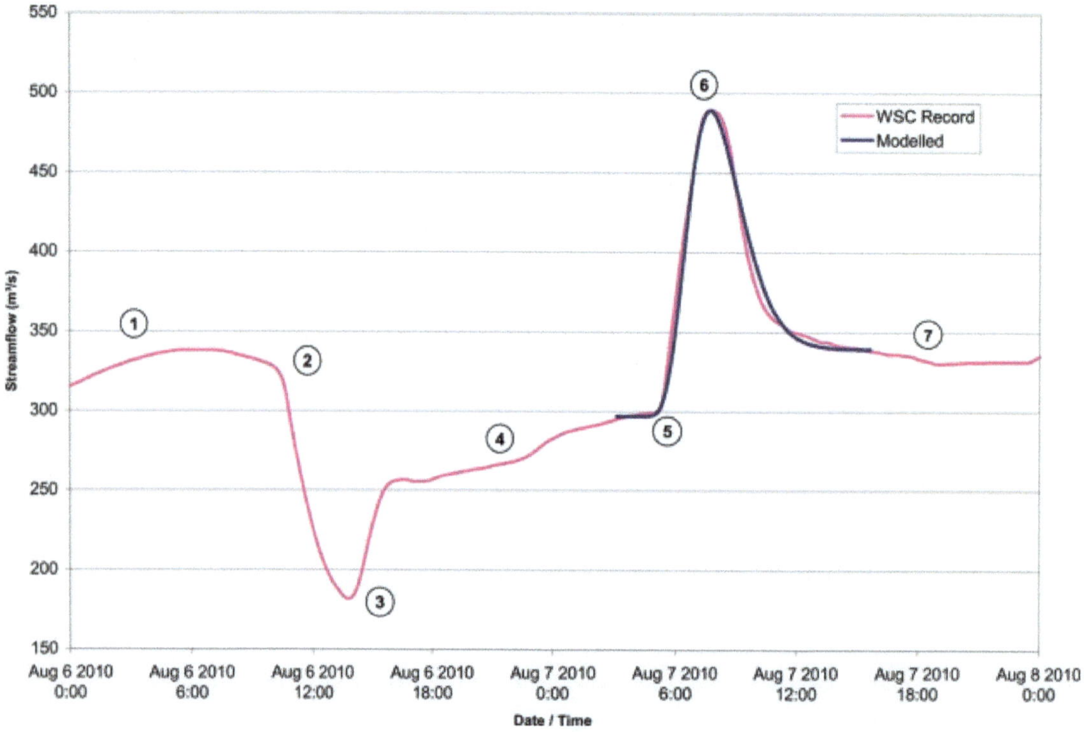

Figure 16. WSC gauge 08MG005, Lillooet River near Pemberton. The gauge records the response of the Lillooet River at Pemberton to the Mount Meager landslide and subsequent damming events. Delay time to the station is approximately 6 h and 45 min. 1, stable base flow (curve is diurnal fluxuation in stream flow); 2, initial drop in flow due to the complete blockage of Meager Creek, and partial blockage of Lillooet at 10:15; flow dropped to 182 $m^3 s^{-1}$; 3, rapid recovery of flow beginning at 13:15 associated with the incision of Lillooet River through the landslide deposit; 4, continued gradual recovery of Lillooet River flow, minus the Meager Creek portion; 5, arrival of the Meager dam breach floodwave at 05:15; 6, arrival of the flood peak at 07:45; 7, re-establishment of stream flow conditions. The blue line represents the modelled dam breach and corroborates the observed stage discharge at Pemberton.

Sediment moving downstream puts pressure on the Pemberton district dyking system, raising the effective flood levels, and the obliteration of Capricorn Creek and inundation of Meager Creek will have long-lasting environmental effects.

In total, direct costs associated with the event are estimated at $10 M CAD. If gravel removal or dike elevation changes become necessary as a result of the increased sediment load, the total long-term costs of this landslide could exceed the direct costs.

Eyewitness Account

Eyewitness accounts, while arguably less scientific, provide important corroboration and context to rarely seen catastrophic events. In the case of the Mount Meager landslide, no lives were lost; however, four witnesses to the event were in extreme jeopardy on several occasions during and immediately following the landslide. K. Kraliz, J. Duffy, J. Tilley, and P. Smith, along with their dog, arrived at upper Lillooet forest campsite at 3:25 a.m. and began unloading their gear. They were surprised by "two large cracks" (loud explosive noises, not physical cracks in the ground) occurring in quick succession, followed by a rumbling that initially sounded like a train or a forest fire, but that grew to a deafening volume in about 20 s (P. Smith, personal communication, 2010). All four campers and their dog got back into their truck and headed for higher ground. Chaos ensued for the next few hours before daylight as they encountered debris flows, mud, falling trees and other hazards at the edge of the landslide deposit in the Lillooet River valley. The most alarming part of their story occurred as they came face to face with a 1.5-m high wall of mud and water that appeared to them like a "turbulent bubbling wedge of black oil" (P. Smith, personal communication, 2010). They turned their truck around, but not before they were overtaken by the hyper-concentrated flow. Accelerating rapidly, they nonetheless escaped, in what P. Smith described as most resembling "the Millennium Falcon escaping the Death Star explosion at the end of Return of the Jedi" (personal communication, 2010).

The account corroborates both the approximate time of the event, the violent nature and noise level of a massive land-slide even at the distal margins, and ongoing activity on the landslide long after the main event occurred. We propose that the two cracks (loud explosive noises) are the subjective memory of the impact of the failing secondary peak.

CONCLUSIONS

In this paper we have documented a large catastrophic land-slide that occurred in August 2010 on the southern flank of Mount Meager, a dissected Quaternary volcanic center within the Mount Meager Volcanic Complex (MMVC) of southwestern British Columbia. The initial failure mass had a volume of 48.5×10^6 m^3 and involved a slope consisting of poor quality volcanic rock. The failed rock mass rapidly transformed into a high-velocity debris flow that swept 7.8 km down Capricorn Creek to dam Meager Creek. At Meager Creek, the debris flow split and traveled 3.7 km upstream, and 4.9 km downstream across the Lillooet River, creating landslide dams on both streams. Our analysis of superelevation in the bends of Capricorn Creek and the dramatic run-up at the confluence of Capricorn Creek and Meager Creek suggest average velocities of 62–73 m s $^{-1}$ and 63 m s^{-1}, respectively.

The 2010 Mount Meager landslide is significant for seven reasons; (1) it is the tenth mass flow involving volumes in excess of 0.5×10^6 m^3 to have occurred in the MMVC since 1850 and the sixth largest mass flow identified in the MMVC over the Holocene; (2) it is one of the largest landslides to have occurred worldwide since 1945; (3) the landslide exhibited dramatic transformation from an initial rock slope failure to a rapidly moving debris flow as a result of the high degree of fragmentation of the initial failure mass; (4) the formation of a significant landslide dam illustrated the importance of landslide damming as a secondary landslide process and as a hazard to distant downstream communities; (5) the seismic trace of the landslide initiation and motion allowed de-tailed reconstruction of the stages and velocity of movement; (6) the event demonstrates the role of rapid post-Little Ice Age deglaciation in destabilizing slopes adjacent to modern glaciers; and (7) it represents a unique opportunity to con-sider hazard and risk from catastrophic failures to mountain communities.

The event illustrates the extreme landslide hazard of glacier-clad dissected Quaternary volcanic centers, which results from the existence of steep slopes in poor quality rock (reflecting such factors as hydrothermal alteration and het-erogeneity of volcanic products). In this geological environment, landslides are a major process of denudation and, through transformation into debris flows, deliver large volumes of debris to river systems.

ACKNOWLEDGEMENTS

The authors gratefully acknowledge logistical support provided by the province of British Columbia, the Centre for Natural Hazard Research at the Simon Fraser University, and the Landslide Research Programme at the University of Waterloo.

REFERENCES

1. Ayotte, D. and Hungr, O.: Calibration of a runout prediction model for debris flows and avalanches, in: Proceedings of the 2nd Inter-national Conference on Debris Flows, Taipei, edited by: Wiec-zorek, G. F. and Naeser, N. D., Balkema, Rotterdam, 505–514, 2000.
2. Berrocal, J., Espinosa, A. F., and Galdos, J.: Seismological and geological aspects of the Mantaro landslide in Peru, Nature, 275, 532–536, 1978.
3. Bovis, M. J. and Jakob, M.: The July 29, 1998, debris flow and land-slide dam at Capricorn Creek, Mount Meager Volcanic Complex, southern Coast Mountains, British Columbia, Can. J. Earth Sci., 37, 1321–1334, 2000.
4. Brodsky, E. E, Gordeev, E., and Kanamori, H.: Landslide basal friction as measured by seismic waves, Geophys. Res. Lett., 30, 2236 I-1–2236 I-5, 2003.
5. Carter, N. M.: Exploration in the Lillooet River watershed, Can. Alpine J., 21, 8–18, 1932.
6. Clague, J. J.: Late Quaternary Geology and Geochronology of British Columbia Part 2: Summary and Discussion of Radiocarbon-dated Quaternary History, Geol. Surv. Can. Pap., 80–35, 41 pp., 1981.
7. Clague, J. J., Evans, S. G., Rampton, V. N., and Woodsworth, G. J.: Improved age estimates for the White River and Bridge River tephras, western Canada, Can. J. Earth Sci., 32, 1172–1179, 1995.
8. Croft, S. A. S.: Stability Assessment of the Capricorn Creek Valley, British Columbia, B.Sc. thesis, University of British Columbia, Vancouver, BC, 1983.
9. Dahlen, F. H.: Single-force representation of shallow landslide sources, Bull. Seismol. Soc. Am., 83, 130–143, 1993.
10. Decker, F., Fougberg, M., and Ronayne, M.: Pemberton: The His-tory of a Settlement, Hemlock Printing Ltd., Burnaby, BC, 1977.
11. Deparis, J., Jongmans, D., Cotton, F., Baillet, L., Thouvenot, F., and Hantz, D.: Analysis of rock-fall and rock-fall avalanche seismo-grams in the French Alps, Bull. Seismol. Soc. Am., 98, 1781– 1796, 2008.
12. Dufresne, A. and Davies, T. R.: Longitudinal ridges in mass move-ment deposits, Geomorphology, 105, 171–181, 2009.
13. Evans, S. G.: A rock avalanche from the peak of Mount Meager, British Columbia, Geol. Surv. Can. Pap. 87-1A, 929–934, 1987.
14. Evans, S. G.: Landslides, in: A synthesis of Geological Hazards in Canada, Geol. Surv. Can. Bull., 548, 43–80, 2001.
15. Evans, S. G.: Single event landslides resulting from massive rock slope failure: Characterising their frequency and impact on soci-ety, in: Landslides from Massive Rock Slope Failure, edited by: Evans, S. G., Scarascia-Mugnozza, G., Strom, A. L., and Her-manns, R. L., NATO Sci. Ser. IV, Earth Environ. Sci., Springer, Dordrecht, 49, 53–73, 2006.
16. Evans, S. G., Hungr, O., and Clague, J. J.: Dynamics of the 1984 rock avalanche and associated debris flow on Mount Cayley, British Columbia, Canada; Implications for landslide hazard as-sessment on dissected volcanoes, Eng. Geol., 61, 29–51, 2001.
17. Evans, S. G., Guthrie, R. H., Roberts, N. J., and Bishop, N. F.: The disastrous 17 February 2006 rockslide-debris avalanche on Leyte Island, Philippines: a catastrophic landslide in tropi-cal mountain terrain, Nat. Hazards Earth Syst. Sci., 7, 89–101, doi:10.5194/nhess-7-89-2007, 2007.
18. Evans, S. G., Roberts, N. J., Ischuck, A., Delaney, K. B., Morozova, G. S., and Tutubalina, O.: Landslides triggered by the 1949 Khait earthquake, Tajikistan, and associated loss of life, Eng. Geol., 109, 195–212, 2009.
19. Friele, P. and Clague, J. J.: Large Holocene landsldies from Py-lon Peak, southwestern British Columbia, Can. J. Earth Sci., 41, 165–182, 2004.
20. Friele, P. and Clague, J.J.: Paraglacial geomorphology of Qua-ternary volcanic landscapes in the southern Coast Mountains, British Columbia, in: Periglacial and Paraglacial Processes and Environments, edited by: Knight, J. and Harrison, S., Geol. Soc. Spec. Publ. London, 320, 219–233, 2009.

21. Friele, P., Clague, J. J., Simpson, K., and Stasiuk, M.: Impact of a Quaternary volcano on Holocene sedimentation in the Lillooet River valley, British Columbia, Sediment. Geol., 176, 305–322, 2005.

22. Friele, P., Jakob, M., and Clague, J. J.: Hazard and risk from large landslides from Mount Meager volcano, British Columbia, Canada, Georisk, 2, 48–64, 2008.

23. Glicken, H.: Sedimentary architecture of large volcanic-debris avalanches, in: Sedimentation in volcanic settings, edited by: Smith, G. A. and Fisher, R. V., SEPM, Tulsa, Special Publica-tion, 45, 99–106, 1991.

24. Guthrie, R. H. and Evans, S. G.: Work, persistence and formative events: The geomorphic impact of landslides, Geomorphology, 88, 266–275, 2007.

25. Hickson, C. J.: Character of volcanism, volcanic hazards, and risk, northern end of the Cascade magmatic arc, British Colubmia and Washington State, in: Geology and Geological Hazards of the Vancouver Region, Southwestern British Columbia, edited by: Monger, J. W. H., Geol. Surv. Can. Bull., 481, 231–250, 1994.

26. Holm, K., Bovis, M., and Jakob, M.: The landslide response of alpine basins to post-Little Ice Age glacial thinning and retreat in southwestern British Columbia, Geomorphology, 57, 201–216, 2004.

27. Huang, C.-J., Yin, H.-Y, Chen, C.-Y., Yeh, C.-H., and Wang, C.-L.: Ground vibrations produced by rock motions and debris flows, J. Geophys. Res., 112, F02014, doi:10.1029/2005JF000437, 2007.

28. Hungr, O.: A model for the runout analysis of rapid flow slides, de-bris flows, and avalanches, Can. Geotech. J., 32, 610–623, 1995.

29. Hungr, O.: Simplified models of spreading flow of dry granular material, Can. Geotech. J., 45, 1156–1168, 2008.

30. Hungr, O. and Evans, S. G.: Rock avalanche runout prediction using a dynamic model, in: Proceedings of the 7th International Sym-posium on Landslides, Trondheim, Norway, 1, 233–238, 1996.

31. Hungr, O. and Evans, S. G.: Entrainment of debris in rock avalanches: An analysis of a long run-out mechanism, Geol. Soc. Am. Bull., 116, 1240–1252, 2004.

32. Hungr, O. and McDougall, S.: Two numerical models for landslide dynamic analysis, Comp. Geosci., 35, 978–992, 2009.

33. Jakob, M.: Morphometric and Geotechnical Controls on De-bris Flow Frequency and Magnitude in Southwestern British Columbia. Ph.D. thesis, University of British Columbia, Vancou-ver, BC, 1996.

34. Jordan, P.: Debris Flows in the Southern Coast Mountains, British Columbia: Dynamic behavior and physical properties, Ph.D. thesis, University of British Columbia, Vancouver, BC, 1994.

35. Julian, B. R., Miller, A. D., and Foulger, G. R.: Non-double-couple Eearthquakes, 1. Theory, Rev. Geophys., 36, 525–549, 1998.

36. Kanamori, H. and Given, J. W.: Analysis of long-period seismic waves excited by the May 18, 1980 eruption of Mount St. He-lens – A terrestrial monopole?, J. Geophys. Res., 87, 5422–5432, 1982.

37. Lipovsky, P. S., Evans, S. G., Clague, J. J., Hopkinson, C., Couture, R., Bobrowsky, P., Ekstrom,¨ G., Demuth, M. N., Delaney, K. B., Roberts, N. J., Clarke, G., and Schaeffer, A.: The July 2007 rock and ice avalanches at Mount Steele, St. Elias Mountains, Yukon, Canada, Landslides, 5, 445–455, doi:10.1007/s10346-008-0133-4, 2008.

38. Mathews, W. H. and McTaggart, K. C.: Hope rockslides, British Columbia Canada, in: Rockslides and Avalanches, Natural Phe-nomona, edited by: Voight, B., Elsevier, New York, 259–275, 1978.

39. McNeely, R. and McCuaig, S.: Geological Survey of Canada Ra-diocarbon Dates XXIX, Geol. Surv. Can. Pap. 89–9, 1991.

40. McSaveney, M. J. and Downes, G.: Application of landslide seis-mology to some New Zealand rock avalanches, in: Landslides, edited by: Rybar, J., Stemberk, J., and Wagner, P., Balkema, Lisse, 649–654, 2002.

41. Mokievsky-Zubok, O.: Glacier caused slide near Pylon Peak, British Columbia, Can. J. Earth Sci., 15, 1039–1052, 1977.

42. Nasmith, H., Mathews, W. H., and Rouse, G. E.: Bridge River ash and some other Recent ash beds in British Columbia, Can. J. Earth Sci., 4, 163–170, 1967.

43. Norris, R. D.: Seismicity of rockfalls and avalanches at three Cas-cade Range volcanoes: Implications for seismic detection of haz-ardous mass movements, Bull. Seismol. Soc. Am., 84, 1925– 1939, 1994.

44. Province of British Columbia: Coast selling price system, average log prices for the 1 month period ending 2010-08-31, Province of British Columbia, Ministry of Forests, Mines and Lands online resource, available at: http://www.for.gov.bc.ca/hva/logreports coast.htm, last access: 4 March 2011.

45. Read, P. B.: Geology, Meager Creek Geothermal Area, British Columbia, Geol. Surv. Can. Open File 603, 1978.

46. Read, P. B.: Mount Meager complex, Garibaldi Belt, southwestern British Columbia, Geosci. Can., 17, 167–170, 1990.

47. Roche, A. D., Guthrie, R. H., Roberts, N. J., Ellis, E., and Friele, P.: Once more into the breach: A forensic analysis of the Au-gust 2010 landslide dam outburst flood at Meager Creek, British Columbia, in: Proceedings of the 5th Canadian Conference on Geotechnique and Natural Hazards, Canadian Geotechnical So-ciety, May 15–17, 2011, Kelowna, BC, 10 pp., 2011.

48. Siebert, L.: Large volcanic debris avalanches: Characteristics of source areas, deposits and associated eruptions, J. Volc. Geo. Res., 22, 163–197, 1984.

49. Simpson, K. A., Stasiuk, M., Shimamura, K., Clague, J. J., and Friele, P. A.: Evidence for catastrophic volcanic debris flows in Pemberton Valley, British Columbia, Can. J. Earth Sci., 43, 679– 689, 2006.

50. Sosio, R., Crosta, G. B., and Hungr, O.: Complete dynamic model-ing calibration for the Thurwieser rock avalanche (Italian Central Alps), Eng. Geol., 100, 11–26, 2008.

51. Stasiuk, M. V., Russell, J. K., and Hickson, C. J.: Distribution, nature, and origins of the 2400 BP eruption products of Mount Meager, British Columbia, Linkages between magma chemistry and eruption behavior, Geol. Surv. Can. Bull., 486, 1–27 1996.

52. Stewart, M.: Dacite Block and Ash Avalanche Hazards in Moun-tainous Terrain: 2360 yr BP Eruption of Mount Meager, British Columbia, M.Sc. thesis, University of British Columbia, 2002.

53. Weichert, D., Horner, R. B., and Evans, S. G.: Seismic signatures of landslides: The 1990 Brenda Mine collapse and the 1965 Hope rockslides, Bull. Seismol. Soc. Am., 84, 1523–1532, 1994.

54. Westgate, J. A. and Dreimanis, A.: Volcanic ash layers of Recent age at Banff National Park, Alberta, Canada, Can. J. Earth Sci., 4, 155–161, 1967.

CHAPTER 14

TECHNICAL NOTE: IMPLEMENTATION OF A GEODATABASE OF PUBLISHED AND UNPUBLISHED DATA ON THE CATASTROPHIC VAIONT LANDSLIDE

L. Superchi, M. Floris, M. Ghirotti, R. Genevois, M. Jaboyedoff, and D. Stead

INTRODUCTION

The Vaiont landslide (Fig. 1) is considered to represent a natural laboratory in which to investigate failure mechanisms and evolution in large rock masses. The catastrophic 1963 landslide demonstrated the paramount importance of detailed geologic investigations. Geological, structural, geo-morphological, hydrogeological and geomechanical features at Vaiont are being reanalyzed using new methods and techniques, such as photogrammetric analyses, terrestrial and aerial laser scanning data acquisition and interpretation, topographic DEM analyses, rock mass characterization and numerical 3-D modelling. Despite that, to date, considerable re-search has been carried out, the Vaiont landslide continues to provide an engineering case study of both high scientific interest and significant technological challenges. The first de-tailed geological studies were carried out by Giudici and Se-menza (1960), who emphasized the existence of a huge pre-historic landslide on the reservoir mountainside, which could be remobilized in the presence of the planned reservoir. Subsequently, following nearly 3 years of intermittent and slow slope movements, a rock mass of about 270 million m^3 suddenly collapsed into the reservoir, generating a huge wave that hit the town of Longarone and other villages, killing almost 2000 people. Several interpretations of this event have been proposed during the last 45 years (e.g., Muller,¨ 1964, 1987a, b; Selli et al., 1964; Skempton, 1966; Hendron and Patton, 1985; Semenza, 2000). Some authors have attempted to explain the sudden acceleration of the mass in some cases suggesting varied mechanisms (Nonveiller, 1992; Tika and Hutchinson, 1999; Kilburn and Petley, 2003). Nevertheless, a comprehensive explanation of both the triggering and dynamics of the phenomenon still remains elusive. The most comprehensive work to date was undertaken by Hendron and Patton (1985) who concluded by emphasizing the need for continued and more in depth research. A complete review of the most important papers related to the 1963 landslide is presented in Genevois and Ghirotti (2005).

Figure 1. Panoramic view of the slip surface of the Vaiont landslide that occurred on 9 October 1963.

Citation: L. Superchi, M. Floris, M. Ghirotti, R. Genevois, M. Jaboyedoff, and D. Stead, "Technical Note: Implementation of a geodatabase of published and unpublished data on the catastrophic Vaiont landslide," Nat. Hazards Earth Syst. Sci., 10, 865–873, 2010. http://www.nat-hazards-earth-syst-sci.net/10/865/2010/nhess-10-865-2010.pdf.

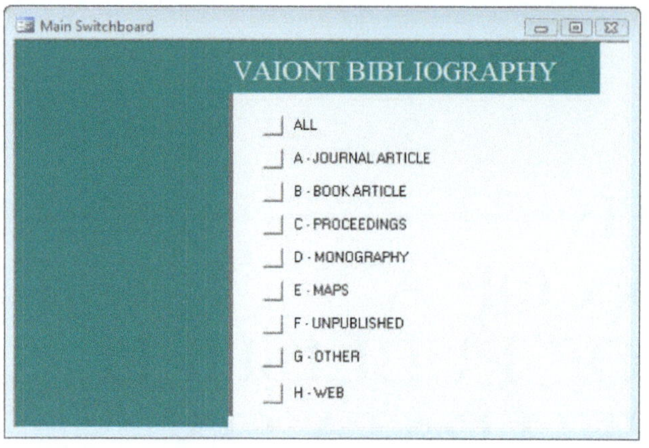

Figure 2. Main switchboard of the electronic bibliographic database.

In this paper, the authors describe the structure and contents of recently developed alphanumeric and spatial databases on the Vaiont landslide. The potential use of these databases in a re-evaluation of the landslide mechanics and dynamics, utilizing current knowledge on rock slides and state-of-the-art methods of data acquisition and numerical modelling is briefly discussed. In addition, all the references to published documents, theses and unpublished technical reports on the Vaiont landslide have been organized and stored in a database and are provided as Supplementary Material (http://www.nat-hazards-earth-syst-sci.net/10/ 865/2010/nhess-10-865-2010-supplement.pdf).

THE ELECTRONIC-BIBLIOGRAPHIC DATABASE

The catastrophic Vaiont landslide has stimulated and generated a significant volume of research, however, most of the available documentation owing to the time at which it was produced is in a non-electronic format. Hence, the first extremely important step was to scan digitally all documentation on the landslide. All the references to published documents, theses, unpublished technical reports and maps on landslide, have subsequently been organized and stored in a database using MS Access software.

The references are contained in a main table and are divided, on the basis of document typology, into Journal article (A), Book article (B), Proceedings (C), Monography (D), Maps (E), Unpublished (F), Other (G), and Web (H), (Fig. 2). Input masks to queries that extract the relevant information for each typology, were created to help in storing and visualizing data (Fig. 3a, b). In order to facilitate data retrieval, all the masks include local (field “LOCAL LINK”) and/or internet (field “LINK”) connections to documents available in a local directory or on a web site. A further important database field is termed “NOTE” where the presence in the document of data from geological and geomechanical surveys, geotechnical data from laboratory tests and scientific contents, is reported (Fig. 4).

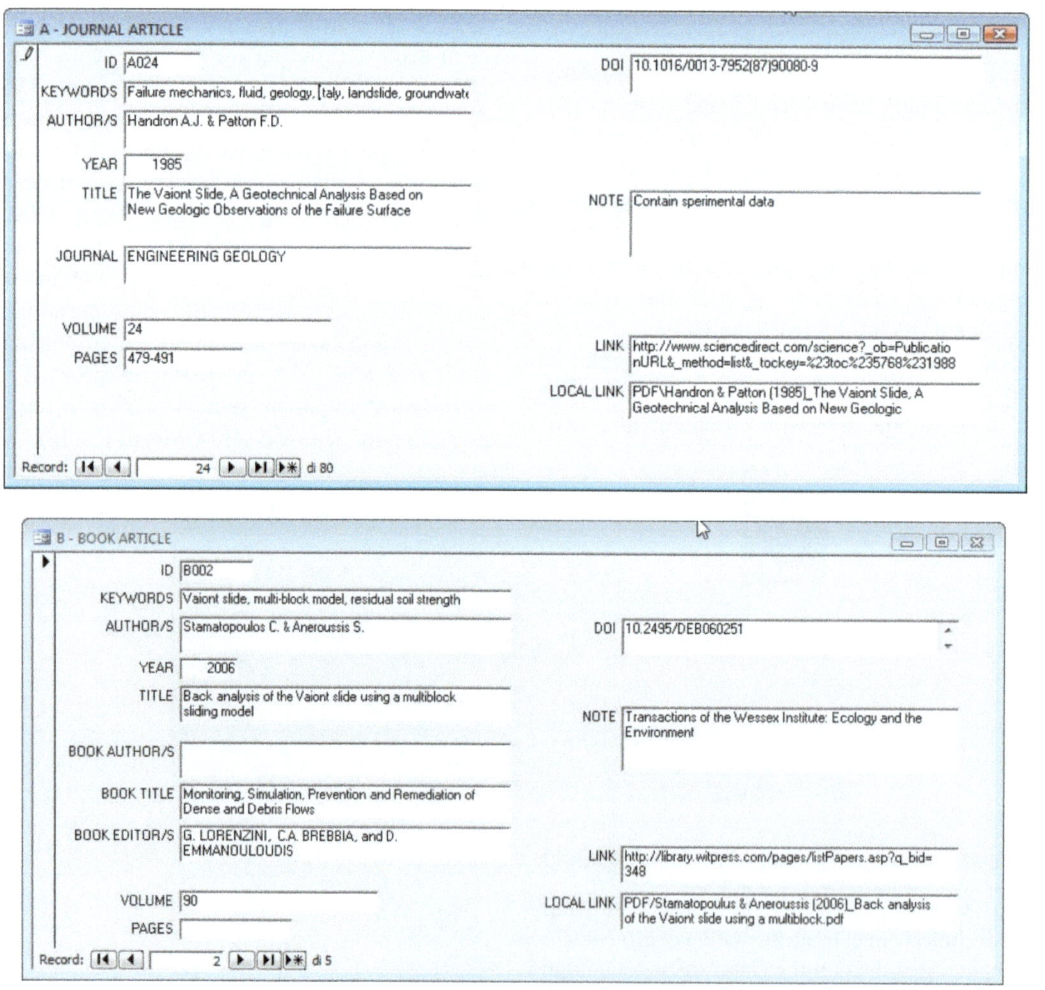

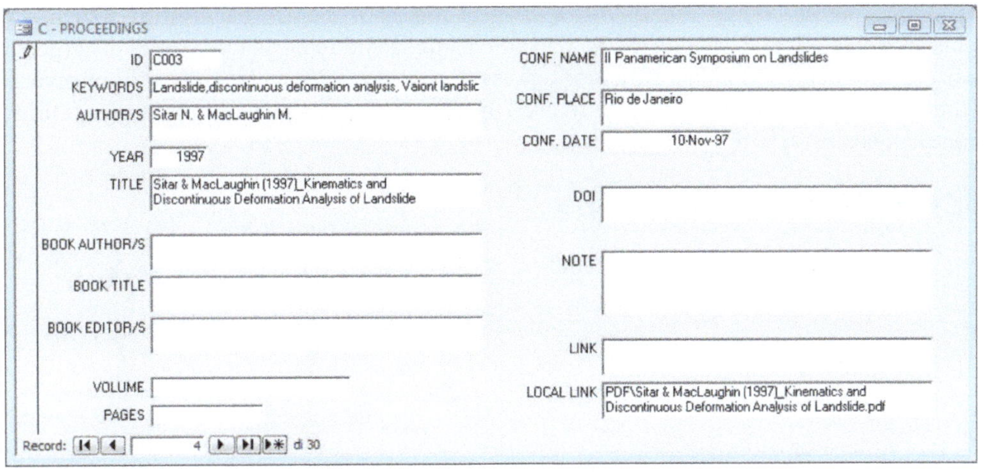

Figure 3a. Input masks created for inserting and visualizing references to journal articles (A), Book articles (B), Proceedings (C). Input mask of Monographs (D) is similar to that for Book article (B).

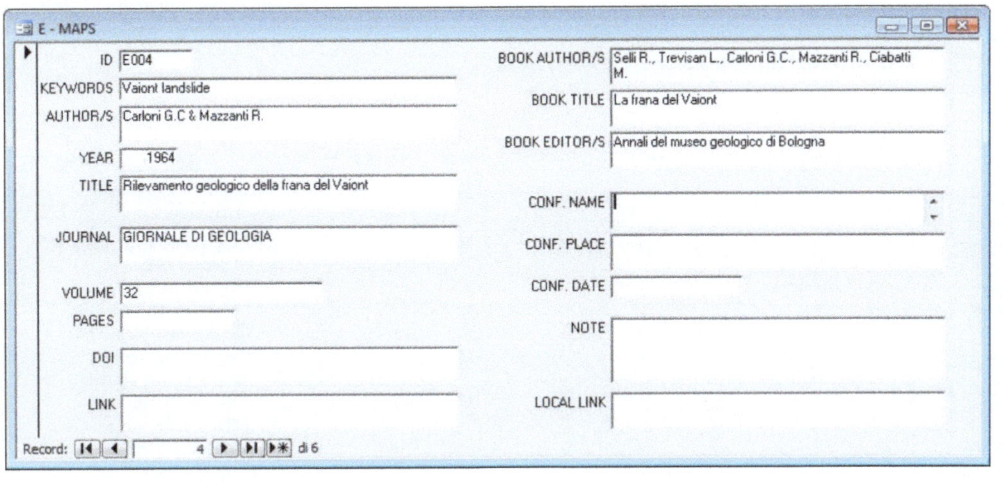

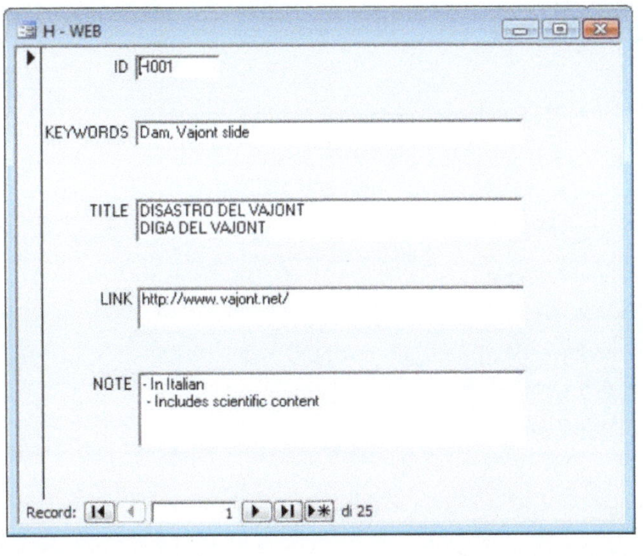

Figure 3b. Input masks created for inserting and visualizing references to maps (E) and web site (H). Input mask of unpublished reports (F) and other (newspaper articles, conference presentations and relevant material) (G) are similar to that for maps (E).

The database currently contains references to 80 journal articles, 5 book articles, 51 monographs, 6 thematic maps, 30 conference proceedings, 5 unpublished reports, 25 web sites and 13 documents classified as "Other" (newspaper articles, conference presentations and relevant material). The chronology of the documentation shows a relatively high production immediately after the landslide (Fig. 5). After a period (1970–1979) of apparent decreasing interest, the documentation increased during 1980–1999, in a large part due to an International Conference on the Vaiont landslide, organized by E. Semenza at the University of Ferrara (Italy) on September 1986. Of note is that most of the documentation, especially journal articles and monographs, have been produced in the last 10 years, as a consequence of both new methods and techniques available for numerical analyses and of the large increase in the electronic tools available for publishing and sharing scientific papers.

STRUCTURE AND CONTENTS OF THE GEODATABASE

A geodatabase is a database designed to store, query, and manipulate geographic information and spatial data. Different types of spatial data, such as vector and raster datasets, and their attributes and location can be stored. In addition, tables and relationships between data can be included. The geodatabase forms the first step in implementing a Geographic Information System (GIS) organized to allow data collection, management and visualization of large slope instabilities and contributing factors (Chacon et al., 2006; Giardino et al., ´ 2004). Vector datasets consist of geometrical primitives such as points, lines, curves and shapes or polygon(s), which are all based on mathematical equations and represent territorial data (geomechanical stations, faults, lithology, etc.). A raster dataset is a data structure representing a generally rectangular grid of pixels, or points of color. In the GIS, raster datasets of varying formats are usually used to represent continuous territorial data (DEM, slope, etc.) and to perform simple to complex analyses.

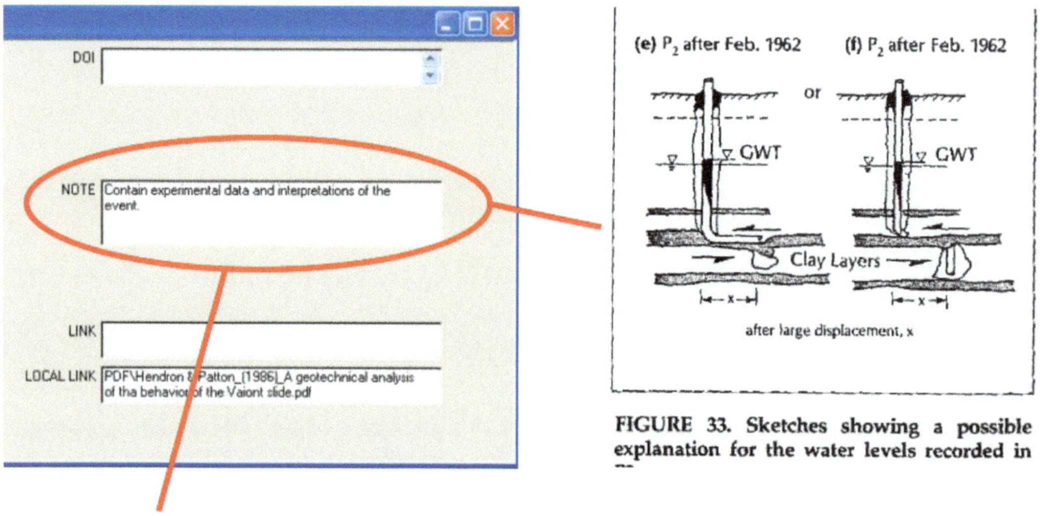

Sample No.	Liquid Limit	Plastic Limit	Plasticity Index	Descriptive Notes
12-6	72	23	49	Clay in debris about 4 m above failure plane
12-6A	35	19	16	Clay on failure plane (10 m from 12-6)
18-6	49	27	22	Clay on failure plain at scarp
18-6A	39	20	19	Clayey debris on rock surface near scarp
18-8	45	32	13	Clay in-situ forms failure plane above
18-9	37	25	12	Clay in-situ forms adjacent failure plane
18-9A	48	33	15	Clay in-situ on failure plane
18-11	38	25	13	Clay in-situ forms failure plane below

Figure 4. Example of data and interpretations reported in the field NOTE of the e-bibliography. Sketches from Hendron and Patton (1986).

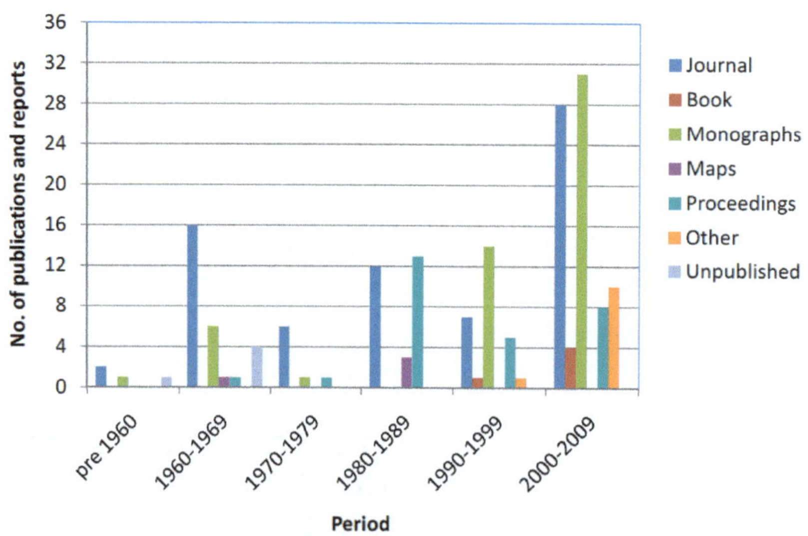

Figure 5. Chronology of the documentation on the Vaiont landslide.

The Vaiont geodatabase was implemented using ESRI ArcGIS Desktop 9.3 software and comprises:

- Feature classes are homogeneous collections of common features, each having the same spatial representation, such as points, lines or polygons.
- Feature datasets are objects that allow to group together related feature classes.
- Tables are data collection of rows (records) and columns (attributes) used to store non-spatial data.
- Relationship classes manage associations between objects in one table and objects in another. Rules to relate feature to feature (spatial relationships), row to row (non spatial relationships), and feature to row (spatial to non spatial relationships), are stored.
- Raster datasets are grid-based representations of spatial data.
- *Raster catalogues* are objects that allow efficient storage and management of multiple spatially-related raster datasets.
- *Terrain datasets* are surfaces that represent three-dimensional space. They use measurements (stored as feature classes) and rules to generate triangular irregular network (TIN) pyramids to represent elevation. From the terrain dataset, it is possible to obtain both a vectorial-based elevation model (TIN) and a raster-based Digital Elevation Model (DEM).

Currently, the Vaiont geodatabase contains the following data.

- Vector datasets representing the attitude of strata, faults, lithology, and geological sections; digitized from pre- and post-landslide geological maps at 1:5000 scale (Rossi and Semenza, 1965).
- Vector datasets representing geognostic boreholes from Broili (1959) and geophysical investigation from Caloi (1960), and their attributes (type, depth, stratigraphy).
- Vector dataset representing geomechanical stations from past and recent surveys. Related attributes consist of rock and rock mass properties (discontinuities at-tributes such as orientation, size, aperture, roughness, etc., Schmidt Hammer testing, Geological Strength Index, Point Load Test, etc.) (see Table 1 for an example).
- Vector datasets representing elevation points and con-tour lines digitized from pre- and post-landslide topographic maps, official regional maps (CTR, Regione Friuli Venezia Giulia) and LiDAR survey.
- Raster datasets representing the Digital Elevation Model (DEM) of the area before and after the landslide. DEMs were calculated from the above mentioned topographic data.
- Raster datasets organized in a Raster Catalog, representing the aerial photos of the area before the landslide (1960).

In addition, the geodatabase contains tables of geotechni-cal and geomechanical properties of the rock mass inside and outside the landslide area, extracted both from the e-bibliography and from current author's field surveys and lab-oratory tests. From simple to complex rules, stored in the Relationships classes, join and relate records in the tables to the location of the collected samples and of the geomechanical stations.

USING THE DATABASES

Modern techniques, such as photogrammetric analyses, ground-based and airborne LiDAR will greatly contribute to geometric and geomechanical rock mass characterization. DEM-based structural analysis, performed by COLTOP-3D software (Derron et al., 2005), on available data before and after the landslide, is an important tool in identifying the structural setting that led to the failure and controlled the direction of the movement. COLTOP-3D uses a color representation merging slope aspect and slope angle in order to obtain a unique color code for each orientation of a topographical element (Jaboyedoff et al., 2007, 2009). Simple analysis of DEMs allow rapid identification of structural features (joints, lineaments, faults) affecting the slope (Derron et al., 2005). The 3-D surface reconstruction is extremely useful as it enables easy identification of the main morphostructural features from which joint set orientations and persistence relevant to the area of interest can be detected. Furthermore, these desktop analysis allow us to explore the area under investigation and thereby provide an aid in planning the field work and the mapping of structural data in inaccessible areas.

Seven main joint sets were detected in the pre- and post-Vaiont landslide slope using the color coding of COLTOP-3D (Fig. 6). The joint surface pole orientations have been measured directly on the DEMs: J1(160°/40°), J2(238°/45°), J3(300°/50°), J4(345°/30°), J5(50°/20°), J6(50°/70°), J7(70°/45°). Figure 6a–b shows that one of the most important sets forming the morphostructure is J4 (light-blue), dipping northward and corresponding to the orientation of the sliding plane. The data, obtained by using COLTOP-3D, agree reasonably well with the field measurements despite the present limited number of field data (Fig. 6e).

CONCLUDING REMARKS

The e-bibliography and the geodatabase represent a powerful tool to extract and select experimental data and scientific con-tents from the extensive documentation on the Vaiont land-slide. The e-bibliography provides an important input for further scientific research not only on the Vaiont landslide but also on large rock landslides in general. Data collected in the electronic bibliographic database show that the majority of the research publications have been authored in the last decade, probably due to recent advances in both the body of knowledge on rock slides and the developments in methods and techniques for rock mass data collection and numerical analyses.

The key objective of the generated geodatabase is to store, manage, visualize and update a large number of different types of data in a central location and to make them avail-able to the scientific community.

3-D geological, hydrogeological and geomechanical reconstruction of rock slope instabilities, such as the Vaiont 1963 event, will be more accurate and comprehensive as it will now be possible to apply new or advanced relationships to collected data, to develop new methods of viewing the available information assets in a interrogative and discriminatory fashion;

in summary, this will allow the leverage of all the collected data in order to optimize their potential for understanding rock slope failure mechanisms. Detailed examination of the geodatabase can provide important data required for numerical modelling of rock slopes i.e.: i) the geological, structural and hydrogeological characteristics prior to the landslide; ii) the geomechanical characterisation of the involved rock masses; iii) the complete definition of the displacement field prior, during and after the landsliding event.

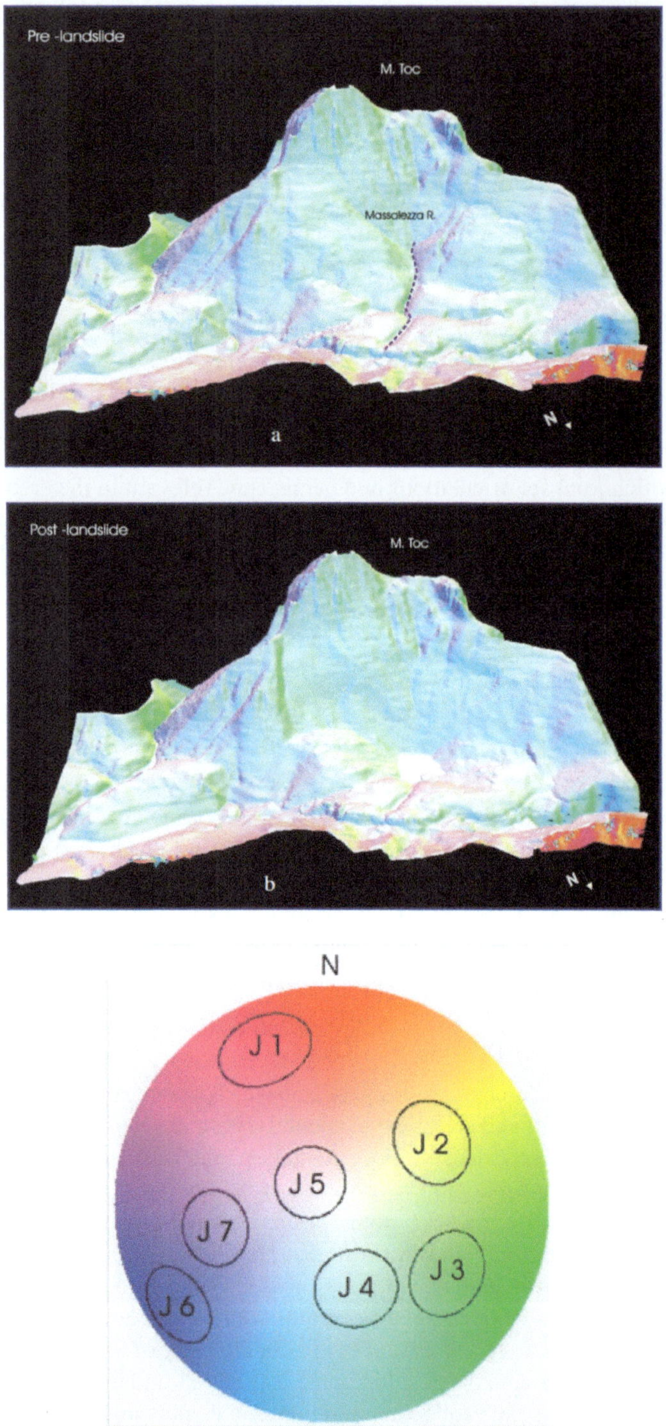

c

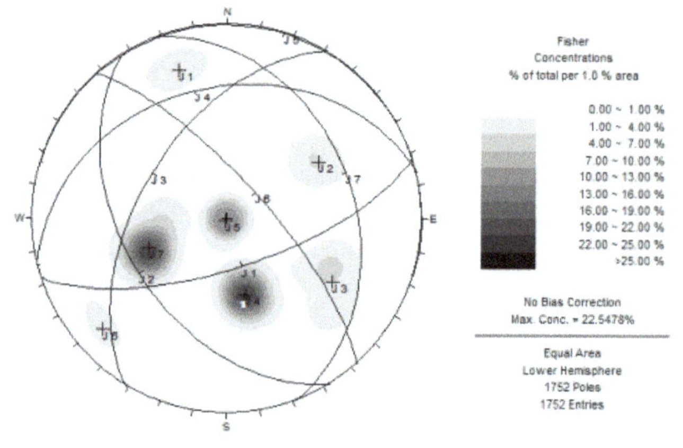

d

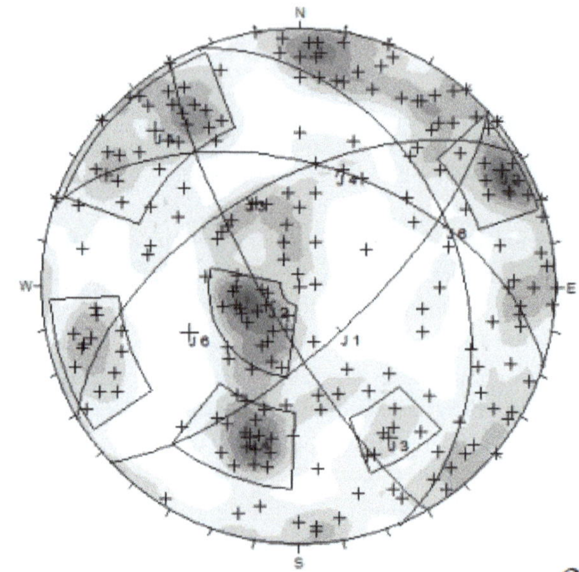

e

Figure 6. COLTOP 3D analysis for the Vaiont area. In the pre- and post-landslide 3-D representation, different colours in the same area show different orientations of the structural features (faults and lineaments) (a–b). Seven morpho-structural sets were detected by COLTOP 3D, evaluated by means of the corresponding Schmidt-Lambert projection (c) and plotted using usual software (d). Obtained results were validated by comparison with the field data (e).

At this stage, the Vaiont geodatabase remains largely incomplete: collected data to date relating mainly to the geological setting, whereas hydrogeological and geomechanical data are somewhat limited, scanty or absent. An example of both the potential of the geodatabase and of its current limitation in available data is given in Table 1. Values and relationships between distinctive properties, such as the Volumetric Joint Count (Jv) and Block Size Index (Ib) (Cai et al., 2004; Kalenchuk et al., 2006; Palmstrom, 1996, 2005), can ¨ aid in defining lithotechnical units, while the Joint Roughness Coefficient (JRC), the Joint Compressive Strength

(JCS) and the Geological Strength Index (GSI) are necessary for continuum or discontinuum modelling of the rock slope. However, the number of data available for each geological formation are insufficient to obtain spatial distributions at a statistically significant level, thus, making it currently impossible to subdivide the rock masses involved in the event into realistic geomechanical units.

Table 1. Geomechanical properties of the material involved in the slide, from past and current surveys. The slip surface has been placed at the bottom of the Calcare di Soccher′ formation (Semenza, 1965). Volumetric Joint Count (Jv), Block Size Index (Ib), Joint Roughness Coefficient (JRC), Joint Compression Strength (JCS), Geological Strength Index (GSI)

Formation	Lithology	Statistics	Jv	Ib	JRC	JCS (MPa)	GSI
	Red and green limestones and marls with red cherts.	No. stations	7	7	7	7	7
		No. of data	3	3	4	4	2
		Minimum	35	3.9	3	44	47
		Maximum	73	11.8	6	104	52
	Fine-grained limestones with various colored cherts.	No. stations	18	18	18	18	18
		No. of data	3	3	17	17	2
		Minimum	22.8	4.3	1	29	47
		Maximum	99.3	8.8	10	180	62
	Red marly, silty limestones interbedded with conglomerate and limestones.	No. stations	8	8	8	8	8
		No. of data	1	1	8	8	1
		Minimum	58	6.4	2	30	40
		Maximum			10	150	
Calcare di Socchér	Compact beds of gray limestones that alternated with a sequence of less resistant thin layers of greenish limestones and calcareous marls	No. stations	17	17	17	17	17
		No. of data	2	2	16	14	2
		Minimum	43	3.2	2	54	52
		Maximum	44.5	8	10	165	52
	conglomerate with pinkish or gray cement	No. stations	3	3	3	3	
		No. of data			3	3	
		Minimum			4	48	
		Maximum			9	130	
	Micrites and e marley-micrites from light green to reddish with red cherts	No. stations	8	8	8	8	8
		No. of data	3	3	7	5	2
		Minimum	58.3	4.11	1	40	40
		Maximum	66.5	9.9	8	125	62
Amm. Rosso							
	Red or nodular micrite, locally containing cherts	No. stations	1	1	1	1	1
		No. of data	1	1	1		1
		Minimum	46.8	7.36	5		57
		Maximum			8		
Fonzaso	Micrites and calcarenites containing cherts, with interbeds of green clay	No. stations	3	3	3	3	3
		No. of data			2	2	
		Minimum			2	44	
		Maximum			6	96	
	Gray cherty limestones with black cherts, which can be nodular						
		No. stations	5	5	5	5	5
		No. of data	3	3	4	4	3
Calcare del Vaiont	Oolitic and crystalline limestones	Minimum	20	3.5	1	53	42
		Maximum	89	14	6	140	50

A re-evaluation of the failure mechanism of the Vaiont landslide requires further research on still poorly constrained features of the area with a collection of additional spatial data. Field surveys, tests and analyses are in progress in order to improve the geodatabase. Field investigations are being extended into areas peripheral to the landslide, in or-der to provide a more comprehensive characterization of the rock masses in the pre- and post-event conditions; this will form the foundation for 2-D/3-D numerical simulation using continuum and discontinuum numerical codes.

Surface based techniques, such as the comparison of pre- and post-event DEMs, and acquisition of deep level seismic tomography, to investigate the 3-D distribution of different rock masses, will allow improved definition of the displacement field of the landslide body.

Completing the geodatabase and performing the re-evaluation of the Vaiont landslide will represent an important contribution to a more comprehensive understanding of large rock slope instabilities.

ACKNOWLEDGEMENTS

This research was financially supported by the University of Padova, research projects CPDA085240 (Principal investigator: Mario Floris, Department of Geosciences) and GEO-RISKS (Principal investigator: Rinaldo Genevois, Department of Geosciences).

REFERENCES

1. Broili, L.: New knowledge on the geomorphology of the Vaiont Slide slip surfaces. Rock Mechanics & Eng. Geol., J. Int. Soc. Rock Mechanics., V(1), 38–88, 1967.
2. Cai, M., Kaiser, P. K., Uno, H., Tasaka, Y., and Minami, M.: Esti-mation of rock mass deformation modulus and strength of jointed hard rock masses using the GSI system. Int. J. Rock Mech. Min-ing Sci., 41, 3–19, 2004.
3. Caloi, P.: L'evento del Vajont nei suoi aspetti geodinamici, Ist. Naz. Geofisica, Roma, 1966.
4. Chacon,´ J., Irigaray, C., Fernandez, T., and El Hamdouni, R.: Engi-neering geology maps: landslides and geographical information systems, B. Eng. Geol. Environ., 65, 341–411, 2006.
5. Derron, M.-H., Jaboyedoff, M., and Blikra, L. H.: Preliminary as-sessment of rockslide and rockfall hazards using a DEM (Opp-stadhornet, Norway), Nat. Hazards Earth Syst. Sci., 5, 285–292, 2005, http://www.nat-hazards-earth-syst-sci.net/5/285/2005/.
6. Genevois, R. and Ghirotti, M.: The 1963 Vaiont Landslide – Giornale di Geologia Applicata, I, 41–53, 2005.
7. Giardino, M., Giordan, D., and Ambrogio, S.: G.I.S. technologies for data collection, management and visualization of large slope instabilities: two applications in the Western Italian Alps, Nat. Hazards Earth Syst. Sci., 4, 197–211, 2004, http://www.nat-hazards-earth-syst-sci.net/4/197/2004/.
8. Giudici, F. and Semenza, E.: Studio geologico del serbatoio del Vajoint. Unpublished report for S.A.D.E., Part A: 21 pp, text, Part B: 68 photos with discussions, 42 pp, Venezia, Italy, 1960.
9. Hendron, A. J. and Patton, F. D.: The Vaiont Slide, a Geotechni-cal Analysis Based on New Geologic Observations of the Failure Surface. I, II, Technical Report GL-85-5, US Army Eng. Water-ways Experiment Station, Vicksburg, MS, 1985.
10. Hendron, A. J. and Patton, F. D.: A geotechnical analysis of tha behavior of the Vaiont slide, Civil Engineering Practice Fall, 65– 130, 1986.
11. Jaboyedoff, M., Metzger, R., Oppikofer, T., Couture, R., Derron, M.-H., Locat, J., and Turmel, D.: New insight techniques to ana-lyze rock-slope relief using DEM and 3D-imaging cloud points: COLTOP-3D, Rock mechanics Vol. 1: Meetings Society's Chal-lenges and demands, chap. 7, 61–68, edited by: Eberhardt, E., Stead, D., and Morrison, T., 2007.
12. Jaboyedoff, M., Couture, R., and Locat, P.: Structural analysis of Turtle Mountain (Alberta) using digital elevation model: toward a progressive failure, Geomorphology, 103, 5–16, 2009.
13. Kalenchuk, K. S., Diederichs, M. S., and McKinnon, S.: Character-izing block geometry in jointed rock masses, Int. J. Rock Mech. Mining Sci., 43(8), 1212–1225, 2006.
14. Kilburn, C. R. J. and Petley, D. N.: Forecasting giant, catastrophic slope collapse: lessons from Vajont, Northern Italy, Geomor-phology, 54(1–2), 21–32, 2003.
15. Muller,¨ L.: The Rock Slide in the Vaiont Valley, Rock Mech. Eng. Geol., 2, 148–212, 1964.
16. Muller,¨ L.: The Vaiont catastrophe - A personal review, in: Dam Failures, edited by: Leonards, G. A., Eng. Geol., 24(1–4), 423– 444, 1987a.
17. Muller,¨ L.: The Vaiont Slide, in: Dam Failures, edited by: Leonards, G. A., Eng. Geol., 24(1–4), 513–523, 1987b.
18. Nonveiller, E.: Vaiont Slide – Influence of Frictional Heat on Slip Velocity, in: Proceedings of the meeting on the 1963 Vaiont land-slide, edited by: Semenza, E. and Melidoro, G., IAEG Italian Section, University of Ferrara, Ferrara 1986, 1, 187–197, 1992.
19. Palmstrom,¨ A.: Characterizing rock masses by the RMi for use in practical rock engineering, Tunn. Undergr. Sp. Tech., 11(2), 175– 188, 1996.
20. Palmstrom,¨ A.: Measurements of and correlations between block size and rock quality designation (RQD), Tunn. Undergr. Sp. Tech., 20, 362–377, 2005.
21. Rossi, D. and Semenza, E.: Carte geologiche del versante setten-trionale del M. Toc e zone limitrofe, prima e dopo il fenomeno di scivolamento del 9 ottobre 1963, Scala 1:5000, Ist. Geologia Universita` di Ferrara, 2 Maps, 1965.
22. Selli, R., Trevisan, L., Carloni, C. G., Mazzanti, R., and Cia-batti, M.: La Frana del Vajont, Giornale di Geologia, serie 20, XXXII(I), 1–154, 1964.
23. Semenza, E.: Sintesi degli studi geologici sulla frana del Vaiont dal 1959 al 1964, Mem. Museo Tridentino Sc. Nat., 16, 1–52, 1965.
24. Semenza, E.: La storia del Vaiont raccontata dal geologo che ha scoperto la frana. Tecomproject Ed., Ferrara, Italy, 2000.
25. Semenza, E. and Ghirotti, M.: History of 1963 Vaiont Slide. The importance of the geological factors to recognise the ancient landslide, B. Eng. Geol. Env., 59, 87–97, 2000.
26. Skempton, A. W.: Bedding-plane slip, residual strength and the Vaiont Landslide, Geotechnique,´ XVI(1), 82–84, 1966.
27. Tika, Th. E. and Hutchinson, J. N.: Ring shear tests on soil from the Vaiont landslide slip surface, Geotechnique, 49(1), 59–74, 1999.

CHAPTER 15

MACRO-ROUGHNESS MODEL OF BEDROCK–ALLUVIAL RIVER MORPHODYNAMICS

L. Zhang, G. Parker, C. P. Stark, T. Inoue, E. Viparelli, X. Fu, and N. Izumi

INTRODUCTION

The pace of river-dominated landscape evolution is set by the rate of downcutting into bedrock across the channel network. The coupled process of river incision and hillslope response is both self-promoting and self-limiting (Gilbert, 1877). Although there are multiple processes that can lead to erosion into bedrock, we here focus on incision driven by abrasion of a bedrock surface as moving particles collide with it. Low rates of incision entail some sediment supply from upstream hillslopes, which provides a modicum of abrasive material in river flows that further facilitates bedrock channel erosion. Faster downcutting leads to higher rates of hillslope sediment supply, boosting the concentration of erosion "tools" and bedrock wear rates, but also leading to greater cover of the bedrock bed with sediment (Sklar and Dietrich, 2001, 2004, 2006; Turowski et al., 2007; Lamb et al., 2008; Turowski, 2009). Too much sediment supply leads to choking of the channels by alluvial cover and the retardation of further channel erosion (e.g., Stark et al., 2009). This competition between incision and sedimentation leads long-term eroding channels to typically take a mixed bedrock–alluvial form in which the pattern and depth of sediment cover fluctuate over time in opposition to the pattern of bedrock wear.

Theoretical approaches to treating the erosion of bedrock rivers have shifted over recent decades (see Turowski, 2012, for a recent review). The pioneering work of Howard and Kerby (1983) focused on bedrock channels with little sediment cover; it led to the detachment-limited model of Howard et al. (1994), in which channel erosion is treated as a power function of river slope and characteristic discharge, and the "stream-power-law" approach, in which the power-law scaling of channel slope with upstream area underpins the way in which landscapes are thought to evolve (Whipple and Tucker, 1999; Whipple, 2004; Howard, 1971, fore-shadows this approach). At the other extreme, sediment flux came into play in the transport-limited treatment of mass removal from channels of, for example, Smith and Brether-ton (1972), in which no bedrock is present in the channel and where the divergence of sediment flux determines the rate of lowering. Whipple and Tucker (2002) blended these approaches, and imagined a transition from detachment limitation upstream to transport-limited behavior downstream. They also discussed, in the context of the stream-power-law approach, the idea emerging at that time (Sklar and Dietrich, 1998) of a "parabolic" form of the rate of bedrock wear as a function of sediment flux normalized by transport capacity. Laboratory experiments conducted by Sklar and Dietrich (2001) corroborated this idea, and they led to the first true sediment flux-dependent model of channel erosion of Sklar and Dietrich (2004, 2006). This saltation–abrasion model was subsequently extended by Lamb et al. (2008) and Chatanantavet and Parker (2009). It was explored experimentally by Chatanantavet and Parker (2008) and Chatanantavet et al. (2013); evaluated in a field context by Johnson et al. (2009), Chatanantavet and Parker (2009), Hobley et al. (2011) and Turowski et al. (2013); adapted to treat alluvial intermittency by Lague (2010); and given a stochastic treatment by Turowski et al. (2007), Turowski (2009) and Lague (2010), the latter of whom introduced several new elements. Howard (1998) presents an alternative formulation for incision that relates bedrock wear to the thickness of alluvial cover rather than sediment supply, in a form that can be thought to be a predecessor of the present work.

At the heart of their saltation–abrasion model lies the idea of a cover factor p corresponding to the areal fraction of the bedrock bed that is covered by alluvium (Sklar and Diet-rich, 2004). This bedrock bed is imagined as a flat surface on which

Citation: Zhang, L., Parker, G., Stark, C. P., Inoue, T., Viparelli, E., Fu, X., and Izumi, N.: Macro-roughness model of bedrock–alluvial river morphodynamics, Earth Surf. Dynam., 3, 113-138, doi:10.5194/esurf-3-113-2015, 2015.

sediment intermittently accumulates and degrades during bedload transport over it. The fraction of sediment cover is assumed to be a linear function of bedload transport relative to capacity. Bedrock wear takes place when bedload clasts strike the exposed bedrock. In the simplest form of the saltation–abrasion model, the subsequent rate of bedrock wear is treated as a linear function of the impact flux and inferred to be proportional to the bedload flux, which leads to the parabolic shape of the cover-limited abrasion curve.

The saltation–abrasion model is considerably more sophisticated and flexible (Sklar and Dietrich, 2004, 2006) than this sketch explanation can encompass. It does, however, have three major restrictions. First, it is formulated in terms of sediment supply rather than local sediment transport. The model is thus unable to capture the interaction between processes that drive evolution of an alluvial bed and those that drive the evolution of an incising of bedrock–alluvial bed. Second, for related reasons, it cannot account for bedrock topography significant enough to affect the pattern of sediment storage and rock exposure. Such a topography is illustrated in Fig. 1 for the Shimanto River, Japan. Third, it is designed for quasi-steady conditions, and thus cannot account for the effects of cyclic variation in sediment supply on channel development downstream of the point of sediment supply.

Here we address all three of these issues in a model that allows both alluvial and incisional processes to inter-act and coevolve. We do this by relating the cover factor geometrically to a measure of the vertical scale of elevation fluctuations of the bedrock topography, here called macro-roughness, rather than to the ratio of sediment supply rate to capacity sediment transport rate. Our model encompasses downstream-advecting alluvial behavior (e.g., waves of alluvium), diffusive alluvial behavior and upstream-advecting incisional behavior (e.g., knickpoint migration). In order to distinguish between the model of Sklar and Dietrich (2004, 2006) and the present model, we refer to the former as the CSA (Capacity-based Saltation-Abrasion) model, and the latter as the MRSAA (Macro-Roughness-based Saltation-Abrasion-Alluviation) model. We point out here that the first and third issues indicated above have also been addressed by Lague (2010), although in a substantially different way than presented here. The notation used in this paper is defined in Table A1.

Figure 1. Views of the Shimanto River, a mixed alluvial–bedrock river in Shikoku, Japan. **(a)** Upstream view at low stage. **(b)** Macroscopic roughness of the bed and alluvial patches. Channel width is about 100 m.

CAPACITY-BASED SALTATION-ABRASION (CSA) GEOMORPHIC INCISION LAW AND ITS IMPLICATIONS FOR CHANNEL EVOLUTION: UPSTREAM-MIGRATING WAVES OF INCISION

CSA Geomorphic Incision Law

Sklar and Dietrich (2004, 2006) present the following model, referred to here as the Capacity-based Saltation-Abrasion (CSA) model, for bedrock incision in mixed bedrock– alluvial rivers transporting gravel. Defining $\boldsymbol{E}$ as the vertical rate of erosion into bedrock, $\boldsymbol{q}_a$ as the volume gravel transport rate per unit width (specified in their model solely in terms of a supply, or feed rate $\boldsymbol{q}_{af}$) and $\boldsymbol{q}_{ac}$ as the capacity volume gravel transport per unit width such that $\boldsymbol{q}_a < \boldsymbol{q}_{ac}$,

$$E = \beta q_a \left(1 - \frac{q_a}{q_{ac}}\right), \tag{1}$$

where β is an abrasion coefficient with dimension L^{-1} . By introducing a cover factor parameter p, this equation can be rewritten as

$$E = \beta q_{ac} p(1-p). \tag{1b}$$

This cover factor p is defined (Sklar and Dietrich, 2006) as the areal fraction of bedrock surface covered with alluvium and is given by

$$p = \begin{cases} \dfrac{q_a}{q_{ac}} \ldots & 0 \le \dfrac{q_a}{q_{ac}} \le 1 \\ 1 \ldots & \dfrac{q_a}{q_{ac}} > 1, \end{cases} \tag{2}$$

where this fraction is calculated by averaging over a window larger than a characteristic macroscale of bedrock elevation variation. We refer to this formulation for cover factor p as "capacity based" because Eq. (2) dictates that p is determined in terms of the ratio of sediment supply to its capacity value in the CSA model.

In the above formulation, it is assumed that the gravel transport rate q_a over a bedrock surface can be estimated by simply multiplying the capacity rate q_{ac} by the areal cover fraction p. While this is the simplest first-order assumption, it should be recognized that the roughness of the bedrock itself can change the flow resistance, leading to a relationship that is more complex than Eq. (1b) (Inoue et al., 2014; Johnson, 2014).

Before introducing the relation of Sklar and Dietrich (2006) for abrasion coefficient β, it is of value to provide an interpretation for this parameter not originally given by Sklar and Dietrich (2004, 2006), but which plays a useful role in the analysis below. The abrasion coefficient has a physical interpretation in terms of Sternberg's law (Sternberg, 1875) for downstream diminution of grain size (Parker, 1991, 2008; Chatanantavet et al., 2010). Consider a clast of material that is of identical rock type to the bedrock being abraded. Sternberg's law is

$$D = D_u e^{-\alpha_d x}, \tag{3}$$

where D is gravel clast size, D_u is the upstream value of D, x is downstream distance and α_d is a diminution coefficient. If all diminution results from abrasion, α_d is related to β by

$$\alpha_d = \frac{\beta}{3}. \tag{4a}$$

In the case of constant β, and therefore constant αd, the distance Lhalf for such a clast to halve in size is given by

$$L_{half} = \frac{\ln(2)}{\alpha_d}. \tag{4b}$$

This interpretation of abrasion coefficient β in terms of diminution coefficient αd allows for comparison of the experimental results of Sklar and Dietrich (2001) with values of αd previously obtained from abrasion mills (Parker, 2008: see Fig. 3-41 therein; Kodama, 1994).

The relations of Sklar and Dietrich (2004, 2006) to compute β and q_{ac} can be cast in the following form:

$$\beta = \frac{0.08\rho_g R g Y}{k\sigma_t^2}\left(\frac{\tau^*}{\tau_c^*} - 1\right)^{-1/2}\left[1 - \frac{\tau^*}{R_f^2}\right]^{3/2}, \tag{5a}$$

$$R_f = \frac{v_f}{\sqrt{RgD}}, \tag{5b}$$

$$q_{ac} = \alpha_a \sqrt{RgD}\, D\left(\tau^* - \tau_c^*\right)^{n_a}. \tag{5c}$$

In the above relations, D corresponds to the characteristic size of the gravel clasts that are effective in abrading the bedrock; ρ_g is the material density of the grains; R is their submerged specific gravity (~ 1.65 for quartz); g is gravitational acceleration, τ* is the dimensionless Shields number of the flow; $\tau*_c$ is the threshold Shields number for the onset of significant bedload transport; α_a and n_a denote, respectively, a relation-specific dimensionless coefficient and an exponent; v_f is the fall (settling) velocity corresponding to grain size D, Y is the bedrock modulus of elasticity; σ_t is the rock tensile strength; and k is a dimensionless coefficient of the order of 10^{-6}. (In the above two relations and the text, several misprints in Sklar and Dietrich (2004, 2006) have been corrected on the advice of the authors.) Equation (5c) corresponds to the bedload transport relation of Fernandez Luque and van Beek (1976) when $\alpha_a = 5.7$ and $n_a = 1.5$; Sklar and Dietrich (2004, 2006) used this relation with the assumed value $\tau*_c = 0.03$.

The relations above define a 0-D formulation. It must be augmented with other parameters and relations, including channel width, relations for hydraulics, quantification of flow discharge or flow duration curve, etc., to allow application at the river reach scale.

It is useful to cast Eq. (5a) in the form

$$\beta = \beta_{ref}\frac{\left(\frac{\tau^*}{\tau_c^*} - 1\right)^{-1/2}\left[1 - \frac{\tau^*}{R_f^2}\right]^{3/2}}{\left(\frac{\tau_{ref}^*}{\tau_c^*} - 1\right)^{-1/2}\left[1 - \frac{\tau_{ref}^*}{R_f^2}\right]^{3/2}}, \tag{5d}$$

where β_{ref} is a reference value of β, either computed from known values of the parameters Y , k, σ_t , R_f , etc., or estimated indirectly.

Embedding of CSA into a Model of Bedrock Surface Evolution

A relation for the evolution of bedrock surface elevation η_b is obtained by substituting the CSA geomorphic law for incision of Eq. (1b) into a simplified 1-D mass conservation equation for bedrock material subjected to piston-style rock uplift or base level fall (Sklar and Dietrich, 2006):

$$\frac{\partial \eta_b}{\partial t} = \upsilon - IE. \tag{6a}$$

Here t denotes time, υ denotes the relative vertical velocity between the rock underlying the channel (which is assumed to undergo no deformation) and the point at which base level is maintained, and I denotes a flood intermittency factor to account for the fact that only relatively rare flow events are likely to drive incision (Chatanantavet and Parker, 2009). Also, I is assumed to be a prescribed constant; a more generalized formulation for flow hydrograph is given in Sklar and Dietrich (2006) and DiBiase and Whipple (2011). In interpreting Eq. (6a), it should be noted that υ denotes a rock uplift rate (in the sense of England and Molnar, 1990) for the case of constant base level, or equivalently a rate of base level fall for rock undergoing neither uplift nor subsidence. Below we use the term "rock uplift" as shorthand for the relative vertical velocity between the rock and the point of base level maintenance. Substituting Eq. (1b) into Eq. (6a) yields (Sklar and Dietrich, 2006)

$$\frac{\partial \eta_b}{\partial t} = \upsilon - I\beta q_{ac} p(1-p). \tag{6b}$$

Character of the CSA Model: Upstream Waves of Incision

The MRSAA model (introduced below) has several new features as compared to CSA. These are best illustrated by first characterizing the mathematical nature of CSA in the context of Eq. (6). Let

$$S_b = -\frac{\partial \eta_b}{\partial x} \tag{7}$$

denote the streamwise bedrock-surface slope. Reducing Eq. (6b) with Eq. (7) the CSA model of Eq. (1) reveals itself as a nonlinear kinematic wave equation with a source term:

$$\frac{\partial \eta_b}{\partial t} - c_b\frac{\partial \eta_b}{\partial x} = \upsilon, \tag{8a}$$

$$c_b = \frac{I\beta q_{ac} p(1-p)}{S_b}. \tag{8b}$$

Here c_b denotes the wave speed associated with bedrock incision. The form of Eq. (8a) dictates that disturbances in bedrock elevation always move upstream. We will see later that these disturbances can take the form of upstream-migrating knickpoints (e.g., Chatanantavet and Parker, 2009).

Any solution of Eqs. (8a) and (8b) subject to the cover relation of Eq. (2) requires specification of a flow model. In mountain streams, backwater effects are likely to be negligible (e.g., Parker, 2004). The normal (steady, uniform) flow assumption allows for simplification. Let Q denote water discharge during (morphodynamically active) flood flow taking place with intermittency I , and let H denote flood depth and g denote acceleration due to gravity. Momentum and mass balance take the forms

$$\tau = \rho g H S_b, \tag{9a}$$

$$Q = UBH, \tag{9b}$$

where τ is boundary shear stress at flood flow, U is the corresponding mean flow speed, B is channel width and ρ is water density. The dimensionless Shields number τ* and dimensionless Chézy resistance coefficient Cz are defined as

$$\tau^* = \frac{\tau}{\rho R g D}, \tag{10a}$$

$$Cz = \frac{U}{\sqrt{\tau/\rho}}. \tag{10b}$$

As shown in Parker (2004) and Chatanantavet and Parker (2009), reducing Eqs. (7), (9) and (10) yields the following relations for H and τ*:

$$H = \left(\frac{Q^2}{Cz^2 g B^2 S_b}\right)^{1/3}, \tag{11a}$$

$$\tau^* = \left(\frac{Q^2}{Cz^2 g B^2}\right)^{1/3} \frac{S_b^{2/3}}{RD}. \tag{11b}$$

A comparison of Eqs. (2), (5c) and (11b) indicates that even for constant values of other parameters, the functional forms for q_{ac} and thus p are such that c_b is in general a nonlinear function of $S_b = -\partial\eta_b/\partial x$.

Limitations of the CSA Model

The CSA model (Sklar and Dietrich, 2004, 2006) was a major advance in the analysis of bedrock incision due to abrasion because it (a) accounts for the effect of alluvial cover and tool availability on the incision rate through the term $\boldsymbol{p}(1 - \boldsymbol{p})$ in Eq. (1b) and (b) provides a physical basis for incision due to abrasion as gravel clasts collide with the bedrock surface. The CSA model been used, modified, adapted and extended by a number of researchers (Crosby et al., 2007; Lamb et al., 2008; Chatanantavet and Parker, 2009; Turowski, 2009; Lague, 2010).

The model does, however, have a significant limitation in that it specifically does not include either alluvial morphodynamics or the morphodynamics of transitions between bedrock and alluvial zones. Here we study this limitation, and how to overcome it, in terms of the highly simplified configuration of a reach (HSR, highly simplified reach) with constant width; fixed, non-erodible banks; constant water discharge; and sediment input only from the upstream end. For simplicity, we also neglect abrasion of the gravel itself, so that grain size $\boldsymbol{D}$ is a specified constant. (This condition, while introduced arbitrarily here, can be physically interpreted in terms of clasts that are much more resistant to abrasion than the bedrock.) The means to relax these constraints is available (e.g., Chatanantavet et al., 2010; DiBiase and Whipple, 2011), and indeed many of them have been implemented in the SSTRIM model of Lague (2010). Such a relaxation, however, obscures the first-order physics under-lying the rich patterns of interaction between completely and partially alluviated conditions illustrated herein.

In the CSA model, the bedload transport rate $\boldsymbol{q}_a$ is specified as a "supply". That is, the bedload transport rate is con-strained so that it cannot change in the downstream direction, and is always equal to the bedload feed rate (supply) $\boldsymbol{q}_{af}$ at the upstream end. When the feed rate $\boldsymbol{q}_{af}$ increases, $\boldsymbol{q}_a$ must increase simultaneously everywhere. That is, a change in bedload supply is felt instantaneously throughout the en-tire reach, regardless of its length.

We illustrate this behavior in Fig. 2. The reach has length $\boldsymbol{L}$. The gravel feed rate at $\boldsymbol{x} = 0$ follows a cyclic "sedimentograph" (in analogy to a hydrograph) with period $\boldsymbol{T} = \boldsymbol{T}_h + \boldsymbol{T}_l$, in which the sediment feed rate has a constant high feed rate $\boldsymbol{q}_{afh}$ for time $\boldsymbol{T}_h$, and a subsequent constant low feed rate $\boldsymbol{q}_{afl}$ for time $\boldsymbol{T}_l$. According to the CSA model, at $\boldsymbol{x} = \boldsymbol{L}$ corresponding to the downstream end of the reach, the temporal variation in bedload transport rate must precisely reflect the feed rate. That is, the model was not designed to route sediment in the downstream direction.

In a more realistic model, the effect of a change in bedload feed rate $\boldsymbol{q}_{af}$ would gradually diffuse and propagate down-stream, so that the bedload transport rate at the downstream end of the reach would show more gradual temporal variation. This effect is illustrated in Fig. 2. This same diffusion and propagation can be expected in the cover fraction $\boldsymbol{p}$, which in general should vary in both $\boldsymbol{x}$ and $\boldsymbol{t}$. The change in cover fraction in turn should affect the incision rate as quantified in Eq. (1a). To capture this effect, however, Eq. (1b) must be coupled with an alluvial formulation that routes sediment downstream over the bedrock.

A second limitation concerns alluviation of the bedrock surface. Consider a wave of sediment moving over this sur-face, as shown in Fig. 3. We characterize the vertical scale of the geometric roughness of the bedrock surface (as seen in Fig. 1) in terms of a vertical macro-roughness $\boldsymbol{L}_{mr}$. For simplicity, we assume that the bottom of the bedrock relief has a specific elevation $_b$, an assumption that will be relaxed later in favor of a probabilistic formulation. The alluvial thickness above this basal elevation $_a$ represents an average value of local bed elevation over an appropriately defined window. It can be seen in Fig. 3 that the surface undergoes both partial ($\eta_a < L_{mr}$) and then complete ($\eta_a \geq L_{mr}$) alluviation, only to be excavated later as the wave passes through.

Bed elevation is given as

$$\eta = \eta_b + \eta_a. \tag{12}$$

Figure 3 shows that, in the case of complete alluviation, the elevation of the bed η can be arbitrarily higher than the elevation η_b of the bedrock, the difference between the two corresponding to the thickness η_a. The CSA model does not describe the variation in bed elevation η when the bed undergoes transitions between partial and complete alluviation; it simply infers that incision is shut down by the complete alluvial cover.

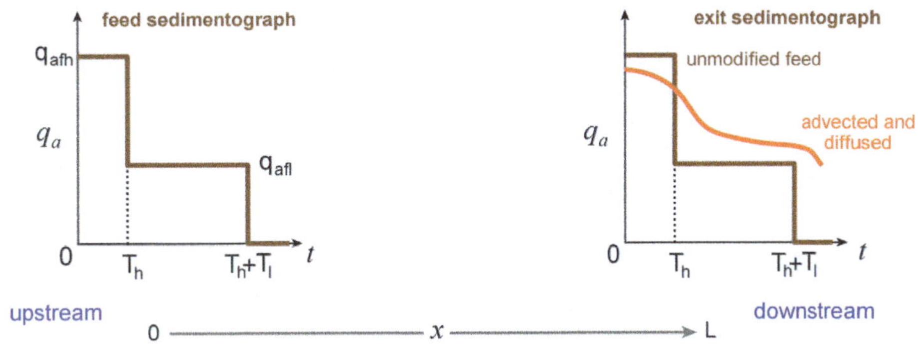

Figure 2. Schematic diagram illustrating downstream modification of a sedimentograph. At the upstream feed point (x =0, left panel), the bedload transport rate q_a takes the high feed value q_{afh} for time T_h and the low feed value q_{afl} time T_l, for a total cycle time of $T = T_h + T_l$. At the downstream end (x = L, right panel), the discontinuous brown line represents the unaltered sedimentograph at the downstream end of the reach, assumed to have propagated instantaneously from the supply point, while the smoother red line represents the sedimentograph as modified by advective–diffusive effects.

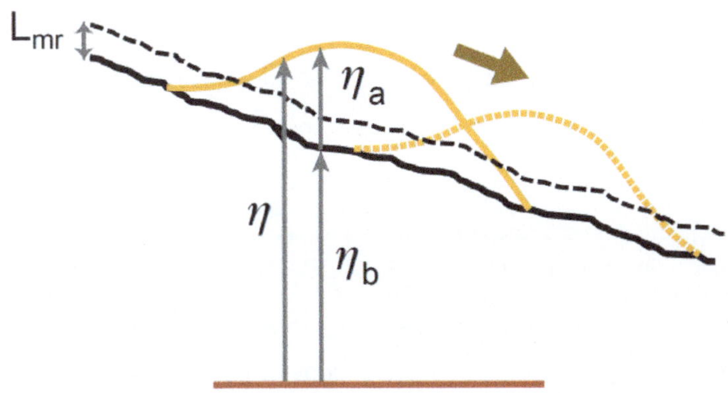

Figure 3. Schematic diagram illustrating the propagation of a wave of sediment over bedrock (orange line shifting to dashed orange line over time). Here η_b denotes the elevation of the bottom of the bedrock relief (black line), L_{mr} denotes the bedrock macroroughness thickness (dashed black line), η_a denotes the thickness of the alluvial cover (which may be less than or greater than L_{mr}) and $\eta = \eta_b + \eta_a$ denotes the elevation of the top of the alluvium.

The goal of this paper is the development and implementation of a model that overcomes these limitations by (a) capturing the spatiotemporal coevolution of the sediment trans-port rate, alluvial cover thickness and bedrock incision rate, and (b) explicitly enabling spatiotemporally evolving transitions between bedrock–alluvial morphodynamics and purely alluvial morphodynamics. The form of the model presented here is simplified in terms of the HSR outlined above, including a constant-width channel and a single sediment source upstream.

MACRO-ROUGHNESS-BASED SALTATION-ABRASION-ALLUVIATION (MRSAA) FORMULATION AND ITS IMPLICATIONS FOR CHANNEL EVOLUTION

Formulation for Alluvial Sediment Conservation and Cover Factor

The geomorphic incision law of the MRSAA model is identical to that of CSA, i.e., Eq. (1b). The essential differences are contained in (a) a formulation for the cover factor p that differs from Eq. (2) and (b) the inclusion of alluvial morpho-dynamics in a way that tracks the spatiotemporal evolution of the bedload transport rate, and allows for smooth spatiotemporal transitions between the bedrock–alluvial state and the purely alluvial state.

The specific case we consider here is one for which (a) the bedrock surface is rough in a hydraulic sense (as opposed to a hydraulically smooth or transitional surface; see Schlichting, 1979), and (b) the characteristic vertical scale of bedrock elevation fluctuation about a mean value based on an appropriately defined window, here denoted as the macro-roughness L_{mr} of the bedrock, is large compared to the characteristic size of the clasts constituting the alluvium. We use the term "macro-roughness" so as to clearly distinguish it from hydraulic roughness, which is specifically defined in terms the logarithmic velocity profile. Inoue et al. (2014) introduced the terms "clast-rough" and "clast-smooth", the former referring to a bedrock surface roughness that is large compared to the characteristic size of the alluvium, and the latter referring to a bedrock surface macro-roughness that is small compared to the size of the alluvium. Here we specifically consider the clast-rough case.

We formulate the problem by considering a conservation equation for the alluvium, in standard Exner form, appropriately adapted to include below-capacity transport over a non-erodible surface. The first model of this kind is due to Struiksma (1999), and further progress has been made by Parker et al. (2009, 2013), Izumi and Yokokawa (2011), Izumi et al. (2012), Tanaka and Izumi (2013) and Zhang et al. (2013). These models are expressed in continuous form; Lague (2010) presents a discrete version based on a series of reaches of finite length that allows for generalization to a continuous form.

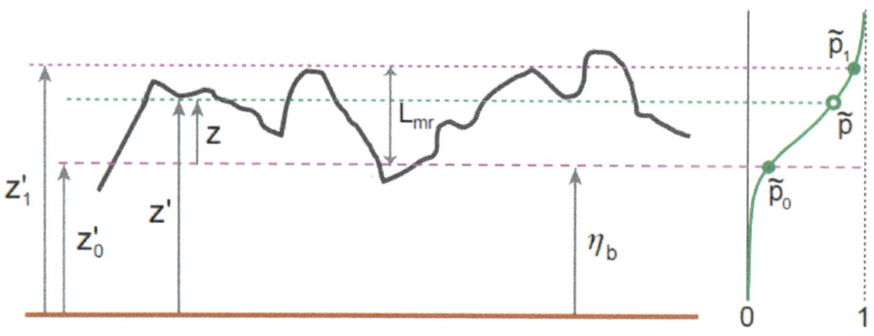

Figure 4. Illustration of the statistical structure or local hypsometry of the bedrock surface topography (dark-grey line). Here z' denotes an elevation above some arbitrary datum (dark-red line) deep in the bedrock and $\tilde{p}(z')$ is the probability (green line, marker) that a point at elevation z' is in water or alluvium rather than bedrock, i.e., above the local bedrock surface. The effective "bottom" of the bedrock relief is located at elevation $z'_0 = \eta_b$, where $\tilde{p}$ takes an appropriately selected low value $\tilde{p}_1$ (e.g., 0.05); the effective "top" of the bedrock relief is located at elevation z'_1, where petakes an appropriately selected high value $\tilde{p}_1$ (e.g., 0.95); and the macro-roughness L_{mr} is given as $z'_1 - z'_0$. The coordinate $z = z' - z'_0$ is referenced to the effective "bottom" of the bedrock relief.

None of the above models is specifically designed to handle the clast-rough case, in particular that shown in Fig. 1, where the elevation of the bedrock roughness has a random element. Here we handle the clast-rough case by first characterizing the statistical nature of the bedrock surface alone. As noted in Fig. 4, z'denotes elevation above an arbitrary datum deep in the bedrock, and $\tilde{p}$ (z') denotes the probability that a point located at elevation z' is located in alluvium or water rather than bedrock. Conversely, 1 − pedenotes the probability that a point at elevation z' is in bedrock (rather than water or alluvium above). As seen on the right-hand side of the figure, $\tilde{p}$ (z')→0 as z' → −∞ (pure bedrock) and $\tilde{p}$ (z')→1 as z'→ +∞. This statistical structure function (a hypsometric curve for local bedrock topography) which we use here to characterize bedrock elevation fluctuations is analogous to that used in Parker et al. (2000) for alluvial beds. It should be noted that "−∞" is shorthand for "far below the bedrock surface" and "+∞" is shorthand for "far above the bedrock surface". It should also be emphasized that the tilde in the parameter peindicates it is not a cover factor, but rather a statistical parameter referring to the bedrock relief itself.

In such a statistical formulation, bedrock relief has neither a precise "bottom" nor a precise "top". Rather, the "bottom" and "top" of the bedrock topography, as well as the macroroughness L_{mr}, are here defined in a statistical sense. This can be done using moments or exceedance probabilities; here we use the latter.

Let $\tilde{p}_0$ denote some low reference value of $\tilde{p}$ (e.g., $\tilde{p}_0$ = 0.05, or deep into the bedrock relief) and $\tilde{p}$ denote a corresponding high reference value of $\tilde{p}$ (e.g., $\tilde{p}_1 = 1 - \tilde{p}_0 = 0.95$, or near the upper portion of the bedrock relief), and z'_0 and z'_1 denote the corresponding bed elevations. An effective "base" of the bedrock relief can be set at z'_0, a macro-roughness height L_{mr} defined as

$$L_{mr} = z'_1 - z'_0 \qquad (13)$$

and an effective "top" of the bedrock specified as $z'_0 + L_m r$. The clast-rough condition considered here satisfies the constraint that $L_{mr}/D >> 1$. The problem can now be rephrased in terms of a vertical coordinate z with its origin located at the effective bottom of the bedrock topography:

$$z = z' - z'_0 . \qquad (14)$$

As noted in Fig. 4, the statistical variable $\tilde{p}$ (z) with shifted coordinate z now has the following properties:

$$\tilde{p}(z' = 0) = \tilde{p}_0, \; \tilde{p}(z = L_{mr}) = \text{pe}1. \qquad (15)$$

Here we define the thickness of the alluvial cover ηa as the elevation difference between the locally averaged top of the alluvium and the elevation z = 0. The cover fraction p associated with any alluvial thickness η_a relative to the macroroughness height L_{mr} is then given as

$$p(\eta_a/L_{mr}) = \tilde{p}\,(z = \eta_a). \qquad (16)$$

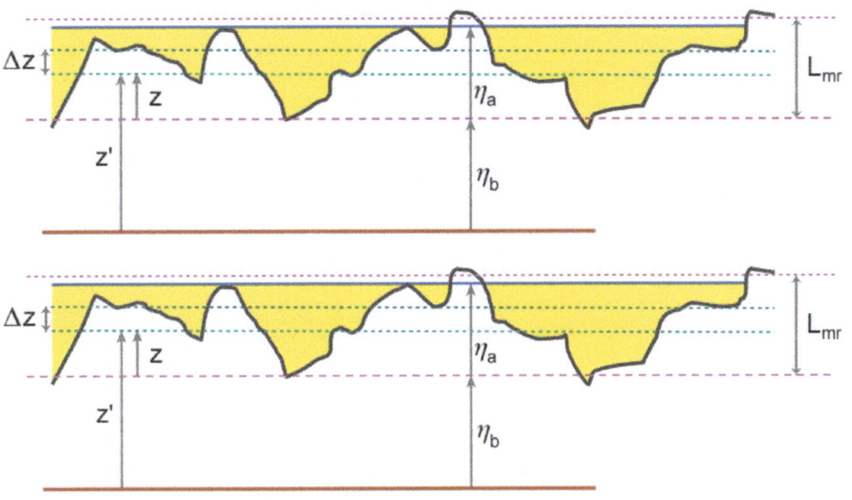

Figure 5. Schematic diagram for derivation of the Exner equation of sediment continuity over a bedrock surface (dark-grey line). As in Figs. (3) and (4), z is the elevation above the effective "bottom" of the bedrock relief, L_{mr} is macro-roughness height, η_b is elevation at base of the bedrock and η_a is the thickness of alluvium (yellow fill). The diagram shows alluvium–water interfaces (blue line) that are at spatially constant elevations. This is for illustrative purposes only; the interfaces should instead be those that result from averaging over an appropriate spatial window, i.e., alluvial fill levels are expected to vary from one pocket in the bedrock relief to another.

Exner Equation of Alluvial Sediment Conservation over a Bedrock Surface

The alluvial sediment is taken to have constant porosity λ. As illustrated in Fig. 5, the volume of alluvial sediment per unit area between elevations z and $z + \Delta z$ is $(1-\lambda)\,\tilde{p}(z)\,\Delta z$, , and the corresponding volume bedload transport rate per unit width q_o is estimated as p q_{ac}, where again, according to Eq. (16), $p = \tilde{p}(\eta_a)$.

For the case of sediment of constant density, the Exner equation for mass balance of alluvial sediment can be expressed as

$$(1-\lambda)\frac{\partial}{\partial t}\int_0^{\eta_a}\tilde{p}\mathrm{d}z = -I\frac{\partial p q_{ac}}{\partial x}, \qquad (17)$$

where the factor I accounts for the fact that morphodynamics are active only during floods. Reducing Eq. (17) using Leibniz's rule,

$$(1-\lambda)p\frac{\partial \eta_a}{\partial t} = -I\frac{\partial p q_{ac}}{\partial x}. \qquad (18)$$

The above formulation for the conservation of alluvial sediment over a bedrock surface differs in one essential way from the earlier forms due to Struiksma (1999), Parker et al. (2009, 2013), Izumi and Yokokawa (2011), Izumi et al. (2012), Tanaka and Izumi (2013) and Zhang et al. (2013). Specifi- cally, in Eq. (18), the cover fraction p is present on the lefthand side of the equation as well as the right-hand side. It is shown below that this feature dictates a strong nonlinearity in the speed of propagation of alluvial waves over a bedrock surface, such that wave speed increases with decreasing wave amplitude. This feature is specifically captured by means of Leibniz's rule, as implemented between Eqs. (17) and (18).

The combination of Eqs. (6b) and (18) delineates a formulation encompassing both mixed bedrock–alluvial rivers and alluvial rivers.

Closure Model for Cover Relation

In the present formulation, the cover fraction p is free to vary in both x and t, i.e., p = p(x, t). In order to complete the problem, however, it is necessary to specify a closure model for p. We characterize the local variation in bedrock elevation in terms of the macro-roughness, i.e., the vertical length scale L_{mr} of Fig. 4. Here we seek a formulation that averages over a window capturing a statistically relevant sample of this local variation. In general, we assume a cover relation that characterizes to what extent the alluvial cover "drowns" the bedrock roughness elements. More specifically, we assume the form

$$p = f(\chi), \qquad (19a)$$

$$\chi = \frac{\eta_a}{L_{mr}} \geq 0, \qquad (19b)$$

$$f(\chi = 0) = \tilde{p}_0, \qquad (19c)$$

$$f(\chi = 1) = \tilde{p}_1. \qquad (19d)$$

The precise details of the relation can be expected to vary from case to case, but the overall characteristics that we hypothesize are illustrated in Fig. 6a–c. It is seen therein that p takes the residual value $p = \tilde{p}_0$ (e.g., 0.05) at $\chi = 0$, increases monotonically to $p = \tilde{p}_1$ (e.g., 0.95) at $\chi = 1$, and then takes the asymptotic value $p \rightarrow 1$ as χ becomes sufficiently large. The first of these conditions corresponds to a bedrock surface that is bare of alluvium except in deep pockets of the macro-roughness elements, the second to a bedrock surface that is nearly completely alluviated but with some parts of the macro-roughness elements exposed, and the third to a bedrock surface that is deeply alluviated.

Note that the cover relation of Fig. 6 and Eq. (19) is based on the macro-roughness height scale L_{mr} rather than the transport capacity q_{ac} of Eq. (2). This is the motivation for referring to the new model presented here as the MacroRoughness-based Saltation-Abrasion-Alluviation (MRSAA) model.

In applying the MRSAA model to general cases, it is useful to delineate the simplest functional form for the closure relation for cover fraction that satisfies the constraints of Eq. (19) and Fig. 6. This relation is the piecewise-linear form

$$p = f(\chi) = \begin{cases} \widetilde{p}_0 + (\widetilde{p}_1 - \widetilde{p}_0)\,\chi & \ldots 0 \le \chi \le \frac{1-\widetilde{p}_0}{\widetilde{p}_1-\widetilde{p}_0} \\ 1 & \ldots \chi > \frac{1-\widetilde{p}_0}{\widetilde{p}_1-\widetilde{p}_0} \end{cases} \tag{20a}$$

or rephrasing to emphasize the dependence of p on the thickness of alluvial cover η_a,

$$p = f(\eta_a/L_{mr}) = \begin{cases} \widetilde{p}_0 + (\widetilde{p}_1 - \widetilde{p}_0)\,\frac{\eta_a}{L_{mr}} & \ldots 0 \le \frac{\eta_a}{L_{mr}} \le \frac{1-\widetilde{p}_0}{\widetilde{p}_1-\widetilde{p}_0} \\ 1 & \ldots \frac{\eta_a}{L_{mr}} > \frac{1-\widetilde{p}_0}{\widetilde{p}_1-\widetilde{p}_0} \end{cases}. \tag{20b}$$

The above relation is illustrated in Fig. 7a with the sample evaluations

$$\widetilde{p}_0 = 0.05, \quad \widetilde{p}_1 = 0.95. \tag{21}$$

Equations (20) and (21) are used in implementations of the MRSAA below. One way to develop forms of Eq. (19) that can show a wider variety of behavior than Eq. (20) would be through the performance of experiments similar to those of Chatanantavet and Parker (2008), but with a specific focus on various forms of macro-roughness that mimic those in the field.

The form for the derivative of Eq. (20) with respect to χ, which is given below, will prove useful in succeeding analysis.

$$\frac{df}{d\chi} = \begin{cases} \widetilde{p}_1 - \widetilde{p}_0 & \ldots 0 \le \chi \le \frac{1-\widetilde{p}_0}{\widetilde{p}_1-\widetilde{p}_0} \\ 0 & \ldots \chi > \frac{1-\widetilde{p}_0}{\widetilde{p}_1-\widetilde{p}_0} \end{cases} \tag{22}$$

Our model is specifically meant to apply to the case for which the characteristic size of the roughness elements is large compared to the size of clasts transported as bedload. With this in mind, it should be noted that in Fig. 6, no bed elevation variations are shown over parts of the bed that are covered with alluvium. This is done only for simplicity, and reflects the condition that in the clast-rough case considered here, grain size is small compared to macro-roughness height. Figure 6 also contains another simplification, in that all pockets are assumed to be filled to the same level by alluvium. While this condition is not likely to be true at the local scale, it is a reasonable first approximation when averaging over an appropriately defined window.

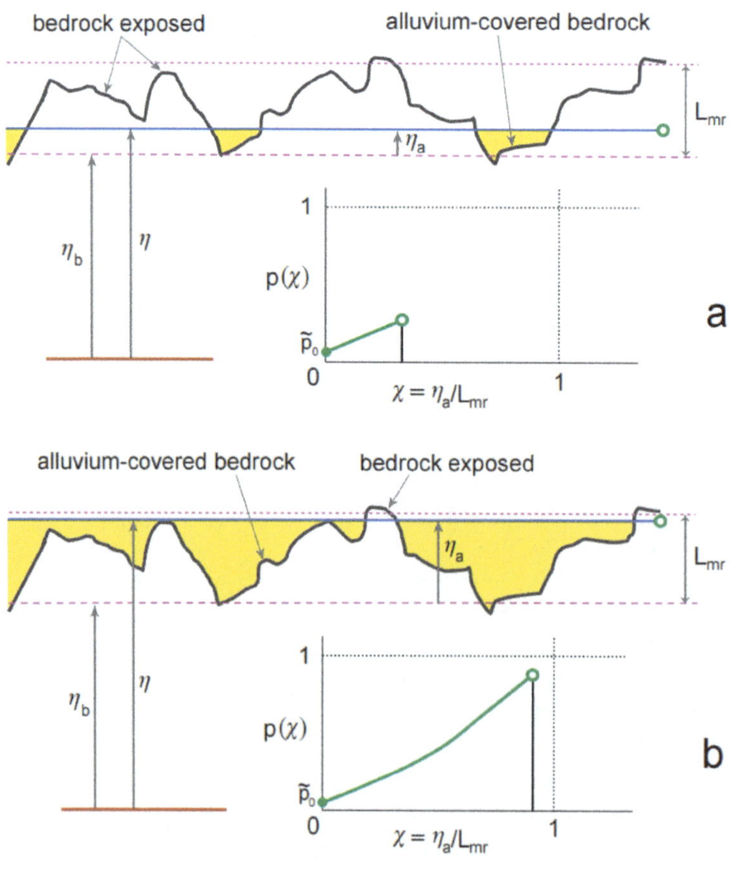

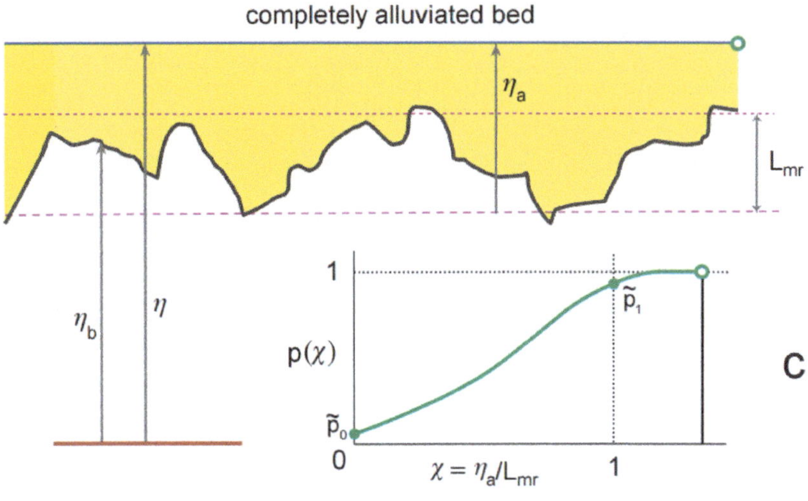

Figure 6. Illustration of the MRSAA model relation between areal fraction of alluvial cover of bedrock p(χ) (green curves) and $\chi = \eta_a/L_{mr}$, where η_a is the thickness of alluvium (yellow fill) and L_{mr} is the macro-roughness height, for (a) low cover, (b) intermediate cover (c) complete alluviation above the top of the bedrock. The diagrams show alluvium–water interfaces at spatially constant elevations (blue lines). This is for illustrative purposes only: the interfaces should instead be those that result from averaging over an appropriate spatial window.

The formulation presented here has an obvious limitation. Since it is a 1-D expression of sediment conservation over a bedrock surface, it cannot capture 2-D variation, which will result in a more complex pattern than that shown in Fig. 6, and in particular will provide more connectivity between adjacent pockets. This two-dimensionality is known to have an effect on the pattern of incision, as illustrated by Johnson and Whipple (2007). The extension of the formulation to the 2-D case represents a future goal; some relevant comments can be found in the "Discussion" section (Sect. 8).

Sections 3.4 and 3.5, immediately below focus, on the mathematical interpretation of the MRSAA problem in terms of diffusion and wave characteristics. The reader whose primary interest is in applications may jump directly to Sect. 3.6, with the two exceptions of Eqs. (27) and (28) in Sect. 3.5. Equation (27) is a version of Eq. (6b) in which incisional morphodynamics are recast into a kinematic wave equation, revealing upstream-migrating waves of incision with wave speed c_b. Equation (28) is a version of Eq. (18) in which alluvial morphodynamics are recast in terms of an advection–diffusion equation, with downstream migrating waves of alluviation with speed c_a, and alluvial diffusion with kinematic coefficient of diffusion κ_a.

Character of the Alluvial Part of the MRSAA Problem: Alluvial Diffusion and Downstream-Migrating Waves of Alluviation

Equation (18) may be reduced to reveal the presence of an alluvial wave speed as follows. The derivative on the righthand side of the equation is expanded using the chain rule, the derivative ∂_p/∂_x is reduced in accordance with the general closure form of Eq. (19), and both sides of Eq. (18) are then divided by p to yield

$$\frac{\partial \eta_a}{\partial t} + c_a \frac{\partial \eta_a}{\partial x} = -\frac{1}{(1-\lambda)} I \frac{\partial q_{ac}}{\partial x}, \tag{23a}$$

where

$$c_a = \frac{I}{(1-\lambda)} \frac{q_{ac}}{L_{mr} p} \frac{\mathrm{d}f}{\mathrm{d}\chi}. \tag{23b}$$

The left-hand side of Eq. (23a) thus takes a kinematic wave form, such that c_a is the wave speed of downstream-directed alluviation.

It is important to realize that alluvial wave speed c_a is a nonlinear function of alluvial thickness η_a. Using the example functions of Eqs. (20) and (22), the speed c_{ai} of an alluvial wave of infinitesimal height $\eta_a = 0$ is given from Eq. (23b) as

$$c_{ai} = \frac{I}{(1-\lambda)} \frac{q_{ac}}{L_{mr}} \frac{(\widetilde{p}_1 - \widetilde{p}_0)}{\widetilde{p}_0}. \tag{24a}$$

The ratio c_a/c_{ai} of the wave speed c_a at height η_a to the corresponding value for an infinitesimal wave is then found to be

$$\frac{c_a}{c_{ai}} = \frac{(\widetilde{p}_1 - \widetilde{p}_0)}{\widetilde{p}_0} \frac{1}{p} \frac{\mathrm{d}f}{\mathrm{d}\chi}. \tag{24b}$$

In Fig. 7b, the ratio c_a/c_{ai} is plotted against $\chi = \eta_a/L_{mr}$ in accordance with the specifications of Eqs. (20), (21), and (22). It is seen therein that alluvial wave speed takes its maximum value for the limit $\eta_a \rightarrow 0$, and decreases to a vanishing value as the bed becomes completely alluviated ($\eta_a/L_{mr} = 1.056$). That is, waves of alluvium run fastest over a nearly bare bed, and wave-like behavior ceases to exist under conditions of complete alluviation. This latter result is in accordance with Lisle et al. (2001) and Cui et al. (2003a, b).

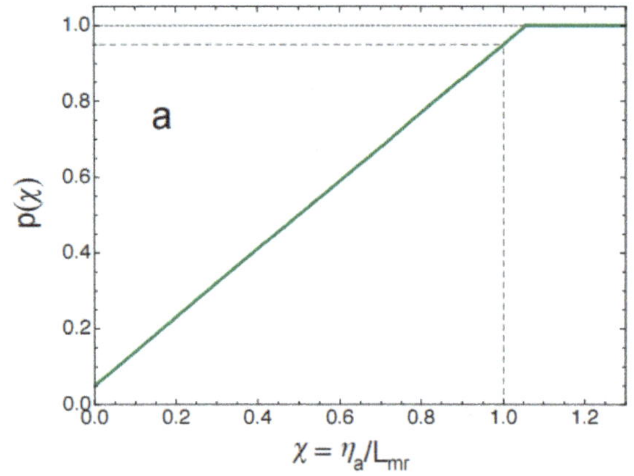

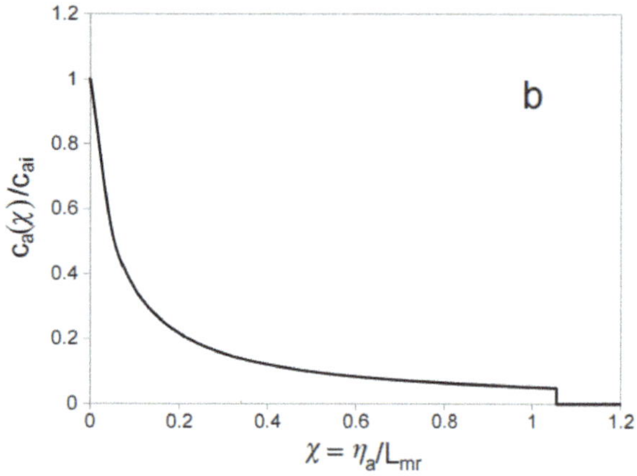

Figure 7. (a) Simplest modified form the MRSAA cover function (green curve) satisfying the conditions p(0) = r, p(1) = 1 − r and p(∞) = 1, where in this case r = 0.05. The parameter on the vertical axis p(χ) denotes the cover fraction, and the parameter on the horizontal axis is $\chi = \eta_a/L_{mr}$, where η_a is the thickness of alluvium and L_{mr} is the macro-roughness height. (b) Plot of the ratio of alluvial wave speed $c_a(\chi)$ at a finite value of $\chi = \eta_a/L_{mr}$ to the corresponding speed c_{ai} for an alluvial wave of infinitesimal height versus χ. The plot shows that alluvial wave speed declines with increasing alluvial thickness, and that it vanishes under complete alluviation.

It is of interest to inquire as to how the model would behave if the clast-rough condition, i.e., $L_{mr}/D >> 1$, were not satisfied here. The limiting case of clast-smooth conditions would correspond to a probability distribution $\tilde{p}(z')$ in Fig. 4 that obeys a step function; $\tilde{p}(z')$ would be vanishing up to a smooth, horizontal bedrock surface, and would take the value unity above it. Consequently, the cover fraction p would converge precisely to zero as $\eta_a \rightarrow 0$, thus resulting, according to Eq. (23b), in an infinite speed of propagation of an alluvial wave of infinitesimal height. This is not entirely unrealistic: the physical realization would consist of clasts rolling rapidly over a smooth bed with no alluviation (Inoue et al., 2014). The presence of such a singularity, would, however, preclude the modeling of the migration of a pulse of alluvium of finite extent over the otherwise bare bed schematized in Fig. 3. In the present clast-rough formulation, the presence of deep pockets within the bedrock relief where alluvium can be stored without transport ensures that the wave speed of alluvium never displays a singularity.

The form of Eq. (23a) can be further clarified by rewriting it as

$$\frac{\partial \eta_a}{\partial t} + c_a \frac{\partial \eta_a}{\partial x} - \frac{\partial}{\partial x}\left(\kappa_a \frac{\partial \eta_a}{\partial x}\right) = \frac{\partial}{\partial x}\left(\kappa_a \frac{\partial \eta_b}{\partial x}\right), \tag{25}$$

where

$$\kappa_a = \frac{I q_{ac}}{(1-\lambda) S}, \tag{26a}$$

$$S = -\frac{\partial \eta}{\partial x} = -\frac{\partial \eta_b}{\partial x} - \frac{\partial \eta_a}{\partial x}. \tag{26b}$$

In the above relation, κ_a has the physical meaning of a kinematic diffusivity. In general, q_{ac}, q_{ac}/S and thus κ_a are nonlinear functions of S. The alluvial problem thus takes the form of a nonlinear advective–diffusive problem with a source term arising from a bedrock term.

Full MRSAA Formulation: Alluvial Diffusion, Upstream-Migration Waves of Incision, Downstream-Migrating Waves of Alluviation

The full MRSAA model consists of the kinematic wave equation with a source term Eq. (8a) for the bedrock part, Eqs. (23b), (25) and (26) for the alluvial part, and the linkage between the two embodied in the cover relation of Eq. (19). Restating these equations for emphasis, Eq. (6a) can be recast as

$$\frac{\partial \eta_b}{\partial t} - c_b \frac{\partial \eta_b}{\partial x} = \upsilon, \tag{27a}$$

$$c_b = I \beta q_{ac} p(1-p)\left(-\frac{\partial \eta_b}{\partial x}\right)^{-1}, \tag{27b}$$

where c_b denotes the speed of upstream-migrating incisional waves. Equation (18) can be cast in conjunction with Eqs. (19) and (20) as

$$\frac{\partial \eta_a}{\partial t} + c_a \frac{\partial \eta_a}{\partial x} - \frac{\partial}{\partial x}\left(\kappa_a \frac{\partial \eta_a}{\partial x}\right) = \frac{\partial}{\partial x}\left(\kappa_a \frac{\partial \eta_b}{\partial x}\right), \tag{28a}$$

$$c_a = \frac{I}{(1-\lambda)} \frac{q_{ac}}{L_{mr} p} \frac{df}{d\chi}, \tag{28b}$$

$$\frac{df}{d\chi} = \frac{dp}{d(\eta_a/L_{mr})}, \tag{28c}$$

$$\kappa_a = I(1-\lambda) q_{ac}\left[-\frac{\partial}{\partial x}(\eta_a + \eta_b)\right]^{-1} \tag{28d}$$

and

$$p = f(\chi) = \begin{cases} \tilde{p}_0 + (\tilde{p}_1 - \tilde{p}_0)\chi & \ldots 0 \le \chi \le \frac{1-\tilde{p}_0}{\tilde{p}_1 - \tilde{p}_0} \\ 1 & \ldots \chi > \frac{1-\tilde{p}_0}{\tilde{p}_1 - \tilde{p}_0}, \end{cases} \tag{29}$$

where c_a denotes the speed of downstream-migrating alluvial waves, and κ_a is the kinematic diffusivity of alluvium. In this way, upstream-migrating incisional waves are combined with downstream-migrating alluvial waves and alluvial diffusion.

In MRSAA, then, the spatiotemporal variation in the cover fraction p(x, t) is specifically tied to the corresponding variation in η_a through Eq. (19), e.g., the specific example of Eq. (29) above. This variation then affects incision through Eq. (27). Consider the simplified case of a wave of alluvium of finite extent illustrated in Fig. 3. There is no incision upstream of the wave because p = 0 and there is no sediment in motion over the bed. At the peak of the wave, $\eta_a > L_{mr}$, so p = 1; the bed is entirely covered with sediment, and again there is no incision. Incision can only occur on the rising and falling parts of the wave, where bedrock is partially exposed and sediment is in motion over it, i.e., $0 < p < 1$. It can thus be expected that the spatiotemporal variation in cover thickness η_a will affect the evolution of the long profile of an incising river that undergoes transitions between alluvial and mixed bedrock–alluvial states.

Amendment of the Flow Component of the MRSAA Model

The flow model, and in particular Eqs. (9a) and (11), must be modified to include the alluvial formulation, so that bedrock slope S_b is replaced with slope S of the top of the bed, where

$$S = -\frac{\partial \eta}{\partial x} = S_b + S_a, \quad (30a)$$

$$S_b = -\frac{\partial \eta_b}{\partial x}, \quad (30b)$$

$$S_a = -\frac{\partial \eta_a}{\partial x}. \quad (30c)$$

Thus Eqs. (9a) and (11a, 11b) are amended to

$$\tau = \rho g H S \quad (31)$$

and

$$H = \left(\frac{Q^2}{Cz^2 g B^2 S}\right)^{1/3}, \quad (32a)$$

$$\tau^* = \left(\frac{Q^2}{Cz^2 g B^2}\right)^{1/3} \frac{S^{2/3}}{RD}. \quad (32b)$$

The purely alluvial case, i.e., p = 1, $df/d\chi = 0$ and η_b = const $< \eta_a$ in Eq. (28), results in the purely diffusional relation

$$\frac{\partial \eta_a}{\partial t} = \frac{\partial}{\partial x}\left(\kappa_a \frac{\partial \eta_a}{\partial x}\right), \quad (33)$$

in which the diffusivity κ_a is a function of $S_a = -\partial\eta_a/\partial x$.

How the Governing Equations Connect to Each Other

In the numerical analysis below, the actual equations used to solve for morphodynamic evolution are not those of Sects. 3.4 and 3.5, but rather the primitive forms presented earlier. The unknowns to be solved are η_b, η_a, η, p, q_{ac}, τ^*, S, S_b and S_a. These nine parameters are connected to each other via nine equations, i.e., Eqs. (5c), (6b), (12), (18), (20), (30a, b, c) and (32b).

Equivalence of the MRSAA and CSA Models at Steady State

In the restricted case of the highly simplified reach (HSR) configuration constrained by (a) temporally constant, belowcapacity sediment feed (supply) rate q_{af}, (b) bedload transport rate q_a everywhere equal to the feed rate q_{af}, and (c) a steady-state balance between incision and rock uplift, η_a, p and S_b become constant and Sa vanishes, so that Eq. (28) is satisfied exactly. Equation (18) integrates to give

$$p = \frac{q_{af}}{q_{ac}}, \quad (34)$$

so that η_a can then be back-calculated from Eq. (20). In this case, then, the MRSAA model reduces to Eqs. (27) and (34), i.e., the CSA model.

THE BELOW-CAPACITY STEADY-STATE CASE COMMON TO THE CSA AND MRSAA MODELS

The steady-state form of Eq. (6) under below-capacity conditions ($p < 1$) can be expressed with the aid of Eq. (2) in the form

$$p_{ss} = 1 - \varphi, \quad (35a)$$

$$\varphi = \frac{v}{I \beta_{ss} q_{af}}, \quad (35b)$$

$$q_{acss} = \frac{q_{af}}{p_{ss}}, \quad (35c)$$

where p_{ss}, β_{ss} and q_{acss} denote steady-state values of p, β and q_{ac}, respectively. Equations (35a)–(35c) describe a balance between the incision rate and relative vertical rock velocity (e.g., constant rock uplift rate at constant base level or constant rock elevation with constant rate of base level fall). CSA and MRSAA yield the same solution for this case, which must be characterized before showing how the models differ.

Equation (35a) has an interesting character. When the value of the dimensionless number j exceeds unity, p falls below zero and no steady-state solution exists. Equation (35b) reveals that j can be interpreted as a dimensionless rock uplift rate. Thus when the rock uplift rate is sufficiently large for j to exceed unity, incision cannot keep pace with rock uplift. The model thus implicitly predicts the formation of a hanging valley. This issue was earlier discussed in Crosby et al. (2007).

In solving for this steady state, and in subsequent calculations, we use the bedload transport relation of Wong and Parker (2006a), a development and correction of the semiempirical relation of Meyer-Peter and Müller (1948), rather than the similar formulation of Fernandez Luque and van Beek (1976); in the case of the former, $\alpha_a = 4$, $n_a = 1.5$ and $\tau^*_c = 0.0495$. We consider two cases: one for which $\beta_{ss} = \beta$ is a specified constant, and one for which only a reference value βref is specified, and β_{ss} is computed from Eq. (5d).

In the case of a specified constant abrasion coefficient β, specification of υ, I and q_{af} allow computation of j, p_{ss} and q_{acss} from Eqs. (35a)–(35c). Further specification of R (here chosen to be 1.65, the standard value for quartz) and D allows the steady-state Shields number τ^*_{ss} to be computed from Eq. (5c). Steady-state bedrock slope S_{bss} can then be computed from Eq. (11b) upon specification of flood discharge Q, Chézy resistance coefficient C_z and channel width B. In the case of β_{ss} calculated according to Eq. (5d) using a specified reference value βref, the problem can again be solved with Eqs. (35), (5c) and (11b), but the solution is implicit.

We performed calculations for conditions loosely based on (a) field estimates for a reach of the bedrock Shimanto River near Tokawa, Japan (Fig. 1), for which bed slope S is about 0.002 and channel width is about 100 m and (b) estimates using relations in Parker et al. (2007) for alluvial gravel-bed rivers with similar slopes, and reasonable choices for otherwise poorly constrained parameters. The input parameters, Cz = 10, Q = 300 m³ s⁻¹, B = 100 m, are loosely justified in terms of bankfull characteristics of alluvial gravel-bed rivers of the same slope (Parker et al., 2007; Wilkerson and Parker, 2011) as shown in Fig. 8a and b. The value D = 20 mm represents a reasonable characteristic size of the substrate (and thus the bedload) for gravel-bed rivers; a typical size for surface pavement is 2 to 3 times this (e.g., Parker et al., 1982). Flood intermittency I is estimated at 0.05, i.e., 18 days per year, and thus a reasonable estimate for a river subject to frequent heavy storm rainfall. Alluvial porosity is λ = 0.35.

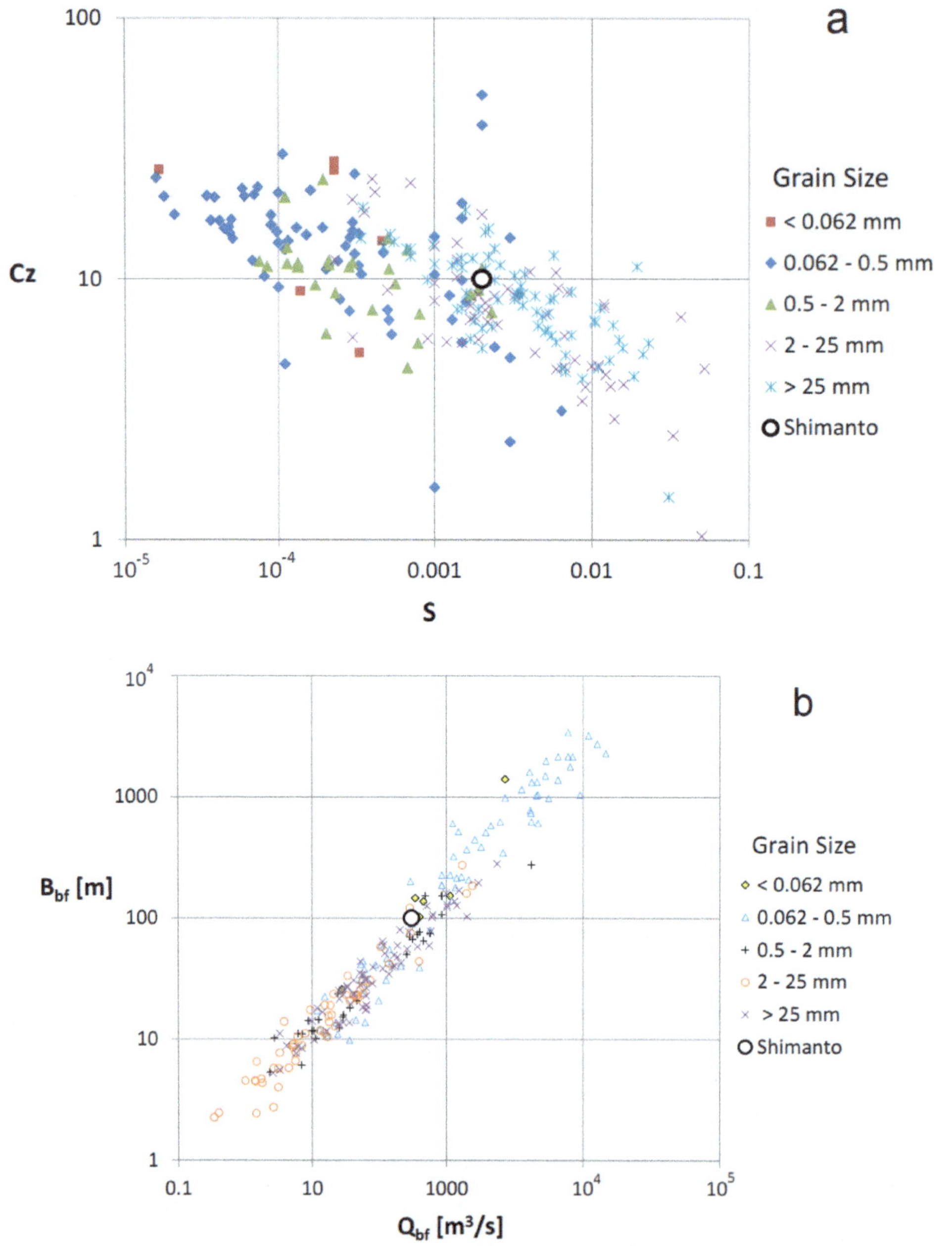

Figure 8. (a) Chézy resistance coefficient Cz plotted against bed slope S for alluvial rivers. Also included are estimated values for the Shimanto River, Japan. (b) Bankfull width B_{bf} versus bankfull discharge Q_{bf} for alluvial rivers. Also included are estimates for bank-to-bank width and characteristic flood discharge in the Shimanto River, Japan. The ranges for characteristic bed material size of the alluvial rivers are denoted in the legends.

Two sediment feed rates were considered. The high feed rate was set at 3.5×10^5 t yr^{-1}, which corresponds to the following steady-state parameters at capacity conditions: Shields number $\tau* = 0.12$, depth $H = 1.5$ m, steady-state alluvial bed slope $S_{ass} = 0.0026$ and Froude number $Fr = 0.51$, where

$$Fr = \frac{Q}{BH\sqrt{gH}}. \quad (36)$$

The low feed rate was set at 3.5×10^4 t yr^{-1}, corresponding to the following parameters at capacity conditions: Shields number $\tau* = 0.064$, depth $H = 2.1$ m, steady-state alluvial bed slope Sass = 0.0010 and Froude number $F_r = 0.32$. The value $\beta_{ss} = 0.05$ km^{-1} was used for the case of a constant steady-state abrasion coefficient. This corresponds to a value of αd of 0.017 km^{-1}, which falls in the middle of the range measured by Kodama (1994) for chert, quartz and andesite (see Fig. 3-41 of Parker, 2008). For the case of a variable abrasion coefficient, Eq. (1a) was used with βref set to 0.05 km−1 and τ^*_{ref} set to 0.12, i.e., the value for the high feed rate. This value of τ^*_{ref} is about 2.5 times the threshold value of Wong and Parker (2006a).

For the high feed, predicted relations for a steady-state abrasion coefficient β_{ss} versus rock uplift rate υ are shown in Fig. 9a; the corresponding predictions for S_{bss} versus υ are shown in Fig. 9b; the corresponding predictions for p_{ss} and j are shown in Fig. 9c. Both the cases of constant and variable βss are shown. There are five notable aspects of these figures: (a) in Fig. 9a, the predictions for variable β_{ss} are very similar to the case of constant, specified β_{ss}, and indeed are nearly identical for $\upsilon \leq 3.3$ mm yr^{-1} (corresponding to $j \leq 0.05$ in Fig. 9c). (b) In Fig. 9b and c, the predictions for S_{bss}, p_{ss} and j for variable β_{ss} are again nearly identical to those for constant β_{ss}, and again essentially independent of υ for $\upsilon \leq 3.3$ mm yr^{-1}. (c) In Fig. 9c, p_{ss} is only slightly below unity (i.e., $p_{ss} \geq 0.95$), and $j \leq 0.05$ for $\upsilon \leq 3.3$ mm yr^{-1}). (d) For $\upsilon > 3.3$ mm yr^{-1}, the predictions for S_{bss} and p_{ss} become dependent on υ, such that S_{bss} increases, and p_{ss} decreases, with increasing υ. The values for constant β_{ss} diverge from those for variable β_{ss}, but are nevertheless close to each other up to some limiting value. (e) This limiting value corresponds to $j = 1$ and thus $p_{ss} = 0$ from Eq. (35a); larger values of j lead to hanging valley formation. Here $j = 1$ for the very high values $\upsilon = 65$ mm yr^{-1} for constant β_{ss} and $\upsilon = 30$ mm for variable β_{ss}.

These results require interpretation. It can be seen from Eqs. (35a–35c) that when $\upsilon/(I\,\beta_{ss}\,q_{af}) = j << 1$, p becomes nearly equal to unity (very little exposed bedrock), in which case q_{af} is constrained to be only slightly smaller than q_{ac}. From Eqs. (5c) and (11), then, S_{bss} is only slightly above the steady-state alluvial bed slope Sass. Note that the steadystate bedrock slope decouples from rock uplift rate under these conditions: the predictions for $\upsilon = 0.2$ mm yr^{-1} are nearly identical to this for $\upsilon = 3.3$ mm yr^{-1}. This behavior is a specific consequence of the condition $j << 1$ corresponding to a low ratio of uplift rate to reference incision rate $E_{ref} = I\,\beta_{ss}\,q_{af}$. They imply a wide range of conditions for which (a) very little bedrock is exposed, and (b) bedrock slope is independent of uplift rate.

The results for the low feed rate are very similar. The values for variable steady-state abrasion coefficient β_{ss} differ from the constant value β_{ss} in Fig. 10a, but this is because the constant value $\beta_{ss} = 0.05$ was set based on the high feed rate. The results in Fig. 10b and c are qualitatively the same for Fig. 9b and c; the uplift rate below which $j < 0.05$ is $\upsilon < 0.33$ mm yr^{-1} for the case of constant β_{ss}, and $\upsilon < 0.73$ mm yr^{-1} for the case of variable β_{ss}. The critical value of υ beyond which a hanging valley forms is $\upsilon \geq 6.8$ mm yr^{-1} for constant β_{ss} and $\upsilon \geq 7.1$ mm yr^{-1} for variable β_{ss}.

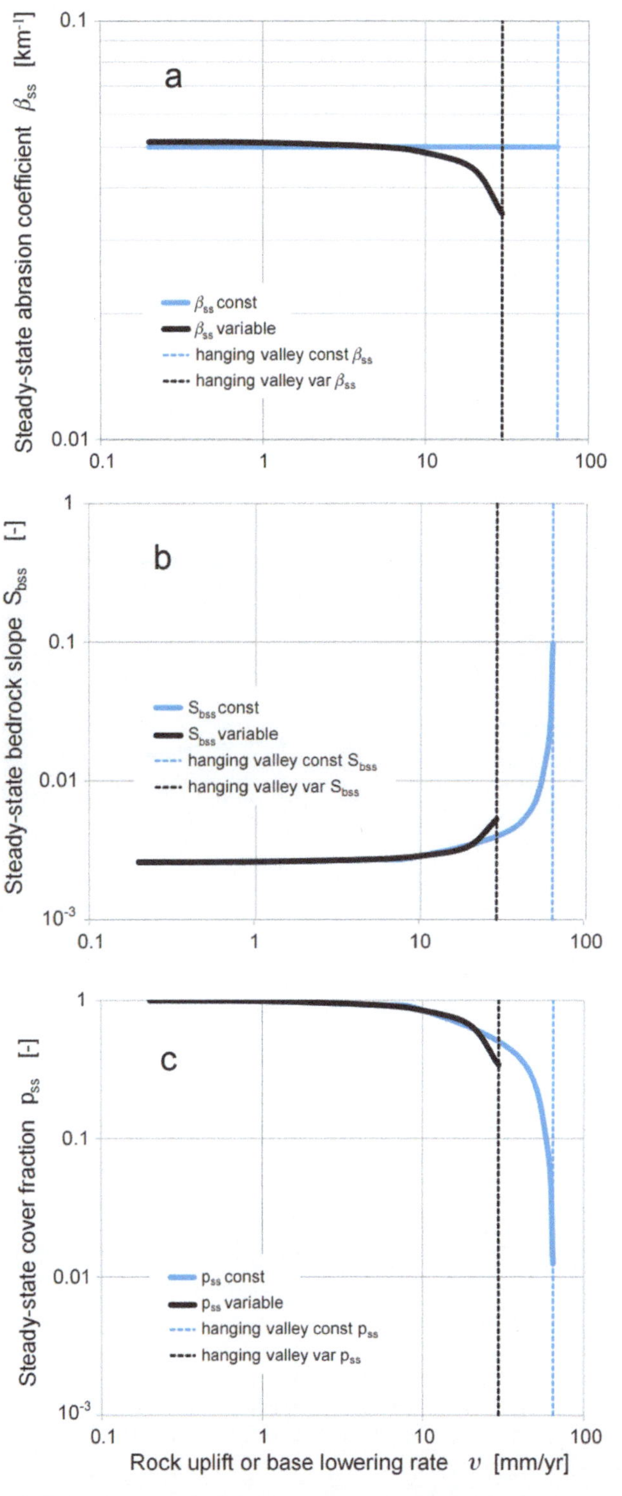

Figure 9. Variation at steady state (black curves) of (a) abrasion coefficient β_{ss}, (b) bedrock slope S_{bss} and (c) cover fraction p_{ss} and parameter j on rock uplift or base lowering rate υ, for a high bedload feed rate of 3.5×10^5 t yr^{-1}. The cases of constant, specified β_{ss} and S_{bss} varying according to Eq. (34) are shown as blue curves. The vertical dashed lines denote the incipient conditions for the formation of a hanging valley. The predictions are the same for the CSA and MRSAA models.

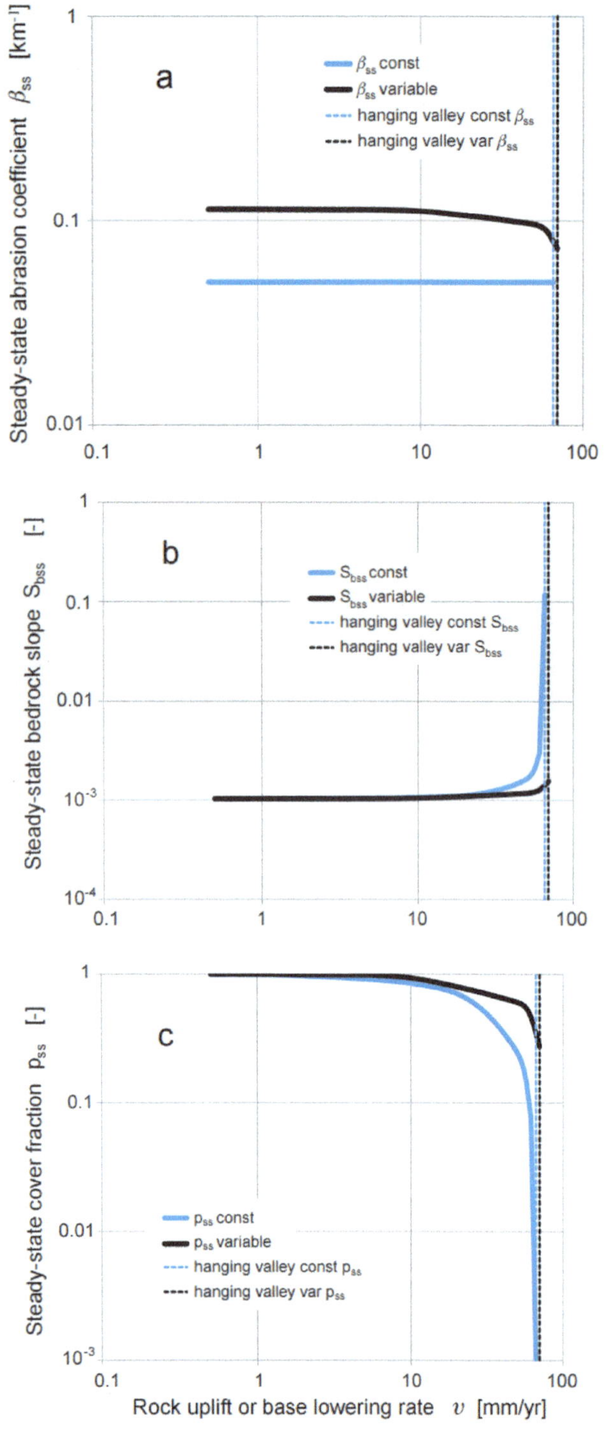

Figure 10. Variation at steady state (black curves) of (a) abrasion coefficient β_{ss}, (b) bedrock slope S_{bs}s and (c) cover fraction p_{ss} and parameter j with rock uplift or base lowering rate υ, for a low bedload feed rate of 3.5×10^4 t yr^{-1}. The cases of constant, specified β_{ss} and S_{bss} varying according to Eq. (34) are shown as blue curves. The vertical dashed lines denote the incipient conditions for the formation of a hanging valley. The predictions are the same for the CSA and MRSAA models.

The lack of dependence of steady-state bedrock slope S_{bss} on rock uplift rate υ below a threshold value for the steadystate solutions of the CSA model (and thus the MRSAA model as well) is in stark contrast to earlier work for which the incision rate E is assumed to have the following dependence on slope S_b and drainage area A ("slope–area" formulation, Howard and Kerby, 1983):

$$E = K S_b^n A^m, \tag{37}$$

where A denotes drainage area, n and m are specified exponents, and K is a constant assumed to decrease with increasing rock hardness.

In order to compare the steady-state predictions of the slope–area relation in Eq. (37) for constant υ with CSA, drainage area A must be taken to be a constant value A_o so as to correspond to the HSR configuration used here. The steady-state slope S_{bss} corresponding to a balance between incision and rock uplift is found from Eq. (37) to be

$$S_{bss} = \frac{\upsilon^{1/n}}{K^{1/n} A_o^{m/n}}. \tag{38}$$

The issue as to the values of m and n has been considered by many researchers, including Whipple and Tucker (2002) and Lague (2014).

In their Table 1, Whipple and Tucker (2002) quote a range of values of n, but their most quoted value is n = 2. We compare the results for CSA for S_{bss} with the predictions from Eq. (38) with n = 2 by normalizing against a reference value Sb_{ref} that corresponds to a reference rock uplift rate υ_{ref} of 0.2 mm yr^{-1}. Equation (38) yields

$$\frac{S_{bss}}{S_{bref}} = \left(\frac{\upsilon}{\upsilon_{ref}}\right)^{1/2}. \tag{39}$$

In Fig. 11, Eq. (39) is compared against the CSA predictions of Figs. 9b and 10b (high and low feed rate, respectively) for both constant and variable βss. In order to keep the plot within a realistic range, only values of υ between 0.2 and 10 mm yr^{-1} (the upper limit corresponding to Dadson et al., 2003) have been used in the CSA results. The remarkable insensitivity of the CSA predictions for steady-state slope S_{bss} on rock uplift rate is readily apparent from the figure.

One more difference between the CSA and slope–area formulations is worth noting. If the slope–area relation is installed into Eq. (6) in place of CSA, it is readily shown that bedrock slope gradually relaxes to zero in the absence of rock uplift. CSA does not obey the same behavior under the constraint of constant sediment feed rate: Figs. 9b and 10b indicate that bedrock slope converges to a constant, nonzero value as rock uplift declines to zero. This is not necessarily a shortcoming of CSA; the sediment feed rate can be expected to decline as relief declines.

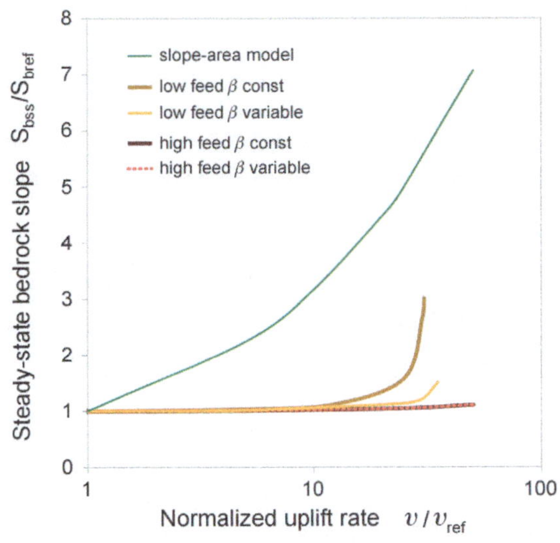

Figure 11. Normalized steady-state bedrock slope S_{bss}/Sb_{ref} versus normalized rock uplift rate υ as predicted by the CSA model for a low feed rate (orange-brown curves) and a high feed rate (black curve and dashed red curve), and for constant and variable abrasion coefficient. The results are the same for the MRSAA model. Also shown is the prediction of a model for which the incision rate is specified in terms of bedrock slope and upstream drainage area (green curve). Note that the predictions for steady-state bedrock slope of the CSA model are insensitive to the rock uplift rate over a wide range.

BOUNDARY CONDITIONS AND PARAMETERS FOR NUMERICAL SOLUTIONS OF THE MRSAA MODEL

Having conducted a fairly thorough analysis of the steady state common to the CSA and MRSAA models, it is now appropriate to move on to examples of behavior that can be captured by the MRSAA model, but are not captured by models that assume a relation for cover based on the ratio of sediment supply to capacity transport rate, i.e., Eq. (2). Before doing so, however, it is necessary to delineate the boundary conditions and other assumptions used in the MRSAA model.

Let L denote the length of the reach. Equation (27a) indicates the formulation for bedrock incision is first order in x and so requires only one boundary condition. The example considered here is that of a downstream bedrock elevation, i.e., base level, set to zero:

$$\eta_b|_{x=L} = 0. \tag{40}$$

According to Eq. (18), or alternatively Eq. (28a), the alluvial formulation is second order in x and thus requires two boundary conditions. The following boundary condition applies at the upstream end of the reach, where $q_{af}(t)$ denotes a feed rate that may vary in time,

$$q_a|_{x=0} = q_{af}(t). \tag{41}$$

At the downstream end, a free boundary condition is applied for $\eta_a/L_{mr} < 1$, and a fixed boundary condition is applied for $\eta_a/L_{mr} \geq 1$ as follows:

$$\left[(1-\lambda)p\frac{\partial \eta_a}{\partial t} + I\frac{\partial p q_{ac}}{\partial x}\right]_{x=L} = 0 \quad \ldots \text{if} \left[\frac{\eta_a}{L_{mr}}\right]_{x=L} < 1, \tag{42a}$$

$$\eta_a|_{x=L} = L_{mr} \qquad \ldots \text{if} \left[\frac{\eta_a}{L_{mr}}\right]_{x=L} \geq 1. \tag{42b}$$

Here, Eq. (42a) specifies a free boundary in the case of partial alluviation, thus allowing below-capacity sediment waves to exit the reach. Equation (42b), on the other hand, fixes the maximum downstream elevation at $\eta = \eta_a = L_{mr}$.

In order to illustrate the essential features of the new formulation of the MRSAA model for the morphodynamics of mixed bedrock–alluvial rivers, it is useful to consider the most simplified case that illustrates its expanded capabilities compared to the CSA model. Here we implement the HSR simplification. In addition, based on the results of the previous section, we approximate β_{ss} as a prescribed constant. Finally, we assume that the clasts of the abrading bedload are sufficiently hard compared to the bedrock so that grain size D can be approximated as a constant. These constraints are easily relaxed.

In the numerical solution of the differential Eqs. (6b) and (18), spatial derivatives have been computed using an upwinding scheme for short timescales (so as to capture downstream-migrating alluvial waves) and a downwinding scheme for long timescales (so as to capture upstreammigrating incisional waves). Time derivatives have been computed using the Euler step method.

SEDIMENT WAVES OVER A FIXED BED: STRIPPING AND EMPLACEMENT OF ALLUVIAL LAYER AND ADVECTION–DIFFUSION OF A SEDIMENT PULSE

Three numerical solutions of the MRSAA model are studied here: (a) stripping of an alluvial cover to bare bed, (b) emplacement of an alluvial cover over a bare bed and (c) advection–diffusion of an alluvial pulse over a bare bed. Reach length L is 20 km. As the time for alluvial response is short compared to incisional response, βs_s and υ are set equal to zero for these calculations. In addition, flood intermittency I is set to unity so as to illustrate the migration from the feed point to the end of the reach under the condition of continuous flow. The macro-roughness L_{mr} is set to 1 m based on observation of the Shimanto River near Tokawa, Japan. The values for Cz, Q, B, D and λ are the same as in Sect. 5, i.e., Cz = 10, Q = 300 $m^3\ s^{-1}$, B = 100 m, D = 20 mm and λ = 0.35. Bedrock slope S_b, which is constant due to the absence of abrasion, is set to 0.004. The above numbers combined with Eqs (5c) (using the constants of the formulation of Wong and Parker, 2006a), (32a) and (32b) yield the following values: depth H = 1.32 m, Froude number Fr = 0.63, Shields number $\tau* = 0.016$ and capacity bedload transport rate $q_{ac} = 0.0017\ m^2\ s^{-1}$.

None of these three cases can be treated using models that assume a relation for cover based on the ratio of sediment supply to capacity transport rate, i.e., Eq. (2). They thus illustrate capabilities unique to MRSAA.

Alluvial Stripping

The case of stripping of an initial alluvial layer to bare bedrock is considered here. In this simulation, the bedload feed rate q_{af} = 0 and the initial thickness of alluvial cover η_a is set to 0.8 m, i.e., 80 % of the macro-roughness length L_{mr}. To drive stripping of the alluvial layer, the feed rate is set equal to zero. Figure 12a shows how the alluvial cover is progressively stripped off from upstream to downstream as a wave of alluvial rarification migrates downstream. The alluvial layer is completely removed (except for residual sediment in deep pockets, as specified by Eq. 21) after a little more than 0.12 years.

Of interest in Fig. 12a is the fact that the wave of stripping maintains constant form in spite of the diffusive term in Eq. (28a) which should cause the wave to spread. The reason the wave does not spread is the nonlinearity of the wave speed c_a in Eq. (28b): since p enters into the denominator on the right-hand side of the equation, wave speed is seen to increase as p decreases, and thus η_a decreases. As a result, the lower portion of the wave tends to migrate faster than the higher portion, sharpening the wave and opposing diffusion.

Emplacement of an Alluvial Layer over an Initially Bare Bed

In this simulation, the initial thickness of alluvium η_a is set to zero and the sediment feed rate is set to 0.0013 m^2 s^{-1}, i.e., 80 % of the capacity value. The result of the calculation is shown in Fig. 12b. Here nonlinear advection and diffusion act in concert to cause the wave of alluviation to spread. The steady-state thickness of alluvium is 0.83 m; by 0.1 years it has been emplaced only down to about 5 km from the source. This steady-state condition, and only this condition, corresponds to a convergence of results from MRSAA and CSA.

Propagation of a Pulse of Alluvium over an Initially Bare Bed

In this example the initial bed is bare of sediment. The sediment feed rate is set equal to 0.0012 m^2 s^{-1}, i.e., 70 % of the capacity value for 0.05 years from the start of the run, and then dropped to zero for the rest of the run. Figure 12c shows the propagation of a damped alluvial pulse through the reach, with complete evacuation of the pulse in a little more than 0.15 years. Nonlinear advection acts against diffusion to suppress the spreading of the upstream side of the pulse, but advection acts together with diffusion to drive spreading of the downstream side of the pulse.

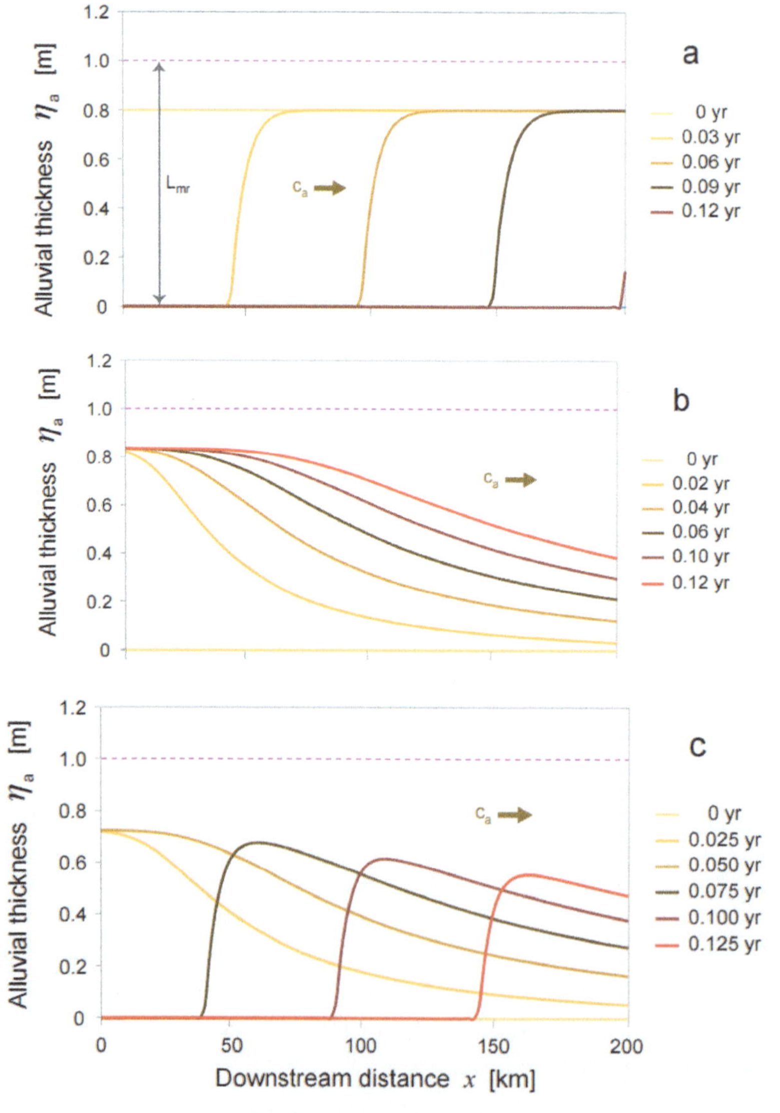

Figure 12. MRSAA model solutions for (a) stripping of an alluvial layer to bare bedrock, (b) emplacement of an alluvial cover over initially bare bedrock and (c) evolution of a pulse of sediment over bare bedrock. Numerical simulations of the evolution of alluvial thickness $\eta_a(x, t)$ over streamwise distance x are shown for a series of time steps t as indicated in the legends.

COMPARISON OF EVOLUTION TO UPLIFT-DRIVEN STEADY STATE FOR THE CSA AND MRSAA MODELS

Here we consider three cases of channel profile evolution to steady state that include both rock uplift and incision. In the first case, the initial bedrock slope is set to a value below the steady-state value, and the sediment feed rate is set to a value that is well above the steady-state value for the initial bedrock slope, causing early-stage massive alluviation. The configuration for the second case is a simplified version of a graben with a horst upstream and a horst downstream. The configuration for the third case is such that there is an alluviated river mouth downstream and a bedrock–alluvial transition upstream. In all cases, MRSAA predicts evolution that cannot be predicted by models that assume a relation for cover based on the ratio of sediment supply to capacity transport rate, i.e., Eq. (2).

Evolution of Bedrock Profile with Early-Stage Massive Alluviation

Here we set Q, B, Cz, D and λ to the same values as Sect. 6. The reach length L is 20 km, the flood intermittency I is set to 0.05, macro-roughness L_{mr} is set to 1 m, initial alluvial thickness $\eta a| t=0 = 0.5$ m, downstream bed elevation $\eta b|_x=L = 0$ and the abrasion coefficient β_{ss} is 0.05 km^{-1} . The initial bed slope is 0.004. The feed rate is set to twice the capacity rate for this slope, i.e., $q_{af} = 0.0033$ m^2 s^{-1} . The uplift rate is set to the very large value of 5 mm yr^{-1} . It should be noted, however, that as shown in Fig. 10b, the steady-state bedrock slope for this feed rate is independent of the uplift rate for $\upsilon \leq 5$ m yr^{-1} . This is because the steady-state value of j is 0.019, i.e., $j << 1$.

The results for the CSA model are shown in Fig. 13a. The bed slope evolves from the initial value of 0.004 to a final steady-state value of 0.0068. Evolution is achieved solely by means of an upstream-migrating knickpoint. Only the first 4000 years of evolution are shown in the figure.

Figure 13b shows the results of the first 400 years of the calculation with MRSAA. By 100 years, the bed is completely alluviated, and by 400 years, the thickness of the alluvial layer at the upstream end of the reach is 52 m. This massive alluviation is, unsurprisingly, not predicted by CSA, which was designed to treat incision only. Figure 13c shows the results of the first 4000 years of evolution. The upstream migrating knickpoint takes the same form as CSA, but it is nearly completely hidden by the alluvial layer. The knickpoint gradually migrates upstream, driving the completely alluviated layer out of the domain, but this process is not complete by 4000 years. A comparison of Fig. 13a and c show that a knickpoint that is exposed in CSA is hidden in MRSAA. Models that assume a relation for cover based on the ratio of sediment supply to capacity transport rate cannot predict the presence of a hidden knickpoint.

Evolution of Horst–Graben Configuration

In this example, Cz, Q, B, D, λ, I , β_{ss}, L_{mr}, L, $\eta_a| t=0$ and $\eta_b|_x=L$ are set to the values used in Sect. 7.1. The sediment feed rate $q_{af} = 0.00083$ m^2 s^{-1} , and the initial bedrock slope S_b is set to the steady-state value for a rock uplift rate of 1 mm yr^{-1} , i.e., 0.0027. The model is then run for a rock uplift rate of 1 mm yr^{-1} for the domains $0 \leq x \leq 8$ km and 12 km $\leq x \leq 20$ km and a rock subsidence rate of 1 mm yr^{-1} for the domain 8 km $< x < 12$ km. This configuration corresponds to a simplified 1-D configuration of a graben bounded by two horsts, one upstream and one downstream.

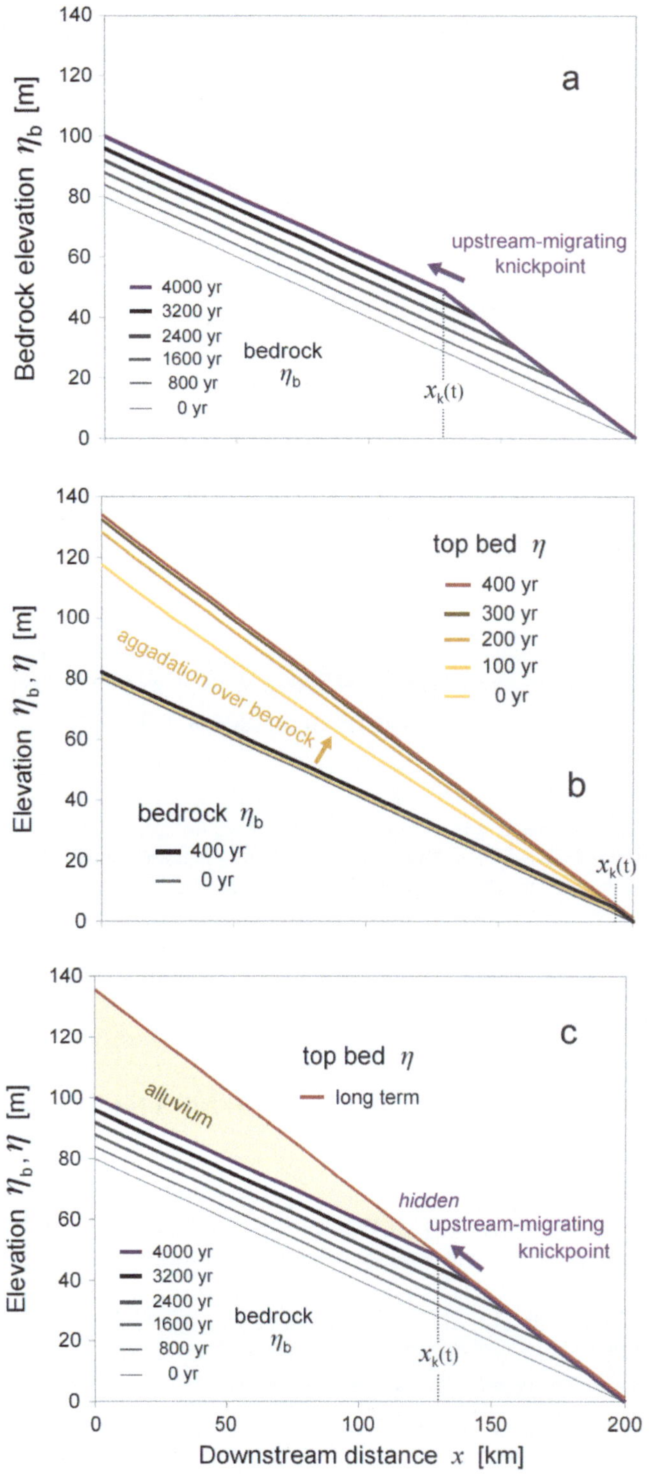

Figure 13. Progression to steady state after an impulsive increase in sediment supply: (a) CSA model; (b) MRSAA model, early stage; and (c) MRSAA model, late stage. Note the knickpoints $x_k(t)$ in bedrock in (a) and (b) and hidden in a migrating alluvial–bedrock transition in (c). Here η denotes elevation of the top bed surface, η_b denotes elevation of the base of bedrock relief, and x denotes streamwise distance. The retreating alluvial wedge is shaded in (c) for emphasis.

This case cannot be implemented in models that assume a relation for cover based on the ratio of sediment supply to capacity transport rate, i.e., Eq. (2). This is because such models are not designed to handle the case of alluvial fill of accommodation space created by subsidence. The results for MRSAA are shown in Fig. 14. By 15 kyr, the uplifting domains evolve to a steady state in terms of both bedrock elevation and alluvial cover. The bedrock elevation of the subsiding domain never reaches steady state because it is completely alluviated. The profile at the top of the alluvium in this domain has indeed reached steady state by 15 000 years, with a bed slope that deviates only modestly from the steadystate bedrock slope driven by a uniform $\upsilon = 1$ mm yr^{-1} .

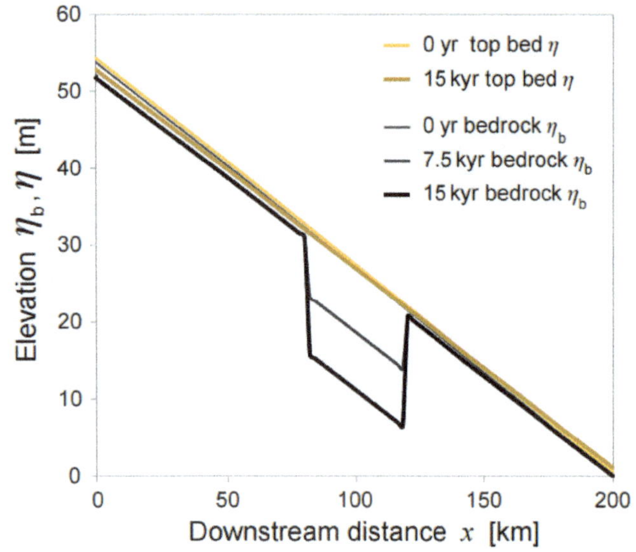

Figure 14. Evolution predicted by the MRSAA model for localized subsidence at a narrow graben superimposed on broader uplift. Note the bedrock–alluvial and alluvial–bedrock transitions at the margins of the graben. By 15 kyr, the bed top has reached steady state, even though the bedrock surface in the graben continues to subside. The regional rock uplift rate and graben subsidence rate are assumed constant for simplicity. Here η denotes elevation at the bed top, η_b denotes elevation of the bottom of bedrock relief, and x denotes streamwise distance.

Evolution of River Profile with Alluviated Zone at River Mouth

In this example Cz, Q, B, D, λ, β_{ss}, L_{mr}, L and $\eta_a|$ t=0 are again set to the values chosen in Sect. 7.1. The bedload feed rate is 0.00083 m^2 s^{-1} ; the steady-state bedrock slope Sb associated with this feed rate is 0.0026 for $\upsilon < 5$ mm yr^{-1} (Fig. 10b). The initial bedrock slope is set, however, to the higher value of 0.004. The rock uplift rate υ for this case is set to zero, for which the steady-state slope is again 0.0026.

The result of CSA for this case, with base level $\eta_b|_x$=L pinned at zero elevation, is shown in Fig. 15. As in the case of Sect. 7.1, the bedrock slope evolves from the initial value of 0.004 to the steady-state value 0.0026 by means of an upstream-migrating knickpoint. Only 4000 years of evolution are shown in the figure, by which time the knickpoint is 4.8 km from the feed point.

MRSAA is implemented with somewhat different initial and downstream boundary conditions in order to model the case of a bed that remains alluviated at the downstream end. This condition thus corresponds to an alluviated river mouth. The initial bedrock slope is again 0.004, and the downstream bedrock elevation $\eta_b|_x$=L is again 0 m. The downstream alluvial elevation $\eta_a|_x$=L, however, is held at 10 m, so that the downstream end is completely alluviated. The initial slope S for the top of the bed is 0.0021, a value chosen so that the bed elevation equals the bedrock elevation at the upstream end. Results of the MRSAA simulation are shown in Fig. 16.

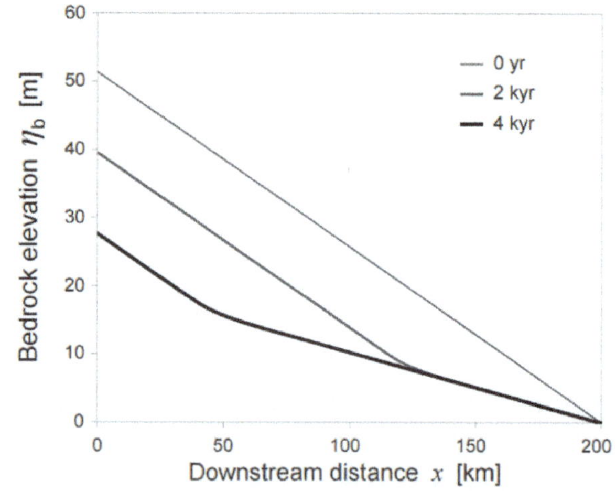

Figure 15. CSA model evolution of an initial bedrock profile towards a steady-state profile. Compare with the MRSAA model behavior in Fig. 16. Here η_b denotes elevation of the bottom of bedrock relief, and x denotes streamwise distance.

Figure 16a–c show the early-stage evolution, i.e., at t = 0, 10 and 100 years. Over this period, a bedrock–alluvial transition (from mixed bedrock–alluvial to purely alluvial) migrates downstream from the feed point to x = 13.6 km, i.e., 6.4 km upstream of the terminus. Bedrock incision is negligible over this period.

Figure 16d–f show the bedrock and top bed profiles for 1000, 2000 and 4000 years. Over this period, the bedrock–alluvial transition migrates upstream. As it does so, the bedrock slope downstream of x = 13.6 km remains alluviated and does not change. The bedrock slope between the transition and x = 13.6 km evolves to the steady-state value of the case in Sect. 7.2, and the top bed slope downstream of the transition evolves to the same slope as the steadystate bedrock slope (because with $\upsilon = 0$, j vanishes). The figures show that the upstream-migrating bedrock knickpoint is located at the bedrock–alluvial transition. By 4000 years, the transition has migrated out of the domain and the bed is completely alluviated. The thickness of the alluvial cover upstream of x = 13.6 km is, however, only 1.05 m, i.e., only slightly larger than the macro-roughness height of 1 m. This means that although the reach is everywhere alluvial at 4000 years, the bedrock is only barely covered.

DISCUSSION

The MRSAA model is a direct descendant of the model of Sklar and Dietrich (2004) in terms of the formulation for bedrock incision, and the model of Struiksma (1999) in terms of the formulation of the conservation alluvium over a partly covered bedrock surface. In terms of its capabilities, however, it shares much in common with the previous work of Lague (2010), and in particular with his SSTRIM model. These include (a) the melding of incision and alluviation into a single model, (b) the inclusion of a cover relation that is based on geometric bed structure, and (c) the ability to track simultaneously the spatiotemporal variation in both incision rate and alluvial cover. Priority should accrue to Lague (2010) in regard to these features. The present model has the following advantages: (a) the Exner equation of sediment conservation is specifically based on a formulation of the statistics of partial and complete cover over a rough bedrock surface; (b) the formulation yields a specific relation for alluvial wave velocity as a function of cover, ranging to a maximum value for minimum cover to 0 for complete alluviation; and (c) it allows for explicit description of the nonlinear advective diffusive physics of the problem in terms of an alluvial diffusivity and two wave celerities, one directed upstream and associated with bedrock incision, and one directed downstream and associated with alluviation.

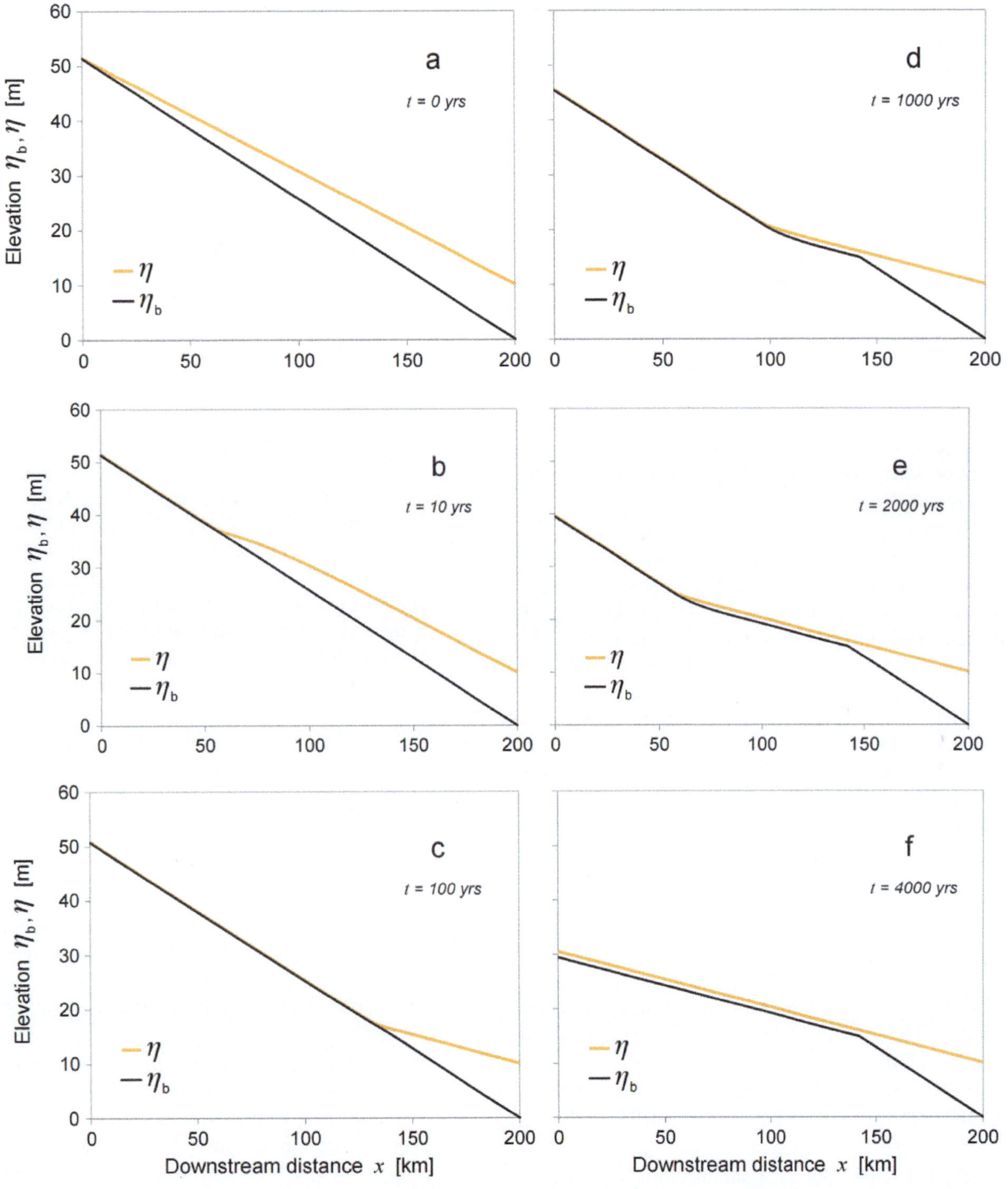

Figure 16. MRSAA model evolution of bed top and bedrock profiles with an imposed alluvial river mouth at the downstream end and an upstream-migrating bedrock–alluvial transition. The results are for (a) t = 0 years, (b) t = 10 years, (c) t = 100 years, (d) t = 1000 years, (e) t = 2000 years and (f) t = 4000 years (steady state). Here (a), (b) and (c) show the early response, and (d), (e) and (f) show the late response. In these plots, η denotes elevation of the bed surface, η_b denotes elevation of the bottom of bedrock relief, and x denotes streamwise distance. Compare with the CSA model solution in Fig. 15.

The form of the MRSAA model presented here has been simplified as much as possible, i.e., to treat a HSR (highly simplified reach) with constant grain size ***D***. This has been done to allow for a precise and complete characterization of the behavior of the governing equations. It can relatively easily be extended to: (a) abrasion of the clasts that abrade the bed, so abrasional downstream fining is captured (Parker, 1991); (b) size mixtures of sediment (Wilcock and Crowe, 2003); (c) multiple sediment sources (Lague, 2010; Yanites et al., 2010); (d) channels with width variation down-stream (Lague, 2010); (e) discharge varying according to a flow duration curve (Sklar and Dietrich, 2006; Lague, 2010), or fully unsteady flow (An et al., 2014); and (f) cyclically varying hydrographs (Wong and Parker, 2006b) or "sedimen-tographs", the latter corresponding to events for which the sediment supply rate first increases, and then decreases cyclically (Zhang et al., 2013). In addition, the model can and should be extended to include the stochasticity emphasized by Lague (2010). Sections 6 and 7 illustrate features captured by MRSAA but not by models that assume a cover relation based on the ratio of sediment supply to capacity transport rate, i.e., Eq. (2).

The MRSAA model presented here is applied to several 1-D cases with spatiotemporal variation. The model can easily be generalized to 2-D simply by expressing Eq. (18) in 2-D form. In any such implementation, however, the effect of 2-D connectivity between deep holes should be considered in the relation to the cover factor.

The MRSAA model in the form presented here has a weakness in that the flow resistance coefficient Cz is a prescribed constant. The recent models of bedrock incision of Inoue et al. (2014) and Johnson (2014) pro-vide a much more detailed description of flow resistance. In addition to characterizing macro-roughness, their models use two micro-roughnesses, one characterizing the hydraulic roughness of the alluvium and the other characterizing the hydraulic roughness of the bedrock surface. Their models can thus discriminate between (a) "clast-smooth" beds, for which bedrock roughness is lower than clast rough-ness and (b) "clast-rough" beds, for which bedrock rough-ness is greater than clast roughness. This characterization al-lows for two innovative features: (a) both bed resistance and fractional cover become dependent on the ratio of bedrock micro-roughness to alluvial micro-roughness and (b) incision can result from throughput sediment passing over a purely bedrock surface with no alluvial deposit. The models of Inoue et al. (2014) and Johnson (2014) use modified forms of the capacity-based form for cover of Eq. (2) in order to capture these phenomena. Their models are thus unable to capture the coevolution of bedrock–alluvial and purely alluvial processes of MRSAA. Amalgamation of their models and the one presented here, however, appears to be feasible and is an attractive future goal.

Because MRSAA tracks the spatiotemporal variation in both bedload transport and alluvial thickness, it is applicable to the study of the incisional response of a river subject to temporally varying sediment supply. It thus has the potential to capture the response of an alluvial–bedrock river to massive impulsive sediment inputs associated with landslides or debris flows. A preliminary example of such an extension is given in Zhang et al. (2013). When extended to multiple sediments sources, it can encompass both the short- and long-term responses of a bedrock–alluvial river to intermittent massive sediment supply due to landslides and debris flows. As such, it has the potential to be integrated into a framework for managing sediment disturbance in mountain rivers systems such as those affected by the 2008 Wenchuan earthquake in Sichuan, China. Over 200 landslide dams formed during that event (Xu et al., 2009; Fu et al., 2011). A similar potential application is the case of drastic sediment supply to, and evacuation from, rivers in Taiwan due to typhoon-induced or earthquake-induced landsliding (e.g., Yanites et al., 2010).

CONCLUSIONS

We present a 1-D model of alluvial transport and bedrock erosion in a river channel whose bed may be purely alluvial, or mixed bedrock–alluvial, or may transition freely between the two morphologies. Our model, which we call the Macro-Roughness-based Saltation-Abrasion-Alluviation (MRSAA) model, specifically tracks not only large-scale bedrock morphodynamics but also the morphodynamics of the alluvium over it. The key results are as follows:

- The transport of alluvium over a bedrock surface cannot in general be described simply by a supply rate that instantaneously affects the entire river reach downstream as it is varied in time. Here we track the alluvium in terms of a spatiotemporally varying alluvial thickness.
- The area fraction of cover ***p*** enters into both incisional and alluvial evolution. The alluvial part allows for the downstream propagation and diffusion of sediment waves, so that at any given time the alluvial bed can be above or below the top of the bedrock. The model thus allows for spatiotemporal transitions between complete cover, under which no incision occurs; partial cover, for which incision may occur; and no cover, for which no incision occurs.
- The MRSAA model captures three processes: downstream alluvial advection at a fast timescale, alluvial diffusion, and upstream incisional advection at a slow timescale. Only the third of these processes is captured by models that assume a relation for cover based on the ratio of sediment supply to capacity transport rate rather than a measure of the thickness of alluvial cover itself. The CSA model can be thought of as a 0-D model that applies locally. The MRSAA model lends itself more directly to application to long 1-D reaches because it embeds the elements necessary to route sediment down the reach.
- The MRSAA model reduces to the CSA model under the conditions of steady-state incision in balance with rock uplift and below-capacity cover. The steady-state bedrock slope predicted by both models is insensitive to the rock uplift rate over a wide range of conditions. This insensitivity is in marked contrast to the commonly used incision model in which the incision rate is a power function of bedrock slope and drainage area upstream. The two models can differ substantially under transient conditions, particularly under those that include migrating transitions between the bedrock–alluvial and purely alluvial state.
- In the MRSAA model, inclusion of alluvial advection and diffusion lead to the following phenomena: (a) a

wave-like stripping of antecedent alluvium over a bedrock surface in response to cessation of sediment supply, (b) advection–diffusional emplacement of a sediment cover over initially bare bedrock and (c) the propagation and deformation of a sediment pulse over a bedrock surface.

- In the case of transient imbalance between rock up-lift and incision with a massive increase in sediment feed, MRSAA captures an upstream-migrating transition between a purely alluvial reach upstream and a bedrock–alluvial reach downstream (here abbreviated as a alluvial–bedrock transition). The bedrock profile shows an upstream-migrating knickpoint, but this knickpoint is hidden under alluvium. Models that assume a relation for cover based on the ratio of sediment supply to capacity transport rate, i.e., Eq. (2), capture only the knickpoint, which is completely exposed, and thus miss the thick alluvial cover predicted by MRSAA.
- MRSAA captures the mixed incisional–alluvial evolution for the case of a simplified 1-D subsiding graben bounded by two uplifting horsts. It captures alluvial filling of the graben, and thus converges to a steady-state top-bed profile with a bedrock–alluvial transition at the upstream end of the graben and an alluvial–bedrock transition at the downstream end.
- In the case studied here of an uplifting bedrock profile with an alluviated bed at the downstream end modeling a river mouth, MRSAA predicts an upstream-migrating bedrock–alluvial transition at which the bedrock undergoes a sharp transition from a higher to a lower slope. MRSAA further predicts a bedrock long profile under the alluvium that has the same slope as the top bed. It also predicts that the cover is thin, so that the purely alluvial reach is only barely so. The steady state for this case is purely alluvial.
- The new MRSAA model provides an entry point for the study of how bedrock–alluvial rivers respond to occasional large, impulsive supplies of sediment from landslides and debris flows. It thus can provide a tool for forecasting river-sedimentation disasters associated with such events. An example application would be treatment of the aftereffects of the 2008 Wenchuan earthquake, which triggered massive alluviation and the formation of over 200 landslide dams.

ACKNOWLEDGEMENTS

The participation of L. Zhang and X. Fu in this work was made possible by the National Natural Science Foundation of China (grant nos. 51379100 and 51039003). The participation of G. Parker was made possible in part by a grant from the US National Science Foundation (grant no. EAR-1124482) The participation of C. P. Stark was made possible in part by grants from the US National Science Foundation (grant nos. EAR-1148176, EAR-1124114 and CMMI-1331499). The participation of T. Inoue was made possible by support from the Hokkaido Regional Development Bureau.

REFERENCES

1. An, C. G., Fu, X. D., and Parker, G.: River morphological evolution in earthquake-hit region: effects of floods and pulsed sediment supply, in: River Flow 2014 Conference, 3–5 September 2014, Lausanne, Switzerland, p. 7, 2014.
2. Chatanantavet, P. and Parker, G.: Experimental study of bedrock channel alluviation under varied sediment supply and hydraulic conditions, Water Resour. Res., W12446, doi:10.1029/2007WR006581, 2008.
3. Chatanantavet, P. and Parker, G.: Physically based modeling of bedrock incision by abrasion, plucking, and macroabrasion, J. Geophys. Res., 114, F04018, doi:10.1029/2008JF001044, 2009.
4. Chatanantavet, P., Lajeunesse, E., Parker, G., Malverti, L., and Meu-nier, P.: Physically-based model of downstream fining in bedrock streams with lateral input, Water Resour. Res., 46, W02518, doi:10.1029/2008WR007208, 2010.
5. Chatanantavet, P., Whipple, K. X., Adams, M. A., and Lamb, M. P.: Experimental study on coarse grain saltation dynamics in bedrock channels, J. Geophys. Res., 118, 1–16, doi:10.1002/jgrf.20053, 2013.
6. Crosby, B. T., Whipple, K. X., Gasparini, N. M., and Wobus, C. W.: Formation of fluvial hanging valleys: theory and simulation, J. Geophys. Res., 112, F03S10, doi:10.1029/2006JF000566, 2007.
7. Cui, Y., Parker, G., Lisle, T., Gott, J., Hansler, M., Pizzuto, J. E., Allmendinger, N. E., and Reed, J. M.: Sediment pulses in mountain rivers, Part 1. Experiments, Water Resour, Res., 39, 1239, doi:10.1029/2002WR001803, 2003a.
8. Cui, Y., Parker, G., Pizzuto, J. E., and Lisle, T. E.: Sediment pulses in mountain rivers, Part 2. Comparison between experiments and numerical predictions, Water Resour. Res., 39, 1240, doi:10.1029/2002WR001805, 2003b.
9. Dadson, S. J., Hovius, N., Chen, H., Dade, W. B. Hsieh, M. L., Willet, S. D., Hu, J. C., Horng, M. J., Chen, M. C., Stark, C. P., Lague, D., and Lin, J. C.: Links between erosion, runoff variability and seismicity in the Taiwan orogeny, Nature, 426, 648–651, 2003.
10. DiBiase, R. A. and Whipple, K. X.: The influence of erosion thresh-olds and runoff variability on the relationships among topography, climate, and erosion rate, J. Geophys. Res., 116, F04036, doi:10.1029/2011JF002095, 2011.
11. England, P. and Molnar, P.: Surface uplift, uplift of rocks and ex-humation rate of rocks, Geology, 18, 1173–1177, 1990.
12. Fernandez Luque, R. and van Beek, R.: Erosion and transport of bedload sediment, J. Hydraul. Res., 14, 127–144, 1976.
13. Fu, X. D., Liu, F., Wang, G. Q., Xu, W. J., and Zhang, J. X.: Neces-sity of integrated methodology for hazard mitigation of quake lakes: case study of the Wenchuan Earthquake, China, Front. Arch. Civ. Eng. China, 5, 1–10, 2011.
14. Gilbert, G. K.: Report on the geology of the Henry Mountains: geographical and geological survey of the Rocky Mountain region, Government Printing Office,

Washington, D.C., 106 pp., 1877.

15. Hobley, D. E. J., Sinclair, H. D., Mudd, S. M., and Cowie, P. A.: Field calibration of sediment flux dependent river incision, J. Geophys. Res., 116, F04017, doi:10.1029/2010JF001935, 2011.

16. Howard, A. D.: Simulation of stream capture, Geol. Soc. Am. Bull., 82, 1355–1376, 1971.

17. Howard, A. D.: Long profile development of bedrock channels: interaction of weathering, mass wasting, bed erosion, and sediment transport, Am. Geophys. Un. Monogr., 107, 297–319, 1998.

18. Howard, A. D. and Kerby, G.: Channel changes in badlands, Bull. Geol. Soc. Am., 94, 739–752, 1983.

19. Howard, A. D., Seidl, M. A., and Dietrich, W. E.: Modeling fluvial erosion on regional to continental scales, J. Geophys. Res., 99, 13971–13986, 1994.

20. Inoue, T., Izumi, N., Shimizu, Y., and Parker, G.: Interaction be-tween alluvial cover, bed roughness and incision rate in purely bedrock and alluvial-bedrock channel, in press, J. Geophys. Res., 119, 2123–2146, doi:10.1002/2014JF003133, 2014.

21. Izumi, N. and Yokokawa, M.: Cyclic steps formed in bedrock rivers, Proceedings of River, Coastal and Estuarine Morphodynamics, RCEM 2011, Tsinghua University Press, Beijing, 2084–2090, 2011.

22. Izumi, N., Yokokawa, M., and Parker, G.: Cyclic step morphology formed on bedrock, J. JSCE Div. B, 68, I_955–I_960, 2012.

23. Johnson, J. P. L.: A surface roughness model for predicting alluvial cover and bedload transport rate in bedrock channels, J. Geophys. Res., 119, 2147–2173, doi:10.1002/2013JF003000, 2014.

24. Johnson, J. P. L. and Whipple, K. X.: Feedbacks between erosion and sediment transport in experimental bedrock channels, Earth Surf. Proc. Land., 32, 1048–1062, doi:10.1002/esp.1471, 2007.

25. Johnson, J. P. L., Whipple, K. X., Sklar, L. S., and Hanks, T. C.: Transport slopes, sediment cover, and bedrock channel incision in the Henry Mountains, Utah, J. Geophys. Res., 114, F02014, doi:10.1029/2007JF000862, 2009.

26. Kodama, Y.: Experimental study of abrasion and its role in produc-ing downstream fining in gravel-bed rivers, J. Sed. Res. A, 64, 76–85, 1994.

27. Lague, D.: Reduction of long-term bedrock incision efficiency by short-term alluvial cover intermittency, J. Geophys. Res., 115, F02011, doi:10.1029/2008JF001210, 2010.

28. Lague, D.: The stream power river incision model: evidence, theory and beyond, Earth Surf. Proc. Land., 39, 38-61, doi:10.1002/esp.3462, 2014.

29. Lamb, M. P., Dietrich, W. E., and Sklar, L. S.: A model for fluvial bedrock incision by impacting suspended and bedload sediment, J. Geophys. Res., 113, F03025, doi:10.1029/2007JF000915, 2008.

30. Lisle, T. E., Cui, Y., Parker, G., Pizzuto, J. E., and Dodd, A. M.: The dominance of dispersion in the evolution of bed material waves in gravel-bed rivers, Earth Surf. Proc. Land., 26, 1409– 1420, doi:10.1002/esp.300, 2001.

31. Meyer-Peter, E. and Müller, R.: Formulas for bed-load transport, in: Proceeding of the 2nd IAHR Meeting, International Association for Hydraulic Research, 7–9 June 1948, Stockholm, Sweden, 39– 64, 1948.

32. Parker, G.: Selective sorting and abrasion of river gravel: theory, J. Hydraul. Eng., 117, 131–149, 1991.

33. Parker, G.: 1-D Sediment Transport Morphodynamics with Ap-plications to Rivers and Turbidity Currents, available at: http: //vtchl.uiuc.edu/people/parkerg/ (last access: 18 January 2015), 2004.

34. Parker, G.: Transport of Gravel and Sediment Mixtures, Sedimenta-tion Engineering: Processes, Measurements, Modeling and Practice, ASCE Manual of Practice 110, ch. 3, edited by: Garcia, M. H., American Society of Civil Engineers, Reston, USA, 165–252, 2008.

35. Parker, G., Klingeman, P., and McLean, D.: Bedload and size distribution in natural paved gravel bed streams, J. Hydraul. Eng., 108, 544–571, 1982.

36. Parker, G., Paola, C., and Leclair, S.: Probabilistic form of Exner equation of sediment continuity for mixtures with no active layer, J. Hydraul. Eng., 126, 818–826, 2000.

37. Parker, G., Wilcock, P., Paola, C., Dietrich, W. E., and Pitlick, J.: Quasi-universal relations for bankfull hydraulic geometry of single-thread gravel-bed rivers, J. Geophys. Res., 112, F04005, doi:10.1029/2006JF000549, 2007.

38. Parker, G., Nittrouer, J. A., Mohrig, D., Allison, M. A., Diet-rich, W. E., and Voller, V. R.: Modeling the morphodynamics of the lower Mississippi River as a quasi-bedrock river, Eos Trans. AGU 90, Fall Meet., John Wiley and Sons, Hoboken, NJ, USA, Suppl., Abstract EP32A-01, 2009.

39. Parker, G., Viparelli, E., Stark, C. P., Zhang, L., Fu, X., Inoue, T., Izumi, N., and Shimizu, Y.: Interaction between waves of alluvi-ation and incision in mixed bedrock-alluvial rivers, in: Advances in River Sediment Research, published by CRC Press/Balkema, Proc. of the 12th Inter. Symp. on River Sedimentation, Kyoto, p. 8, 2013.

40. Schlichting, H.: Boundary layer theory, 7th Edn. 1979, McGraw-Hill, New York, 1979.

41. Sklar, L. S. and Dietrich, W. E.: River longitudinal profiles and bedrock incision models: stream power and the influence of sediment supply, in: Rivers over Rock: Fluvial Processes in Bedrock Channels, edited by: Tinkler, K. and Wohl, E. E., Am. Geophys. Un. Geophys. Monogr., 107, 237–260, 1998.

42. Sklar, L. S. and Dietrich, W. E:. Sediment and rock strength controls on river incision into bedrock, Geology, 29, 1087–1090, 2001.

43. Sklar, L. S. and Dietrich, W. E.: A mechanistic model for river inci-sion into bedrock by saltating bed load, Water. Resour. Res., 40, W06301, doi:10.1029/2003WR002496, 2004.

44. Sklar. L. S. and Dietrich, W. E.: The role of sediment in controlling steady-state bedrock channel slope: implications of the saltation– abrasion incision model, Geomorphology, 82, 58–83, 2006.

45. Smith, T. R. and Bretherton, F. P.: Stability and the conservation of mass in drainage basin evolution, Water Resour. Res., 8, 1506– 1529, 1972.

46. Stark, C. P., Foufoula-Georgiou, E., and Ganti, V.: A nonlocal theory of sediment buffering and bedrock channel evolution, J. Geo-phys. Res., 114, F01029, doi:10.1029/2008JF000981, 2009.

47. Sternberg, H.: Untersuchungen über Längen- und Querprofil geschiebeführender Flüsse, Z. Bauwesen, 25, 483–506, 1875.

48. Struiksma, N.: Mathematical modelling of bedload transport over non-erodible layers, in: Vol. 1, Proceeding of IAHR Conference on River, Coastal and Estuary Morphodynamics, Genova, Italy, 89–98, 1999.

49. Tanaka, G. and Izumi, N., The bedload transport rate and hydraulic resistance in bedrock channels partly covered with gravel, J. JSCE Div. B, 69, I_1033–I_1038, 2013.

50. Turowski, J. M.: Stochastic modeling of the cover effect and bedrock erosion, Water Resour. Res., 45, W03422, doi:10.1029/2008WR007262, 2009.

51. Turowski, J. M.: Semi-alluvial channels and sediment-flux-driven bedrock erosion, in: Gravel Bed Rivers: Processes, Tools, Envi-ronments, 1st Edn., ch. 29, edited by: Church, M., Biron, P. M., and Roy, A., John Wiley & Sons, Chichester, UK, 401–416, 2012.

52. Turowski, J. M., Lague, D., and Hovius, N.: Cover effect in bedrock abrasion: a new derivation and its implications for the mod-eling of channel morphology, J. Geophys. Res., 112, F04006, doi:10.1029/2006JF000697, 2007.

53. Turowski, J. M., Badoux, A., Leuzinger, J., and Hegglin, R.: Large floods, alluvial overprint, and bedrock erosion, Earth Surf. Proc. Land., 38, 947–958, doi:10.1002/esp.3341, 2013.

54. Whipple, K. X.: Bedrock rivers and the geomorphology of active orogens, Ann. Rev. Earth Pl. Sci., 32, 151–185, 2004.

55. Whipple, K. X. and Tucker, G. E.: Dynamics of the stream-power river incision model: implications for height limits of mountain ranges, landscape response timescales, and research needs, J. Geophys. Res., 104, 17661–17674, 1999.

56. Whipple, K. X. and Tucker, G. E.: Implications of sediment-flux-dependent river incision models for landscape evolution, J. Geo-phys. Res., 107, 2039, doi:10.1029/2000JB000044, 2002.

57. Wilcock, P. R. and Crowe, J. C.: Surface-based transport model for mixed-size sediment, J. Hydraul. Eng., 129, 120–128, 2003.

58. Wilkerson, G. V. and Parker, G.: Physical basis for quasi-universal relations describing bankfull hydraulic geometry of sand-bed rivers, J. Hydraul. Eng., 137, 739–753, 2011.

59. Wong, M. and Parker, G.: Reanalysis and correction of bed-load relation of Meyer-Peter and Müller using their own database, J. Hydraul. Eng., 132, 1159–1168, 2006a.

60. Wong, M. and Parker, G.: One-dimensional modeling of bed evolu-tion in a gravel bed river subject to a cycled flood hydrograph, J. Geophys Res., 111, F3018, doi:10.1029/2006JF000478, 2006b.

61. Xu, Q., Fan, X. M., Huang, R. Q., and Van Westen, C.: Landslide dams triggered by the Wenchuan Earthquake, Sichuan Province, south west China, Bull. Eng. Geol. Environ., 68, 373–386, 2009.

62. Yanites, B. J., Tucker, G. E., Mueller, K. J., and Chen, Y. G.: How rivers react to large earthquakes: Evidence from central Taiwan, Geology, 38, 639–642, doi:10.1130/G30883.1, 2010.

63. Zhang, L., Fu, X. D., Stark, C. P., Fernandez, R., and Parker, G.: Modeling of incision of bedrock rivers subject to temporally varying sediment supply, Proceedings of the 2013 IAHR World Congress, Chengdu, Tsinghua University Press, China, p. 11, 2013.

INDEX

T

V